AF740161

Incomputable Earth

Theory in the New Humanities

Series Editor: Rosie Braidotti

Theory is back! The vitality of critical thinking in the world today is palpable, as is a spirit of insurgency that sustains it. Theoretical practice has exploded with renewed energy in media, society, the arts and the corporate world. New generations of critical 'studies' areas have grown alongside the classical radical epistemologies of the 1970s: gender, feminist, queer, race, postcolonial and subaltern studies, cultural studies, film, television and media studies. This series aims to present cartographic accounts of emerging critical theories and to reflect the vitality and inspirational force of on-going theoretical debates.

Other titles in the series:

Hormonal Theory, edited by Andrea Ford, Roslyn Malcolm, Sonja Erikainen, Lisa Raeder and Celia Roberts
Posthuman Glossary, edited by Rosi Braidotti and Maria Hlavajova
Conflicting Humanities, edited by Rosi Braidotti and Paul Gilroy
General Ecology, edited by Erich Hörl with James Burton
Philosophical Posthumanism, Francesca Ferrando
The Philosophy of Matter, Rick Dolphin
Materialist Phenomenology, Manuel DeLanda
From Deleuze and Guattari to Posthumanism, edited by Christine Daigle and Terrance H. McDonald
Vibrant Death, Nina Lykke
Visceral Prostheses, Margrit Shildrick

Incomputable Earth

Technology and the Anthropocene Hypothesis

Edited by Antonia Majaca

BLOOMSBURY ACADEMIC
LONDON • NEW YORK • OXFORD • NEW DELHI • SYDNEY

BLOOMSBURY ACADEMIC
Bloomsbury Publishing Plc, 50 Bedford Square, London, WC1B 3DP, UK
Bloomsbury Publishing Inc, 1385 Broadway, New York, NY 10018, USA
Bloomsbury Publishing Ireland, 29 Earlsfort Terrace, Dublin 2, D02 AY28, Ireland

BLOOMSBURY, BLOOMSBURY ACADEMIC and the Diana logo are trademarks of Bloomsbury Publishing Plc

First published in Great Britain 2025

Copyright © Antonia Majaca, and Contributors 2025

Antonia Majaca has asserted her right under the Copyright, Designs and Patents Act, 1988, to be identified as Editor of this work.

Cover design by Rafaela Drazic

All rights reserved. No part of this publication may be: i) reproduced or transmitted in any form, electronic or mechanical, including photocopying, recording or by means of any information storage or retrieval system without prior permission in writing from the publishers; or ii) used or reproduced in any way for the training, development or operation of artificial intelligence (AI) technologies, including generative AI technologies. The rights holders expressly reserve this publication from the text and data mining exception as per Article 4(3) of the Digital Single Market Directive (EU) 2019/790.

Bloomsbury Publishing Plc does not have any control over, or responsibility for, any third-party websites referred to or in this book. All internet addresses given in this book were correct at the time of going to press. The author and publisher regret any inconvenience caused if addresses have changed or sites have ceased to exist, but can accept no responsibility for any such changes.

A catalogue record for this book is available from the British Library.

A catalog record for this book is available from the Library of Congress.

ISBN: HB: 978-1-3502-6497-7
PB: 978-1-3502-6498-4
ePDF: 978-1-3502-6499-1
eBook: 978-1-3502-6500-4

Series: Theory in the New Humanities

Typeset by Deanta Global Publishing Services, Chennai, India

For product safety related questions contact productsafety@bloomsbury.com.

To find out more about our authors and books visit www bloomsbury.comand sign up for our newsletters.

Contents

Illustrations

Contributors

Mari Bastashevski is an artist, writer, and a lecturer at the Trondheim Academy of Fine Art. Her work currently focuses on future imaginaries that transcend the culture of techno-optimism and collapsism. She works extensively with snails and other animals while contemplating the tripartite relationship between animals, humans, and technology. Her research examines the potential of emerging visual technologies as sites of field research into these relationships. Previously, she has exhibited with Bonniers Konsthall, Stockholm; Maison Populaire, Montreuil; Photo Elysée, Lausanne; HKW, Berlin; Art Souterrain, Montreal; and Noorderlicht International Photo Festival. Her work has been published with Rutgers University Press, EPFL Press, as well as *Time Magazine*, *New York Times*, *Courrier International*, *Le Monde*, and *e-flux*, among others. Bastashevski currently lives in Amsterdam, where she continues to collaborate in various capacities with the Academy of Architecture, KABK–Royal Academy of Art, Sandberg Instituut, and the Amsterdam School of Cultural Analysis.

Tega Brain is an Australian-born artist and environmental engineer, exploring issues of ecology, data, automation, and infrastructure. She has created digital networks that are controlled by environmental phenomena, schemes for obfuscating personal data, and a wildly popular smell-based online dating service. She exhibits internationally, having recently shown work at the Arts and Industries Building, Smithsonian; ZKM, Karlsruhe; Whitney Museum of American Art, New York; Australian Centre for Contemporary Art, Melbourne; and Haus der Elektronischen Künste, Münchenstein. Her first book, *Code as Creative Medium* (2021), is coauthored with Golan Levin. Brain is an Industry Associate Professor of Integrated Design and Media at NYU–Tandon School of Engineering and also serves on the board of the School for Poetic Computation in New York.

Imani Jacqueline Brown is an artist, activist, writer, and researcher from New Orleans, based in London. Her work investigates the "continuum of extractivism," which spans from settler-colonial genocide and slavery to fossil fuel production and climate change. Brown combines artistic and archival research, ecological philosophy, legal theory, people's history, and counter-cartographic strategies to map the spatial logics that make geographies, unmake communities, and break Earth's geology. Her research is disseminated through films and art installations, public actions and performance, and reports and testimony delivered to courts and organs of the United Nations. Her artistic output has been presented internationally, including in the United States, the UK, Poland, Germany, and the UAE. Among other things, Brown is currently a PhD candidate at Queen Mary, University of London, a research fellow with Forensic Architecture, and an associate lecturer in MA Architecture at the Royal College of Arts.

Michaela Büsse is a postdoctoral researcher at the Chair of Digital Culture, Technische Universität Dresden, and Associated Investigator at the cluster of excellence "Matters of Activity: Image Space Material" at Humboldt-Universität, Berlin. In her research, she focuses on sociomaterial transformations in the context of speculative urbanism, climate change mitigation, and energy transition. Drawing on elemental anthropology, feminist science, and technology studies, Büsse investigates how design practices and technologies govern environments and define who and what is being rendered inhuman.

Jingru (Cyan) Cheng works across architecture, anthropology, and filmmaking. Her practice follows drifting bodies—from rural migrant workers to forms of water—to confront intensified social injustice and ecological crisis. Cheng was awarded the Harvard GSD's 2023 Wheelwright Prize for *TRACING SAND*; received two commendations from the RIBA (Royal Institute of British Architects) President's Awards for Research in 2020 and 2018; and won the Architecture Short Film Award at the Milano Design Film Festival in 2024 and the Best Short Film at the Venice Architecture Film Festival in 2023, all as part of *RIPPLE RIPPLE RIPPLING*. Her work has appeared in solo shows at Storefront for Art and Architecture in New York (2025) and the Architectural Association in London (2024), among others. Together with Elise Misao Hunchuck and Marco Ferrari, she co-led the architectural design studio ADS7: Politics of the Atmosphere at the Royal College of Art in London (2019–22).

Kate Crawford is a leading international scholar of the social implications of artificial intelligence. She is a research professor at USC Annenberg in Los Angeles, a senior principal researcher at Microsoft Research in New York, and the inaugural visiting research chair on Social Justice and Artificial Intelligence at the École normale supérieure in Paris. Her latest book, *Atlas of AI* (2021), won the Sally Hacker Prize from the Society for the History of Technology, the ASSI&T Best Information Science Book Award, and was named one of the best books of 2021 by *New Scientist* and the *Financial Times*. Over her twenty-year research career, she has also produced groundbreaking creative collaborations and visual investigations. Crawford has advised policy makers in the United Nations, the White House, and the European Parliament, and she currently leads Knowing Machines, an international research collaboration that investigates the foundations of machine learning.

Arturo Escobar is an activist-researcher from Cali, Colombia, working on territorial struggles against extractivism, ecosocial transitions, and ontological design. He was professor of anthropology and political ecology at the University of North Carolina, Chapel Hill, until 2018 and is currently affiliated with the PhD program in environmental sciences at Universidad del Valle, Cali. Over the past thirty years, he has worked closely with Afro-descendant, environmental, and feminist organizations in Colombia. His most well-known book is *Encountering Development: The Making and Unmaking of the Third World* (1995; 2nd ed., 2011). His most recent books are *Designs for the Pluriverse: Radical Interdependence, Autonomy, and the Making of Worlds* (2018); and *Pluriversal Politics: The Real and the Possible* (2020). In 2021, he was inducted into the American Academy of Arts and Sciences.

Marco Ferrari is the co-founder of Studio Folder, an interdisciplinary design practice based in Milan, working both on commissions in the fields of culture and the arts and on research projects examining the politics and visualization of spatial data. Together with Elisa Pasqual and Andrea Bagnato, he is the co-author of *A Moving Border: Alpine Cartographies of Climate Change*, a book based on Studio Folder's long-term project Italian Limes (2019). He has been teaching Methods and Tools for Representation at ISIA Urbino (2010–22) and co-led ADS7 at the Royal College of Art in London (2019–22), together with Elise Misao Hunchuck and Jingru (Cyan) Cheng. Currently, he is an adjunct assistant professor at Columbia GSAPP and the head of MA Information Design at Design Academy Eindhoven.

Victor G. García-Castañeda is a researcher focused on the fields of philosophy of technology and the sociology of digital cultures. He has degrees in philosophy, social sciences, and sociology from ITESO–Instituto Tecnológico y de Estudios

Superiores de Occidente in Guadalajara and Universidad Iberoamericana in Mexico City. He is currently a PhD candidate in philosophy at the Universitat Autònoma de Barcelona and an affiliate member of the Research Network for Philosophy and Technology. He has presented his work in academic journals, conferences, and institutions across different countries in Latin America and Europe.

Sara Garzón, PhD, is a curator and art historian, specializing in modern and contemporary Latin American art with a focus on issues relating to decoloniality, temporality, and Indigenous ecocriticism. Sara has contributed to several exhibition catalogs, anthologies, peer-reviewed journals, and art magazines. Her recent editorial project, "Worldmaking Practices: A Take on the Future," was published with the support of the Colección Patricia Phelps de Cisneros. In addition to art exhibitions, she has served as guest curator of the experimental summer school on art and technology Materia Abierta, Mexico City, and has taught courses on art, ecology, and care at the Dutch Art Institute's Roaming Academy. She also participated in the residency "Creative Ecologies and Decolonial Futurities," Chiapas (2019); the Workshops for Emerging Curators at Para Site, Hong Kong (2019); and the program *science_technology_society* at Delfina Foundation, London (2020).

fields harrington is an artist who lives and works in Brooklyn. He works across disciplines and media to investigate the social and political dimensions of race, value, and the complex history of science. harrington received his BFA from the University of North Texas and his MFA from the University of Pennsylvania. He was a participant in the Whitney Independent Study Program (2019–20). He has presented solo exhibitions at David Salkin Creative (2020) and Y2K group (2021). He has also exhibited in group shows at MIT List Projects, KAJE, Parsons School of Design, and Recycled Artist in Residence, among others, and was a LAB researcher in residence at the Kitchen in collaboration with the School for Poetic Computation (2023). harrington teaches at Eugene Lang College of Liberal Arts, Parsons School of Design, and Cooper Union.

Yuk Hui obtained his PhD from Goldsmiths, University of London, and his Habilitation in philosophy from Leuphana Universität, Lüneburg. Hui is the author of several monographs that have been translated into a dozen languages, including *On the Existence of Digital Objects* (2016), *The Question Concerning Technology in China: An Essay in Cosmotechnics* (2016), *Recursivity and*

Contingency (2019), and *Art and Cosmotechnics* (2021). Hui is the convenor of the Research Network for Philosophy and Technology and sits as a juror of the Berggruen Prize for Philosophy and Culture since 2020. He is currently a professor of philosophy of technology and media at the City University of Hong Kong.

Elise Misao Hunchuck (she/her) is a transdisciplinary researcher, editor, writer, and educator. Based in Berlin and Milan, her practice brings together architecture, landscape architecture, and media studies to sites in Canada, Japan, China, Ukraine, and the United States. She is the journal editor and curator of transmediale, the annual festival for art and digital culture in Berlin. She is an editorial board member for the journal *Scapegoat: Architecture, Landscape, Political Economy* and Adjunct Assistant Professor at the Graduate School of Architecture, Planning and Preservation, Columbia University. Alongside Marco Ferrari and Jingru (Cyan) Cheng, she co-led the architectural design studio ADS7: Politics of the Atmosphere at the Royal College of Art in London (2019–22). Hunchuck has previously taught at the John H. Daniels Faculty of Architecture, Landscape, and Design at University of Toronto, the Azrieli School of Architecture and Urbanism at Carleton University, and the Bartlett Faculty of the Built Environment at University College London.

Vladan Joler is an academic, researcher, and artist whose work blends data investigations, counter-cartography, investigative journalism, writing, data visualization, critical design, and numerous other disciplines. He explores and visualizes different technical and social aspects of algorithmic transparency, digital labor exploitation, invisible infrastructures, and many other contemporary phenomena at the intersection between technology and society. He has authored and published many projects with the independent research collective Share Lab, based in Novi Sad. His work is in the permanent collections of the Museum of Modern Art, New York; Victoria and Albert Museum, London; Design Museum, London; and the permanent exhibition of Ars Electronica Center. Joler's work has been exhibited in more than one hundred international exhibitions.

Antonia Majaca is a curator and theorist living between Venice and Berlin. Her work incorporates political philosophy, critical media studies, the political epistemology of technology and intellectual history. She was the principal investigator for the research project "The Incomputable" (2015–21) at IZK–Institut für Zeitgenössische Kunst, Technische Universität Graz, funded by the

Austrian Science Fund FWF. She was one of the research curators in the long-term project "Kanon Fragen" at HKW–Haus der Kulturen der Welt, Berlin, where in 2017, she also co-curated the large-scale research exhibition *Parapolitics: Cultural Freedom and the Cold War*. In 2015, she founded "Feminist Takes," a long-term research project considering the relation between the non-Western avant-garde cinema and feminist theory, with sessions held so far at Kunsthalle Wien, Nottingham Contemporary, and Tranzit Display Prague. Majaca edited the book *Feminist Takes: "Early Works" by Želimir Žilnik* (2021) with Rachel O'Reilly and Jelena Vesić.

Conor McKeown (they/them) explores how video games help us understand complex concepts. Whether ecologies, quantum theory, and human extinction, they use cutting-edge philosophy—like Karen Barad's agential realism—to reframe games into lenses. Through their published work and lectures available on YouTube, their hope is that players and scholars alike can view the world through these lenses and find something new. Their research relies on creative practice, making games, and creative coding experiments to test concepts and find meaning in game mechanics. In 2022–23, McKeown also worked as Co-PI on the "Making a Difference" project, which engages with the impact of maker spaces in Scotland.

Katherine McKittrick is Professor of Gender Studies and Canada Research Chair in Black Studies at Queen's University in Kingston, Canada. She authored *Demonic Grounds: Black Women and the Cartographies of Struggle* (2006) and edited and contributed to *Sylvia Wynter: On Being Human as Praxis* (2015). McKittrick's recent monograph, *Dear Science and Other Stories* (2021), is an exploration of black methodologies.

Margarida Mendes holds a PhD in Research Architecture from the Department of Visual Cultures, Goldsmiths, University of London. She is a researcher, curator, and educator, exploring the overlap between systems thinking, experimental film, sound practices, and ecopedagogy. She consults for Sciaena, an NGO working on marine policy and deep-sea mining, and has co-directed several educational platforms such as *Escuelita* (CA2M, 2018), *The World in Which We Occur* (Matter in Flux, 2018), and the Barber Shop (Lisbon, 2009–16). Mendes was part of the curatorial team of the 11th Liverpool Biennial, 4th Istanbul Design Biennial, and 11th Gwangju Biennale, and is a member of Natural Contract Lab, a transdisciplinary collective of lawyers and artists working on restorative justice and nature rights across Europe.

Konstantin Mitrokhov's practice spans across image-making, writing, and academic research, focusing on open-ended modes of knowledge production that are enabled by computation. He works with moving images as a vehicle for practice-based inquiry. Mitrokhov also collaborates with other researchers as a cinematographer and video editor.

Juaniko Moreno is a curator, researcher, and artist based in Bogotá, with interests in cosmotechnics, the nature/culture divide, planetarity, and digital art. He holds a BA from Colombia National University, Bogotá, and a MA in Contemporary Art and Curatorial Studies from China Academy of Art, Hangzhou. He was part of the research and think-tank program Terraforming 2021, Strelka Institute, Moscow. He currently works as a professor of visual arts at Universidad El Bosque and as a curator for MAMBO–Bogotá Museum of Modern Art.

Conrad Moriarty-Cole is a senior lecturer at the Bath School of Art, Film and Media. His research sits at the intersection of philosophy, cultural studies, and computer science, and considers questions about computational infrastructures, ecology, decolonialism, and aesthetics. The core focus of his current research is a critical reinterpretation of the philosophical legacy of phenomenology, developing a post-phenomenological process metaphysics of technology, nature, and society. Outside of academia, Moriarty-Cole is Board Chair at the Centre for Progressive Change, London.

Michal Osterweil earned a PhD in cultural anthropology with a certificate in cultural studies and is Teaching Professor in the Curriculum in Global Studies, University of North Carolina at Chapel Hill. She leads a campus-wide initiative on Transformative Pedagogy in Times of Crisis that seeks to create communities of practice for bringing anti-racist, anti-oppression practices into pedagogical practices in ways that address and honor the full selves of faculty and students alike. These are based on her recent training and research in mindfulness, trauma-informed leadership, and other somatic modalities. Since 2011, she has co-curated and led the Carolina Seminar on the Theory and Politics of Relationality and was involved with the Social Movements Working Group, which began in 2003. With Arturo Escobar and Kriti Sharma, Osterweil is co-author of *Relationality: An Emergent Politics of Life and the Human* (2024).

Luciana Parisi works across three main areas of study: cybernetics, philosophies of nature, and philosophies of difference; computation, aesthetics, and media; and computational logic, philosophies of technology, and political theories. She is currently a professor in the Program in Literature and core faculty in the graduate program of Computational Media, Art & Culture at Duke University. She was a member of the CCRU (Cybernetic Culture Research Unit) and is currently a co-founding member of CCB (Critical Computation Bureau) and co-organized the symposium "Recursive Colonialism, Artificial Intelligence & Speculative Computation: A Manifesto" (2020). Her philosophical writings are published in the book *Abstract Sex: Philosophy, Biotechnology and the Mutations of Desire* (2004) and *Contagious Architecture: Computation, Aesthetics, and Space* (2013). Parisi is currently writing a monograph titled *Automating Philosophy, Instrumentality and Critique in Technoculture* (forthcoming).

Matteo Pasquinelli is Professor in Media Philosophy at Karlsruhe University of Arts and Design, where he is coordinating the research group KIM–Künstliche Intelligenz und Medienphilosophie. His research focuses the intersection of philosophy of mind, political economy, and the automation of knowledge and cultural production. He edited the anthology *Alleys of Your Mind: Augmented Intelligence and Its Traumas* (2015) and wrote, with Vladan Joler, the visual essay "The Nooscope Manifested: AI as Instrument of Knowledge Extractivism" (2020). For Verso he wrote the monograph *The Eye of the Master: A Social History of Artificial Intelligence* (October 2023). Among others, he has published essays for the *Journal of Interdisciplinary History of Ideas*, *Qui Parle*, *Radical Philosophy*, *Les Mondes du Travail*, *South Atlantic Quarterly*, *Parrhesia*, *Theory, Culture & Society*, *Multitudes*, and *e-flux*.

James Phillips is an independent researcher based in London whose work focuses on ecology, decolonialism, and speculation. He holds a PhD in Cultural Studies from Goldsmiths, University of London.

Kriti Sharma is a microbial ecologist and philosopher. She is Assistant Professor of Critical Race Science and Technology Studies at University of California Santa Cruz and the author of *Interdependence: Biology and Beyond* (2015) and *Relationality: An Emergent Politics of Life and the Human* (2024) with co-authors Michal Osterweil and Arturo Escobar. Dr. Sharma completed her PhD in Ecology, Evolution, and Organismal Biology at University of North Carolina at Chapel Hill and her postdoctoral research in geobiology at California Institute of Technology. Her field- and laboratory-based research on microbes in deep-sea sediments of seagrass meadows elucidates how these carbon-sequestering organisms can mitigate climate change and what they can teach us about making brave transitions into the unknown. Dr. Sharma's transdisciplinary laboratory brings together multispecies collectives around the question of how to transition past the limits of the dominant bioeconomic view of life and the human.

Olivia Sullivan is a visual artist from the UK whose practice is based on storytelling and poetry through comics, animation, and experimental design. Her topics of interest include psychogeography and surreal exploration. Sullivan has published comics and anthologies, including *ESCAPADES*, which was nominated for a Broken Frontier Award in 2020. Recently, she was commissioned to create a short comic for the *Tori Amos: Little Earthquakes—The Graphic Album* (2023), an anthology that has been nominated for numerous Eisner Awards. All of her artwork is created with traditional tools and digital applications such as Procreate, and no AI is used or involved in her creative process. Sullivan graduated from the Royal College of Art in 2018 and is currently working in commercial video game design as a narrative designer and animator.

Ana Teixeira Pinto is a writer and cultural theorist based in Berlin. She is a professor of art theory at the Hochschule für Bildende Künste Braunschweig and a theory tutor at the Dutch Art Institute. Her writings have appeared in publications such as *Third Text*, *Afterall*, *e-flux*, *Artforum*, and *Texte zur Kunst*. She is the editor of the book series *On the Antipolitical*, published by Sternberg Press. Together with Anselm Franke, she organized the conference and podcast series *The White West: Whose Universal?* that took place at Haus der Kulturen der Welt, Berlin.

Marina Vishmidt (1976–2024) was an American writer, editor, and critic born in Kharkiv in the former Soviet Union. She was Professor of Art Theory in the Department of Art Theory, University of Applied Arts, Vienna until April 2024. She was a fellow at the Leuphana Institute of Advanced Studies in Culture and Society, Lüneburg, 2024, held the Rudolph Arnheim Guest Professorship in Art History, Humboldt University, Berlin, April–September 2022 and held a Visiting Professorship at the Department of Critical and Cultural Theory, Södertörn University, Stockholm, September 2020. She was a lecturer at the Centre for Cultural Studies at Goldsmiths University of London 2016–2023 and taught art theory at the Netherlands Art Institute in Arnhem 2014–2018.

Kevin Walker is an artist and a researcher at Coventry University, where he leads a research team on AI and Algorithmic Cultures, working across anthropology, technology, and the arts. Prior to that, he launched and led the Information Experience Design program at the Royal College of Art in London. He develops hardware and software, and his artistic practice involves sound, moving image, and drawing. He has made commissioned work for Centre Pompidou, Kensington Palace, Transport for London, and the British Council. He has also curated exhibitions and events and written about art, AI, and related topics.

Brian Kuan Wood is a writer based in New York, an editor of e-flux's book series and monthly journal, and a co-curator of the 2023 Taipei Biennial. He has taught and lectured in the MA Curatorial Practice program at the School of Visual Arts in New York (where he was Director of Research from 2017 to 2022), Ashkal Alwan in Beirut, Moderna Museet in Stockholm, Para Site in Hong Kong, Inside-Out Art Museum in Beijing, and China Academy of Art in Hangzhou, among other places. He edited Natascha Sadr Haghighian's *relearning bearing witness* (2021), Yuk Hui's *Art and Cosmotechnics* (2021), as well as the 2017 Sharjah Biennial publication *Tamawuj* (with Amal Issa, Omar Berrada, and Kaelen Wilson-Goldie).

Gary Zhexi Zhang is an artist and writer whose work explores connections between cosmology, technology, and economy. He was born in Suzhou and works in London. He edited *Catastrophe Time!* (2023), a book of fiction, essays, and interviews about finance and time. The opera he co-created with Waste Paper Opera (Klara Kofen and James Oldham), *Dead Cat Bounce*, premiered at Somerset House in 2022. His last solo exhibition, *Cycle 25*, documented events that blur the boundaries between speculative belief and the material world, like natural disasters, scam nations, and cosmic economies. Zhang is a lecturer in critical studies at Goldsmiths and taught previously at Parsons School of Design. Recent books and chapters include *Against Reduction* (2021), *Future Art Ecosystems: Art x Decentralised Tech*, coauthored with Victoria Ivanova (2022), and *Interspecies: A Primer* (2024).

Daniela Zyman is the artistic director of TBA21 Thyssen-Bornemisza Art Contemporary, a position she has held since 2003, where she has played a crucial role in shaping the organization's exhibitions and commissions program. Prior to this, from 1995 to 2001, she held the position of chief curator at the MAK–Museum of Applied Arts in Vienna, during which she also established and curated the MAK Center for Art and Architecture in Los Angeles. In addition to her curatorial work, Zyman is also an adjunct lecturer at the University of Art and Industrial Design in Linz and contributes essays to various art publications. Since 2022, she has curated a series of exhibitions at the Centro de Creación Contemporánea de Andalucía in the city of Córdoba on behalf of TBA21, including *Abundant Futures in Troubled Times* (2022–23) and *Remedios: Where New Land Might Grow* (2023–24).

1

Incomputable Earth

Reclaiming Abstraction

Antonia Majaca

Beginning in the early 2000s, Inuit elders across the Canadian Arctic reported something extraordinary: The sun was setting many kilometers further west along the horizon, and the stars were no longer where they should be. For generations trained from childhood to observe the most minute details of Arctic skies—sent out every morning to study weather patterns—these communities possessed unmatched expertise in celestial navigation developed across thousands of years of Arctic habitation. As Susan Schuppli narrates in her film *Can the Sun Lie?*, when the Inuits brought their observations to climate scientists, they were dismissed as impossible. The Earth had not shifted on its axis, the researchers insisted. Yet elders from across Nunavut—from Pangnirtung to Iqaluit, from Resolute Bay to Igloolik—were independently reporting identical phenomena. Community after community offered the same precise observations about altered celestial positions. The confrontation between Indigenous knowledge and Western scientific authority reached international attention when these observations were presented at COP15 in 2009. The resolution came only when scientists finally investigated Arctic atmospheric conditions and discovered that warming air was creating temperature inversions—literally refracting sunlight and shifting its apparent position across the horizon. The Inuit had been tracking planetary-scale atmospheric changes with a precision that exceeded sophisticated satellite monitoring and computational systems. Even more remarkably, they already had a term for this optical phenomenon: *qapirangajuq*, meaning "spear strangely," describing how hunters must adjust for refraction when spearing fish through water.[1]

What manifests at Earth's poles as cryospheric breakdown and dramatic epistemological confrontation saturates the entire planetary system: from soil microbiomes resisting algorithmic prediction to monsoon patterns defying computational modeling. As the ice sheets fracture into warming oceans, the political epistemological foundations of planetary climate "management" dissolve alongside the ancient ice, exposed as catastrophically inadequate. It is the time of "great derangement,"[2] a fundamental epistemological crisis wherein conventional frameworks for understanding human-nature relationships become inadequate to our planetary predicament. What if Earth is not a broken machine awaiting algorithmic repair, not a chaotic system demanding computational control, not a wayward spaceship requiring technological course correction? What if climate breakdown reveals, alongside the deadly pathogens emerging from thawing permafrost, the constitutive epistemological pathology and a fundamental cosmological error? What if it shows us that earthly complexity will always burst through the computable data streams that cybernetic logic has amplified into algorithmic control—streams that nonetheless cannot grasp, contain, or heal climate breakdown?

This book posits that the climate crisis is an irreducibly incomputable problem[3]—yet precisely as with incomputable problems in mathematics and

[1] Zacharias Kunuk and Ian Mauro, "Inuit Knowledge and Climate Change," documentary film, 2010; "New Documentary Recounts Bizarre Climate Changes Seen by Inuit Elders," *The Globe and Mail*, October 19, 2010; "Could the Inuit Steer Us Through Climate Change?" *The Daily Beast*, May 4, 2019; Susan Schuppli, "Can the Sun Lie?" HD video, 12:51 minutes, 2014, https://susanschuppli.com/CAN-THE-SUN-LIE.

[2] Amitav Ghosh, *The Great Derangement: Climate Change and the Unthinkable* (Chicago: University of Chicago Press, 2016). Ghosh argues that climate change represents a crisis not only of the physical world but of culture and imagination, examining how modern literary and artistic forms prove inadequate to representing the scale and temporal complexity of planetary transformation.

[3] Incomputability refers to problems that cannot be solved by any algorithm, even given unlimited time and computational resources. This concept emerged from Alan Turing's foundational work on computability theory, particularly his proof that the "halting problem"—determining whether a given program will eventually halt or run forever—is undecidable. See Alan M. Turing, "On Computable Numbers, with an Application to the Entscheidungsproblem," *Proceedings of the London Mathematical Society* 42, no. 2 (1937): 230–65; Alonzo Church, "An Unsolvable Problem of Elementary Number Theory," *American Journal of Mathematics* 58, no. 2 (1936): 345–63. Incomputable problems include not only formally undecidable questions but also those requiring infinite computational resources, such as certain optimization problems in complex systems. Gregory Chaitin's work on algorithmic information theory demonstrates that most real numbers are incomputable and that randomness itself cannot be algorithmically compressed, revealing fundamental limits to what computation can capture about reality. See Chaitin, *The Limits of Mathematics* (Singapore: Springer-Verlag, 1998). In the context of complex systems, incomputability manifests through emergence, where system

computer science, this incomputability opens potentials for new, transversal thinking. What makes incomputable problems uniquely perplexing yet generative is that they contain undecidable properties or require infinite computational resources and, as such, have no immediate application but often lead to the creation of new hypotheses. This means that incomputability destabilizes, from within, the very mode of solving problems that can be posed, described, or named. It points to the inherent, structural contradictions embedded within the technology reproducing the cybernetic paradigms of the climate crisis.

The concept of incomputability in this book operates across three interrelated dimensions that structure our analysis. *Epistemological excess* identifies what escapes algorithmic capture within dominant frameworks—those relational, temporal, and cosmological dimensions that resist reduction to computational variables yet remain constitutive of planetary relationships. *Political remainder* reveals the excluded knowledges and foreclosed forms of labor upon which cybernetic systems depend through ongoing processes of systematic appropriation—particularly women's reproductive work, place-based ecological knowledge, and collective care practices. And finally, *generative potential* enables critical-creative practices that prefigure alternative knowledge production oriented toward ecological regeneration rather than continued extraction on a world scale.

These three dimensions of planetary incomputability expose not only the limits of algorithmic capture but also the fictive nature of the knowing subject these systems presuppose. This incomputability demands we abandon not merely cybernetic frameworks but the entire edifice of Man as planetary manager—that subject position that claims mastery over planetary processes

behaviors arise from interactions between components in ways that cannot be predicted from understanding the components separately—what complexity theorists identify as "irreducible" properties. See Melanie Mitchell, *Complexity: A Guided Tour* (New York: Oxford University Press, 2009), 12–35, 298–312.

For contemporary uses of the term, see especially: Luciana Parisi, *Contagious Architecture: Computation, Aesthetics, and Space* (Cambridge, MA: MIT Press, 2013). Parisi's foundational work argues that algorithmic computation constitutes "a mode of thought in its own right" that operates through "forms of abstraction that lie beyond direct human cognition and control," including "modes of infinity, contingency, and indeterminacy, as well as incomputable quantities underlying the iterative process of algorithmic processing." See also Antonia Majaca and Luciana Parisi, "The Incomputable and Instrumental Possibility," *e-flux Journal* 77 (November 2016), where we develop the concept of "the incomputable subject" as one that comes into being by way of reclaiming the contingent as a mode of reasoning and as the field of the political. Instrumentality is thereby theorized within the possible against the probable and as a way to repoliticize and reclaim unknown unknowns from the apparatuses of capture and prediction.

through fantasies of steering "Spaceship Earth"[4] toward technospheric salvation. Drawing on Sylvia Wynter's critique of "Overrepresentation of Man" as the universal human figure,[5] this analysis proceeds through what might be termed decapitalized humanism—simultaneously the human without capital H and humanism unbound from capitalocentric frameworks. This reconceptualization proves essential because, as Ana Teixeira Pinto demonstrates in this volume,[6] the evolutionary schemes haunting contemporary technological discourse position each being as a "relay point in the Spirits' search for its optimal vessel," reproducing the same transcendence fantasies that legitimated imperial expansion through technological moral imperatives. Against the formation of human agency as transcendence from Earth's processes, decapitalized humanism positions human transformative capacity as emerging through rather than against the material conditions of planetary reproduction. *Incomputable Earth* is thus also, crucially, an invitation to consider the human as an incomplete and unfinished praxis of instantiating noninstrumental relations. The Faustian fantasies of mastering time and space cannot be undone by expanding and adapting Promethean instrumentality or by "repurposing" technology (as it has been conceptualized by Western productivism) for the advancement of the dispossessed. Instead, the undoing of the damage caused by the Anthropos must also be based on the recognition of the myriad obscured forms of subjectivity, labor, and social praxis resistant to the reductive logic of Western technoscientific reason.

To counteract the Anthropocene hypothesis' narrative of human mastery turned destructive—which inevitably accelerates techno-solutionist fantasies of planetary management—this book argues that the incomputable nature of our crisis demands a genuinely different hypothesis. We need one that centers the material practices of regeneration constitutive of all world-making and life-

[4] The metaphor of "Spaceship Earth," popularized by R. Buckminster Fuller in the 1960s, encodes cybernetic logics of planetary management. See Fuller, *Operating Manual for Spaceship Earth* (Carbondale: Southern Illinois University Press, 1969).

[5] Sylvia Wynter, "Unsettling the Coloniality of Being/Power/Truth/Freedom: Towards the Human, After Man, Its Overrepresentation—An Argument," *The New Centennial Review* 3, no. 3 (Fall 2003): 257–337; "On How We Mistook the Map for the Territory, and Re-Imprisoned Ourselves in Our Unbearable Wrongness of Being, of Désêtre: Black Studies Toward the Human Project," in *Not Only the Master's Tools: African-American Studies in Theory and Practice*, ed. Lewis R. Gordon and Jane Anna Gordon (Boulder: Paradigm Publishers, 2006), 107–69; For critical engagements with Wynter's work, see Katherine McKittrick, ed., *Sylvia Wynter: On Being Human as Praxis* (Durham: Duke University Press, 2015); Denise Ferreira da Silva, "Before *Man*: Sylvia Wynter's Rewriting of the Modern Episteme," in *Sylvia Wynter*, ed. McKittrick, 90–105; Alexander G. Weheliye, *Habeas Viscus: Racializing Assemblages, Biopolitics, and Black Feminist Theories of the Human* (Durham: Duke University Press, 2014), 8–12, 130–35.

[6] Ana Teixeira Pinto, "Nature, Estranged from the Idea: Gendered Metaphors and Evolutionary Allegories in the Long Nineteenth Century." This volume, 375–91.

making, one that recovers reciprocal ways of knowing: sciences that listen to Earth's own teachings; technologies that become not instruments of extraction but practices of relation, woven into the communities of land, water, wind, and root systems that constitute the material basis through which human-nature relations are produced and reproduced. This hypothesis must attend to the gendered political economy of ecological reproduction—recognizing that the climate crisis emerges from the systematic devalorization of life-sustaining labor and the foreclosure of intricate knowledges of terran metabolisms. Rather than the Anthropocene's computational epistemic paradigm, we need a hypothesis grounded in ways of knowing that the irreducibly incommensurable dimensions of terrestrial life and the incomputable nature of the climate crisis are already forcing into existence.

Incomputable Earth undertakes this epistemological transformation through a methodology that operates on two levels: first, as a critical-creative critique revealing how computational and cybernetic epistemologies become naturalized through apparently neutral technoscientific forms; second, as a radical epistemological proposition exploring how different ways of knowing, brought into mutually supportive relation, can generate alternative hypotheses for collective response to climate breakdown—invoking Earth sciences fundamentally different from those undergirding the Anthropocene hypothesis. Drawing from companion planting practices found in agricultural traditions worldwide, this epistemology weaves together three strands: feminist critiques of Western science, decolonial epistemologies, and historical materialist analysis. Together, these become tools for reclaiming abstraction from computational capture. This proposal is feminist in the broadest sense: It seeks to fundamentally transform the technological ordering of the world while reconfiguring our understanding of human, nature, technology, value, knowledge, and intelligence. This demands denaturalizing Western conceptions of technology to salvage it from the violent reductions of the Anthropos.

*

The Earth is not a globe, nor a system, nor a spaceship hurtling through space demanding steersmen who know how to navigate through the cosmic debris of their own making. Norbert Wiener's 1940s antiaircraft gun control systems, requiring prediction and response to moving targets through continuous feedback adjustment, established foundational principles of cybernetics, a framework for understanding mechanical and biological systems as goal-directed entities operating through information feedback loops.[7] Wiener derived his term for this new science from the ancient Greek *kybernētikós*—the helmsman who reads wind and tide, adjusting rudder in endless feedback loops between intention and drift. This, of course, represents only one navigational concept and practice among many. Pacific wayfinders, for instance, read swells through the vessel's motion, memorize star houses that divide the horizon into thirty-two bearings, and follow the flight paths of golden plovers and bristle-thighed curlews—practicing a form of navigation where sailor and ocean become part of the same knowledge system rather than controller and controlled.[8] Yet, over the course of the twentieth century, the very metaphor of planetary navigation has conjured the specter of control over what was once simply home. When the vessel becomes planetary, the cybernetic art of navigation mutates into something far more troubling: a fantasy of total system control where Earth itself becomes the ship, its living processes reduced to variables in a computational helm. Wiener's *Cybernetics: Or Control and Communication in the Animal and the Machine* and Claude Shannon's "A Mathematical Theory of Communication" both appeared in 1948, emerging from wartime research into control systems and signal processing.[9] While Wiener investigated information theory independently through his work on antiaircraft guns and developed the Wiener filter, Shannon's theory quantified information as statistical patterns that could be encoded, transmitted, and decoded.[10] For Shannon, Wiener's work already contained "the first clear-cut formulation of communication theory as a statistical problem."[11] Shannon's mathematical formalization enabled the translation of circular causality—the recursive dynamic where systems influence each other in continuous feedback

[7] Norbert Wiener, *Cybernetics: Or Control and Communication in the Animal and the Machine* (Cambridge, MA: MIT Press, 1948). Wiener's framework became foundational to subsequent developments in systems theory, artificial intelligence, and computational approaches to understanding complex systems. For a historical analysis of cybernetics' development, see Peter Galison, "The Ontology of the Enemy: Norbert Wiener and the Cybernetic Vision," *Critical Inquiry* 21, no. 1 (1994): 228–66; N. Katherine Hayles, *How We Became Posthuman: Virtual Bodies in Cybernetics, Literature, and Informatics* (Chicago: University of Chicago Press, 1999), 50–83.

[8] David Lewis, *We, the Navigators: The Ancient Art of Landfinding in the Pacific*, 2nd ed. (Honolulu: University of Hawaii Press, 1994).

[9] Claude E. Shannon and Warren Weaver, *The Mathematical Theory of Communication* (Urbana: University of Illinois Press, 1949).

[10] Rudolf Seising, "Cybernetics, System(s) Theory, Information Theory and Fuzzy Sets and Systems in the 1950s and 1960s," *Information Sciences* 180, no. 3 (2010): 459–76.

[11] Shannon and Weaver, *Mathematical Theory of Communication*, 85.

loops—into computable algorithms that could process these interdependencies through discrete computational steps.[12]

The development of digital computers in the 1950s, particularly systems like SAGE (Semi-Automatic Ground Environment), operationalized these theoretical advances. These machines could implement Shannon's information-processing algorithms at scale, transforming continuous environmental signals into discrete data points that could be stored, analyzed, and fed back into control systems. If Shannon's information theory provided the mathematical foundation for translating Wiener's circular causality into algorithmic operations, then the development of digital computers enabled the processing of vast quantities of environmental data.[13] Over the following two decades, early command-and-control systems evolved from military applications into civilian scientific infrastructure, as scientists recognized that the same computational architectures designed for tracking missiles could model planetary atmospheric dynamics. By the 1970s, these computational extensions of cybernetic logic became institutionalized in climate science through global circulation models that treated atmospheric and oceanic systems as information-processing problems.[14] This same period witnessed the emergence of what would become known as Earth System Science, formalized through NASA's Earth System Science Committee in 1983, which promoted a "holistic" approach to understanding Earth as an integrated system of interacting components.[15]

The transformation from Wiener's relatively simple mechanical systems to contemporary algorithmic applications represents a fundamental shift from circular causality between system and environment to computational processes that promise planetary optimization through real-time data processing and automated response systems.[16] This enabled the standardization of observational protocols, coordination of data collection efforts, and development of shared computational modeling platforms—particularly

[12] Karl Johan Åström and Richard M. Murray, *Feedback Systems: An Introduction for Scientists and Engineers* (Princeton: Princeton University Press, 2008).

[13] Paul N. Edwards, *The Closed World: Computers and the Politics of Discourse in Cold War America* (Cambridge, MA: MIT Press, 1996), 75–111.

[14] Paul N. Edwards, *A Vast Machine: Computer Models, Climate Data, and the Politics of Global Warming* (Cambridge, MA: MIT Press, 2010), 184–225.

[15] NASA Earth System Science Committee, *Earth System Science: A Closer View* (Washington, DC: NASA, 1988); Clark A. Miller and Paul N. Edwards, (eds.), *Changing the Atmosphere: Expert Knowledge and Environmental Governance* (Cambridge, MA: MIT Press, 2001), 120–45.

[16] Andrew Pickering, *The Cybernetic Brain: Sketches of Another Future* (Chicago: University of Chicago Press, 2010), 18–44.

general circulation models—that became foundational to contemporary climate science. The requirement for "global" data necessitated observational protocols that could be standardized across diverse ecological and cultural contexts—a standardization process that systematically privileged forms of environmental knowledge amenable to quantitative representation while marginalizing those that resist such reduction.[17]

Contemporary climate modeling exemplifies the computational evolution of cybernetic principles, transforming planetary monitoring into algorithmic governance. Vast networks of environmental sensors feed data streams to supercomputers processing quintillions of calculations per second while Earth System Models containing millions of lines of code simulate atmospheric, oceanic, and biogeochemical processes to predict future climate states and evaluate intervention strategies. The planetary steersman's wheel has become the algorithm itself—sophisticated mathematical operations promising to navigate ecological crises through computational mastery. Machine learning algorithms accelerate these models tenfold while cloud-resolving simulations at kilometer-scale resolution aspire to capture Earth's complexity in ever-finer computational grids. These simulations promise more accurate predictions of regional precipitation, extreme weather events, and climate sensitivity but require enormous energy and computational resources. This proliferation of predictive modeling and automated control systems as instruments of planetary governance creates what Matteo Pasquinelli identifies in this volume as a feedback loop between the very systems designed for planetary control and the acceleration of planetary disequilibrium they purport to manage.[18] The institutional architecture and political economy behind this cybernetic management are perhaps best exemplified by NASA's strategic rebranding of "Mission to Planet Earth" as the "Earth Science Enterprise"—a shift that marks the inherent relation between computational and market logic and exposes

[17] I am referring to the establishment of the Intergovernmental Panel on Climate Change in 1988 and the development of coordinated international research programs: The World Climate Research Programme, established in 1980, and the International Geosphere-Biosphere Programme, launched in 1986. These new institutions created organizational frameworks for coordinating global environmental research across national boundaries and disciplinary divisions. See Bert Bolin, *A History of the Science and Politics of Climate Change: The Role of the Intergovernmental Panel on Climate Change* (Cambridge: Cambridge University Press, 2007), 54–78; and World Climate Research Programme, *WCRP Strategic Plan 2019–2028* (Geneva: World Meteorological Organization, 2018), 8–15.

[18] Matteo Pasquinelli, "The Automaton of the Anthropocene: On Carbosilicon Machines and Cyberfossil Capital." This volume, 65–79.

how neoliberal governance colonizes even the conceptualization of Earth itself.[19] This trajectory, which positions Earth as an external object available for scientific observation and technological intervention, results in the systematic augmentation of what Gayatri Chakravorty Spivak identified as the "globe"—Earth transformed into a computational object existing only "on our computers" where "no one lives."[20]

Across the second half of the twentieth century and into the twenty-first, planetary steersmen clutched ever more sophisticated instruments while Earth systems spiraled beyond correction. Every satellite measurement and climate model becomes another attempt to navigate what has already been lost to the logic of cybernetic management itself. The contemporary kybernētikós stands at the helm of Spaceship Earth with instruments capable of measuring everything—carbon parts per million, ocean acidification, biodiversity loss—yet cannot see that the very act of steering perpetuates the shipwreck. This becomes the ultimate tragic irony of planetary management: The helmsman is obviously lost but keeps steering in the same direction, because to stop would mean confronting the fact that there was never any ship, only the ocean, and the wheel was always an illusion of mastery over forces that were never meant to be steered.[21]

*

This genealogy of planetary computational governance culminates in what this volume takes as its central object of critique and point of departure for thinking the incomputable: the Anthropocene hypothesis, the defining concept of Earth System Science and its most influential claim after the millennium. It traces the hypothesis not as a geological stratigraphic designation but as a paradigmatic

19 See Edwards, *A Vast Machine*; Sheila Jasanoff, "Technologies of Humility: Citizen Participation in Governing Science," *Minerva* 41, no. 3 (2003): 223–44.

20 Gayatri Chakravorty Spivak, *Death of a Discipline* (New York: Columbia University Press, 2003), 72; see also Spivak, "Imperatives to Re-Imagine the Planet," in *An Aesthetic Education in the Era of Globalization* (Cambridge, MA: Harvard University Press, 2012), 335–50; and Spivak, "Planetarity," in *Dictionary of Untranslatables: A Philosophical Lexicon*, ed. Barbara Cassin (Princeton: Princeton University Press, 2014).

21 In "The Oikos of the Earth, the Nomos of the Black Hole," I argue that contemporary planetary governance operates through what I term the "nomos of the Black Hole"—a condition where computational capitalism creates a totalizing system that perpetually consumes and regenerates itself while rendering invisible the reproductive labor that sustains it. Through analysis of three paradigmatic cases—the Earthrise image, the Black Hole photograph, and the Anthropocene hypothesis—I examine how these operate as "fetish objects of techno-positivist ideology" that create an illusion of computational mastery over planetary systems while excluding the "incomputable" dimensions of planetary relationships that resist algorithmic capture. Antonia Majaca, "The Oikos of the Earth, the Nomos of the Black Hole," *e-flux Journal*, no. 143 (2024), https://www.e-flux.com/journal/143/589526/the-oikos-of-the-earth-the-nomos-of-the-black-hole/.

discursive and ideological apparatus working across multiple disciplinary and cultural domains.[22]

By positing "humans" as an undifferentiated geological force, the Anthropocene reproduces hegemonic knowledge paradigms through what appears as radical periodization. This universalizing gesture obscures the differential power relations, colonial histories, and racialized hierarchies through which specific communities and political-economic systems have devastated planetary systems. As Kathryn Yusoff demonstrates, this enables a "geophysics of race," to remain invisible within ostensibly neutral geological designation, naturalizing colonial violence as species-wide evolutionary trajectory.[23] While grounded in these critiques of racial and colonial erasure, this volume focuses specifically on how the Anthropocene's prescriptive dimension—its proposition of enhanced technological intervention as the rational response to ecological breakdown—reveals it not merely as scientific diagnosis but also as an ideological apparatus that reproduces particular understandings of the human as planetary manager, nature as standing reserve, and technology as salvation, thereby legitimating the acceleration of extractive modernity under the banner of technological salvation.

The ideological apparatus of the Anthropocene becomes starkly visible when examining its genealogy within specific scientific networks. Emerging from atmospheric chemist Paul Crutzen and marine biologist Eugene Stoermer's 2000 publication "The Anthropocene"[24] and gaining institutional momentum through Crutzen's 2002 "Geology of Mankind,"[25] the hypothesis was subsequently developed through Crutzen's extensive collaboration with Earth System scientist Will Steffen. Their joint work—including the influential 2007 paper "The Anthropocene: Are Humans Now Overwhelming the Great Forces of Nature?"[26]

[22] Haus der Kulturen der Welt, "The Anthropocene at HKW," HKW Mediathek, https://mediathek.hkw.de/video/das-anthropozaen-am-hkw; Max Planck Institute for the History of Science, "The Anthropocene at Haus der Kulturen der Welt," https://www.mpiwg-berlin.mpg.de/research/projects/anthropocene-haus-der-kulturen-der-welt; Max Planck Institute for the History of Science, "IV. Anthropocene Formations," https://www.mpiwg-berlin.mpg.de/project/knowledge-anthropocene; Max Planck Institute for the History of Science, "Announcement of the Anthropocene Working Group's Proposed GSSP Candidate Site of the Anthropocene," https://www.mpiwg-berlin.mpg.de/video/announcement-anthropocene-working-groups-proposed-gssp-candidate-site-anthropocene.

[23] Kathryn Yusoff, *A Billion Black Anthropocenes or None* (Minneapolis: University of Minnesota Press, 2018), 25–53.

[24] Paul J. Crutzen and Eugene F. Stoermer, "The Anthropocene," *IGBP Global Change Newsletter* 41 (May 2000): 17–18.

[25] Paul J. Crutzen, "Geology of Mankind," *Nature* 415, no. 6867 (January 3, 2002): 23.

[26] Will Steffen, Paul J. Crutzen, and John R. McNeill, "The Anthropocene: Are Humans Now Overwhelming the Great Forces of Nature?" *Ambio* 36, no. 8 (2007): 614–21.

and Steffen's concept of the "Great Acceleration"[27]—proposed "planetary stewardship" through enhanced technological management as the rational response to Earth System disruption.

Consider Crutzen and Stoermer's revealing conclusion from the original paper:

> Without major catastrophes like an enormous volcanic eruption, an unexpected epidemic, a large-scale nuclear war, an asteroid impact, a new ice age, or continued plundering of Earth's resources by partially still primitive technology (the last four dangers can, however, be prevented in a real functioning noösphere) mankind will remain a major geological force for many millennia, maybe millions of years, to come.[28]

This passage explicitly positions human demiurgic planetary influence as a condition to be perpetuated through enhanced application of *noöspheric* knowledge—what they characterize as the "knowledge or information society."[29] Crutzen's subsequent iteration made the technological prescription even more explicit, suggesting that "environmentally sustainable management during the era of the Anthropocene" may well involve "large-scale geo-engineering projects, for instance to 'optimize' climate."[30] These formulations clearly demonstrate how the Anthropocene hypothesis functions as a normative apparatus that legitimizes particular forms of techno-political governance through an apparently geological designation.

The trust that Crutzen and Stoermer place in the noösphere as a mechanism for preventing ecological catastrophe is particularly telling, as it points to the concept that had undergone systematic institutional transformation prior to its deployment within the Anthropocene hypothesis. Originally conceived in the 1920s by biogeochemist Vladimir Vernadsky, the noösphere was formulated as an evolutionary apex wherein human cognitive capacities would catalyze unified global consciousness oriented toward "the right distribution of wealth" and "the consciousness of the unity and equality of all people."[31] Vernadsky's materialist framework envisioned the noösphere as a pathway toward harmonious

[27] Will Steffen et al., "The Trajectory of the Anthropocene: The Great Acceleration," *The Anthropocene Review* 2, no. 1 (2015): 81–98.

[28] Crutzen and Stoermer, "The Anthropocene," 17–18.

[29] Crutzen and Stoermer, "The Anthropocene," 17–18.

[30] Crutzen, "Geology of Mankind," 23.

[31] Vladimir I. Vernadsky, *Scientific Thought as a Planetary Phenomenon* (written 1938, published Moscow: Nauka, 1991). English translation: Moscow: Nongovernmental Ecological V.I. Vernadsky Foundation, 1997.

relationships between technological development and ecological processes through egalitarian social transformation.[32]

Vernadsky's understanding of the noösphere emerged from his decades-long empirical engagement with biogeochemical processes.[33] Working with soil samples and mineral specimens in his laboratories from the 1920s through the 1940s, he observed how "civilized humanity has introduced changes into the structure of the film on land which have no parallel in the hydrosphere . . . the systematic destruction during human history of forests, the most powerful parts of the film."[34] But more importantly, his vision of the noösphere required transformation of social relations because, as he put it, "the right distribution of the wealth is now on the agenda. This question is tied up with the consciousness of the unity and equality of all people, with the consciousness of the unity of the noösphere."[35] For Vernadsky, consciousness could only function as planetary force through eliminating the social relations that concentrated wealth and separated humans from Earth's processes, since "profound social changes, giving support to the broad masses, advanced their interests into the first rank, and the question of eliminating malnutrition and famine became a realistic option that can no longer be ignored."[36] This demanded an egalitarian transformation, not through moral imperative but through scientific recognition that human intelligence operates as a biogeochemical phenomenon requiring specific material and social conditions for its just planetary deployment.

This egalitarian vision underwent systematic reconfiguration through strategic appropriation by Cold War institutions. In 1999, one year before the Anthropocene hypothesis' formal articulation, the RAND Corporation published "The Emergence of Noopolitik: Toward an American Information

[32] Vladimir I. Vernadsky, "The Biosphere and the Noösphere," *American Scientist* 33, no. 1 (1945): 1–12.

[33] Giulia Rispoli, "Planetary Environing: The Biosphere and the Earth System," in *Environing Media*, ed. Adam Wickberg and Johan Gärdebo (London and New York: Routledge, 2023), 54–74. Rispoli traces the fundamental historical bifurcation between biosphere and Earth system concepts and contrasts Vernadsky's early twentieth-century understanding of the biosphere as a dynamic, human-reconfigured system experienced "from the inside" with the post-1970s Earth System Science approach that renders the planet as a techno-ecological system observable "from the outside." Through analyzing the role of environing media—including geographical expeditions, cybernetic models, and satellite imagery—Rispoli demonstrates how the contemporary Earth system paradigm obscures human agency and fails to adequately address the predicament of the Anthropocene. She advocates for recovering Vernadsky's biosphere-noösphere theory to better understand humanity's transformative role in planetary processes and develop more effective responses to ecological crisis.

[34] Ian Angus, "Vladimir Vernadsky and the Disruption of the Biosphere," *Climate & Capitalism*, June 5, 2018.

[35] Vernadsky, *Scientific Thought as a Planetary Phenomenon*, written 1936–1938.

[36] Vernadsky, as quoted in Ian Angus, "Vladimir Vernadsky and the Disruption of the Biosphere."

Strategy."[37] This document transformed Vernadsky's egalitarian concept into a framework for geopolitical management through informational technologies, articulating the noösphere as "the realm of the mind," constituting an intelligence infrastructure enabled by the digital information revolution. The noösphere's journey from Vernadsky's laboratories in 1920s Leningrad to RAND Corporation's strategic documents in 1999 California represents a particular kind of strategic appropriation, where emancipatory concepts return as their own opposites, maintaining the original terminology while serving precisely those power relations they were designed to transform. Rather than a pathway toward social and ecological transformation, the noösphere became a mechanism for maintaining existing power relations. It was a transformation of the spherical to the global epistemology of earthly transcendence by technological means. Pinto's analysis reveals the deeper genealogy of this transformation, tracing how imperial expansion became legitimated as moral imperative through technological transcendence fantasies rooted in thermodynamic thought. Her insight that "mind, masculinity, immateriality, and value converge on one side; nature, biology, femininity, and the past on the other" exposes the gendered metaphysics underlying Western technoscience, where both *élan vital* and cybernetic information seek to transcend death while locating life's essence in an "immaterial elsewhere."[38] This shift fundamentally alienates planetary knowledge from lived experience, requiring an imaginative flight beyond the world in order to witness it as complete totality, with, as Tim Ingold puts it, "vast scientific and technological resources devoted to actualizing this fantasy of exteriorized observation."[39]

This epistemological trajectory—whereby planetary totality becomes visible only through radical exteriorization—materializes today through the convergence of techno-optimist ideology and ecomodernist frameworks that explicitly embrace planetary abandonment as a solution to ecological crisis.[40]

[37] John Arquilla and David Ronfeldt, *The Emergence of Noopolitik: Toward an American Information Strategy* (Santa Monica: RAND Corporation, 1999).

[38] Pinto, "Nature, Estranged from the Idea," 375–91.

[39] Tim Ingold, "Globes and Spheres: The Topology of Environmentalism," in *The Perception of the Environment: Essays on Livelihood, Dwelling and Skill* (London: Routledge, 2000), 209–18.

[40] Stewart Brand's book *Whole Earth Discipline: An Ecopragmatist Manifesto* represents a pivotal work in the ideology of technological salvation. Brand, founder of the *Whole Earth Catalog* and co-founder of The Long Now Foundation, advocates for "nuclear energy, genetic modification, mass urbanization, and geoengineering" as technological solutions to climate change. Ecomodernism, as articulated in "An Ecomodernist Manifesto," co-authored by Brand, argues that modernization processes, including urbanization, agricultural intensification, nuclear power, and technological development can reduce humanity's environmental footprint while improving living standards. See Stewart Brand (ed.), *Whole Earth Catalog* (Menlo Park: Portola Institute, 1968–1985); Stewart Brand,

Contemporary Silicon Valley techno-solutionism operates through the conviction that "there is no material problem—whether created by nature or by technology—that cannot be solved with more technology," while ecomodernist manifestos promote "decoupling" human development from environmental impacts through technological intensification that enables the wealthy to escape ecological consequences entirely.[41] The rise of bunker capitalism, space colonization projects, and libertarian seasteading initiatives represents the material actualization of Ingold's global perspective: a mode of planetary apprehension that can only comprehend Earth's totality by abandoning it entirely, transforming ecological crisis into entrepreneurial opportunity. The systematic conversion of the Earth into extractable data enables intensified material extraction that drives socio-ecological metabolism toward collapse. While the destruction of the actual life on Earth might seem irrational, it is entirely rational from capital's perspective: Within this technomessianic scheme, algorithmic optimization enables both the final extraction of value from a dying planet and the pedestrian earth beings while simultaneously enabling the construction of computationally managed bunkers where the planetary oligarchs imagine they can weather the storms they have summoned. The globe of the *computable* Earth thus becomes the *killable* Earth through this nihilistic acceleration: Faced with the contradiction between infinite growth imperatives and finite planetary boundaries, capital resolves the impasse by rendering the planet itself expendable. The fantasies of planetary abandonment represent the apotheosis of this death drive—the ultimate expression of capital's willingness to destroy its own material foundation while rationalizing continued primitive accumulation on world-scale and unlimited expansion beyond even the Earth itself. These techno-oligarchic fantasies of exodus reveal how technological solutionism functions as epistemic apparatus for planetary desertion and as technological escape routes for those with sufficient capital to purchase their way out of ecological collapse, thus actualizing what Ingold describes as the globe epistemology's fundamental requirement that the world can only truly exist by leaving it.[42]

The Anthropocene becomes thinkable only through technologies that fulfill the flight from lived experience—satellite imagery, ice core data, atmospheric

Whole Earth Discipline: An Ecopragmatist Manifesto (New York: Viking, 2009); "An Ecomodernist Manifesto," Ecomodernism.org, April 2015, http://www.ecomodernism.org/manifesto-english.

41 Marc Andreessen, "The Techno-Optimist Manifesto," *Andreessen Horowitz*, October 16, 2023, https://a16z.com/the-techno-optimist-manifesto/.

42 Ingold, "Globes and Spheres," 211.

measurements, and geological surveys that render the planet visible as a complete object available for scientific observation and management. From what we know today, it is hard to believe anymore that these are the tools of planetary stewardship. Rather, they seem like instruments of organized abandonment. RAND Corporation's noösphere provides the conceptual infrastructure for this departure by transforming Vernadsky's embodied consciousness emerging from within ecological relationships into information systems that coordinate planetary control. The ghostly presence of Vernadsky's original, egalitarian spherical vision haunts the Anthropocene hypothesis.

The systematic reproduction of globe epistemology and cybernetic logic within the Anthropocene narrative becomes even more apparent through examination of its temporal architecture. The Anthropocene Working Group's 2019 consensus definitively displaced the chronological parameters initially proposed by Crutzen and Stoermer, situating the epoch's commencement in the mid-twentieth century rather than the industrial revolution—precisely coincident with what Earth System Science designates as the "Great Acceleration."[43] This temporal revision reveals that the most measurable transformations in planetary hydrological, atmospheric, glacial, and terrestrial systems converge exactly with the historical moment when these geological processes became integrated into computational infrastructures designed to render them as quantifiable, extractable information. Earth System Science's capacity to identify the Great Acceleration as the definitive Anthropocene marker depends entirely upon the post-1950 proliferation of global monitoring networks, satellite systems, and digital databases that transformed planetary processes into computational objects amenable to algorithmic analysis and mathematical modeling. The constitutive relationship between material planetary transformation and epistemic reconfiguration operates as the foundational logic of both the Anthropocene narrative and the cybernetic totalization embedded within Earth System Science. The Great Acceleration periodization further naturalized computational governance as planetary necessity. The geological evidence of human planetary dominance coincides with the intensified ecological extraction and the technological capacity to measure that extraction through the same computational systems that enable its acceleration—a recursive relationship

[43] Anthropocene Working Group, "Results of Binding Vote by AWG Released 21st May 2019," Subcommission on Quaternary Stratigraphy, May 21, 2019, http://quaternary.stratigraphy.org/working-groups/anthropocene/; Steffen et al., "The Trajectory of the Anthropocene: The Great Acceleration," 81–98.

that positions cybernetic measurement as both symptom and instrument of the planetary crisis it purports to objectively document.

The planetary-scale measurement necessarily excludes forms of environmental labor and knowledge that operate according to different temporal rhythms, spatial scales, and social relationships than those amenable to digital processing and statistical aggregation. These constitutive exclusions have so far generated substantial alternative periodizations that expose the Anthropocene's erasure of differential histories and ongoing violences constitutive of planetary crises. Among these critical interventions, Jason Moore's theorization of the "Capitalocene" most directly challenges how the Anthropocene hypothesis obscures the historical specificity of capitalist accumulation by converting fundamentally social relations into a species-wide evolutionary narrative.[44] Moore's analysis reveals how the "web of life" encompasses not generic human activity but the particular operations of capital's appropriation of what he terms "cheap nature"—the systematic undervaluation of women's work, racialized labor, and ecological processes that enables accumulation through the ongoing transformation of social and ecological relations into zones of capitalist production of surplus value.[45]

Similarly, Donna Haraway's concept of the "Chthulucene," emerging through her engagement with feminist science studies networks, operates through what she characterizes as "tentacular temporality" that refuses the linear progressivism embedded within both Anthropocene optimism and apocalyptic pessimism.[46] The theorization of the "Plantationocene"—a concept that emerged from collaborative interdisciplinary conversations between Anna Tsing, Haraway, and others[47]—foregrounds the colonial plantation as an originary site of modern ecological devastation, emphasizing how racialized labor exploitation and environmental extraction are co-constitutive rather than separate processes.[48]

Fundamental to these critiques is the understanding that the process of rendering the Earth computable was not initiated, as the Anthropocene thesis

[44] Jason W. Moore, "The Capitalocene, Part I: On the Nature and Origins of Our Ecological Crisis," *Journal of Peasant Studies* 44, no. 3 (2017): 594–630.

[45] Jason W. Moore, *Capitalism in the Web of Life: Ecology and the Accumulation of Capital* (London: Verso, 2015), 53–96.

[46] Donna J. Haraway, *Staying with the Trouble: Making Kin in the Chthulucene* (Durham: Duke University Press, 2016), 30–57.

[47] Donna Haraway et al., "Anthropologists Are Talking—About the Anthropocene," *Ethnos* 81, no. 3 (2016): 535–64; Gregg Mitman, "Reflections on the Plantationocene: A conversation with Donna Haraway and Anna Tsing," *Edge Effects Magazine* (June 18, 2019): 2–19, https://edgeeffects.net/haraway-tsing-plantationocene.

[48] Donna Haraway, "Anthropocene, Capitalocene, Plantationocene, Chthulucene: Making Kin," *Environmental Humanities* 6, no. 1 (2015): 159–65.

originally proposed, with the invention of the steam engine in late eighteenth-century England—nor, for that matter, during the Great Acceleration of the twentieth century and the development of computational technologies—but rather with the European imperial colonization of the world and the subsequent development of the capitalist world economy, which in turn incorporated and built upon the racialized and gendered computational logics introduced by colonialism. It was already the colonial plantation, with its discipline and abject alienation of the human and natural world, that introduced paradigms of Earth "optimization" and computational thinking into the processes of agricultural labor so as to maximize the productivity of both enslaved humans and land, establishing templates for organizing labor and landscapes through what Mari Bastashevski discusses in this volume as the "relentless classification" that marks modernity's computational origins.[49] These colonial technologies established templates for organizing labor and landscapes that were later adapted to optimize industrial techniques for maximum profit.

These important critical takes generated rich debates across disciplines. Yet the fundamental epistemological question remains: How does the Anthropocene narrative maintain its constitutive authority over legitimate planetary knowledge? Which political epistemologies enable technoscientific frameworks to continue positioning themselves as uniquely capable of addressing planetary crises while simultaneously marginalizing the relational knowledge systems, place-based ecological practices, and embodied ways of knowing that these and other alternative periodizations demonstrate are essential for understanding planetary transformation? And how do these exclusions operate at the historical conjuncture when planetary emergency demands precisely the kinds of epistemological diversification that dominant Anthropocene discourse renders illegitimate or invisible?

Addressing computational capitalism's epistemological foreclosures requires articulation of what Marisol de la Cadena and Mario Blaser call "A World of Many Worlds" and theorize as a "cosmos composed through divergent political practices that do not need to become the same," where what intersects at unpredictable junctures are worlds, each of which recognizes its own environment.[50] This connects to what Arturo Escobar, Michal Osterweil, and Kriti Sharma in this

[49] Mari Bastashevski, "Pending Xenophora." This volume, 345–71.

[50] Marisol de la Cadena and Mario Blaser, "Introduction: Pluriverse Proposals for a World of Many Worlds," in *A World of Many Worlds*, ed. Marisol de la Cadena and Mario Blaser (Durham: Duke University Press, 2018). See also Marisol de la Cadena, *Earth Beings: Ecologies of Practice across Andean Worlds* (Durham: Duke University Press, 2015), 155–89.

volume describe as "pluriversal horizons"—an onto-epistemic reorientation that recognizes how "we are all within the pluriverse."[51]

Building on this foundation, the following section of this introduction develops a methodological approach oriented toward what remains necessarily irreducible, incommensurable, and incomplete within dominant frameworks of planetary computation. This approach attempts to hold historical materialist and feminist science methodologies in productive tension with place-based and Earth-oriented knowledge. It seeks to cultivate analytical practices that affirm the irreducibly relational dimensions of planetarity. Such practices find powerful articulation in Robin Wall Kimmerer's work, which illuminates the critical epistemological crisis at the heart of contemporary knowledge production through the braiding of Indigenous wisdom and scientific understanding.[52]

*

Kimmerer's writing, which in great part inspired this research and publishing project, reveals the profound limitations of an intellectual landscape dominated by a singular, reductive approach to understanding the world. The fundamental methodological divergence between Western scientific frameworks and Indigenous knowledge systems lies not in their pursuit of understanding but in their fundamental epistemological premises. While both represent sophisticated methods of "reading the land," Western scientific epistemology constrains itself to measurable, observable phenomena, effectively excising the spiritual, relational, and unseen dimensions of knowledge. In contrast, Indigenous ecological knowledge integrates spiritual relationships, holistic understanding, and interconnected ways of perceiving that transcend the narrow empirical boundaries of Western scientific methodology.[53] Kimmerer's critical intervention advocates for an epistemological pluralism that dismantles the false dichotomy between scientific rigor and traditional wisdom and suggests a transformative approach: a symbiotic engagement between scientific methodological tools and Indigenous philosophical frameworks. This is not a simplistic syncretism but a fundamental epistemological reconstruction that challenges the fundamental

[51] Arturo Escobar, Michal Osterweil, and Kriti Sharma, "Pluriversal Horizons: Notes for an Onto-epistemic Reorientation of Technology." This volume, 127–46.

[52] Robin Wall Kimmerer, *Braiding Sweetgrass: Indigenous Wisdom, Scientific Knowledge, and the Teachings of Plants* (Minneapolis: Milkweed Editions, 2013).

[53] Kimmerer's work is situated within the broader Indigenous scholarship critically engaging with dominant epistemological frameworks, including Linda Tuhiwai Smith's critical analysis of decolonizing methodologies and Vine Deloria Jr.'s critiques of Western scientific epistemologies.

assumptions of knowledge production under colonial and capitalist regimes of understanding.

Kimmerer explains how Indigenous ecological knowledge might guide Western science within what she terms a "knowledge garden." She explains this relationship through the agricultural wisdom of planting corn and beans together: The corn serves as lattice for bean plants to climb, while the beans absorb nitrogen from the air and add it to the soil needed for the corn's growth. Yet the bean, as Kimmerer observes, is inherently curious and, powered by nitrogen fixation, keeps wandering in new directions—if unguided, it becomes destructive to the garden, taking over and causing chaos. In this agricultural metaphor, Western science operates as the nitrogen-fixing bean: enormously generative yet desperately requiring guidance from what Kimmerer characterizes as "emotion and spirit and the compass of empathy and compassion" embedded within traditional ecological knowledge systems.[54]

From the *milpa* polycultures of Mesoamerica to the intercropping traditions of East Africa and Asia, from the forest gardens that have flourished across Te Moananui Oceania to the complex mixed cropping systems of West Africa, from the Mediterranean polycultures where olives, grapes, and cereals create layered agroecological systems to the countless other regional traditions of agricultural relationality, communities worldwide have developed sophisticated practices that demonstrate how different species can provide mutual nourishment. These global traditions of companion planting offer profound lessons about relationality, reciprocity, and interdependence that extend far beyond agricultural practice[55]—they constitute embodied epistemologies that challenge the logics of monoculture and separation. They are an example of what Denise Ferreira da Silva calls "difference without separability," where "difference is not a manifestation of an unresolvable *estrangement* but the expression of an elementary *entanglement.*"[56]

Among the global companion planting traditions, the Three Sisters planting method central to many Indigenous North American agricultural traditions,

[54] Kimmerer, *Braiding Sweetgrass*, 89.

[55] In postwar socialist Yugoslavia, women returning from partisan struggle against fascism—among them my paternal grandmother—planted corn and beans to feed starved communities.

[56] Ferreira da Silva's concept of "difference without separability" operates through her sustained critique of modern thought's reliance on what she identifies as the "three ontological pillars"—determinacy, separability, and sequentiality—that legitimate the analytics of raciality. See Denise Ferreira da Silva, *Toward a Global Idea of Race* (Minneapolis: University of Minnesota Press, 2007); Denise Ferreira da Silva, "Toward a Black Feminist Poethics: The Quest(ion) of Blackness Toward the End of the World," *The Black Scholar* 44, no. 2 (2014): 81–97; and Denise Ferreira da Silva, "On Difference Without Separability," in *Incerteza Viva: 32nd Bienal de São Paulo*, ed. Jochen Volz and Júlia Rebouças (São Paulo: Fundação Bienal de São Paulo, 2016), 57–65.

including those of the Haudenosaunee (Iroquois), Cherokee, and numerous other Indigenous nations across Turtle Island, offers perhaps the most inspiring framework for methodological intervention. In this practice, corn provides vertical structure for beans to climb while the beans fix atmospheric nitrogen that enriches the soil for all three plants, and squash spreads across the ground to retain moisture and suppress weeds. This practice, based on reciprocity, interdependence, and mutual aid—might provide a foundation for thinking of a new, symbiotic epistemology that can resist the mono-epistemologies of the globe where nobody lives.

Gratefully inspired by thinking with the three sisters, this volume attempts to plant some seeds for braiding together three distinct yet complementary epistemological traditions as a counterhegemonic response to the monocultural and mono-epistemic totalization of the "world that has granted itself the right to assimilate all other words,"[57] the abstract universalism of the globe. Through the methodological symbiosis of feminist philosophy of science, decolonial and Indigenous epistemologies, and historical materialist analysis,[58] this framework invokes a different Earth and earthly science, in affinity with Wynter's call for a decolonial *scientia* and a new science of being human.[59]

[57] John Law, "What's Wrong with a One-World World?" *Distinktion: Journal of Social Theory* 16, no. 1 (2015): 126–39.

[58] Maria Mies's theorization of "capitalist-patriarchy" demonstrates how women's labor becomes invisible through "housewifization," constituting the foundation of capital accumulation (Maria Mies, *Patriarchy and Accumulation on a World Scale: Women in the International Division of Labour* [London: Zed Books, 1986]). Mariarosa Dalla Costa and Selma James launched the "domestic labor debate" by redefining housework as reproductive labor necessary to capitalism's functioning, rendered invisible by its removal from the wage-relation (Mariarosa Dalla Costa and Selma James, *The Power of Women and the Subversion of the Community* [Bristol: Falling Wall Press, 1972]). Silvia Federici's analysis extends this critique by demonstrating how European witch-hunts served as mechanisms of primitive accumulation, enforcing the separation of production from reproduction through extreme violence against women (Silvia Federici, *Caliban and the Witch: Women, the Body and Primitive Accumulation* [New York: Autonomedia, 2004]). Vandana Shiva's ecofeminist scholarship reveals how "the marginalization of women and the destruction of biodiversity go hand in hand," particularly through development projects that displace Indigenous and women's ecological knowledge systems (Vandana Shiva, *Staying Alive: Women, Ecology and Development* [London: Zed Books, 1988]; Maria Mies and Vandana Shiva, *Ecofeminism* [London: Zed Books, 1993]). Ariel Salleh's concept of "meta-industrial labor" identifies a global class of "housewives, peasants, indigenous peoples—whose reproductive labors minimize risk and hold complex living systems together," creating "metabolic value" that sustains capitalism's bio-infrastructure while remaining unacknowledged (Ariel Salleh, *Ecofeminism as Politics: Nature, Marx and the Postmodern* [London: Zed Books, 1997]). These theoretical interventions collectively identify the constitutive externalities that enable cybernetic systems to maintain apparent coherence through the systematic invisibilization of the regenerative labor that sustains both human societies and ecological systems.

[59] Sylvia Wynter's call for a decolonial scientia and "new science of being human" operates through multiple theoretical interventions across her oeuvre. See Wynter, "Unsettling the Coloniality of Being/Power/Truth/Freedom," 257–337, which establishes her foundational critique of "Man's" overrepresentation as the universal human"; Sylvia Wynter and Katherine McKittrick, "Unparalleled Catastrophe for Our Species? Or, to Give Humanness a Different Future: Conversations," in *Sylvia Wynter: On Being Human as Praxis*, ed. Katherine McKittrick (Durham: Duke University Press,

Each epistemological approach supports and strengthens the others, creating collaborative correspondence that Katherine McKittrick and fields harrington,[60] in this volume, demonstrate through poetics of science capable of maintaining analytical rigor while remaining accountable to collective liberation: Feminist science studies provide methodological frameworks for understanding how knowledge is situated and embodied; decolonial and Indigenous epistemologies offer alternative cosmologies and relational ways of knowing that challenge Western scientific universalism; and historical materialist analysis reveals the economic and social relations that shape both knowledge production and ecological crisis. Each methodological framework both nourishes and draws analytical sustenance from its corresponding incomputable dimension, creating what Sara Garzón, in her contribution to this volume, identifies in the Nahua milpa as "a set of processes that are fully informed by the social fabric created between human, other-than-human, ancestral, spiritual, and historical actants."[61]

Within this symbiogenetic framework, each epistemological tradition assumes a specific role precisely because it operates according to different relationalities and temporal rhythms[62] and what de la Cadena describes as "partial connections"

2015), 9–89, where Wynter articulates her vision for "a new science, a hybrid science: a science of the Word" that understands consciousness and experience as neurochemically determined by social systems rather than genetics or naturalism; Walter D. Mignolo, "Sylvia Wynter: What Does It Mean to Be Human?" in *Sylvia Wynter: On Being Human as Praxis*, ed. Katherine McKittrick (Durham: Duke University Press, 2015), 106–23, which analyzes Wynter's decolonial scientia as a practice of epistemic disobedience that delinks from colonial knowledge systems; and Jennifer D. Adams and Matthew Weinstein, "Sylvia Wynter: Science Studies and Posthumanism as Praxes of Being Human," *Cultural Studies of Science Education* 15, no. 4 (2020): 1–31, which explores how Wynter's "new science" or scientia challenges the ideological construction of "Man" as humanity's paradigm. This theoretical framework enables what Wynter terms the movement "towards the human, after Man" through the development of hybrid knowledge practices that refuse the colonial separation of nature/culture and reason/emotion.

60 fields harrington and Katherine McKittrick, "Poetics of Science/Dialogic Curiosities/ Incomputabilities." This volume, 295–310.

61 Sara Garzón, "The Rise of the Coyote: Towards a Socio-Technological Approach to Worldmaking." This volume, 189–99.

62 Indigenous cosmologies frequently operate through circular or cyclical rather than linear temporal frameworks. See Kyle P. Whyte, "Indigenous Science (Fiction) for the Anthropocene: Ancestral Dystopias and Fantasies of Climate Change Crises," *Environment and Planning E: Nature and Space* 1, no. 1–2 (2018): 224–42; Samantha Chisholm Hatfield et al., "Indian Time: Time, Seasonality, and Culture in Traditional Ecological Knowledge of Climate Change," *Ecological Processes* 7, no. 25 (2018): 1–11. "spiralic time," as theorized in Indigenous contexts, refers to the "cyclical, but transformed for the moment rather than mere repetition," as discussed in Laura De Vos, "Spiralic Time and Cultural Continuity for Indigenous Sovereignty: Idle No More and The Marrow Thieves," *Transmotion* 6, no. 2 (2020): 1–42. Indigenous relational thinking centers "natural law, ethics and protocols rather than human well-being; a focus on a collective beyond humans; and relationality that involves more than ontology," as analyzed in Rachelle K. Gould et al., "Exploring Indigenous Relationality to Inform the Relational Turn in Sustainability Science," *Ecosystems and People* 19, no. 1 (2023): 1–9. On temporality, see Nancy D. Munn, "The Cultural Anthropology of Time: A Critical Essay," *Annual Review of Anthropology* 21 (1992): 93–123; for the prevalence of relationality across diverse cosmologies including "dharma, din, dao, advaita, Buddhism, Confucianism and Sikhi," where "different notions of temporality co-exist and constitute different subjects within the

between knowledge worlds.[63] This represents not the abandonment of empirical precision but its transformation through acknowledgment that objectivity emerges through systematic incorporation of standpoint perspectives, as argued by Sandra Harding's analysis of "strong objectivity."[64]

cosmos," see Tamara Trownsell et al., "Recrafting International Relations through Relationality," *E-International Relations*, January 8, 2019. These relationalities are transmitted intergenerationally through "contextual narratives of the self and the collective, and symbolic, mythic relationships and imagination" that extend care and personhood to landscapes and waters. See Yuria Celidwen and Dacher Keltner, "Kin Relationality and Ecological Belonging: A Cultural Psychology of Indigenous Transcendence," *Frontiers in Psychology* 14 (2023): 994508; Gregory Cajete, *Native Science: Natural Laws of Interdependence* (Santa Fe: Clear Light Publishers, 2000); Margaret Kovach, *Indigenous Methodologies: Characteristics, Conversations, and Contexts* (Toronto: University of Toronto Press, 2009); Linda Tuhiwai Smith, *Decolonizing Methodologies: Research and Indigenous Peoples*, 2nd ed. (London: Zed Books, 2012).

63 The concept of "partial connections" as de la Cadena employs it, refers to "how social (in the broadest sense) worlds may not be commensurable, but communication is possible between them," originally developed by anthropologist Marilyn Strathern. De la Cadena extends this concept to explore possibilities for communication across different onto-epistemic formations, particularly between Indigenous runakuna-earth being relations and the Peruvian state. The concept is central to de la Cadena's methodology for understanding how different worlds can communicate despite fundamental incommensurabilities. De la Cadena, *Earth Beings*.

64 Sandra Harding's concept of "strong objectivity" represents a fundamental reconfiguration of scientific objectivity through feminist standpoint epistemology. See Sandra Harding, "Rethinking Standpoint Epistemology: What Is 'Strong Objectivity'?" in *Feminist Epistemologies*, ed. Linda Alcoff and Elizabeth Potter (New York: Routledge, 1993), 49–82; Sandra Harding, *Whose Science? Whose Knowledge? Thinking from Women's Lives* (Ithaca: Cornell University Press, 1991), particularly chapter 6, "'Strong Objectivity' and Socially Situated Knowledge," 138–63; Sandra Harding, *The Science Question in Feminism* (Ithaca: Cornell University Press, 1986). Strong objectivity builds on the insights of feminist standpoint theory, which argues for the importance of starting from the experiences of those who have been traditionally left out of the production of knowledge. Harding argues that such reflexive analysis, coupled with an active attempt to engage in research from the perspectives of the oppressed, will result in a stronger form of objectivity than the "weak objectivity" available through a neutral approach to knowing that ignores the role of social location and cultural assumptions in shaping one's perspective. This approach acknowledges that "a researcher's life experiences will always be a lens through which they view" their research, making researcher reflexivity essential. This kind of research process starts from questions that arise in the daily lives of people in oppressed groups. To answer such questions, it "studies up," examining the principles, practices, and cultures of dominant institutions, from the design and management of which oppressed groups have been excluded. "Starting off research from women's lives will generate less partial and distorted accounts not only of women's lives but also of men's lives and of the whole social order." For Harding, the prevailing knowledge production community is "too weak" to uncover and eliminate all of its "social values, interests, and agendas," thus to "maximize objectivity" in research, the scientific community needs to incorporate the standpoint of marginalized groups. See also Sandra Harding, *Objectivity and Diversity: Another Logic of Scientific Research* (Chicago: University of Chicago Press, 2015). For critical engagements with strong objectivity, see Susan Hekman, "Truth and Method: Feminist Standpoint Theory Revisited," *Signs: Journal of Women in Culture and Society* 22, no. 2 (Winter 1997): 341–65, and the responses in the same issue by Nancy Hartsock, "Comment on Hekman's 'Truth and Method: Feminist Standpoint Theory Revisited': Truth or Justice?" 367–74; Patricia Hill Collins, "Comment on Hekman's 'Truth and Method: Feminist Standpoint Theory Revisited': Where's the Power?" 375–81; and Sandra Harding, "Comment on Hekman's 'Truth and Method: Feminist Standpoint Theory Revisited': Whose Standpoint Needs the Regimes of Truth and Reality?" 382–91.

Wynter's insistence that humans exist as a "hybrid bios/logos or bios/mythoi level of existence" directly challenges the presumption that the climate crisis can be addressed through purely technical solutions divorced from the mythological frameworks that structure our relationships to Earth systems. Her concept of "autopoiesis"—the self-instituting nature of human orders through narrative—reveals how current climate science, despite its claims to objectivity, remains trapped within the cosmological assumptions of Western Man's extractive relationship to nature. This monoepistemological enclosure prevents recognition of what she calls "sociogenic" rather than purely "ontogenic" knowledge—those collective intelligences and reciprocal cosmologies that understand planetary systems through registers entirely outside technoscientific rationality's reductive logic. The incomputable Earth thus represents both a critique of the computational logic embedded within the Anthropocene narrative and an opening toward a counterhegemonic epistemology of the planetary, in line with the critique advanced by Spivak, Wynter, and de la Cadena. The Earth and its messy complexity is irreducible, as is the climate "problem." As much as there is no one instrument, there is also not one single episteme or one cosmology that can be employed to solve it.

Technomessianic narratives—from Silicon Valley's digital transcendence fantasies to geoengineering and planetary computational management schemes—are based on the temporal enclosure of futurity. The reclamation of the future can only happen by way of mobilizing political imagination, beyond the mere critique of technological solutionism and interrogation of the epistemological assumptions that render such solutions conceivable. Indeed, revolutionary transformation requires both critique of dominant ideas and construction of alternative "common sense," as Antonio Gramsci so pertinently warned. That dual imperative is the mode of critical-creative praxis that must be a part of any counterhegemonic epistemology. This means that critical-creative art-based research and transversal praxis and thinking are not mere cultural interventions but forces supporting the development of collective imagination capacities necessary for what Gramsci called the war of position. In that sense, symbiotic epistemological thinking and the critical-creative praxis it necessarily integrates in the production of new hypotheses might also create a space for imagining radically different forms of "integrative" Earth science grounded in alternative common sense. Indeed, as a speculative leap of political imagination, we might ask: Which historically and materially grounded solutions might emerge from millions of incomputable Earths? Could we imagine "people's Earth science" that centers community-led research over expert technocracy; "decolonial Earth science" that grounds analysis in place-based cosmologies and relational ontologies;

or feminist Earth science that foregrounds care work and reproductive labor as foundational to planetary survival?

*

Throughout the long durée of what Earth System Science designates as the Anthropocene, Western technoscience—despite individual scientists' intentions—has been instrumentalized as apparatus for extraction, militarization, and control. To state this in a volume dealing with climate breakdown and the blind spots of the dominant concepts of human, nature, and technology may read as redundant to many, but it also seems necessary in the face of an avalanche of actor-network-influenced theories that increasingly blur hierarchies of power and chains of responsibilities. Despite the urgent need for new narratives and genuinely transformative hypotheses, this volume diverges from certain tendencies within posthumanism and new materialism that, while offering valuable critiques of anthropocentrism, privilege ontological speculation over concrete analysis. Rather than engaging primarily with questions of being and materiality in abstract terms, we focus on the specific historical and institutional processes through which particular subjectivities and knowledges achieve dominance. The challenge lies in developing conceptual tools that can acknowledge the more-than-human dimensions of ecological relationships without losing sight of the colonial, capitalist, and patriarchal structures that organize both environmental destruction and the knowledge systems deployed to manage it. The problem with posthumanism emerges when recognizing "the intelligence of plants and molds" becomes a substitute for rather than complement to analyzing how capitalism appropriates and commodifies both human and more-than-human forms of labor and collective intelligence.

What, on the other hand, contemporary Marxian ecological critique often overlooks is that the political project of reversing ecological catastrophe cannot succeed without first radically reconceptualizing the very figure of the actionable human as the knower and the agent of revolutionary transformation.[65] This

[65] The debates within ecological Marxism reveal fundamental tensions about knowledge production and political epistemology. See John Bellamy Foster, *Marx's Ecology: Materialism and Nature* (New York: Monthly Review Press, 2000); Andreas Malm, *Fossil Capital* (London: Verso, 2016); *The Progress of This Storm* (London: Verso, 2018); Alf Hornborg, *Nature, Society, and Justice in the Anthropocene* (Cambridge: Cambridge University Press, 2019). Jason Moore's intervention is particularly significant epistemologically because he attempts to move beyond what he identifies as the "Cartesian dualism" embedded in traditional Marxist approaches by drawing on feminist science studies (especially Haraway's "situated knowledges"), world-systems analysis, and attention to the "web of life" that integrates human and extra-human natures (*Capitalism in the Web of Life*). Moore's

reconceptualization requires anthropological deuniversalization: the systematic recognition that categories of human, nature, and technology operate as anthropologically non-universal formations that achieve apparent stability only through ongoing processes of colonial and patriarchal naturalization. In other words, climate breakdown demands not merely political and economic analysis but profound epistemological transformation—recognition that capitalism's instrumental organization of human-nature relationships depends structurally upon the "overrepresentation of Man" as universal human figure. In this navigation between the residual productivist anthropocentrism that continues to affirm the Anthropocene narrative and the post humanist depoliticized ontological speculation, this volume identifies an epistemological position that remains stubbornly committed to human agency and responsibility, through what I referred to as decapitalized humanism. This project is an attempt at an epistemological orientation that neither reproduces the Promethean subject of Western modernity nor dissolves political agency into distributed networks but instead positions human transformative capacity as emerging through, rather than despite, metabolic embeddedness within Earth.

In the context of ecological crisis, capital continues to operate as what Alfred Sohn-Rethel theorized as "real abstraction" through the systematic transformation of qualitatively different ecological relationships into quantitatively equivalent units amenable to calculation, extraction, and exchange. The reduction of forests to "carbon credits," watersheds to "ecosystem services," and people's livelihoods to "natural resources" represents the material reorganization of Earth's systems according to the abstract logic of value posed by capital. Contemporary climate governance operates precisely through such real abstraction: Complex bioregional relationships become standardized data inputs for algorithmic models. While it may not be possible to dismantle the master's house with the master's tools, as Audre Lorde famously put it, it is necessary to consider the genealogy of these abstractions and their tools in order to identify the blind spots within the scaffolding of the master's techno-political epistemology. This book mobilizes incomputability simultaneously as conceptual tool, analytical object, and epistemological proposition. At the same time, thinking from the perspective of the incomputable ultimately advances by thinking through what is capable of exploding the logic of instrumentality from within by considering the political subjectivity and epistemology of those who have historically been

epistemological framework offers important resources for integrating insights from decolonial and feminist standpoint theory in historical materialism.

consigned to the role of the instrument. This is necessary not only in order to locate the monoepistemological mode of thought that we should abandon but also to think the planetary climate breakdown incomputably, to reclaim abstraction, and to reclaim the future. This is what counterhegemonic planetary epistemology in this volume invokes: reclaiming abstraction and making it incomputable.

The Political Economy of Anthropocene Technologies

2

Externality and Necessity between Materialism and Ecology

Marina Vishmidt

Setting off from some of my previous work concerning art and exceptionality, this text will survey some of the implications of thinking in terms of exceptions and externalities, both for the economic paradigms that seek to "value" nature and the political ecology that would take a critical stance on such projects. If we look at a spectrum spanning from feminist theorizations of social reproduction to climate investment models, the relationship between an activity or entity and a recognizable, that is to say economic, form of value must be established. This visibility to value can be thought of as the hallmark of the purported necessity of the activity or entity. This equation of visibility and value has been disputed as a commitment to functionalism in some variants of social reproduction theory, such as the less-dialectical interpretations of the Wages for Housework campaign. A divergence can, however, be charted between an understanding of something as in principle or practice "external" to commodification, or to capitalist value more systemically, and the location of it as nonetheless "necessary" in a way that silently overrides its location in a social relation determined by value. An example that frequently occurs here would be the suggestion that "care" is necessary in any form of society while at the same time bringing the historically specific social relation back in as a further step articulating how that care is functional to its reproduction. I have consistently analyzed this (perhaps unavoidable) vacillation in my writing on social reproduction theory as a definitive tendency in contemporary Marxist feminism and recently sought to frame it in terms of the two notions of "reproduction," which are both operative and elided in it: the reproduction of labor power and the reproduction of capitalism.[1]

[1] Marina Vishmidt, "The Two Reproductions in (Feminist) Art and Theory since the 1970s," *Third Text* 31, no. 1 (2017): 49–66. For a reframing of "social reproduction" in feminist theory as necessarily "capitalist social reproduction," see Martha E. Giménez, *Marx, Women, and Capitalist*

This juxtaposition also allowed me, in specific contexts, to develop a related discussion of art as something that is external to both value and use, thus, in a sense, "nonfunctional" but, if approached from the standpoint of the critique of ideology and institution adopted by Louis Althusser, perfectly functional as an element of the reproduction of capitalist social relations.[2]

In parallel with this, the creation of exceptions and externalities can be seen to be fundamental to the functioning of capitalist economies and the societies they shape: from the centrality of unpaid labor to "surplus value" in Karl Marx; to the role of "cheap nature" in Jason W. Moore's analysis of capitalism in the ecological "web of life"; to other contemporary critical theories of capitalist society that focus on the "ex-" of capital beyond the long-recognized fundamentals of exchange and exploitation (extraction, expropriation, . . .). At the same time, this question of visibility needs to be complicated by the fact that most often nowadays we understand visibility, not to say credibility, in terms of numbers, data, and statistics, signaling the saturation of many social and political debates with the truth criteria of the economic that Michel Foucault diagnosed a long time ago as the hallmark of neoliberalism. Although this is a subject that merits a far more extensive and historically informed discussion than there is room for in this essay, in many ways the connection between the visibility of the subject and the verification provided by numbers inflects normative liberal and oppositional political grammars, as with the widespread vocabulary of "accountability" and "transparency" used to refer to behaviors that inspire trust.

Assessing this dynamic of inclusion and exclusion from a strategic, as well as a theoretical, standpoint makes it possible to see ways of moving away from "accounting" paradigms in emancipatory politics more generally, whether the question is framed in terms of socialist ecological transition or critical aesthetics. Here, the concept of accounting, or accounting for, should be linked more broadly to the question of representation, though that is of course a much larger question than can be engaged with here. A touchstone for this inquiry recently came into view during a lecture by Christoph Menke, where he made a distinction between critique in general and the kind of genealogical critique pursued by Marx, which, in his terms, is the difference between exposing the

Social Reproduction (Leiden: Brill, 2018). Giménez's work draws substantively on Althusser, whom I also looked to in elaborating the second of the "two reproductions." See Louis Althusser, *On the Reproduction of Capitalism: Ideology and Ideological State Apparatuses*, trans. G. M. Goshgarian (London: Verso Books, 2014).

[2] See, indicatively, Marina Vishmidt, "Anomaly and Autonomy: On the Currency of the Exception in the Value Relations of Contemporary Art," *Zeitschrift für Kunstgeschichte* 81, no. 4 (2018): 588–600; Marina Vishmidt, *Speculation as a Mode of Production: Forms of Value Subjectivity in Art and Capital* (Leiden: Brill, 2018/Chicago: Haymarket, 2019).

content behind a specific form and asking why a specific form appears as it does. This would not be to say, "Here is the labor behind value!" but rather, "Why does labor take the form of value?"[3]

I would like to start with a suggestion, which, though somewhat counterintuitive, has clear ecological implications. This is that externality is intrinsic to the capitalist mode of production in general and to a phase of production dominated by extraction in particular. While it is clearly visible that the process of value being converted into waste is constantly being condensed in time, political ecology tells us that the construction and devaluation of nature, from not-so-human life to "naturalized" gendered and racialized labor, is a precondition of commodification. This has been analyzed in terms such as "permanent primary (primitive) accumulation" or, latterly, "accumulation by dispossession." It can likewise be found in the "world-ecological surplus"[4] discussed by Moore, which he uses to refer to the exhaustion of resources by the expansion of surplus capital. Such a concept of the parallelism between capitalist production and waste aligns with Marx's identification of the "surplus population," which is corollary with the capitalist mode of production; in other words, the expulsion of workers from production as unnecessary to the goal of valorization. Thus, exhaustion, expulsion, and extraction without reproduction (of labor power, of nonhuman life) are as key to the reproduction of capital as the reproduction of labor power. As it is often understood, the contradiction identified by Marx between absorption and expulsion of labor power at the basis of his theory of the "industrial reserve army" is that every act of production performed by the worker is simultaneously the production of their own superfluity.[5] This close link between the production of economic surplus and the production of a surplus population is necessary to the reproduction of capital as a class relation. Such a focus has lately been revisited, if not always in the vocabulary of the critique of political economy, in projects looking at racialization and how it facilitates the "surplusing" of specific communities (mass incarceration, murderous policing), as well as the role such surplusing plays in regimes of urban accumulation and the pushing of resource frontiers (land expulsions, housing displacement, gentrification, "sacrifice zones").

This thinking of how externality is internal to the value cycle entails a reflection on the relationship between extraction and exploitation and whether

[3] Christoph Menke, "The Critique of Law and the Law of Critique," Annual Philosophy Lecture, Goldsmiths, University of London, London, May 29, 2019.

[4] Jason W. Moore, "Nature in the Limits to Capital (and Vice Versa)," *Radical Philosophy*, no. 193 (2015): 9–19.

[5] Karl Marx, *Capital: A Critique of Political Economy, Volume One*, trans. Ben Fowkes (1867; London: Penguin Books, 1976), 793.

extraction marks a new configuration of absolute surplus value or simply a new analytic emphasis. Likewise, an understanding of how race and racialization figure into capitalist social relations can depart from an analysis defined around the three "exes," as recently proposed by Chris Chen and Sarika Chandra in their contribution to the *Totality Inside Out* reader: expulsion, exploitation, and expropriation, or, as they write, "how capitalism produces relational interlinkages among different domains of social life through a general measure of capitalist value."[6] They go on to define the three categories. Expropriation is the structural centrality to the capital of processes of dispossession; structural rather than initial, as in the concept of "so-called primitive" (primary/originary) accumulation that kickstarts capitalism but then somehow disappears into more routine forms of exploitation. These processes of dispossession are not just about theft of resources but their subsumption to specifically capitalist circuits of value generation. Thus, expropriation and exploitation can be seen as not always really distinct. Expulsion, on the other hand, refers to the making of surplus populations for capital's cycles of valorization and the displacement and removal of communities from rural and urban spaces, which are seen to be more valuable/controllable without their presence.[7] The triple lens (of expulsion, expropriation, and exploitation), it is contended, precludes the need for identifying a primary determination in race, class, or any other social location or ascriptive identity while positing material grounds for solidarity. The approach proposed by Chen and Chandra underlines the structural weight of the expulsion, expropriation, and exploitation of differentiated and stratified human labor power and nonhuman natures. It also outlines how more ontological inquiries have applied such lenses to the category of the "human" itself and not only the reproduction of capital, both logically and historically. There is a further basic principle here that Moore reminds us of, namely that the externality to value is key for the reproduction of value because the appropriation of "free resources" is not only necessary to the production of value that happens through commodification but may actually be *more* productive of value than the exchange of commodities.[8]

As Liam Campling and Alejandro Colas review this point in their book *Capitalism and the Sea*, "appropriation in new commodity frontiers is greater

[6] Chris Chen and Sarika Chandra, "Remapping the Race/Class Problematic," in *Totality Inside Out: Rethinking Crisis and Conflict under Capital*, ed. Kevin Floyd, Jen Hedler Phillis, and Sarika Chandra (New York: Fordham University Press, 2022), 157.

[7] They write, "We can understand the dynamics of contemporary comparative racial formation then as inscribed within processes of expulsion and expropriation constituted in relation to economic exploitation." Chen and Chandra, "Remapping the Race/Class Problematic," 156.

[8] Jason W. Moore (ed.), "The Rise of Cheap Nature," in *Anthropocene or Capitalocene? Nature, History, and the Crisis of Capitalism* (Oakland: PM Press, 2016), 78–115.

than in existing zones of commodification, allowing for the competitive increase in relative surplus value through productivity gains."[9] While this need for there to be an externality to value in order for value to be created is exemplified in their work on industrial fishing and the extraction of aquatic life-forms as exchange value on global markets, it is clear that unvalued or devalued labor and "resources" more generally manifest at the frontiers of extraction as direct appropriation. Other illustrations could be imagined as well, such as the commodification of networked data in platform economies as a challenge to an orthodox Marxist analysis of commodification versus direct appropriation in these terms.[10] Nonetheless, if, as it has been recognized from Marx to Rosa Luxemburg (in her theory of imperialism, for example, which elaborates this process on a geopolitical scale), the appropriation of free stuff—or at least stuff that capital does not pay for but then valorizes further in the cycle of production or expels as waste—is key to the accumulation of capital per se, then it can still be suggested that capital, in an era characterized by extraction rather than production, becomes dependent on this dynamic of devaluation and extension of the commodity frontier as never before.[11]

In other words, we can see how shifting the locus of critical analysis from the disposability of human labor, as it has been so radically underlined by the effects of the pandemic, to the fabric of nature shows us how the transition from value to waste, or the direct conversion to waste, is unfolding on an unprecedented scale. At the same time, the engagement with categories such as "surplus population" (as already noted), "wageless life," "wasted life," and many of the discussions around precarity index the topicality of this shift in terms of labor and social reproduction.[12] The phenomenon of waste can also be tracked in a more ecological register, as seen in accelerating extraction for cheap energy purposes, which can be illustrated by tracing the continuum from mining to artificial intelligence, or from fracking to cryptocurrency and nonfungible tokens (NFTs). In the meantime,

[9] Liam Campling and Alejandro Colas, *Capitalism and the Sea: The Maritime Factor in the Making of the Modern World* (London: Verso Books, 2021), 183.

[10] Jathan Sadowski, "When Data Is Capital: Datafication, Accumulation, and Extraction," *Big Data & Society* 6, no. 1 (2019): 1–12.

[11] The question here, however, could be whether alongside the commodity frontier should be considered the "labor frontier," human and not, given that nature is also put to work in this process, and that naturalization, often enacted through the vectors of gender and race, is a way of cheapening labor power for capital. This point has also been made by a number of Marxist ecofeminists whom Moore is drawing on, such as Silvia Federici, Ariel Salleh, Vandana Shiva, Maria Mies, and others.

[12] See Michael Denning, "Wageless Life," *New Left Review* 66 (2010): 79–97; Neferti X. M. Tadiar, *Remaindered Life* (Durham: Duke University Press, 2022); as well as the emergent disciplines of "waste studies" and "discard studies," with a useful publication here being Max Liboiron and Josh Lepawsky, *Discard Studies: Wasting, Systems, and Power* (Cambridge, MA: MIT Press, 2022).

labor as an economic or political actor is steadily eclipsed, turning "wageless life" into an endlessly mutable field of accumulation and extraction subject to the marginalizing forces of criminalization and abandonment, and the already cited extension of the commodity form moves further and further into zones of human activity in ways determined by the logic of racialization. This means that it is not the commodification of human labor but the growth of profit extraction from the maintenance of sections of humanity as a surplus (from prison labor to microcredit, predatory mortgages to court fees).[13] As the economic geographer Arthur Scarritt demonstrates in his work on neoliberalism as neofeudalism, these extractive logics actually reproduce class relations more successfully than they produce profit, which contemporary low-growth, stagnant economies in the West would seem to confirm: "With the neoliberal turn away from exploitation the question centers on controlling a population superfluous to the functioning of the system. And a central part of this question focuses on how to extract rents from this population to support the overall aristocratic order."[14] All this suggests that when it comes to arguments about externalities to value, it is important to break down at what point the external becomes not only internal but also even comes to constitute a hegemonic influence in a historical and systemic phase of the capital relation. In the background of this already quite substantial question, a revised understanding of economic rationality in capitalism must be developed to replace the reductive analysis on the Left that has often led to correspondingly ineffective forms of organizing. Parenthetically, this could be related to, but is not identical with, the question of how the division between the economic and the political is drawn in many formulations of anticapitalist critique.

Prior to the more expansive inquiry sketched above, I have long been interested in how that which is considered exceptional to value production—including activities as different in their relation to the market and to prestige, as art making and housework—is actually constitutive of it and how the normativity of waged labor is just a small fraction of a broader "hidden continent," as Silvia Federici called it, of unwaged labor.[15] At the same time, I was not sure whether

[13] An important recent synthesis of these trends, albeit focused on the US, can be found in Jackie Wang, especially in the first two essays, "Racialized Accumulation by Dispossession in the Age of Finance Capital: Notes on the Debt Economy," 99–150 and "Policing as Plunder: Notes on Municipal Finance and the Political Economy of Fees and Fines," in *Carceral Capitalism* (Los Angeles: Semiotext(e), 2018), 151–92.

[14] Arthur Scarritt, "Tripartheid: How Global White Supremacy Triumphs through Neoliberalism," in *Totality Inside Out: Rethinking Crisis and Conflict under Capital*, ed. Kevin Floyd, Jen Hedler Phillis, and Sarika Chandra (New York: Fordham University Press, 2022), 117.

[15] Silvia Federici, "Precarious Labor: A Feminist Viewpoint," Bluestockings Cooperative Bookstore, New York, October 28, 2006.

the solution was the recognition or analysis of all these activities as labor, in line with a dominant Marxian "ontology" of labor perspective that informs a great deal of superlatively important work in social reproduction theory. This seemed to presuppose a certain focus on productivity that was inadequate for making a distinction between capitalist production and production in general or, as I have put it elsewhere, social reproduction and capitalist social reproduction with the suggestion, originally proposed by Martha Giménez, that the idea of social reproduction is always capitalist social reproduction and has no coherence outside of this social and historical determination.[16] With the question of art, it is often considered exceptional to both the metrics of value (in production rather than in circulation) and the productivity of labor, so it is a good crystallization of how the dynamic of exceptionality operates to both undermine and confirm the form of value as the matrix of capitalist social life.[17] What became increasingly clear over years of working at this juncture is that labor itself was getting more and more marginalized or "surplused," with value increasingly being extracted directly via processes of expropriation, expulsion, and exploitation than it was when I began. The point, therefore, is perhaps less about exceptionality per se or the integrative and reproductive work the ideology of autonomy was doing in both glorifying and repudiating the heteronomy of value relations in the dialectical template of art encapsulated by Theodor Adorno. Rather, by thinking about the shifting balance between exploitation and extraction, with capital's tendencies toward the latter supercharged by the deadly biopolitics of the pandemic, I found myself moving toward figuring out where art, as an institution and as a set of material and social practices, was situated in a landscape organized around extraction rather than production as capital's hegemonic norm. While I was trying to move in this direction, NFTs happened with their premise of using a specific, highly energy-consuming technology—the proof-of-work architecture of the blockchain—to turn literally anything that could be made into a digital image on the Internet into an equivocally legal and often unenforceable property claim.

In my work on financialization and art in the speculative mode of production, speculation was understood as a key cultural logic of neoliberal subjectivity, here figured as competitive self-capitalization and the forms of social organization that enforce it, chief among them privatization and debt. It was also an exploration of how modernist notions about the autonomy of the artistic subject

[16] Giménez, *Marx, Women, and Capitalist Social Reproduction.*
[17] Vishmidt, "Anomaly and Autonomy."

as exceptional to value relations prefigured the "creative" as the archetype of capitalist subjectivity, showing how autonomy is the feedstock of precarity. In this logic, art is that which makes labor and its claims disappear, be it in specialized genius or the generalized creativity of human capital. With the NFT, we are seeing something closer to the literalization of this "logic," the far-right dream of the monetization of everything that exists, or at least every form of its mediation. Can we consider this automated enclosure of externality—beyond its perhaps more empirical register of pyramid scam/bubble that characterizes the "crypto space"—as a kind of terminal point, for now, in capital's dream of making labor disappear, an externality that is always and already waste?

With my previous work, I have also sought to counter arguments about the "real subsumption" of art by showing that art is mediated by capital in such a way that it needs to remain partially external in its mode of production and valorization in order to be integrated in the specific way that it is, namely, as both "autonomous" and a "social fact," in Adorno's terms. Unlike other luxury goods, to which at least the art market dimension of art production is functionally similar, art is positioned as de jure ideologically critical of capitalist society, as standing for a different set of values. In this, art production is related to, although different from, a number of other nonvalue-producing or "indirectly market-mediated" sectors, especially the "nonprofit" and publicly funded institutional sectors. This was always an argument for the significance of externality to the reproduction of the capital relation, logically and historically. This significance had to be highlighted as a critical method in addition to, and perhaps in skepticism of, the centrality of those who took integration and subsumption—that is, the production of homogeneity and closure—as their touchstone and thus downplayed not only the importance of exceptionality to those dynamics but also the critical and political salience of negation and struggle. Likewise, there was a skepticism about the traction of establishing visibility and value, often through the category of labor, as adequate understandings of, and challenges to, the capitalist norm. The political implications of this were, of course, dialectical ones. Because productive or waged labor (and there is no necessary equation between the two) is not universally definitive of the capital relation, the appropriation of life and free labor, and organizing against this appropriation (workers' organizing, for example) is crucial. Conversely, the analysis of subsumption, rather than extraction, externalization, and waste, seemed to have idealist consequences, as already noted earlier in reference to social reproduction. In Marx's terms, this idealism encourages contemplation rather than action, so that the extension of the category of subsumption of labor to all experience becomes an ontological move rather than a critical one, and it tends toward the consolidation of domination in the sphere of analysis rather than developing tools for transformation.[18] Subsumption can be considered a diagnosis of integration and thus is affirmative, while extraction and the other exes, as in the Chen and Chandra discussion referenced above, have no legitimation apart from that of capital expanding itself—if not necessarily as profit, at least as simple domination—and thus are more generative of antagonism in theoretical discourse and in the lived reality of social life, which includes art. If I am now turning to a more direct engagement with externalities, or extraction, it is because it is unavoidably clear that rent-seeking, extractive, and ecocidal forms of accumulation have come to dominate our world to such an obvious extent that it is no longer a critical proposition to argue for the centrality of waste and destruction to the capitalist value cycle.

Having established this context for my inquiry into art and the externalities of value, I would now like to return to the discussion of climate investment models and feminist theories of social reproduction insofar as they relate to thinking about the broad paradigm of valuing the invisible and the socially naturalized, albeit for very different ends. The analyses I have undertaken of the registration, or occlusion, of labor in art connect to the discussion to follow via the proposition that the recognition of labor, where before there was only nature, or free activity, is the first step in inscribing an activity, a subjectivity, or a space into value relations. It thus becomes possible to analyze them through a lens of systemic reproduction and totality, as well as differentiated composition and the perpetuation of inequality. There is, however, a problematic link between this critical recognition and the strategic affirmation of such elements externalized by the capitalist value form, or capitalist markets, depending on the level of abstraction of the inquiry. This link must, therefore, be fully comprehended, in all its dialectical and structural necessity, in order to avoid the generation of solutions to capital's valorization crisis from the Left. This should be undertaken while recognizing, as Ed Graham writes in his review of Kate Soper's latest book, "the economic necessity behind environmentally damaging consumption patterns."[19] Thus, a focus on populations as consumers should be filled out with a focus on the valorization of value at the level of global capital and the capital in terms of labor relations as a whole, as well as the externalization of needs that

[18] Karl Marx, "Theses on Feuerbach," Thesis 1, 1845.

[19] Ed Graham, review of *Post-Growth Living: For an Alternative Hedonism* by Kate Soper, *Marx & Philosophy Review of Books*, March 29, 2022, https://marxandphilosophy.org.uk/reviews/20107_post-growth-living-for-an-alternative-hedonism-by-kate-soper-reviewed-by-ed-graham.

creates profitable pools of low-value and decommodified labor and life for the spiral of devaluation in which capital is currently trapped.

The problem of recognition and its connection to the prospects of emancipatory change can be approached theoretically in different ways. There is, for example, an extensive discourse in contemporary post-Frankfurt School critical theory that centers on recognition. A counter-discourse claims that this move abandons the "critical materialist" framework in critical theory in favor of an ethical and idealist approach whose presupposition is equality and homogeneity rather than an antagonistic society of ever more intensified forms of exploitation and oppression, all of which are exacerbated, globally as well as locally, by climate breakdown.[20] There is an analogous discourse targeting assertions of "human rights" and, latterly perhaps, the rights of nature, insofar as these are positioned as given, primary, and rationally administered, rather than a conditional territory gained in asymmetrical struggles between, for example, state power and capital vis-à-vis affected communities. Both of these critiques are informed by a resistance to the naturalization or ontologization (and there is also, of course, an anthropological register in which these claims are made) of political configurations that are actually fundamentally social and historical in nature and derive their political character from this in the first place. The critical framework of these objections can be informed by one or a combination of historical materialist, poststructuralist, feminist, and critical race theory, as well as paradigms linked to queer and Indigenous scholarship and epistemologies.

I would add another lens to this collection that is concerned with repudiating claims, which are ontologically rather than historically grounded and come from the specific trajectory of Marxist value theory, and the critiques of race, gender, and nature that emerge from this trajectory. An example of this would be critiques of social reproduction theory, which suggest that its focus on gendered labor can take on dimensions of naturalizing both labor and gender, including "The Logic of Gender" by the Endnotes collective (Maya Gonzalez and Jeanne Neton), the work of Amy De'Ath on the loss of focus in this theory on capitalist abstraction, some of the writing I have been doing with Zöe Sutherland, or Beverley Best's recent piece, "Wages for Housework Redux: Social Reproduction and the Utopian Dialectic of the Value-Form," which I will focus on here.[21]

[20] See the introduction to the three-volume SAGE handbook of Frankfurt School critical theory for a good summary of this debate: Beverley Best, Werner Bonefeld, and Chris O'Kane (eds.), *The SAGE Handbook of Frankfurt School Critical Theory* (London: SAGE, 2018), 1–16.

[21] Endnotes, "The Logic of Gender: On the Separation of Spheres and the Process of Abjection," *Endnotes*, no. 3 (September 2013), https://endnotes.org.uk/issues; Amy De'Ath, "Reproduction," in *The Bloomsbury Companion to Marx*, ed. Jeff Diamanti, Andrew Pendakis, and Imre Szeman

The point I would like to take from Best's essay is her assertion that social reproduction "doesn't count" for the value metric ensuring the reproduction of capital because it is, in other terms, an "externality" and this is precisely why capitalism is a life-annihilating mode of production—that which does not count for it can be destroyed. As she writes,

> The unwaged activities of social reproduction, potentially the very definition of wealth in an associated mode of production, cannot be objectified (or, therefore quantified) as wealth in capitalist terms. Neither can photosynthesis, mineralization, the water cycle, or the mating of polar bears. None of these activities/processes "count" for capital, because none can be quantified by capital. Value is an *accounting category*; capital is a system of domination by math. It is not that social reproductive activity doesn't "count" as significant in Marx's analysis; it is that, quite literally, unwaged social reproductive activity doesn't "count" for capital. A careful accounting of value-generation is a matter of grasping how and why a capitalist mode of production is anathema to human and planetary well-being.[22]

What is clarified here is that the problem is not a faulty mode of accounting or the lack of an emancipatory theory that can diagnose blind spots in that mode of accounting with a view to compensate for them (as, for example, in feminist economics that look at household economies in particular); it is rather the necessity of taking up the challenge of the dialectic between the call for "Wages for Housework" and "Wages Against Housework" that initiated the social reproduction perspective in Marxist feminism and come to grips with how valuation works in capital's accounting. That is, to understand valuation as signifying both value relations and what they structurally exclude, as well as the attempt to capture different aspects of those relations in the metric of value as being bound to the commodity form. This includes the dimension of fetishism, which internalizes this metric as the most direct channel to recognition and thus compensation of an activity; a fetishism that is socially necessary and thus to be understood dialectically rather than as ideologically deluded or morally compromised. What Best's analysis also elucidates, if only implicitly in the passage quoted, is that this externalization of activities as nonlabor, that is, noneconomic,

(London: Bloomsbury, 2019), 395–404; Zöe Sutherland and Marina Vishmidt, "(Un)making Value: Reading Social Reproduction through the Question of Totality," in *Totality Inside Out: Rethinking Crisis and Conflict under Capital*, ed. Kevin Floyd, Jen Hedler Phillis, and Sarika Chandra (New York: Fordham University Press, 2022), 67–90; Beverley Best, "Wages for Housework Redux: Social Reproduction and the Utopian Dialectic of the Value-Form," *Theory and Event* 24, no. 4 (2021): 896–921.

22 Best, "Wages for Housework Redux," 901, emphasis original.

is symptomatic of the aforementioned accelerating crisis tendency of capital in which the more it expels as waste, or irrelevance, and the less value it produces from what it can subsume, the less value is produced and the more extraction (the desperate part of the cycle of valorization) comes to predominate, undercutting the sustainability of an already principally unsustainable mode of production. Best writes "that exploitation is capital's only means of growth is ultimately its fatal dysfunction. Characterized by Marx as capital's crisis tendency, it is an internal contradiction, unfurling slow-motion car-crash-style, particularly as it mediates various contemporary forms of oppression and group-differentiated vulnerability."[23]

By way of moving on from this question of what results are yielded in recognizing externalities of value as internal to value in capital-critical paradigms—as well as other critical paradigms that diagnose the centrality of racialized, gendered, and illegalized labor to capitalist accumulation and capitalist social reproduction—I would propose its recontextualization by looking at the forms of capitalist accounting that, at the most general level, try to price various aspects of the environment in the ostensible attempt to promote robust ecological improvement outcomes by making them market-immanent rather than external to the market. The most obvious examples of this that we have witnessed over the past decade, apart from the general vocabulary of ecosystem services or natural capital, are carbon trading schemes based on a dynamic of displacement of, or compensation for, pollution between richer and poorer nations or carbon budgets, which can be defined as the total amount of carbon dioxide emissions that human activities can be allowed to generate in order to avoid excessive global warming. These measures, which can be adopted at global, national, or other more local scales, are profoundly indeterminate because their guidelines depend on the sensitivity of the metrics adopted and the range of climate sensitivity to carbon emissions that these metrics aim to capture, while timelines and local variations induce uncertainty. It is indicative of these technocratic or "solutionist" approaches to ecological problem-solving that, whatever they appear to have in common with the critical approaches to unrecognized or externalized labor, they assume that if the market can recognize something as a commodity, it can value it, and then steps will be taken to sustain it as a source of value. Anticapitalist critics of such notions, however, claim that it is instead a matter of *decommodifying* carbon, which may include tactics such as disinvestment, aimed at destroying the value proposition

[23] Best, "Wages for Housework Redux," 901.

of polluting "resources," but on a longer-term, strategic level, requires a shift to a noncapitalist mode of social production and reproduction.

On this point, Peter Somerville, a UK-based climate change policy analyst, notes in his recent essay "Revisiting Connections between Capital and Nature II: The Case of Climate Change" that not only must current polluting energy resources be side-lined, the "green growth" capitalism that is responsible for land grabs, expulsions, and further forms of degradation of the natural and social world in the drive for an ecological transition is also not tenable.[24] He writes, citing Sarah Knuth,

> this categorical decommodification of fossil fuels contradicts the tendency of ecological economists and others to treat nature as if it holds or should hold capitalist value. However, some kinds of anti-fossil capital action, such as on divestment, may not be sufficient to achieve such decommodification (they could just mean reducing the amount of fossil fuels being produced), and may also be compatible with a shift to green capitalism based on renewables. There is a risk that commodified fossil fuels will simply be replaced by commodified renewable energy, so that capitalism continues more or less as before.[25]

This is a problem that, he suggests, few programmatic formulations of the Green New Deal adequately take into account. Therefore, the problem here is perhaps, as with social reproduction theory, the problem of recognition of value. More precisely, the problem is that such a recognition is necessary as an initial basis for the reconfiguration of a mode of production but works by means of making political claims on the basis of some of the structurally necessary elements that it wishes to avoid supporting or accounting for. Taken to its terminal point, this is an argument that would overwhelm capitalist rationality by making it pay and account for everything. However, we see that this is insufficient as an emancipatory critique, or a program for social transformation away from crisis capitalism, in the example of carbon accounting, which artificially sustains fossil fuel markets and actually delays transitions to other energy sources while also, secondarily, preserving the infrastructures of market valuation in a posttransition centering of renewables with the schema of extraction, expropriation, and exploitation—to return to the earlier exes—that this perpetuates. A substitution where not

[24] Peter Somerville, "Revisiting Connections between Capital and Nature II: The Case of Climate Change," *Capitalism Nature Socialism* 32, no. 2 (2021): 40–55.

[25] Somerville, "Revisiting Connections between Capital and Nature II," 44; Sarah Knuth, "Green Devaluation: Disruption, Divestment, and Decommodification for a Green Economy," *Capitalism Nature Socialism* 28, no. 1 (2017): 98–117.

only many things stay the same, but the substitution is undertaken precisely in order to ensure this outcome, as per Giuseppe Tomasi di Lampedusa.[26]

So where does that leave us in light of the foregoing attempts to parse the intricate and often twisted relations between the politics of visibility, which strives to render an account and thus go some way toward normalization, and the structural centrality of the exception to capitalist valorization? This is both a conceptual and a political problem. There are a number of projects in critical theory that have undertaken to disclose and situate these exceptions, but the question remains open whether, without a determinate horizon of negation of the present with emancipatory effects, any project of making visible does more than consolidate the intractability of the present.[27] There is also a sense that the disclosure and debunking of exceptions as constitutive needs to emphasize—and poignantly so in our current era of geophysical collapse—systematic reproduction rather than value production and the interconnections between the living and dead, concrete and abstract, that perpetuate the deadly system keeping us only questionably alive, and one which is nevertheless fragile.

The phenomenal intensity of violence that we are currently witnessing inflicted by state forces on defenseless populations is a symptom of the intractable colonial histories that are neither faced nor altered in their robotic persistence in converting all human and nonhuman nature into waste. The fact that coloniality is structured by the eliminationist instinct of property means that psychosis and rationality are intertwined as capital's most basic infrastructure of feeling. Likewise, at a time when public feelings are roused by notions of "taking back control," it is imperative to subsume neither the present, in all its contradictory particularity, to the business as usual of capital nor the exceptionality of a moment of escalating political and ecological chaos, as both engender sacrificial forms of thinking and agency—sacrificial when it comes to designated human and not-so-human others. Rather, the changes to be contemplated and enacted should be drawn from the negativity of our situations, which can emerge in situations of antagonism (direct contestation of the complacency and cynicism of power), appropriation (the "destituent" emptying it of the resources it extracts from us in order to do violence to us), and abolition (a determinate negation not only of a harmful institution, or practice, but of all the ways of accommodating it

[26] The so-called "di Lampedusa strategy": "If we want things to stay as they are, things will have to change" is voiced by a character in Giuseppe Tomasi di Lampedusa, *The Leopard*, trans. Archibald Colquhoun (1958; London: Collins and Harvill Press, 1960), 31. The Palme d'Or-winning 1963 Luchino Visconti film adaptation is better known than the novel.

[27] Endnotes, "Error," *Endnotes*, no. 5 (October 2019), https://endnotes.org.uk/issues.

that have become socially habitual while placing other possibilities into reality). These are, of course, fragmented and intuitive hints rather than propositions, much less arguments. However, their direction of travel is, as already noted, that of negation, but also negation as a process that is constantly differentiating itself and creating other possibilities. How to draw the boundary between this restless negativity as a guiding rope for emancipation and the hopeless mutability of a society committing suicide-by-commodity, remains an enormous array of questions for another day—a day that cannot be postponed.

3

Between the Planet and the Market

Gary Zhexi Zhang

The Future Has Already Been Distributed, It's Just Not Here Yet

> *Financial activity creates bonds. It opens positions, engages deals, and generates exposure. But, at a crucial point in time, obligations are settled, payments are concluded, counterparties are detached, contracts are terminated, assets are transferred and markets are cleared. Finance is, in part, an art of finalization.*
>
> —Fabian Muniesa[1]

The construction and distribution of planetary futures depend not only on geoengineering but also on temporal engineering: the social, technical, and economic infrastructures that govern time and expectation and the narrative processes by which they become systemically legible. Put another way, the politics of the Anthropocene are grounded in a temporal conflict between the time of species, fossilized carbon, sea-level rises, generational debts, political tenures, economic cycles, earnings reports, news cycles, and price signals. Whereas modernity and its attendant economic models were predicated on universal time and infinite resources, the Anthropocene reverses the equation, revealing a contested world of finite and fragmented temporalities in which futures have been hoarded, constrained, and stolen. This planetary state of spatiotemporal disorientation is difficult to narrate, because it is weird. As the novelist Elvia Wilk

[1] Fabian Muniesa, *The Provoked Economy: Economic reality and the performative turn* (Oxon: Routledge, 2014), 48–49, with reference to Yuval Millo, Fabian Muniesa, Nikiforos S. Panourgias, and Susan V. Scott, "Organised Detachment: Clearinghouse Mechanisms in Financial Market," *Information and Organization* 15, no. 3 (2005): 229–46.

asks, "How can a person *feel* a planetary-scale system?"[2] The future is not only radically uncertain but also encountered as a concatenation of disparate epistemic communities, spatial distributions, and perceptual scales that are fragmented and cacophonous in their simultaneity. If the literary subject attempting to sense the planetary is confronted by the weird and the irreconcilable, then the capitalist subject cuts through the quagmire with a delusional method of highly selectively reading. As the Stern Review on climate economics reported in 2006, "Climate change presents a unique challenge for economics: it is the greatest and widest-ranging market failure ever seen."[3] Mark Carney, then governor of the Bank of England, warned in 2019 that the climate crisis threatened the markets with a "Minsky Moment," whereby massive and sudden adjustments in expectations would cause a fire sale in assets and cause a global economic collapse.[4] Greenhouse gases are, in economic terms, an externality: "Those who produce greenhouse-gas emissions are bringing about climate change, thereby imposing costs on the world and on future generations, but they do not face the full consequences of their actions themselves."[5] As US treasury secretary Larry Summers put it in 1991: "I think the economic logic behind dumping a load of toxic waste on the lowest-wage country is impeccable and we should face up to that."[6] The reality of the climate crisis presents a powerful dissonance that markets are structurally unable to compute. Investments kick the can down the road, trading increasingly catastrophic prospects for business as usual. And yet, climate change represents a singular threat to capitalism, whose value remains largely grounded in land, real estate, and securities linked to physical assets. What it would take for the planet to survive the market?

The planet's de facto temporal infrastructure is finance, which emerged in the seventeenth century to open up a probabilistic world by which time could be engineered to mitigate the insecurities of society and nature. Statistical inventions like the law of large numbers made way for the socialization of risk, making it possible to insure against everyday catastrophes, from fires to illness to sunken merchant ships. In his *Essays Upon Projects*, Daniel Defoe waxed

[2] Elvia Wilk, "Toward a Theory of the New Weird: Elvia Wilk on a Feminist Understanding of Eerie Fiction," *Literary Hub*, August 5, 2019, https://lithub.com/toward-a-theory-of-the-new-weird.

[3] Nicholas Stern, "Stern Review on the Economics of Climate Change - Executive Summary (full)," 2006, i, https://webarchive.nationalarchives.gov.uk/ukgwa/20100407172811/https:/www.hm-treasury.gov.uk/stern_review_report.htm.

[4] "Open Letter on Climate-related Financial Risks," *Bank of England*, April 19, 2019, https://www.bankofengland.co.uk/news/2019/april/open-letter-on-climate-related-financial-risks.

[5] Stern, "Stern Review," xviii.

[6] Larry Summers quoted in "Furor on Memo At World Bank," *New York Times*, February 7, 1992, https://www.nytimes.com/1992/02/07/business/furor-on-memo-at-world-bank.html.

lyrical on "friendly societies," a type of collectively managed insurance fund that emerged in the seventeenth century: "All the contingencies of life might be fenced against by this method (as fire already is), as thieves, floods by land, storms by sea, losses of all sorts, and death itself."[7] Defoe's image of insurance suggests a spirit of mutualism—of collectivizing risk as well as reward—by which collectives may protect themselves from uncertainties held in common. While this imaginary mastery of time is a far cry from the systemic instabilities of today's financial capitalism, the translocal and intertemporal affordances of finance and insurance—that is, risk modeling, different forms of credit, monetary infrastructure—remain critical mechanisms for resilience and redistribution in the engineering of viable collective futures.

In theory, markets were supposed to be adequate to this task of planetary coordination. In "The Use of Knowledge in Society," the seminal essay of modern market economics from 1945, Friedrich A. Hayek argues that a market economy was the only model adequate to the problems of large-scale economic coordination. For Hayek, the price system is an information system, a "machinery for registering change," which aggregates an unthinkably complex "economy of knowledge" into a single flickering figure, the truthful state of the economy at any given moment. But as social theorist Evgeny Morozov has argued, Hayek did not mean by this that an unfettered market was a miraculous oracle of economic truth, as his neoliberal inheritors would have it: "Such an elegant and information-light arrangement as the price mechanism can only work because much of the actual complexity of competition is handled and reduced elsewhere in the economic system."[8] The price system functions "efficiently" because of what economic agents do not need to know in order to operate in a market that nonetheless conceals all manner of informational asymmetries, deceptions, and fabulations instrumentalized in the name of competitive advantage. This is acutely reflected in the history of climate change, in which fossil fuel corporations such as Exxon and Shell have notoriously funded novel climate science only to obfuscate its findings while also supporting climate-skeptic think tanks in order to shape the narrative and policy environment in which they operate. More recently, a 2021 report found that investments held by eighteen US banks and asset managers were responsible for almost two billion

[7] Daniel Defoe, *An Essay upon Projects* (London: Printed by R.R. for Tho. Cockerill, at the Three Legs in the Poultrey, 1697). Edition from 1887, https://www.gutenberg.org/files/4087/4087-h/4087-h.htm.

[8] Evgeny Morozov, "Digital Socialism? The Calculation Debate in the Age of Big Data," *New Left Review* 116/117 (2019): 33–67, here 46, https://newleftreview.org/issues/ii116/articles/evgeny-morozov-digital-socialism.pdf.

tonnes of CO_2 emissions, noting that if it was a nation-state, it would be the fifth largest emitter in the world.[9] Given the voluntary nature of most emissions disclosure frameworks such as the G20's Task Force on Climate-related Financial Disclosures (TCFD), most publicly available figures are considered significant underestimates,[10] notwithstanding the systemic cascade of indirect, second-order risks, and negative externalities that are both difficult to quantify and structurally obscured. Voluntary carbon markets have suffered similar failures of knowledge through inadequate transparency and a lack of unified standards or a trusted way to prove additionality—evidence that financial machinations have actually reduced emissions in the world. Over the past decade, corporate climate efforts have increasingly revolved around "environmental, social and governance" (ESG) policies and investment frameworks. As Bloomberg recently reported on MSCI, the leading provider of ESG ratings, its assessments ignore real environmental effects and serve to "gauge the risk the world poses to a company, not the other way around."[11] The company's CEO, Henry Fernandez, explained during an interview: "Remember, they get paid. They're fiduciaries, you know. They're not as concerned about the risk to the world."[12]

Far from a system of rational economic agents, the competitive market is more like a narrative engine for shaping an asymmetrical system of belief and expectation. Capitalism has evolved significantly since Hayek's time, but the dream of perfect competition in "complete markets"[13] continues to animate the neoliberal economics that has conditioned the global economy during the past half-century. The deindustrialization of the US economy and the rise of financial capitalism since the doctrine of shareholder primacy and agency theory of the 1970s powerfully swung the balance of power from corporate owners and managers to the interests of investors. As economic historian Greta Krippner observes, "it is difficult to escape the impression that we live in the world of finance."[14] In a planetary future largely distributed through programs of investment, finance resembles a labyrinthine supply chain of speculations—

[9] Tim Quinson, "Wall Street Is Close to Triggering a Climate Financial Crisis," *Bloomberg*, December 14, 2021, https://www.bloomberg.com/news/articles/2021-12-14/wall-street-may-trigger-climate-financial-crisis-green-insight.

[10] Quinson, "Wall Street Is Close to Triggering a Climate Financial Crisis."

[11] Cam Simpson, Akshat Rathi, and Saijel Kishan, "The ESG Mirage," *Bloomberg*, December 10, 2021, https://www.bloomberg.com/graphics/2021-what-is-esg-investing-msci-ratings-focus-on-corporate-bottom-line.

[12] Simpson, Rathi, and Kishan, "The ESG Mirage."

[13] A complete market in economics is a market with perfect information, in which there is a price for every asset in every possible state of the world.

[14] Greta R. Krippner, "The Financialization of the American Economy," *Socio-Economic Review* 3, no. 2 (2005): 173–208.

fictions that *function*—oiled by an apparently endless supply of global liquidity. Within this highly speculative imaginary, risk takes on a central role in the shaping of social and economic organization, as seen in the meteoric rise of asset managers, hedge funds, as well as the subordination of public policy to the interests of corporate bondholders. Echoing Hayek's "competition for reputation," the philosopher Michel Feher posits that the center of balance of contemporary capitalism has shifted from the extraction of profit to the attribution of credit. As such, the nature of the capitalist game is not to make rational, calculative choices about the profitability of a firm but to anticipate the appreciation of assets: "to speculate on the speculations of others."[15] Within this performative arena of collective belief and manifestation (consider SoftBank's gargantuan bets on loss-making would-be monopolies or the rise of cryptocurrencies), the stakes and stakeholders of the speculative economy are everywhere, entangling everyone as investor and investee. This happens not only conventionally through home ownership, insurance, and pension funds but also more diffusely as purveyors of social credit, sentiment, and reputation. Feher compellingly argues that this state of affairs surfaces a different kind of politics in which investees must "participate, for its own purposes, in this game of self-fulfilling prophecies"[16] and "occupy the time that creditors have conquered."[17] The framing of "investee politics" brings the conflicts between the Anthropocene and financial capitalism—between narrative failure and market failure—full circle to a question of aligning and redistributing temporalities on the horizons of collective speculation.

As philosopher Achille Mbembe observes, "It may be that we must let go of the dream of reconcilability. It may be that those dreams are so antagonistic that they will never be reconciled. The question then is: Is it at all possible to build anything in common in the face of such agonism? How do we live with irreconcilability?"[18] If the time of irreconcilability cannot be experienced as a coherent totality, other than in the market's delusions of unity, perhaps it can be encountered in its systemic components as a supply chain of speculations and positions by which myriad future risks are transformed into present-day causes. As Ulrich Beck writes of the "risk society": "The single, undivided truth has fractured into hundreds of relative truths," a condition in which uncertainty "often cannot be overcome by more knowledge but is instead the result of more

[15] Michel Feher, *Rated Agency: Investee Politics in a Speculative Age*, trans. Gregory Elliott (New York: Zone Books, 2018), 54.

[16] Feher, *Rated Agency*, 60.

[17] Feher, *Rated Agency*, 111.

[18] Achille Mbembe, "How To Develop A Planetary Consciousness," *Noēma*, January 11, 2022, https://www.noemamag.com/how-to-develop-a-planetary-consciousness.

knowledge."[19] Just as a heatwave in Russia contributes to an uprising in Egypt, the contemporary planetary crisis of technological and ecological uncertainty is characterized by increasingly complex and compounded risks.[20]

Risk is the substance of financial capitalism, the fuel of a speculative engine—is it also a resource to be leveraged? As a multiplicity of risk-fictions exert themselves upon the uncertainties of the present, how does social and scientific knowledge arrive intact as signals rather than externalities? How do desires and threats congeal into collective expectations and demands? And how are these future claims constrained, legitimized, or foreclosed? It is clear that dominant market players cannot be left to determine the response to these questions, yet a certain financial realism demands that planetary crises will be negotiated within and against markets. As is often observed, infrastructures remain invisible until they malfunction; perhaps those that distribute our times are no different.

The Case of Catastrophe Insurance

How does the planet encounter the market? Until the end of the twentieth century, insurance for natural disasters such as hurricanes and droughts were calculated without regard for underlying natural systems. The reinsurance market was largely based at Lloyd's of London, where underwriters (known internally as "Names") would calculate premiums using actuarial analyses of previous years' losses, just as they had done since the origins of their profession in marine insurance. In 1985, a young American insurance analyst named Karen Clark proposed that the entire sector was highly overexposed to catastrophic risks, since insurers took little account of how and where hurricanes actually occurred. When Clark made her point two years later at Lloyd's, arguing for the use of computer models to estimate where hurricanes might occur, she was largely ignored by her audience of British underwriters, most of whom had never used a computer in their work. The Names felt little cause for alarm: After all, the Atlantic hurricane season of the 1970s and 1980s had been unusually quiet, and the catastrophic losses of earlier decades had been largely forgotten. Clark recalls: "People thought they had already figured it out. They were comfortable

[19] Ulrich Beck, *World at Risk*, trans. Ciaran Cronin (2007; Cambridge: Polity, 2009), 5.

[20] Jakob Zscheischler, Seth Westra, Bart J. J. M. van den Hurk, et al., "Future Climate Risk from Compound Events," *Nature Climate Change* 8, no. 6 (2018): 469–77.

with their own subjective judgment."[21] These judgments were undermined when the Category 5 storm Hurricane Andrew tore through small-town Florida in 1992, narrowly missing the urban centers of Miami and Fort Lauderdale. The industry had anticipated insignificant damage, whereas Clark's model indicated an unprecedented $13 billion in losses. As the hurricane progressed, leaving catastrophic storm surges and spin-off tornadoes in its wake, the eventual losses came to just under $16 billion. Hurricane Andrew bankrupted eleven US insurance companies, leaving a million people without coverage and 180,000 people homeless. As a result, the state picked up much of the damage, setting up the Florida Windstorm Underwriting Association and the Florida Hurricane Catastrophe Fund as the insurer and reinsurer of last resort.

Catastrophes are fickle things. Because natural disasters are rare, the available historical sample sizes for estimating devastating outliers are extremely limited. Clark's 1986 paper titled "A Formal Approach to Catastrophe Risk Assessment and Measurement"[22] centered around a catastrophe model for the hurricane insurance industry, which combined statistical simulations of hurricane paths with the insured property values associated with US zip codes. Using hurricane data dating back to the 1850s, Clark ran archival hurricanes through her model to produce thousands of "synthetic storms," fictional variations of real catastrophes, which could populate a viable sample size to represent the risk of a future hurricane season. While modeling techniques have grown considerably more granular, the reinsurance industry uses largely the same techniques Clark proposed at Lloyd's to estimate seasonal losses. Two firms have since emerged as the dominant fact-brokers of the catastrophe insurance market: AIR Worldwide founded by Karen Clark, and RMS–Risk Management Solutions.[23] Both specialize in the development of modeling software that combines windstorm simulations, geographic building codes, insurance industry data, and so forth, chewing up the figures in a giant spreadsheet before spitting out what the financial system can understand: a single risk value.

The formalization of catastrophe models after Hurricane Andrew led to the fearful realization that insurance companies and their global reinsurers simply did not have the capital to underwrite another "big one." In search of greater capital pools, "insurance-linked securities" (ILS) and "catastrophe bonds" (or

[21] Michael Lewis, "In Nature's Casino," *New York Times Magazine*, August 26, 2007, https://www.nytimes.com/2007/08/26/magazine/26neworleans-t.html.

[22] Karen M. Clark, "A Formal Approach to Catastrophe Risk Assessment and Management," *Proceedings of the Casualty Actuarial Society* 73, no. 140 (1986): 69–92.

[23] Their data is not only used by the insurance industry, but it also influences state regulators and international organizations such as the IPCC and OECD.

"cat bonds") were invented through Bermuda-based special purpose vehicles to package insurance risks into special financial securities that could tap into the vast liquidity of global securities markets. Cat bonds were high-risk, high-return products with a unique selling point: They were uncorrelated with the wider financial market because hurricanes do not care about the stock market. In contracts typically lasting a year, investors could effectively "sponsor" a hurricane season. If an agreed loss threshold was surpassed in a particularly disastrous year, the money invested would be used to pay back insurance claims. But if the season was quiet, investors would receive a significant return, typically in the double digits. Over time, cat bonds and ILS markets expanded beyond hurricanes to include a plethora of disasters,[24] both "man-made" and "natural."

A recent study by the sociologist Ian Gray showed how since the mid-2000s, the risk modeling firm RMS sought to bring a more acute understanding of climate change into the purview of catastrophic risk calculations. American hurricane risk models were typically based on a database known as HURDAT2 that contained an archive of all observed North Atlantic storms since the 1850s, but it could not account for the near-term effects of climate change. In 2005, RMS proposed a new model, based not on statistical samples but on an expert elicitation from leading climatologists, that is similar to the RAND Corporation's methodology known as the "Delphi method." As Gray writes, RMS acted on "a credible scientific concern that future weather could no longer be inferred by past trends."[25] When RMS surmised that climate risks would require a significant increase in insurance prices, its conclusions were met with a furious response from consumer groups and politicians alike, who accused the modelers of colluding in price fixing with its insurance industry clients. As a result, not only was the approach roundly rejected but also the South Carolina legislature even passed a law forbidding the use of catastrophe models that selectively accounted for near-term climatic changes, mandating that only simulations based on the full historical record would be permitted.[26]

The absurdity of this episode highlights the fragile political and temporal conflicts by which the market's negotiations with climate uncertainties are constrained, as well as the intricate epistemic and financial machinations required to link them in the first place. Catastrophe models enact an interface between two

24 Cat bonds have been issued for pandemics, wars, droughts, and even "riots and civil unrest."
25 Ian Gray, "Hazardous Simulations: Pricing Climate Risk in US Coastal Insurance Markets," *Economy and Society* 50, no. 2 (2021): 196–223, here 198.
26 Gray, "Hazardous Simulations," 211.

planetary narratives, the environmental and the financial; that insurance-based investments such as bonds are "uncorrelated" with the market is indicative of the chasm which separates the two. Geographer Zac Taylor notes that instruments such as ILS play the role of a "transformer" through which "capital drawn largely from US and European institutional investors is transfigured into reinsurance, or insurance for insurers"[27] while, as Gray observes, catastrophe models translate "new uncertainties about hurricanes into the traditional insurance logic."[28] By generating fictional futures, firms like RMS play a key role in signaling and brokering consensus around climate risks while also opening up new financial channels for realizing them in the market. South Carolina's insurance legislation recalls a choose-your-own-adventure narrative over which a fictional timeline—connecting historical data to future threat—is deemed acceptable to the present political-economic calculus. As Bruno Latour writes, "A conflict will be called, from now on, 'of planetary relevance' not because it has the planet for a stage, but because it is about *which planet* you are claiming to inhabit and defend."[29] At the same time, it also reveals the conflicts of interest at play when risk and reward are determined by the logic of the market. While ostensibly serving as scientific fact-makers, catastrophe modeling firms ultimately work on behalf of financial markets, creating a technical basis that allows investors, reinsurers, and insurers to profit. Their connection to the ultimate subjects of catastrophe—populations facing climate risks and states as insurers of last resort—is tenuously mediated by a chain of private financial interests, leaving their legitimacy open to suspicion in the market for planetary futures.

On Last Resorts

The crisis of the Anthropocene demands a reconceptualization of central banks, the implicit backstop of the global financial, and monetary system toward an active role in the governance of economic temporalities. Since the 1970s, a cornerstone of the global neoliberal consensus has been that of "central bank independence," which is the notion that central banks are apolitical institutions

[27] Zac J. Taylor, "The Real Estate Risk Fix: Residential Insurance-linked Securitization in the Florida Metropolis," *Environment and Planning A: Economy and Space* 52, no. 6 (2020): 1131–49, here 1132.

[28] Gray, "Hazardous Simulations," 204.

[29] Bruno Latour, "'We Don't Seem to Live on the Same Planet'—A Fictional Planetarium," in *Designs for Different Futures*, ed. Kathryn B. Hiesinger and Michelle Millar Fisher (New Haven: Yale University Press, 2019), 193–39.

whose key mandate is maintaining price stability.[30] This notional independence was designated to segregate fiscal policy (the state's power to manage finances through taxation and regulation) from monetary policy (the central bank's power to increase money supply and interest rates for the banking system). This separation of powers between capricious politicians and stoic central bank technocrats became status quo following the inflation crises of the 1970s, which were triggered by Richard Nixon's pursuit of "cheap money" and low interest rates. Moreover, as economic sociologist Onur Özgöde shows, the rise of "monetary government" is fundamentally intertwined with the concept of "systemic risk" that emerged in the 1960s to characterize a deeply interdependent financial system expanding beyond the Federal Reserve's control. This period triggered a shift in the role of the central bank from a risk manager (For example through credit control tools) to the lender of last resort. Systemic risk was understood as a foundational and innate property of this new financial architecture. As Özgöde writes, citing the Federal Reserve Board, "The Fed had finally come to see itself as 'the ultimate source of liquidity to the economy' and thus responsible for the *entire* financial system's security."[31]

The idea that central banks are politically independent has long been unanchored from a reality in which they serve as both the first responder and ultimate backstop to the financial system. This was made painfully evident following the financial crisis of 2007–08, during which the Federal Reserve purchased over a trillion dollars of mortgage-backed securities from the market in order to provide it with the liquidity it needed to survive. The paradox of this dynamic, social theorist Robert Meister notes, "is that the financial system became politically invulnerable at the very moment when its survival was otherwise in doubt."[32] The "independent" central bank thus always steps up to assume systemic financial risk while refusing to meet eyes with the state. After the Covid-19 pandemic began in 2020, the Federal Reserve once again embarked on a massive program of asset purchases, doubling its balance sheets to almost $9 trillion with wealthy nations worldwide following suit. Such expansive "fiscal quantitative easing" was supposed to be anathema to any central bank's

[30] Rodolfo Dall'Orto Mas, Benjamin Vonessen, Christian Fehlker, and Katrin Arnold, "The Case for Central Bank Independence: A Review of Key Issues in the International Debate," European Central Bank, Occasional Paper Series, no. 248 (October 2020), https://www.ecb.europa.eu//pub/pdf/scpops/ecb.op248~28bebb193a.en.pdf.

[31] Onur Özgöde, "The Emergence of Systemic Risk: Federal Reserve, Bailouts, and Monetary Government at the Limits," SocArXiv, January 3, 2019, 12, emphasis original, https://osf.io/preprints/socarxiv/ga5w7.

[32] Robert Meister, *Justice Is an Option: A Democratic Theory of Finance for the Twenty-First Century* (Chicago: The University of Chicago Press, 2021), ix.

disciplined and disciplinary mandate, yet despite mainstream economic logic, US and global inflation rates since 2008 have fallen to historic lows.[33]

The illusion of central bank independence constrains the significant structural leverage it holds to shape climate risk throughout the financial system. Moreover, as economists Michel Aglietti and Étienne Espagne point out, echoing Carney's "Breaking the tragedy of the horizon,"[34] the elephant in the room is that financial systemic risk *is* climate systemic risk.[35] The economic historian Adam Tooze proposes that central bank "independence" should be replaced by a "more direct and openly political approach."[36] As it is, there is no fundamental incentive structure for the hypothetically efficient markets to respond adequately to climate risk because they know that the state will always step in, lest it wants to preside over the systemic collapse of capitalism. To reengineer temporal conflicts between the market and the planet, central banks could broaden their mandate to the long-term (economic) viability of the Anthropocene, rather than serve as an enabler of business as usual. One such scenario is outlined in Kim Stanley Robinson's 2020 science fiction novel, *Ministry for the Future*. Mary Murphy, the leader of the titular ministry, convinces the heads of international central banks to establish a joint program of "carbon quantitative easing" through CarbonCoin, a fiat reserve currency for planetary climate mitigation modeled on a nonfictional proposal by civil engineer and geo-hydrologist Delton Chen. The role of CarbonCoin is to stabilize climate and financial incentives on a long (100-year) horizon. In his paper, Chen proposes a metric called Risk Cost of Carbon (RCC) to complement the existing Social Cost of Carbon (SCC), which calculates the economic cost of a single tonne of CO_2 emissions. Where the SCC targets individual polluting firms, RCC is a holistic system-wide cost borne by the "world economy and its interconnections with people and the environment."[37] In doing so, it invents a planetary framework that integrates

[33] Tooze attributes this to the dramatic weakening in the bargaining power of labor today, as compared to the 1970s. Adam Tooze, "Chartbook Newsletter #22: How Do You Count Inflation? Tracking Weimar's Hyperinflation," *Substack*, June 12, 2021, https://adamtooze.substack.com/p/chartbook-newsletter-22.

[34] Mark Carney, "Breaking the Tragedy of the Horizon – Climate Change and Financial Stability – Speech by Mark Carney," *Bank of England*, September 29, 2015, https://www.bankofengland.co.uk/speech/2015/breaking-the-tragedy-of-the-horizon-climate-change-and-financial-stability.

[35] Michel Aglietta and Étienne Espagne, "Climate and Finance Systemic Risks, More than an Analogy? The Climate Fragility Hypothesis," CEPII Working Paper, no. 2016-10 (2016), 24, http://www.cepii.fr/PDF_PUB/wp/2016/wp2016-10.pdf.

[36] Adam Tooze, "The Death of the Central Bank Myth," *Foreign Policy*, May 13, 2020, https://foreignpolicy.com/2020/05/13/european-central-bank-myth-monetary-policy-german-court-ruling.

[37] Delton B. Chen, Joel van der Beek, and Jonathan Cloud, "Climate Mitigation Policy as a System Solution: Addressing the Risk Cost of Carbon," *Journal of Sustainable Finance & Investment* 7, no. 3 (2017): 233–74.

climate risk into monetary policy and uses the structural leverage of the latter to align the behaviors of the former with the survivability of the Anthropocene. Taking the legal personhood attributed to US corporations, as well as Bolivia's 2010 Ley de Derechos de la Madre Tierra (Law of the Rights of Mother Earth) as a precedent, Chen argues that the world economy and its planetary substrate could be enshrined in a form of legal personhood. This would create a mandate for multilateral central bank coordination to redirect financial market actors toward resilience, sequestration, and mitigation while simultaneously pulling the rug of their "implicit guarantee" from beneath the legacy system dependent on central bank de-risking. Risk is a two-way street: While systemic risk casts central banks as the custodian of the financial system, it also offers them systemic leverage over their protectorate, if only they would use it.

Shifting the macroeconomic stakes of climate risk makes clear the asymmetries of the "market" as co-constituted by governments, central banks, and private financial actors. As observed in catastrophe insurance, where the state plays the first responder and insurer-lender of last resort to the planetary financial complex, the extant economic mantra of the Anthropocene is *socialized risk and privatized reward*. As financial behemoths such as Blackrock acknowledge the climate crisis and throw their weight behind the ESG agenda, and financial centers such as the City of London promise to "unleash" trillions in private finance toward green investment, they do so with the caveat that they will need the state to "de-risk" private investments—in other words, for the public to ensure green infrastructural investment are both profitable and safe. As macroeconomics scholar Daniela Gabor argues, the "de-risking state" is central to the doctrine of what she calls the "Wall Street Consensus," which combines fiscal discipline, central bank independence, and public-private partnerships to turn green infrastructure into a viable asset class.[38] For the likes of vast fixed-income asset managers like Blackrock, whose interests are linked to expanding their assets-under-management (AUM) and not direct investment returns,[39] reconciling the planet with the market means engineering a world in which the climate crisis offers no barriers to private accumulation. Where dissonances form, it is up to public finances to hoist investors onto safer ground. This gesture of double expropriation from the public sphere—the privatization of crucial public infrastructure and the undertaking of transitional risks by the

[38] Daniela Gabor, "The Wall Street Consensus," *Development and Change* 52, no. 3 (2021): 429–59.

[39] Benjamin Braun and Adrienne Buller, "Titans: Tracing the Rise and the Politics of Asset Manager Capitalism," *Phenomenal World*, November 6, 2021, https://www.phenomenalworld.org/analysis/blackrock-asset-manager-capitalism.

state—demonstrates both the delusions and realities of contemporary financial power. As Gabor writes, "This is a *political choice*."[40]

Writing the Planet from Below

In reading the Anthropocene as a financialized planetary complex, the task of temporal engineering is not of mobilizing against the market so much as acknowledging that all forms of activity beyond highly localized phenomena already operate within its supply chains of speculation. It is therefore a task of realizing collective risks in order to bend the distributive arrows of time. As Feher argues, a politics of counter-speculation resembles a cartel within the economy of reputation, expectation, and credit, just as trade unions function like worker cartels over the market price of labor. However asymmetrically, the financial order binds together the fates of corporations and governments, lenders and debtors, consumers and workers. As sociologist of finance Fabian Muniesa writes, "financial activity creates bonds,"[41] it produces networks of dependency among certain claims upon the future. Crisis inevitably emerges when its knots cannot be untied, as when investors discover that they cannot disentangle themselves from their stakes, or borrowers from their loans. Always already entangled, the chains linking investors and investees pull both ways. But to rewrite this narrative from "below" requires a tentacular approach with the understanding that markets are shaped by "nonmarket" forces and are aimed at manifesting the exposures and correlations that bind the market to the planet.

Within the temporality of the Anthropocene, the right of private property is not abstractly enshrined in perpetuity so much as conditioned by the existence of species, society, and nations by which such legal regimes are upheld. As David Graeber shows in *Debt*, the modern view of liberty acquired from the likes of Adam Smith is "essentially the right to do what one likes with one's own property."[42] This concept of private property, in turn, was inherited from Roman law as right of absolute and perpetual power of a person over a "thing," which was highly unusual insofar as it was an abstract relation to an object rather than a relation between people.[43] The modern concept of private property was

[40] Gabor, "The Wall Street Consensus," preprint: 31 emphasis original.
[41] Muniesa, *The Provoked Economy*, 48.
[42] David Graeber, *Debt: The First 5,000 Years* (Brooklyn: Melville House, 2011), 205.
[43] Following Orlando Patterson, Graeber suggests that this concept of *dominium* emerged from the Roman relation between master and slave: "In Roman law, property, or *dominium*, is a relation

developed influentially by John Locke, who argued that states and individuals possessed sovereignty over their property just as one has over one's own body. The idea that private property should be absolute and perpetual rests upon a distinction between person (or corporation-as-person) and nature/world, which flounders in the Anthropocene's conflation of human and natural temporality. One could say that all property is temporary in the Anthropocene. The entangled natural, legal, and sociopolitical environment in which the future is shaped by the present leads all property to be temporally contested—a fact that is obvious to those affected by rising sea levels and increasing natural disasters. The dissonance between the notion of private property and the consciousness of changing planetary conditions is particularly stark in the context of financial assets, which are in essence future-oriented. This can be seen in an ongoing lawsuit filed by five energy conglomerates against four European governments for imposing climate regulations under the Energy Charter Treaty (ECT) such as the phasing out of coal power, which turned their investments into so-called "stranded assets." The fossil fuel companies are seeking €4 billion in "damages" from the accused governments to compensate their shareholders for "the disruption to the company's property."[44] In actuality, if not necessarily in law (one report[45] noted that such cases are commonly settled in favor of investors), fossil fuel investments with globally damaging consequences are not forms of property (subject to deprivation) but an exposure to uninsured risks (subject to loss). As the *Financial Times* noted, "If profits are to be private, so too should losses."[46] To follow this logic, one could interpret the energy companies' dispossession as a counter-claim on the right of the public *not* to bear its negative externalities. The legal-epistemic absurdity of the case demonstrates the Anthropocene's collapse of figure and ground, between the individual's imagined right to ownership and the radical dispossession of the planet and the societies belonging to it. Both conceptually and pragmatically, the Anthropocene renders almost all property contingent—correlated and interdependent—and held together not by natural law but by unnatural risk.

between a person and a thing, characterized by absolute power of that person over that thing." Graeber, *Debt*, 198.

44 Camilla Hodgson, "European Energy Groups Seek €4bn Damages Over Fossil Fuel Projects," *Financial Times*, February 21, 2022.

45 Lea Di Salvatore, "Investor–State Disputes in the Fossil Fuel Industry," *International Institute for Sustainable Development*, December 2021, https://www.iisd.org/system/files/2022-01/investor%E2%80%93state-disputes-fossil-fuel-industry.pdf.

46 Editorial board, "Governments Should Not Foot the Bill for Stranded Assets," *Financial Times*, February 21, 2022.

The recent resurgence of civic climate movements as well as resource-specific mobilizations such as the Gilet Jaunes (over rising fuel prices) and the 2020–21 Indian Farmers' Protests (against neoliberal agricultural reforms) are likely to have been only the beginnings of global upheavals linked to climate instability.[47] To bring these forces to bear at the crucial intersection of climate risk and market vulnerability—in order to turn material dispossession and political grievance into financial risk and temporal redistribution—demands the engineering of encounters between disparate vectors of struggle: in order to "recognize," as Fred Moten reminds us, "that this shit is killing you, too, however much more softly, you stupid motherfucker, you know?"[48] The Dakota Access Pipeline (DAPL) protests beginning in 2016 saw a hopeful confluence of direct action and financial activism. Beginning with protest camps established by Sioux tribe leaders to oppose the desecration of Native land, the DAPL protest evolved into the largest gathering of Native American peoples in a century, and went on to draw mass protest movements worldwide, engendering lasting solidarities between Indigenous rights, climate activism, and antiracism. It did not ultimately succeed: As of 2022, the oil continues to flow, with financial markets taking the case as a legal precedent to ramp up fossil fuel investments.[49] However, through the DefundDAPL initiative, the movement did achieve considerable financial goals, setting a precedent for intangible social risks to enter into investors' fiduciary horizons.[50] As a leading Indigenous DefundDAPL activist proclaimed, "We have the economic power to show companies that when they finance an environmentally racist project and hire private security and collude with law enforcement, their bottom line will suffer."[51] A subsequent analysis shows that the movement incurred an estimated $7.5 billion in costs to its owners and stakeholders and a further $4.4 billion in account closures to affiliated banks. The pipeline's owners, meanwhile, saw a 20 percent fall in

[47] Wilson Center Staff, "Climate Change in the Indian Farmers' Protest," *New Security Beat*, February 4, 2021, https://www.newsecuritybeat.org/2021/02/climate-change-indian-farmers-protest.

[48] Stefano Harney and Fred Moten, *The Undercommons: Fugitive Planning & Black Study* (New York: Minor Compositions, 2013), 10.

[49] Jordan Blum, "Dakota Access Pipeline Will Remain Open After Long Legal Fight, Judge Rules," *S&P Global Commodity Insights*, May 21, 2021, https://www.spglobal.com/commodity-insights/en/market-insights/latest-news/oil/052121-dakota-access-pipeline-will-remain-open-after-long-legal-fight-judge-rules.

[50] Carla Fredericks, Mark Meaney, Nick Pelosi, and Kathleen Finn, "Social Cost and Material Loss: The Dakota Access Pipeline," University of Colorado Law School Legal Studies Research Paper Series, no. 19-1 (2018), 1–66, https://www.colorado.edu/program/fpw/DAPL-case-study.

[51] Julia Carrie Wong, "Private Investor Divests $34.8m from Firms Tied to Dakota Access Pipeline," *Guardian*, March 1, 2017, https://www.theguardian.com/us-news/2017/mar/01/dakota-access-pipeline-storebrand-norway-divest-standing-rock.

stock prices over the period (even as the S&P 500 grew by a third), an outcome attributed to unexamined social risk.[52]

DefundDAPL showed that within the polarized and memetic global culture enmeshed by social networks, financial actors find themselves subject to increasingly volatile forces of collective speculation, recalling Hayek's observation that "competition is in a large measure competition for reputation or goodwill."[53] As the insurance firm CNA Hardy observes, "The inevitable consequence of [. . .] struggles to manage intangible, interconnected risks in this new world economy is reputation damage."[54] Arguably, the intangibility of social risk represents a liminal space in the financial order susceptible to break and entry by outside activism, an arrangement that mirrors the legal membrane that insulates the firm from its externalities in the wider world. For financial activists, generating social risk means manifesting the return of repressed socialized risk and operationalizing financial actors' exposure to the world in which they are enmeshed. Increasingly, high-profile cases of corporate catastrophe have led to a compounding of social and financial risk, in which shareholders sue companies for failing to anticipate harmful consequences as a failure of fiduciary disclosure. As the influential financial columnist Matt Levine observes: "It is a strange combination: Generally speaking the companies do the bad things *on behalf of shareholders*—to make more money for them—but then the securities regulators come in and fine them for *defrauding* shareholders."[55] Moreover, so-called "event-driven security litigations" exact damages on firms where other channels struggle to do so, and are more likely to involve government investigation.[56] This tendency opens a window of exposure through which social risks can be mobilized by animating the contradiction between shareholders' immediate interests in (harmful) corporate activities and their fear of exposure to the consequences of that harm. While shareholders' interests are seldom aligned with those of the wider public, by strategically engineering the environment of social risks—whether through informational campaigns, peaceful demonstrations, strikes, riots, or direct disruption of industrial activities—movements can

[52] Fredericks et al., "Social Cost and Material Loss," 27.

[53] Friedrich A. Hayek, "The Meaning of Competition," in *The Collected Works of F. A. Hayek Volume 15: The Market and Other Orders*, ed. Bruce Caldwell (1948; Chicago: The University of Chicago Press, 2014), 109.

[54] CNA Hardy, "Taking the Pulse of Global Business: Global Risk and Confidence Survey," May 2019, https://www.cnahardy.com/~/media/Files/C/CNA-Hardy/cna-hardy-risk-and-confidence-may-2019.pdf.

[55] Matt Levine, "Snap Earnings and Emissions Fraud," *Bloomberg*, May 11, 2017, emphasis original, https://www.bloomberg.com/opinion/articles/2017-05-11/snap-earnings-and-emissions-fraud.

[56] Emily Strauss, "Is Everything Securities Fraud?," Duke Law School Public Law & Legal Theory Series, no. 2021-04 (2020): 1359, https://ssrn.com/abstract=3664132.

translate negative externalities into prospective losses, making once-attractive investments unviable. It is taken for granted that financial actors operate under uncertainty, but uncertainty is not a passive rule, it can be created by other players in the game of systemic risk.

Ultimately, manifesting social risk through conventional means may be insufficient to challenge the climate-financial complex. It needs to travel upstream in order to become political risk and to force the hand of sovereign power. While financial pressures (even from shareholders themselves) may succeed in forcing public companies' decisions and drive down prices, they also create cheaply traded fossil fuel assets. As Levine points out, the idea that "companies with shareholders who are humans living on Earth can try to make Earth livable" seems commonsensical, but it is actually a "somewhat strange corporate finance argument."[57] For every dollar of impact investment, there are several more amoral dollars to flush through the system. For example, pressure on publicly listed companies to divest from new coal mines has created what one mining executive has called a "bonanza of a lifetime" for private equity markets outside of public scrutiny, giving them the opportunity to cash in on high coal prices and cheap production costs.[58] For as long as assets are not stranded by force, companies may as well divide themselves into more and less high-minded subsidiary factions, taking advantage of ESG premiums and rising fuel prices at the same time. As one investor pointed out, "The quickest way to shrink emissions as a major company is to shed assets so you can hit climate-related targets, [. . .] you're just moving emissions from one hand to another."[59] The goal must therefore be to discredit fossil fuel investments while simultaneously removing the legal and financial backstops that make these assets salvageable. Intervention strategies in the market do not need to involve primarily financial strategies: They might also resemble the rehearsal of future catastrophes through direct action, which is a counter-speculative strategy par excellence. Historian and activist Andreas Malm argues in *How to Blow Up a Pipeline* that climate action cannot limit itself to non-violent or legal actions if it is to unsettle business as usual: "If states cannot on their own initiative open up the fences, others will

[57] Matt Levine, "Not Everyone Likes ESG," *Bloomberg*, January 12, 2022.

[58] LinkedIn post by Vale Executive Vice President Luciano Siani Peres (https://www.linkedin.com/posts/luciano-siani-pires_a-few-days-before-christmas-vale-signed-activity-6883150109136224256-MFW4), quoted in James Attwood, "Private Equity Lines Up for Coal 'Bonanza' Left by Public Miners," *Bloomberg*, January 3, 2022.

[59] Anjli Raval, "A $140bn Asset Sale: The Investors Cashing in on Big Oil's Push to Net Zero," *Financial Times*, July 6, 2021.

have to do it for them."[60] While the term "stranded assets" evokes a passively unrealized profits, Malm proposes it as a condition to be actively imposed through the dismantlement and destruction of CO_2-emitting infrastructure. As R. H. Lossin writes, sabotage is "a sort of prefigurative, if temporary, seizure of property,"[61] a notion of property, which, as argued above, is already temporary in the Anthropocene. One could read Beck's description of risk as the "thematization of future threats"[62] as a prescription for temporal engineering: By destroying fossil fuel assets, one threatens their future vulnerability. Within the informational modality of the market, tactical violence is a discursive act, a narrative intervention that bonds privatized bodies of power to a planetary community of fate. To use Muniesa's term, it is not an act of "finalization"—as Malm notes, saboteurs could hardly blow up every pipeline, nor would that be desirable—but the generation of an (speculative) exposure and to force more powerful actors, namely states, to act. To paraphrase Graeber, to commit climate sabotage is to act as if one is already free (of fossil fuels).

Engineering Catastrophe

I once asked a researcher in the field of existential risk whether he was in touch with the insurance industry, since they were both in the business of quantifying catastrophes. "When they stop caring is where we start caring," he told me, "they do not care if there is no-one to pay out. After a certain level of catastrophe, there will be no more society, there will no longer be capitalism."[63] With its collapse of relations between figure and ground, subject and world, knowledge and risk, the climate-financial complex of the Anthropocene undermines dominant constructions of both narratives and markets. Both are predicated on a fantasy of completion.[64] Where neoclassical markets disguise the myriad contradictions, asymmetries, and inefficiencies of its participants under the totalizing sign of price, conventional narrative abstracts an infinitely expanding world into a

[60] Andreas Malm, *How to Blow Up a Pipeline: Learning to Fight in a World on Fire* (London: Verso Books, 2021), 75.

[61] R. H. Lossin, "Sabotage as Environmental Activism: Why Sabotage of Pipelines is a Justifiable and Effective Form of Resistance to Capitalism and Climate Change," *Public Seminar*, July 3, 2018, https://publicseminar.org/essays/sabotage-as-environmental-activism.

[62] Beck, *World at Risk*, 4.

[63] Conversation with author.

[64] As Hayek wrote, "Competition is essentially a process of the formation of opinion: by spreading information, it creates a unity and coherence of the economic system which we presuppose when we think of it as one market." Hayek, "The Meaning of Competition," 115.

beginning, middle, and end that unfolds in causal, linear time. In many ways, the financial paradigm offers a more truthfully systemic account of a world of multiple and simultaneous processes, held together not by causes but by bonds linking times, spaces, and expectations. As sociologist Elena Esposito writes, "What is sold on financial markets is the possibility of the creation of constraints in the course of time."[65] In practice, their deep and deregulated interdependency produces increasing and profound instability. The ever-impending catastrophe of financial markets in the Anthropocene is not so much a failure of markets to function properly as it is a failure of collective imagination in the coordination of temporal horizons. Nonetheless, to discover a world beyond financial capitalism demands that we think through financialization, at least because at present there is no outside (except for certain enclaves within). Today, the system of global finance is coextensive with the planetary. As Aglietti and Espagne note in their conception of "climate systemic risk," while climate and financial risk are intensifying independently, the real story is the positive feedback loop between them.[66] An irony of these combined catastrophes is that the market is perhaps the only infrastructure in which virtually all humans engage, whose fluctuations we experience as habitually and constantly as the atmosphere we breathe.

The necessary conditions of bringing the planetary to bear on the market are powerfully uncertain, but they seem likely to involve the activity of disparate and accidental alliances, many of which are not directly channeled toward climate nor finance: from mass uprisings and strikes to shareholder activism, from ecological post-Keynesian monetary policy to ecoterrorism, from geoengineering to resource nationalism. By framing these processes within the financial framework, their combined counter-speculative effects are needed for the radical redistribution of climate risk and to engineer the conditions for planetary-scale collective insurance. The notion of "compound events," the co-occurrence of many dependent "hazards" and "drivers," offers an understanding of systemic catastrophe not through isolated causal chains but through multiple and simultaneous effects. The writing of this multivocal planetary narrative means instrumentalizing the many tailwinds of catastrophic time—including the speculative dynamics of finance capital—in order to engineer new temporal bonds. If it can be called a strategy, it is one oriented toward catastrophe in the original sense, an overturning of the narrative that enfolds everything—a plot to steal the horizon before it is eclipsed.

[65] Elena Esposito, *The Future of Futures: The Time of Money in Financing and Society*, trans. Elena Esposito and Andrew K. Whitehead (2009; Cheltenham: Edward Elgar, 2011), 2.
[66] Aglietta and Espagne, "Climate and Finance Systemic Risks," 14–15.

4

The Automaton of the Anthropocene

On Carbosilicon Machines and Cyberfossil Capital[1]

Matteo Pasquinelli

The Bicephalous Machine

The history of industrial civilization can be depicted as a bicephalous chimera whose heads grew out of the same machine, innervated each other, and, after further metamorphoses, still attempted to hegemonize each other. The two heads are Energy and Information and they bifurcated out of the industrial machine of the nineteenth century, although at different tempos. They initiated and extended two technological lineages: the civilizations of Carbon and Silicon, respectively, the one of energy as a medium of motion and the one of energy as a medium of control and communication. The two regimes carried different *entropic costs* and also quite different *colonial costs*, having been developed at different historical stages and latitudes of the planet. Although, for instance, Charles Babbage's Analytical Engine was potentially ready to replace "the mental division of labor"[2] in the industrial factory, only the microchips and labor composition of the twentieth century would be able to trigger the corresponding information revolution. And although Karl Marx registered the "metabolic rift"[3] caused by the pollution of "carboniferous capitalism"[4] on the English landscape, the input and output of the industrial apparatus set in motion a gigantic web of economic relations and supply chains that enslaved populations beyond the borders of the British Empire.

[1] This text was originally published by *The South Atlantic Quarterly* 116, no. 2 (April 2017): 311–26.
[2] Charles Babbage, *On the Economy of Machinery and Manufactures* (London: Charles Knight, 1832), 153.
[3] John Bellamy Foster, "Marx's Theory of Metabolic Rift: Classical Foundations for Environmental Sociology," *American Journal of Sociology* 105, no. 2 (September 1999): 366–405.
[4] Lewis Mumford, *Technics and Civilization* (Chicago: University of Chicago Press, 2010), 156.

The thermodynamic engine is correctly identified as the central axis of the industrial revolution, but the flows of primitive accumulation that prepared its terrain are today finally recognized: the significant contribution of agricultural enclosures, resource expropriation, colonial invasions, unpaid domestic labor, and slavery. Since Marx's formulation, the industrial machine is perceived specifically as the diagram of surplus value, in which machinery is *dead labor* that dialectically absorbs workers' *living labor*. After cybernetics[5] and Gilles Deleuze and Félix Guattari's philosophy of the assemblage,[6] the machine is described in a less dialectical way as a conurbation of flows of money, energy, matter, and information. The unruly technosphere responsible for the Anthropocene should be analyzed with both approaches in mind: the one that sees the machine as a diagram of surplus, accumulation, and crisis, and the one that sees it as a network of flows that responds to a larger social ecology.

Rather than engaging in metaphysical debates on the opposition between Nature and Society, this essay looks for an *empirical assemblage* where the connection between the two, through the paradigms of energy and information, can be studied. This essay illustrates the industrial machine as the forgotten bifurcation of energy and information and follows such a bifurcation along three stages: industrial factory, cybernetic society, and planetary computation. Labor is composed of energy and information, and so is capital. By defining labor as the composition of energy and information, labor can be woven back into the fabric of the Anthropocene paradigm, which itself emerged as a complex architecture of energy and information. Within this picture, labor remains the collective agency that is socially and politically separated by technology and that appears then to be "encrypted" (or, more interestingly, "outcrypted") in all subsequent regimes of production.[7] In this respect, the Anthropocene paradigm seems complicit with a mode of governance that attempts to dissolve labor conflicts into the fabric of information and energy, thus mystifying labor into technological forms so as to render it invisible (as argued in the case of automation by Sigfried Giedion[8]).

The intimate relation between labor and the energy economy has been investigated since the energy crisis of the seventies, which George Caffentzis

[5] Norbert Wiener, *Cybernetics: or Control and Communication in the Animal and the Machine* (Cambridge, MA: MIT Press 1948).

[6] Manuel Delanda, *Assemblage Theory* (Edinburgh: Edinburgh University Press, 2016).

[7] The term "outcryption" refers to something that is invisible and inaccessible for being encoded, paradoxically, in public procedures, common habits, and social techniques: It is historically the nonconscious yet very empirical power of any ideology.

[8] Sigfried Giedion, *Mechanization Takes Command: A Contribution to Anonymous History* (New York: W. W. Norton & Company, 1948).

rightly renames a "work/energy crisis."[9] Caffentzis notices that information is required by capital to allocate resources and workers in the most efficient way against entropy; information is the economic intelligence of energy. But more importantly, both the information revolution and the energy crisis are responses to the refusal of labor and the social movements of the 1960s. Similarly, this essay tries to weave together the *energy theory of labor* (labor as a manual activity) with the *information theory of labor* (labor as a source of information that gives form to energy).

A genealogical study of information that goes back to the industrial age is worthwhile. The global technosphere responsible for the Anthropocene still resembles, in its form and function, the automaton of the industrial age, which was also described by Marx as a central axis of production running nonstop and orchestrating the overall division of mental and manual labor in the factory.[10] Sadly, the *automaton* of the technosphere and its comfort narratives (such as singularity) appear to mirror and capture today the *autonomy* of social movements theorized and practiced in the last decades.

Confronting the Anthropocene paradigm with Guattari's *machinic ecology* (that includes the inorganic, organic, technological, economic, and psychic spheres within the same *Umwelt*), this essay attempts to recompose the *epistemic rift* between energy and information that was provoked by industrial capitalism and then amplified by cybernetics and the digital revolution. The recombination of *labor's intelligence*,[11] which is a novel assemblage of energy and information at a higher scale of labor, will be proposed as a necessary passage toward the machinic ecology that Guattari envisioned also as a political ecology of the mind.

As much as the political economy has discovered the substrate of energy and labor in the diagram of information capitalism too late (by focusing on frictionless paradigms such as knowledge economy and network society for too long), ecology has overlooked the role of information in the *Bildung* of its own cognitive map. If the critique of industrialism helped to also recognize the metabolism of energy and matter in the regime of information machines (see the idea of media geology in Jussi Parikka[12]), likewise, a new critique of cybernetics should help to remind us of the role of information in the growth crisis of the old industrial apparatus. The two regimes of industrialism and informationalism will hopefully

9 George Caffentzis, "The Work/Energy Crisis and the Apocalypse," *Midnight Notes*, no. 3 (1980).

10 Karl Marx, *Capital: A Critique of Political Economy, Volume 1*, trans. Ben Fowkes (1867; London: Penguin Books, 1976), 544.

11 Simon Schaffer, "Babbage's Intelligence: Calculating Engines and the Factory System," *Critical Inquiry* 21, no. 1 (Autumn 1994): 203–27.

12 Jussi Parikka, *A Geology of Media* (Minneapolis: University of Minnesota Press, 2015).

be described one day according to a paradigm that is capable of comprehending their continuum, intersections, and bifurcations, that is, their coevolution.

If labor is reframed according to the composition of the flows of energy and information, a new theory of machines is also necessary. At the end of the essay, the sketch of the *carbosilicon machine* (the infoenergetic assemblage that emerged with the coupling of the Turing machine and the thermodynamic engine) will hopefully cast a different light on the politics of the Anthropocene and the division of labor engendered by the age of planetary computation and logistics. In the last part of the essay, the two paradigms of "fossil capital"[13] and "control revolution"[14] will be united into the exploratory idea of *cyberfossil capital*, the ultimate assemblage of the perennial flows of energy and information.

Coal, or the Fuel of Abstract Labor

It was Gilbert Simondon who noticed that the industrial machine was already an *infomechanical relay*, as it was separating, for the first time, the traditional form of labor into a source of *energy* (propelled by natural resources such as water or coal) and a source of *information* (the conscious movements and instructions of workers supervising the machine).[15] In this view, the traditional *tool* is a design in which energy and information are still united: With the hammer, for example, the preindustrial artisan was providing both energy and form with the same gesture. It was thanks to their separation (bifurcation) that the flows of energy and information could be governed and exponentially multiplied by capital.

The industrial revolution was the reorganization of the labor power of the manufacturing age around the gigantic master axis of the factory—of which workers and flows of natural resources became mere prostheses. The Scottish business theorist Andrew Ure (who Marx humorously called "the Pindar of the automatic factory" for his extravagant prose) described the industrial apparatus as "a vast automaton, composed of various mechanical and intellectual organs, acting in uninterrupted concert for the production of a common object, all of them

[13] Andreas Malm, *Fossil Capital: The Rise of Steam Power and the Roots of Global Warming* (London: Verso Books, 2016).

[14] James R. Beniger, *The Control Revolution: Technological and Economic Origins of the Information Society* (Cambridge, MA: Harvard University Press, 1986).

[15] Gilbert Simondon, "Technical Mentality," trans. Arne De Boever, *Parrhesia*, no. 7 (November 2009): 20.

being subordinate to a self-regulated moving force."[16] In a similar way, Babbage recognized a division of manual and mental labor within the management of the factory and imagined two different forms for their mechanization: Whereas thermodynamic machines were replacing manual labor, his Analytical Engines, prototypes of modern calculators (yet never finalized during his life), were supposed to automate, for instance, the intellectual labor of the factory's accountants.[17] The automation of mental labor (information) takes hold through a more profound relation with the metabolism of energy.

Andreas Malm has illustrated how the motion of the rising industrial automaton had to be propelled by a stable and versatile form of energy, which happened to be found in coal.[18] The physical properties of coal (lightness, homogeneity, measurability, and calorific potential) crucially contributed to the acceleration of industrial capitalism. Steam engines replaced water mills not because coal was cheaper and more abundant than water but because it provided a more stable flow of power than rainfall and allowed factories to move close to urban areas where most of the workers were living at the time. Malm registers in this way the energetic reason for the slow emergence of the industrial mode of production out of the manufacturing age: Indeed, it took roughly forty years for the steam engine to be adopted in place of the water mill. Coal came to be used across the full spectrum of production as it was the most adequate source of *abstract energy*—where abstract means easily computable in terms of cost, transport, stock, and performance. Coal could be transformed into a systemic component of capital only via a technological innovation, the thermodynamic engine.

> For coal to be universalised as a fuel for all sorts of commodity production, it had to be turned into a source of mechanical energy—and, more precisely, of rotary motion. Only by coupling the combustion of coal to the rotation of a wheel could fossil fuels be made to fire the general process of growth: increased production—and transportation—of all kinds of commodities. This is why James Watt's steam engine is widely identified as the fatal breakthrough into a warmer world.[19]

[16] Andrew Ure, *The Philosophy of Manufactures: or, An Exposition of the Scientific, Moral, and Commercial Economy of the Factory System of Great Britain* (London: Charles Knight, 1835); quoted by Marx, *Capital*, 544.

[17] Babbage, *On the Economy of Machinery and Manufactures.*

[18] Andreas Malm, "The Origins of Fossil Capital: From Water to Steam in the British Cotton Industry," *Historical Materialism* 21, no. 1 (January 2013): 15–68.

[19] Malm, "The Origins of Fossil Capitalism," 18.

What is recognized in the gears of such an industrial artifact is also the coupling of *abstract energy* and *abstract labor*.[20] Malm spotlights, in particular, the subtle relation between the energetic versatility of coal and the consolidation of the new spatiotemporal abstractions of capital, namely urban factories and their clock-based labor discipline. Coal provided the energetic continuum that was necessary for the disciplinary abstractions of industrial time and industrial space to emerge.[21]

Extending Malm's genealogy, it may be added that the abstract properties of information emerged thanks to the nature of fossil energy, to its homogenous carbon chains, which made coal easier to quantify and compute than traditional sources, such as water or animal power. If coal could be turned into *abstract energy* and labor into *abstract labor*, this happened specifically thanks to two new technologies of control at the center of the industrial apparatus: "closed-loop feedback devices like James Watt's steam governor (1788) and preprogrammed open-loop controllers like those of the Jacquard loom (1801)."[22] The steam governor was a device to maintain the constant output of an engine by regulating its fuel input in real time (retrospectively, it is considered the first cybernetic device). The punched card was a data device to store instructions for textile patterns for the Jacquard loom (its data format would be adopted by IBM, almost unchanged, throughout the twentieth century). More precisely, Watt's governor was turning the engine impulses into *abstract movement*, that is, constant rotary motion, and Joseph-Marie Jacquard's punched cards were turning manual instructions into *abstract form*, that is, information. Watt's governor and Jacquard's punched cards—control of motion and control of information—can be considered, *in embryo*, the first two anatomical components of the upcoming cybernetic system.[23] Throughout the industrial revolution, the bifurcating lineages of energy and information were already affecting one another and composing novel assemblages.

[20] Orthodox Marxists will pardon, for once, the ambivalent use of the notion of abstract labor in this passage. In Marx, abstract labor refers to human activity that is calculated and valorized by capital as a universal equivalent. Here it points also to the cognitive and informational component of labor in general. Sohn-Rethel would find that the two dimensions are genealogically related. Alfred Sohn-Rethel, *Intellectual and Manual Labour: A Critique of Epistemology* (Atlantic Highlands: Humanities Press, 1978).

[21] The fossil fuel economy will be further "abstractified" by capital with the introduction of Carbon Credit trading. See Emanuele Leonardi, "Carbon Trading Dogma: Theoretical Assumptions and Practical Implications of Global Carbon Markets," *Ephemera* 17, no. 1 (February 2017): 61–87.

[22] Beniger, *The Control Revolution*, 17.

[23] Norbert Wiener (*Cybernetics*, 11) coined the term cybernetics from the Greek *kybernetes* (governor/steersman), also drawing on James Clerk Maxwell's article "On Governors," *Proceedings of the Royal Society of London* 16, no. 16 (December 1868): 270–82.

One may say that somehow both Marxism and environmentalism address the energetic component of capitalism: The former identifies this in the exploitation of human labor, the latter in the exploitation of natural resources. The "autonomy" of both labor and nature is used sometimes to unify the ground of "red" and "green" politics, but this *energy theory of labor* overlooks the role of information in the definition of both labor and nature. Whereas the current section attempted to uncover the role of information within the industrial apparatus and the traditional definition of labor, the following one will show the hidden function of information in the constitution of the paradigm of ecology. Interestingly, both ecology and cybernetics will appear to be the interweaving of the very same flows of energy and information, but outside the factory.

Information, or the Government of Surplus

Conceptually, both ecology and cybernetics share roots in the notion of organism that is found in the German *Naturphilosophie* of the nineteenth century, where any "form of life" (from the animal to the nation-state) was understood as self-centered and in antagonism with the surrounding world (Umwelt). It was the zoologist and popular illustrator Ernst Haeckel who introduced the term "ecology" (*Ökologie*) as the study of the relation between the organism and the environment.[24] The biologist Jakob von Uexküll described the relation between the animal's nervous system (*Innenwelt*) and the outside world (*Außenwelt*) as a "functional circle" (*Funktionskreis*)—a scheme that would later be repeated in the feedback loop of cybernetics.[25] Similar to the Funktionskreis, the feedback loop of cybernetic systems was conceived as a circulation of information and response to an external stimulus. Uexküll viewed the organism as an information-processing system struggling to adapt to the environment, similar to the adaptive model that influenced the early design of the "cybernetic brain."[26] Yet one should remember that Uexküll (as much as Marx) did not possess a notion of information; the mathematical definition of information would be formulated only by Claude Shannon.[27]

[24] Ernst Haeckel, *Generelle Morphologie der Organismen* (Berlin: Georg Reimer, 1866).
[25] Jakob von Uexküll, *Theoretische Biologie* (Berlin: Gebrüder Paetel, 1920).
[26] Andrew Pickering, *The Cybernetic Brain: Sketches of Another Future* (Chicago: The University of Chicago Press, 2010).
[27] Claude E. Shannon, "A Mathematical Theory of Communication," *Bell System Technical Journal* 27, no. 3 (July 1948): 379–423.

Another family trait common to ecology and cybernetics is the idea of conservative equilibrium and self-regulation (later on, this would be further consolidated in the notion of *homeostasis*).[28] There is a distinction to be made though: In ecology, the medium of self-regulation appears to be the energy metabolism itself, whereas in cybernetics, the medium of self-regulation is strictly assigned to information. The two paradigms converged from time to time and formed what is called "cybernetic ecology." The *Whole Earth Catalog*, published in California between 1968 and 1972, was a culminating example of this coevolution and, interestingly, a cultural pioneer of the following regime of production, the *network society*.[29] For stressing the role of the infosphere in the control of the technosphere, the Anthropocene paradigm can also be considered part of the history of cybernetic ecology.

Historically, cybernetics originated from a mix of information theory and cognitive sciences that was heavily sponsored by military research (including the Manhattan Project and the construction of the first nuclear bomb). This essay illustrates cybernetics only in its coupling with the industrial apparatus: The *information flow* bifurcating out of the industrial machine encountered cybernetics and mainframe computers just after the Second World War. As James Beniger shows in his book *The Control Revolution*, the paradigm of informationalism emerged through the continuous pressure of industrial production—in fact, out of a "crisis of control" of Western capitalism. A more and more abstract definition of information (as measurable, computable, and transmissible knowledge) had to be introduced to manage the economic and commodity boom of the United States after the Second World War.

The cybernetic lineage that germinated out of the *information terminal* of the industrial machine aspired to control factories, national economies, and even the whole planet as its new self-reflexive organ, or *world brain*. Douglas Engelbart advanced the idea of machine-aided *augmented intellect* for problem-solving, even at the geopolitical scale.[30] Stafford Beer would apply cybernetics

[28] It is worth noting that Haeckel and von Uexküll embraced reactionary political positions like a good part of the German *Lebensphilosophie*, not to mention Martin Heidegger. See Ann Harrington, *Reenchanted Science: Holism in German culture from Wilhelm II to Hitler* (Princeton: Princeton University Press, 1999).

[29] William Bryant, "Whole System, Whole Earth: The Convergence of Ecology and Technology in Twentieth Century American Culture" (PhD thesis, University of Iowa, 2006); Fred Turner, *From Counterculture to Cyberculture: Stewart Brand, the Whole Earth Network, and the Rise of Digital Utopianism* (Chicago: The University of Chicago Press, 2006).

[30] Douglas Engelbart, "Augmenting Human Intellect: A Conceptual Framework," Stanford Research Institute, Summary Report AFOSR-3233 (Stanford, Menlo Park, CA, October 1962).

to factory management with utopian enthusiasm[31]: Salvador Allende's socialist government would invite him to develop Project Cybersyn with the purpose of regulating Chile's economy (which was, by the way, heavily based on copper extraction).[32] There is a lineage of cybernetics that was progressive: It is sometimes called Social Cybernetics and influenced anti-psychiatry movements and French philosophy too. Deleuze and Guattari took the idea of *plateaus*,[33] for instance, from the work of the English cybernetician Gregory Bateson on Balinese culture.[34]

With the original nucleus of ecology, cybernetics shared the idea of a self-regulating system based on information loops but applied this scheme to the design of intelligent machines. After the Second World War, during the so-called "Great Acceleration,"[35] industrial cybernetics was supposed to contain the overgrowth of production flows as a control apparatus. With the microchip revolution, the technologies of communication and control grew and transformed into a new vast nervous system, a *sentient technosphere* that today is escalating to the size of global datacenters and the sophistication of machine learning algorithms. Cybernetics was also supposed to transform the economy into an ecology of feedback loops in order to control social unrest and potential revolutions. But homeostasis is a troublesome category when transplanted from biological systems to economic and institutional ones; in fact, capitalism keeps on expanding the use of fossil fuels and crunches ever-growing databases, feeding on metabolic surplus. As Beniger noted, the information revolution grew up (and keeps on growing) by feeding itself on the industrial and energetic surplus that it was supposed to measure and control.[36] Equilibrium is rarely seen.

Cybernetics was thus the first *technopolitics*, that is, the first attempt to develop a normative project of power via technology and technological protocols. More exactly, cybernetics was the normative project of power in the age of information

31 Stafford Beer, *Brain of the Firm: A Development in Management Cybernetics* (London: Allen Lane, 1972).

32 On the Cybersyn project see: Eden Medina, *Cybernetic Revolutionaries: Technology and Politics in Allende's Chile* (Cambridge, MA: MIT Press, 2011). It is must be noted that Cybersyn was contemporary to Arpanet (progenitor of the Internet) that was developed by the US Department of Defense. Arpanet was conceptually far more advanced than Cybersyn for implementing a decentralized architecture based on packet-switching communication.

33 Gilles Deleuze and Félix Guattari, *A Thousand Plateaus: Capitalism and Schizophrenia*, trans. Brian Massumi (1980; Minneapolis: University of Minnesota Press, 1987), 21.

34 In general, Deleuze and Guattari's notion of *machinic* is indebted to the open framework inaugurated by cybernetics, which aimed to dissolve the border between organic systems and technical systems (and between vitalism and mechanicism).

35 Will Steffen, Wendy Broadgate, Lisa Deutsch, Owen Gaffney, and Cornelia Ludwig, "The Trajectory of the Anthropocene: The Great Acceleration," *Anthropocene Review* 2, no. 1 (April 2015): 81–98.

36 Beniger, *The Control Revolution*.

machines—a shift that Michel Foucault, but not Deleuze and Guattari, failed to record in his epistemology of power, although French philosophy (since the work of Simondon) was among the few early critical voices of the control paradigm of cybernetics. Marxism developed a critique of cybernetics too late, and Italian *operaismo* started its inquiry on cognitive labor, not by chance, only after 1989. The only exemption may be the pioneering and forgotten work of Romano Alquati, who studied the division of labor at the Italian computer factory Olivetti as early as 1961 and attempted to merge the notion of Marx's surplus value and cybernetic information under the concept of *valorizing information*. Alquati was probably the first to sketch an information theory of labor.

> The *productive labour* is defined by the quality of *information* elaborated and transmitted by the worker to the *means of production* via the mediation of *constant capital*, in a way that is tendentially *indirect*, but completely *socialized*. [. . .] Cybernetics recomposes globally and organically the functions of the general worker that are pulverized into individual microdecisions: the "bit" links up the atomized worker to the figures of the Plan.[37]

Autonomist Marxists like Alquati often stressed how social struggles and the refusal of labor accelerated industrial automation and the dissemination of information technologies. Labor resistance pushed the information revolution in the passage from Fordism to post-Fordism. But post-Fordism is not only the regime of the "hegemony of immaterial production"[38]: it rose as a massive concentration of information, that is, knowledge and intelligence, on the side of capital, in fact as a "control revolution" over industrial production.[39] Post-Fordism is Fordism plus the databases of labor.

Computation, or the Encryption of Labor

Paul Edwards has illustrated how climate science and the computation of global warming are possible only thanks to a planetary network of sensors, data centers, and institutions that conceived and implemented mathematical models

[37] Romano Alquati, "Composizione organica del capitale e forza-lavoro alla Olivetti, Part 2," *Quaderni Rossi*, no. 3 (June 1963): 119–85. Translated by the author in Matteo Pasquinelli, "Italian *Operaismo* and the Information Machine," *Theory, Culture & Society* 32, no. 3 (May 2015): 55.

[38] On the hegemony of immaterial production see Michael Hardt and Antonio Negri, *Multitude: War and Democracy in the Age of Empire* (New York: Penguin Books, 2004), 103–15.

[39] Beniger, *The Control Revolution*.

for data mining and forecasting.[40] Surprisingly (or maybe not), the first picture of the "vast machine" of meteorological computation by John Ruskin resembled closely the "vast automaton" of the industrial factory described by Ure.[41] Ruskin's Meteorological Society appeared to be designed to mirror and second the central technological axis of the time, that is, the giant *automaton* that was orchestrating the division of manual and mental labor in the industrial factory.

> The [M]eteorological Society, therefore, has been formed, not for a city, nor for a kingdom, but for the world. It wishes to be the central point, the moving power, of a vast machine, and it feels that unless it can be this, it must be powerless; if it cannot do all, it can do nothing. It desires to have at its command, at stated periods, perfect systems of methodical and simultaneous observations; it wishes its influence and its power to be omnipresent over the globe, so that it may be able to know, at any given instant, the state of the atmosphere on every point on its surface.[42]

The perception of the whole Earth as an ecosystem (like in the Gaia hypothesis) and the measurement of the Anthropocene are possible only through the most sophisticated information technologies. As much as the Royal Meteorological Society in the UK imitated the automaton of industrial capitalism qua control apparatus, today climate science institutions mirror the datacenters of computational capitalism. With almost identical techniques, global datacenters accumulate information and intelligence, not just about the world's climate but also about financial markets, logistical chains, international terrorism, and more importantly, social networks of billions of individuals. Is the similarity between climate science and control apparatuses just a coincidence or does it point to a more general form of governance?

The vast network of climate science appears like an extended cybernetic loop, with big institutions taking on the role of the nervous system of a pretty large organism—planet Earth. The "vast machine" of early climate science should be considered a prototype of the *governance machine* of the Anthropocene, in which more and more metabolic flows and infrastructures are integrated and computed.[43] Climate science infrastructure and the Anthropocene technosphere

[40] Paul N. Edwards, *A Vast Machine: Computer Models, Climate Data, and the Politics of Global Warming* (Cambridge, MA: MIT Press, 2010).

[41] John Ruskin, "Remarks on the Present State of Meteorological Science," *Transactions of the Meteorological Society* 1 (1839): 56–59; Ure, *The Philosophy of Manufactures*.

[42] Ruskin, "Remarks on the Present State of Meteorological Science," 58–59.

[43] In fact, the metabolism of the global technosphere is incredibly complex: It comprises the cycles of chemical compounds such as agricultural nitrogen and rare earth, for instance, and not just fossil carbon. Haff describes the technosphere as a humungous automaton and proposes six rules to

both emerged like the late twin of computational capitalism, in which computation appears to be oriented to the calculus of the planet's *surplus energy* rather than the calculus of *surplus labor*. Computation gives form to surplus, but one wonders if such a computation of surplus energy is just a way to mystify surplus labor. Since the "work/energy crisis" of the 1970s, we know that any definition and measure of energy affects the governance of labor.[44] More generally, it looks like we have surrendered the antagonism between labor, energy, and information to the Cybernetic Hypothesis[45] on one side and the Anthropocene Hypothesis on the other. The former postulates that life on the planet *is already* under the control of a totalitarian cybernetic apparatus, the latter that life on the planet *should be* under the control of a benevolent cybernetic apparatus. In both scenarios, computation is the adequate form of the paradoxical disappearance of labor, that is, *the ideological encryption of labor* within technology. It is necessary, then, to reveal labor again in the diagram of technology and, conversely, technology in the diagram of labor. The limit of current Marxism is its inability to recognize the new forms of *technified labor* and *technified subjectivities* that have lost any resemblance to the labor struggles of the past. In the mesh of global logistics and algorithmic division of labor, new assemblages of labor must be recognized.

Carbosilicon Assemblages and Cyberfossil Capital

Any bifurcation is the birth of a new assemblage. And vice versa, any new assemblage expands previous bifurcations. In 1989, the same year as the fall of the Berlin Wall and a decade before the much-celebrated rise of the network society, Guattari pictured "the age of planetary computerization" in his book *Schizoanalytic Cartographies*.[46] This age was prophetically marked by a polyphony of technologies, including new chemical compounds and even nuclear fusion energy, but more importantly, artificial intelligence and large databases. According to Guattari, new subjectivities would be based on the computation of "enormous quantities of data"[47] and biological engineering would remodel

frame the fatal destiny of the human outclassed by the metabolism of technology: inaccessibility, impotence, control, scale, performance, and provision (curiously grounding in this way the principles of *anti-cybernetics*, as this looks like a theory of noncontrol). Peter Haff, "Humans and Technology in the Anthropocene: Six Rules," *Anthropocene Review* 1, no. 2 (August 2014): 126–36.

44 Caffentzis, "The Work/Energy Crisis and the Apocalypse."

45 Tiqqun, "L'hypothèse cybernétique," *Tiqqun* 2 (2001): 40–83.

46 Félix Guattari, *Schizoanalytic Cartographies*, trans. Andrew Goffey (1989; London: Bloomsbury, 2013), 11.

47 Guattari, *Schizoanalytic Cartographies*, 11.

living forms anew. In the same year, Guattari also published *The Three Ecologies* and recognized, in parallel, the ecological catastrophe driven by the hubris of technoscience.

> The Earth is undergoing a period of intense techno-scientific transformations. If no remedy is found, the ecological disequilibrium this has generated will ultimately threaten the continuation of life on the planet's surface.[48]

The contrast between the potentiality of computation and the damages of the technosphere has become manifest today, with global datacenters accelerating networks of logistics, the extraction of natural resources (often in the Global South), and fossil fuel emissions worldwide. The incestuous relation between planetary control and planetary disequilibrium is the riddle at stake in the hiatus between the Cybernetic Hypothesis and the Anthropocene Hypothesis, the civilization of Silicon and the one of Carbon, the lineage of Information and the one of Energy, as illustrated throughout this essay. The relation between the chimera's two heads of Energy and Information happened to be a turbulent double bind: of mutual amplification (in the game of capital) but also of containment (in the game of politics). Rather than reiterating the opposition of monotonic paradigms, it may be better to try and consolidate the assemblage of energy and information into new systemic notions.

The idea of the carbosilicon machine is proposed to describe the historical assemblage of industrial and information apparatuses: the grafting of the Turing machine onto the governor of the thermodynamic engine. The carbosilicon machine is the cipher of the technosphere, which seals the molecular imbrication of manual and mental labor that is often overlooked. If Babbage's Analytical Engine, now acknowledged as the first stored-program computer, was "a projection of a more perfect factory,"[49] any Turing machine should be considered an overall *dispositif* for the logistics of mental and manual labor, as well as that of matter and energy. At a planetary scale, the molar coupling of energy and information is obvious in the colonial relation between the data centers for logistics of Western companies and the extractive industries in the Global South. "The Technosphere of the Anthropocene" is therefore the name given to the globalization of the old colonial factory, still waiting to find the present-day Babbage and Marx.

48 Félix Guattari, *The Three Ecologies*, trans. Ian Pindar and Paul Sutton (London: Athlon, 2000), 27.
49 Philip Mirowski, *Machine Dreams: Economics Becomes a Cyborg Science* (Cambridge: Cambridge University Press, 2002), 34.

The notion of the carbosilicon machine may help to decouple and repurpose technology from its colonial and monopolistic destiny and, more importantly, to illuminate new forms of struggle and resistance. Such a clarification is especially hard nowadays due to the double crisis of the Carbon and Silicon regimes: The environmental and energetic crisis, on the one hand, and the crisis of valorization triggered by digital technologies, on the other, have galvanized political fronts that strive to merge. Critical thought (and specifically Marxism) has never tried, in this respect, to unite the lineages of energy and information into a synthetic definition of labor. Everything can be easily described under the hegemony of *financial capitalism*, but *fossil capitalism* and *cognitive capitalism* are still waiting to be integrated. Such a theoretical weakness is mirrored by a sort of "bifurcation" that happens to social struggles too. The disconnect between information-related struggles (from the hacker movement to the digital precariat, from Anonymous to media activism in the post-Snowden age) and energy-related struggles (from antinuclear movements to climate justice, from urban ecology to Indigenous struggles on land and sovereignty) is evident. To use an old topos of the Autonomist thought, a new *political composition* of energy and information must be thought against the *technical composition* that has bifurcated them since the industrial age.

How might we address social autonomy in the age of the planetary automaton? Ironically, the *automaton* of the technosphere[50] appears to absorb and reverse the *autonomy* of social movements and workers' struggles of the period 1960–1970s, as much as the network cultures of the period 1990–2010s (themselves easily captured by the new social monopolies of the Internet). The technological form absorbs features that once belonged to the social form. It may be wise to clarify here that "automaton" means capable of independent motion, whereas "autonomy" means self-governing and able to invent new laws, rules, and habits. Automation is the imitation of old rules; autonomy is the invention of new ones. This is the reason why Langdon Winner says that the expression "autonomous technology"[51] is ironic, because technical objects seem to supplant the free will of subjects. To rethink social autonomy today, one has to see what the autonomy of energy and the autonomy of information mean together in an expanded (and technified) notion of labor.

Eventually, the designation of carbosilicon machine summons the demons of its historical proliferation and logically bespeaks the birth of the regime of

[50] Haff, "Humans and Technology in the Anthropocene."
[51] Langdon Winner, *Autonomous Technology: Technics-out-of-Control as a Theme in Political Thought* (Cambridge, MA: MIT Press, 1977), 16.

cyberfossil capitalism—a regime that has implemented energy and information qua abstract equivalents as much as labor and money. The imbrication of energy and information flows is not new to philosophy. In their reading of the fragment on machines in *Grundrisse*,[52] Deleuze and Guattari recognized a *machinic surplus value* that was distinguished into a *surplus value of flow* (labor, energy) and a *surplus value of code* (information, knowledge).[53] Accordingly, late computational capitalism should be defined as an abstract machine that divides labor into flows of energy and information and manages their synthesis qua real abstractions. Cyberfossil capitalism is the metabolism of the most archaic biosphere and the most abstract technosphere, united by capital.

[52] Karl Marx, *Grundrisse: Foundations of the Critique of Political Economy*, trans. Martin Nicolaus (1939; London: Penguin, 1973), 690–712.

[53] Gilles Deleuze and Félix Guattari, *Anti-Oedipus: Capitalism and Schizophrenia*, trans. Robert Hurley, Mark Seem, and Helen R. Lane (Minneapolis: University of Minnesota Press, 1983), 232.

5

Anatomy of an AI System

The Amazon Echo as an Anatomical Map of Human Labor, Data, and Planetary Resources[1]

Kate Crawford and Vladan Joler

I

A cylinder sits in a room. It is impassive, smooth, simple, and small. It stands 14.8 cm high, with a single blue-green circular light that traces around its upper rim. It is silently attending. A woman walks into the room, carrying a sleeping child in her arms, and she addresses the cylinder.

"Alexa, turn on the hall lights."

The cylinder springs into life. "OK." The room lights up. The woman makes a faint nodding gesture and carries the child upstairs.

[1] This text was originally published by the AI Now Institute and Share Lab, September 7, 2018, https://anatomyof.ai. All images courtesy of Vladan Joler and Kate Crawford.

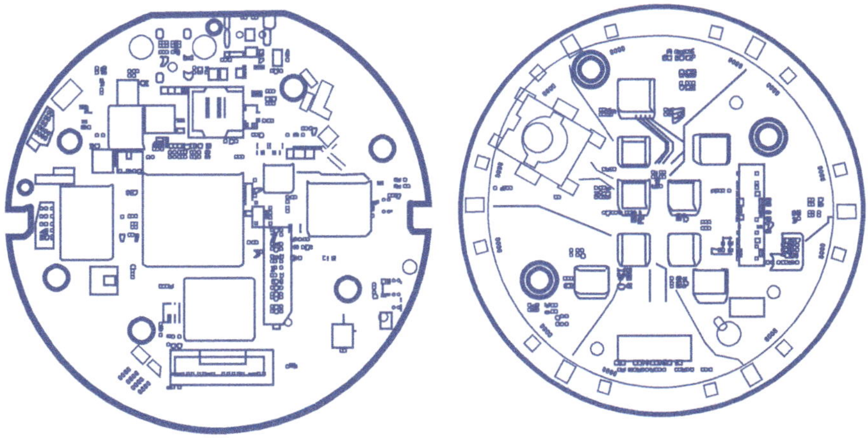

Figure 5.1 Amazon Echo Dot schematics.

This is an interaction with Amazon's Echo device.[2] A brief command and a response is the most common form of engagement with this consumer voice-enabled AI device. But in this fleeting moment of interaction, a vast matrix of capacities is invoked: interlaced chains of resource extraction, human labor, and algorithmic processing across networks of mining, logistics, distribution, prediction, and optimization. The scale of this system is almost beyond human imagining. How can we begin to see it, to grasp its immensity and complexity as a connected form? We start with an outline: an exploded view of a planetary system across three stages of birth, life, and death, accompanied by an essay in twenty-one parts. Together, this becomes an anatomical map of a single AI system.

II

The scene of the woman talking to Alexa is drawn from a 2017 promotional video advertising the latest version of the Amazon Echo. The video begins, "Say hello to the all-new Echo" and explains that the Echo will connect to Alexa (the artificial intelligence agent) in order to "play music, call friends and family, control smart home devices, and more." The device contains seven directional microphones, so the user can be heard at all times even when music is playing. The device comes in several styles, such as gunmetal gray or basic beige, designed to either "blend in or stand out." But even the shiny design options maintain a

[2] Amazon advertising campaign, "All-New Amazon Echo," was released on September 27, 2017, https://youtu.be/IyvZ41XjUjY.

kind of blankness: Nothing will alert the owner to the vast network that subtends and drives its interactive capacities. The promotional video simply states that the range of things you can ask Alexa to do is always expanding. "Because Alexa is in the cloud, she is always getting smarter and adding new features."

How does this happen? Alexa is a disembodied voice that represents the human-AI interaction interface for an extraordinarily complex set of information-processing layers. These layers are fed by constant tides: the flows of human voices being translated into text questions, which are used to query databases of potential answers and the corresponding ebb of Alexa's replies. For each response that Alexa gives, its effectiveness is inferred by what happens next:

Is the same question uttered again? (*Did the user feel heard?*) Was the question reworded? (*Did the user feel the question was understood?*) Was there an action following the question? (*Did the interaction result in a tracked response: a light turned on, a product purchased, a track played?*)

With each interaction, Alexa is training to hear better, to interpret more precisely, to trigger actions that map to the user's commands more accurately, and to build a more complete model of their preferences, habits, and desires. What is required to make this possible? Put simply: Each small moment of convenience—be it answering a question, turning on a light, or playing a song—requires a vast planetary network, fueled by the extraction of nonrenewable materials, labor, and data. The scale of resources required is many magnitudes greater than the energy and labor it would take a human to operate a household appliance or flick a switch. A full accounting for these costs is almost impossible, but it is increasingly important that we grasp the scale and scope if we are to understand and govern the technical infrastructures that thread through our lives.

III

The Salar, the world's largest flat surface, is located in southwest Bolivia at an altitude of 3,656 meters above sea level. It is a high plateau, covered by a few meters of salt crust which is exceptionally rich in lithium, containing 50 to 70 percent of the world's lithium reserves.[3] The Salar alongside the neighboring Atacama regions in Chile and Argentina are major sites for lithium extraction. This soft, silvery metal is currently used to power mobile connected devices as a crucial material used for the

[3] Emily Achtenberg, "Bolivia Bets on State-Run Lithium Industry," NACLA–North American Congress on Latin America, November 15, 2010, https://nacla.org/news/bolivia-bets-state-run-lithium-industry.

production of lithium-ion batteries. It is known as "gray gold." Smartphone batteries, for example, usually have less than eight grams of this material.[4] Each Tesla car needs approximately 7 kilograms of lithium for its battery pack.[5] All these batteries have a limited lifespan, and once consumed they are thrown away as waste. Amazon reminds users that they cannot open up and repair their Echo, because this will void the warranty. The Amazon Echo is wall-powered and also has a mobile battery base. This also has a limited lifespan and then must be thrown away as waste.

According to the Aymara legends about the creation of Bolivia, the volcanic mountains of the Andean Plateau were creations of tragedy.[6] Long ago, when the volcanoes were alive and roaming the plains freely, Tunupa—the only female volcano—gave birth to a baby. Stricken by jealousy, the male volcanoes stole her baby and banished it to a distant location. The gods punished the volcanoes by pinning them all to the Earth. Grieving for the child that she could no longer reach, Tunupa wept deeply. Her tears and breast milk combined to create a giant salt lake: Salar de Uyuni. As Liam Young and Kate Davies observe, "We power our technologies from the tears and breast milk of sacred volcanos."[7] This landscape is connected to the entire planet via the phones in our pockets and links everyone with invisible threads of commerce, science, politics, and power.

[4] Christine Negroni, "How to Determine the Power Rating of Your Gadget's Batteries," *New York Times*, December 26, 2016, https://www.nytimes.com/2016/12/26/business/lithium-ion-battery-airline-safety.html.

[5] Jessica Shankleman, Tom Biesheuvel, Joe Ryman, and Dave Merrill, "We're Going to Need More Lithium," *Bloomberg*, September 7, 2017, https://www.bloomberg.com/graphics/2017-lithium-battery-future.

[6] Nicola Clark and Simon Wallis, "Flamingos, Salt Lakes and Volcanoes: Hunting for Evidence of Past Climate Change on the High Altiplano of Bolivia," *Geology Today* 33, no. 3 (May 2017): 104.

[7] Unknown Fields, http://www.unknownfieldsdivision.com/summer2015bolivia+atacama-lithium dreams.html; Unknown Fields, *Tales from the Dark Side of the City: The Breastmilk of the Volcano Bolivia and the Atacama Desert Expedition* (London: Architectural Association, 2016).

IV

Our exploded view diagram combines and visualizes three central, extractive processes that are required to run a large-scale artificial intelligence (AI) system: material resources, human labor, and data. We consider these three elements across time—represented as a visual description of the birth, life, and death of a single Amazon Echo unit. It is necessary to move beyond a simple analysis of the relationship between an individual human, their data, and any single technology company in order to contend with the truly planetary scale of extraction. Vincent Mosco has shown how the ethereal metaphor of "the cloud" for offsite data management and processing is in complete contradiction with the physical realities of the extraction of minerals from the Earth's crust and dispossession of human populations that sustain its existence.[8] Sandro Mezzadra and Brett Nielson use the term "extractivism" to name the relationship between different forms of extractive operations in contemporary capitalism,[9] which we see repeated in the context of the AI industry. There are deep interconnections between the literal hollowing out of the materials of the earth and biosphere and the data capture and monetization of human practices of communication and sociality in AI. Mezzadra and Nielson note that labor is central to this extractive relationship, which has repeated throughout history: from the way European imperialism used slave labor, to the forced work crews on rubber plantations in Malaya, to the Indigenous people of Bolivia being driven to extract the silver that was used in the first global currency. Thinking about extraction requires thinking about labor, resources, and data together. This presents a challenge to critical and popular understandings of artificial intelligence: It is hard to "see" any of these processes individually, let alone collectively. Hence, the need for a visualization that can bring these connected but globally dispersed processes into a single map.

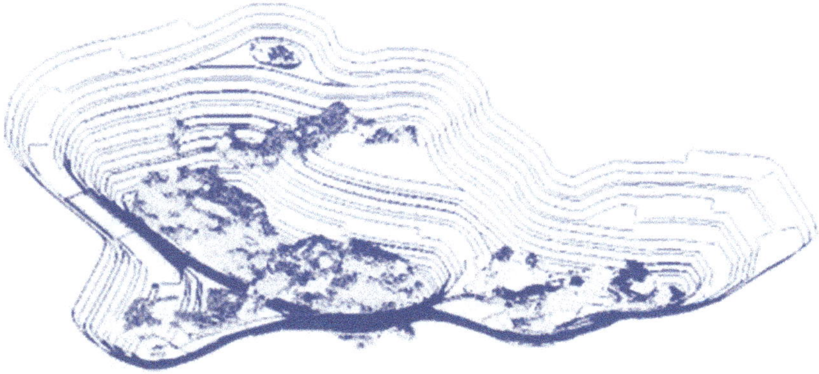

[8] Vincent Mosco, *To the Cloud: Big Data in a Turbulent World* (Boulder: Paradigm, 2014).

[9] Sandro Mezzadra and Brett Neilson, "On the Multiple Frontiers of Extraction: Excavating Contemporary Capitalism," *Cultural Studies* 31, nos. 2–3 (May 2017): 185–204.

V

If you read our map from left to right, the story begins and ends with the Earth and the geological processes of deep time. But read from top to bottom, we see the story as it begins and ends with a human. The top is the human agent, querying the Echo and supplying Amazon with the valuable training data of verbal questions and responses that they can use to further refine their voice-enabled AI systems. At the bottom of the map is another kind of human resource: the history of human knowledge and capacity, which is also used to train and optimize artificial intelligence systems. This is a key difference between artificial intelligence systems and other forms of consumer technology: These systems rely on the ingestion, analysis, and optimization of vast amounts of human-generated images, texts, and videos.

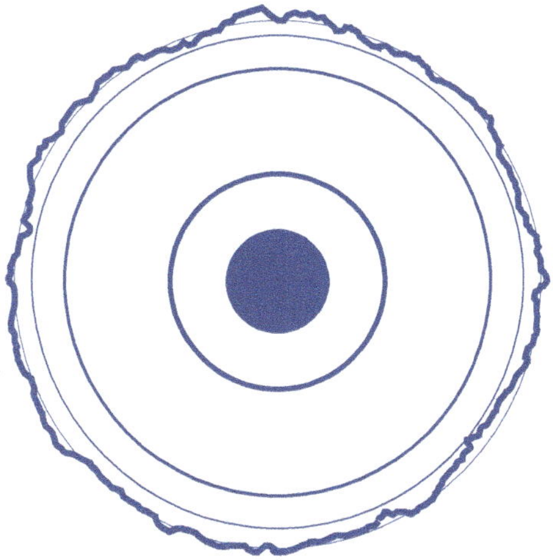

VI

When a human engages with an Echo, or another voice-enabled AI device, they are acting as much more than just an end-product consumer. It is difficult to place the human user of an AI system into a single category: Rather, they deserve to be considered as a hybrid case. Just as the Greek *chimera* was a mythological animal that was part lion, goat, snake, and monster, the Echo user is simultaneously a consumer, a resource, a worker, and a product. This multiple identity recurs for human users in many technological systems. In the specific case of the Amazon Echo, the user has purchased a consumer device for which they receive a set of convenient affordances. But they are also a resource, as their voice commands are collected, analyzed, and retained for the purposes of building an ever-larger corpus of human voices and instructions. And they provide labor, as they continually perform the valuable service of contributing feedback mechanisms regarding the accuracy, usefulness, and overall quality of Alexa's replies. They are, in essence, helping to train the neural networks within Amazon's infrastructural stack.

VII

Anything beyond the limited physical and digital interfaces of the device itself is outside of the user's control. It presents a sleek surface with no ability to open it, repair it, or change how it functions. The object itself is a very simple extrusion of plastic representing a collection of sensors—its real power and complexity lies somewhere else, far out of sight. The Echo is but an "ear" in the home: a disembodied listening agent that never shows its deep connections to remote systems.

In 1673, the Jesuit polymath, Athanasius Kircher, invented the *statua citofonica*, the "talking statue." Kircher was an extraordinary, interdisciplinary scholar and inventor. In his lifetime, he published forty major works across the fields of medicine, geology, comparative religion, and music. He invented the first magnetic clock, many early automatons, and the megaphone. His talking statue was a very early listening system: essentially a microphone made from a huge spiral tube, which could convey the conversations from a public square, up through the tube, and then piped through the mouth of a statue kept within an aristocrat's private chambers. As Kircher wrote:

> This statue must be located in a given place, in order to allow the end section of the spiral-shaped tube to precisely correspond to the opening of the mouth. In this manner it will be perfect, and capable to emit clearly any kind of sound: in fact the statue will be able to speak continuously, uttering in either a human or animal voice: it will laugh or sneer; it will seem to really cry or moan; sometimes with great astonishment it will strongly blow. If the opening of the spiral-shaped tube is located in correspondence to an open public space, all human words pronounced, focused in the conduit, would be replayed through the mouth of the statue: [. . .].[10]

[10] Athanasius Kircher, *Phonurgia Nova*, 1673, cited in Lamberto Tronchin, "The 'Phonurgia Nova' of Athanasius Kircher: The Marvellous Sound World of 17th Century," *Proceedings of Meetings on Acoustics* 4, no. 1 (June 2008): 015002.

Figure 5.2 Athanasius Kircher, *Statua citofonica*, 1673.

The listening system could eavesdrop on everyday conversations in the piazza and relay them to the seventeenth-century Italian oligarchs. Kircher's talking statue was an early form of information extraction for the elites; people talking in the street would have no indication that their conversations were being funneled to those who would instrument that knowledge for their own power, entertainment, and wealth. People inside the homes of aristocrats would have no idea how a magical statue was speaking and conveying all manner of information. The aim was to obscure how the system worked: An elegant statue was all they could see. Listening systems, even at this early stage, were about power, class, and secrecy. But the infrastructure for Kircher's system was prohibitively expensive—available only to the very few. And so the question remains, what are the full resource implications of building such systems? This brings us to the materiality of the infrastructure that lies beneath.

VIII

In his book *A Geology of Media*, Jussi Parikka suggests that we try to think of media not from Marshall McLuhan's point of view—in which media are extensions of human senses[11]—but rather as an extension of Earth.[12] Media technologies should be understood in the context of a geological process, from the creation and the transformation processes to the movement of the natural elements from which media are built. Reflecting upon media and technology as geological processes enables us to consider the profound depletion of nonrenewable resources required to drive the technologies of the present moment. Each object in the extended network of an AI system, from network routers to batteries to microphones, is built using elements that require billions of years to be produced. Looking from the perspective of deep time, we are extracting Earth's history to serve a split second of technological time, in order to build devices that are often designed to be used for no more than a few years. For example, the Consumer Technology Association notes that the average smartphone lifespan is 4.7 years.[13] This obsolescence cycle fuels the purchase of more devices, drives up profits, and increases incentives for the use of unsustainable extraction practices. From a slow process of elemental development, these elements and materials go through an extraordinarily rapid period of excavation, smelting, mixing, and logistical transport—crossing thousands of kilometers in their transformation. Geological processes mark both the beginning and the end of this period, from the mining of ore to the deposition of material in an electronic waste dump. For that reason, our map starts and ends with the Earth's crust. However, all the transformations and movements we depict are only the barest anatomical outline: Beneath these connections lie many more layers of fractal supply chains, exploitation of human and natural resources, concentrations of corporate and geopolitical power, and continual energy consumption.

IX

Drawing out the connections between resources, labor, and data extraction brings us inevitably back to traditional frameworks of exploitation. How is value being generated through these systems? A useful conceptual tool can be found in the work of Christian Fuchs and other authors examining and defining digital labor. The notion of digital labor, which was initially linked with different forms of nonmaterial labor, precedes the life of devices and

[11] Marshall McLuhan, *Understanding Media: The Extensions of Man* (New York: McGraw-Hill, 1964).
[12] Jussi Parikka, *A Geology of Media* (Minneapolis: University of Minnesota Press, 2015), vii–viii.
[13] Chris Ely, "The Life Expectancy of Electronics," Consumer Technology Association, September 16, 2014, accessed September 7, 2018.

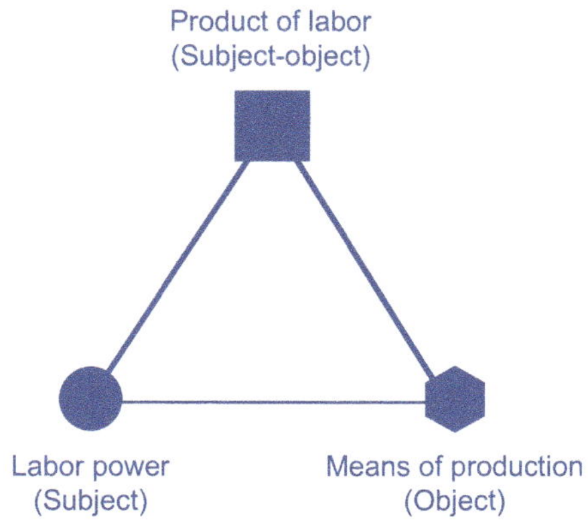

Figure 5.3 Marx's dialectic of subject and object in economy.

complex systems such as artificial intelligence. Digital labor—the work of building and maintaining the stack of digital systems—is far from ephemeral or virtual but is deeply embodied in different activities.[14] The scope is overwhelming: from indentured labor in mines for extracting the minerals that form the physical basis of information technologies; to the work of strictly controlled and sometimes dangerous hardware manufacturing and assembly processes in Chinese factories; to exploited outsourced cognitive workers in developing countries labeling AI training datasets; to the informal workers physically cleaning up toxic waste dumps. These processes create new accumulations of wealth and power, which are concentrated in a very thin social layer.

X

This triangle of value extraction and production represents one of the basic elements of our map, from birth in a geological process, through life as a consumer AI product, and ultimately to death in an electronics dump. Like in Fuchs's work, our triangles are not isolated but linked to one another in the production process. They form a cyclic flow in which the product of work is transformed into a resource, which is transformed into a product, which is transformed into a resource, and so on. Each triangle represents one phase in the production process. Although this appears on the map as a

[14] Christian Fuchs, *Digital Labor and Karl Marx* (London: Routledge, 2014).

linear path of transformation, a different visual metaphor better represents the complexity of current extractivism: the fractal structure known as the Sierpiński triangle.

A linear display does not enable us to show that each next step of production and exploitation contains the previous phases. If we look at the production and exploitation system through a fractal visual structure, the smallest triangle would represent natural resources and means of labor, that is, the miner as labor and ore as product. The next larger triangle encompasses the processing of metals, and the next would represent the process of manufacturing components, and so on. The ultimate triangle in our map, the production of the Amazon Echo unit itself, includes all of these levels of exploitation from the bottom to the very top of Amazon Inc., a role formerly inhabited by Jeff Bezos as CEO of Amazon. Like a pharaoh of ancient Egypt, he stands at the top of the largest pyramid of AI value extraction.

XI

To return to the basic element of this visualization—a variation of Karl Marx's triangle of production—each triangle creates a surplus of value for generating profits. If we look at the scale of average income for each activity

Figure 5.4 Sierpiński triangle, a self-similar fractal.

in the production process of one device, which is shown on the left side of the "Anatomy of an AI system" map, we see the dramatic difference in income earned. According to research by Amnesty International, during the excavation of cobalt, which is also used for lithium batteries of sixteen multinational brands, workers are paid the equivalent of one US dollar per day for working in conditions hazardous to life and health, and were often subjected to violence, extortion, and intimidation.[15] Amnesty has documented children as young as seven working in the mines. In contrast, former Amazon CEO Bezos, at the top of our fractal pyramid, made an average of $275 million a day during the first five months of 2018, according to the Bloomberg Billionaires Index.[16] A child working in a mine in the Congo would need more than 700,000 years of nonstop work to earn the same amount as a single day of Bezos's income.

Many of the triangles shown on this map hide different stories of labor exploitation and inhumane working conditions. The ecological price of transformation of elements and income disparities is just one of the possible ways of representing a deep systemic inequality. We have both researched different forms of "black boxes" understood as algorithmic processes,[17] but this map points to another form of opacity: The very processes of creating, training, and operating a device like an Amazon Echo is in and of itself a kind of black box that is very hard to examine and track *in toto* given the multiple layers of contractors, distributors, and downstream logistical partners around the world. As Mark Graham and Håvard Haarstad write, "Contemporary capitalism conceals the histories and geographies of most commodities from consumers. These consumers [are usually only able to see commodities in the here and now of time and space, and] rarely have opportunities to gaze backward through the chains of production [in order] to gain knowledge about the sites of production, transformation, and distribution of products."[18]

One illustration of the difficulty of investigating and tracking the contemporary production chain process is that it took Intel more than four years to understand

[15] Amnesty International, "'This is What We Die for': Human Rights Abuses in the Democratic Republic of the Congo Power the Global Trade in Cobalt," January 2016, https://www.amnesty.org/en/documents/afr62/3183/2016/en. For an anthropological description of these mining processes, see Jeffrey W. Mantz, "Improvisational Economies: Coltan Production in the Eastern Congo," *Social Anthropology* 16, no. 1 (February 2008): 34–50.

[16] Julia Glum, "The Median Amazon Employee's Salary Is $28,000. Jeff Bezos Makes More Than That in 10 Seconds," *Time*, May 2, 2018.

[17] Frank Pasquale, *The Black Box Society: The Secret Algorithms That Control Money and Information* (Cambridge, MA: Harvard University Press, 2016).

[18] Mark Graham and Håvard Haarstad, "Transparency and Development: Ethical Consumption Through Web 2.0 and the Internet of Things," *Information Technologies & International Development* 7, no. 1 (Spring 2011): 1.

its supply lines well enough to ensure that no tantalum from the Congo was in its microprocessor products. As a semiconductor chip manufacturer, Intel supplies Apple with processors. In order to do so, Intel has its own multitiered supply chain of more than 19,000 suppliers in over 100 countries providing direct materials for their production processes, tools and machines for their factories, and logistics and packaging services.[19] That it took over four years for a leading technology company *just to understand its own supply chain*, reveals just how hard this process can be to grasp from the inside, let alone for external researchers, journalists, and academics. Dutch-based technology company Philips has also claimed that it was working to make its supply chain "conflict-free." Philips, for example, has tens of thousands of different suppliers, each of which provides different components for their manufacturing processes.[20] Those suppliers are themselves linked downstream to tens of thousands of component manufacturers that acquire materials from hundreds of refineries that buy ingredients from different smelters, which are supplied by unknown numbers of traders that deal directly with both legal and illegal mining operations. In *The Elements of Power*, David S. Abraham describes the invisible networks of rare metals traders in global electronics supply chains: "The network to get rare metals from the mine to your laptop travels through a murky network of traders, processors, and component manufacturers. Traders are the middlemen who do more than buy and sell rare metals: they help to regulate information and are the hidden link that helps in navigating the network between metals plants [. . .] and the components in our laptops."[21] According to the computer manufacturing company Dell, complexities of the metal supply chains pose almost insurmountable challenges.[22] The mining of these minerals takes place long before a final product is assembled, making it exceedingly difficult to trace the minerals' origins. In addition, many of the minerals are smelted together with recycled metals, by which point it becomes all but impossible to trace the minerals to their sources. So we see that the attempt to capture the full supply chain is a truly gargantuan task: revealing all the complexity of the twenty-first century global production of technology products.

19 Intel Corporation, "Intel's Efforts to Achieve a 'Conflict Free' Supply Chain," May 2018, accessed September 7, 2018.

20 Philips, "We Are Working to Make Our Supply Chain 'Conflict-Free,'" 2018, accessed September 7, 2018.

21 David S. Abraham, *The Elements of Power: Gadgets, Guns, and the Struggle for a Sustainable Future in the Rare Metal Age*, reprint edition (New Haven: Yale University Press, 2017), 90.

22 Dell, "Responsible Minerals Sourcing Report," June 2018, https://i.dell.com/sites/csdocuments/Corporate_corp-Comm_Documents/en/sourcing-report18.pdf.

XII

Supply chains are often layered on top of one another, in a sprawling network. Apple's supplier program reveals there are tens of thousands of individual components embedded in their devices, which are in turn supplied by hundreds of different companies. In order for each of those components to arrive on the final assembly line where it will be assembled by workers in Foxconn, different components need to be physically transferred from more than 750 supplier sites across 30 different countries.[23] This becomes a complex structure of *supply chains within supply chains*, a zooming fractal of tens of thousands of suppliers, millions of kilometers of shipped materials and hundreds of thousands of workers included within the process even before the product is assembled on the line.

Visualizing this process as one global, pancontinental network through which materials, components, and products flow, we see an analogy to the global information network. Where there is a single network packet traveling to an Amazon Echo, here we can imagine a single cargo container.[24] The dizzying spectacle of global logistics and production would not be possible without the invention of this simple, standardized metal object. Standardized cargo containers allowed the explosion of modern shipping industry, which made it possible to model the planet as a massive, single factory. In 2017, the capacity of container ships in seaborne trade reached nearly 250,000,000 dead-weight tons of cargo, dominated by giant shipping companies such as Maersk of Denmark, the Mediterranean Shipping Company of Switzerland, and CMA CGM, each owning hundreds of container vessels.[25] For these commercial ventures, cargo shipping is a relatively cheap way to traverse the vascular system of the global factory, yet it disguises much larger external costs.

In recent years, shipping boats produce 3.1 percent of global yearly CO_2 emissions, more than the entire country of Germany.[26] In order to minimize their internal costs, most of the container shipping companies use very low-grade fuel in enormous quantities, which lead to increased amounts of sulfur in the air,

[23] Apple, "Supplier Responsibility 2018 Progress Report," 2018, https://www.apple.com/supplier-responsibility/pdf/Apple_SR_2018_Progress_Report.pdf.

[24] Alexander Klose, *The Container Principle: How a Box Changes the Way We Think*, trans. Charles Marcum II (Cambridge, MA: MIT Press, 2015).

[25] UNCTAD–United Nations Conference on Trade and Development, "Review of Maritime Transport," 2017, http://unctad.org/en/PublicationsLibrary/rmt2017_en.pdf.

[26] Zoë Schlanger, "If Shipping Were a Country, It Would be the Sixth-biggest Greenhouse Gas Emitter," *Quartz*, April 17, 2018, https://qz.com/1253874/if-shipping-were-a-country-it-would-the-worlds-sixth-biggest-greenhouse-gas-emitter.

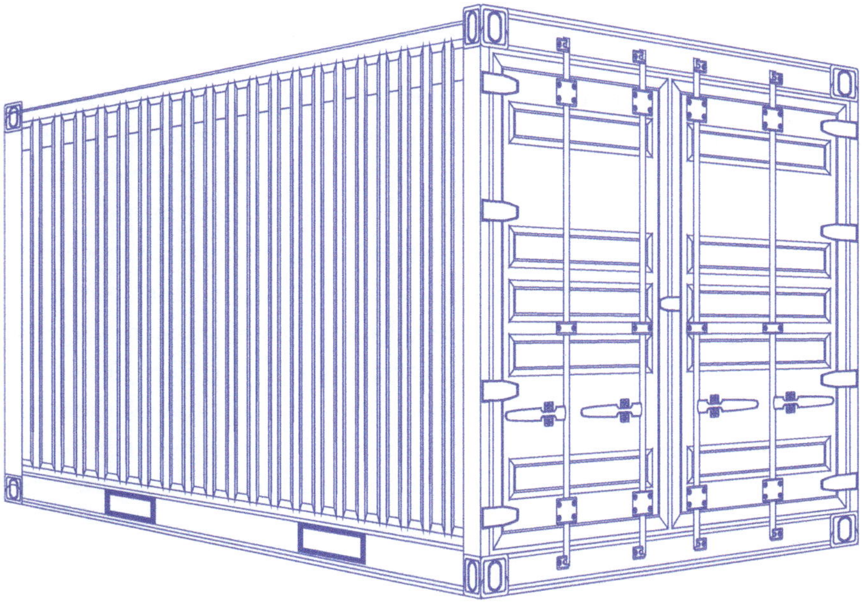

Figure 5.5 Cargo container.

among other toxic substances. It has been estimated that one container ship can emit as much pollution as fifty million cars, and 60,000 deaths worldwide are attributed indirectly to cargo ship industry pollution-related issues annually.[27] Even industry-friendly sources like the World Shipping Council admit that thousands of containers are lost each year, on the ocean floor or drifting loose.[28] Some carry toxic substances which leak into the oceans. Typically, workers spend nine to ten months in the sea, often with long working shifts and without access to external communications. Workers from the Philippines represent more than a third of the global shipping workforce.[29] The most severe costs of global logistics are born by the atmosphere, the oceanic ecosystem and all it contains, and the lowest-paid workers.

[27] John Vidal, "Health Risks of Shipping Pollution Have Been 'Underestimated,'" *Guardian*, April 9, 2009, http://www.theguardian.com/environment/2009/apr/09/shipping-pollution.

[28] World Shipping Council, "Containers Lost at Sea – 2017 Update," July 10, 2017, https://ratical.org/collapsologie/ContainersLostAtSea-2017Jul10.pdf.

[29] Rose George, *Ninety Percent of Everything: Inside Shipping, the Invisible Industry That Puts Clothes on Your Back, Gas in Your Car, and Food on Your Plate* (New York: Metropolitan Books, 2013), 22. Similar to our habit to neglect materiality of Internet infrastructure and information technology, shipping industry is rarely represented in popular culture. George calls this condition, "Sea Blindness" (2013, 4).

XIII

The increasing complexity and miniaturization of our technology depends on the process that strangely echoes the hopes of early medieval alchemy. Where medieval alchemists aimed to transform base metals into "noble" ones, researchers today use rare earth metals to enhance the performance of other minerals. There are seventeen rare earth elements, which are embedded in laptops and smartphones, making them smaller and lighter. They play a role in color displays, loudspeakers, camera lenses, GPS systems, rechargeable batteries, hard drives, and many other components. They are key elements in communication systems from fiber-optic cables and signal amplification in mobile communication towers, to satellites and GPS technology. But the precise configuration and use of these minerals is hard to ascertain. In the same way that medieval alchemists hid their research behind ciphers and cryptic symbolism, contemporary processes for using minerals in devices are protected behind NDAs and trade secrets.

The unique electronic, optical, and magnetic characteristics of rare earth elements cannot be matched by any other metals or synthetic substitutes discovered to date. While they are called "rare earth metals," some are relatively abundant in the Earth's crust, but extraction is costly and highly polluting. David S. Abraham describes the mining of dysprosium and terbium used in a variety of high-tech devices in Jianxi, China. He writes, "Only 0.2 percent of the mined clay contains rare earth elements. This means that 99.8 percent is discarded waste called 'tailings' that are dumped back into the hills and streams,"[30] creating new pollutants like ammonium. In order to refine one ton of rare earth elements, "the Chinese Society of Rare Earths estimates that the process produces 75,000 liters [. . .] of acidic wastewater and one ton of radioactive residue per ton of rare earths."[31] Furthermore, mining and refining activities consume vast amount of water and generate large quantities of CO_2 emissions. In 2009, China produced 95 percent of the world's supply of these elements, and it has been estimated that the single mine known as Bayan Obo contains 70 percent of the world's reserves.[32]

30 David S. Abraham, *The Elements of Power: Gadgets, Guns, and the Struggle for a Sustainable Future in the Rare Metal Age* (New Haven: Yale University Press, 2015), 175.
31 Abraham, *The Elements of Power*, 176.
32 Chris Lo, "The False Monopoly: China and the Rare Earths Trade," *Mining Technology*, August 19, 2015, https://www.mining-technology.com/features/featurethe-false-monopoly-china-and-the-rare-earths-trade-4646712.

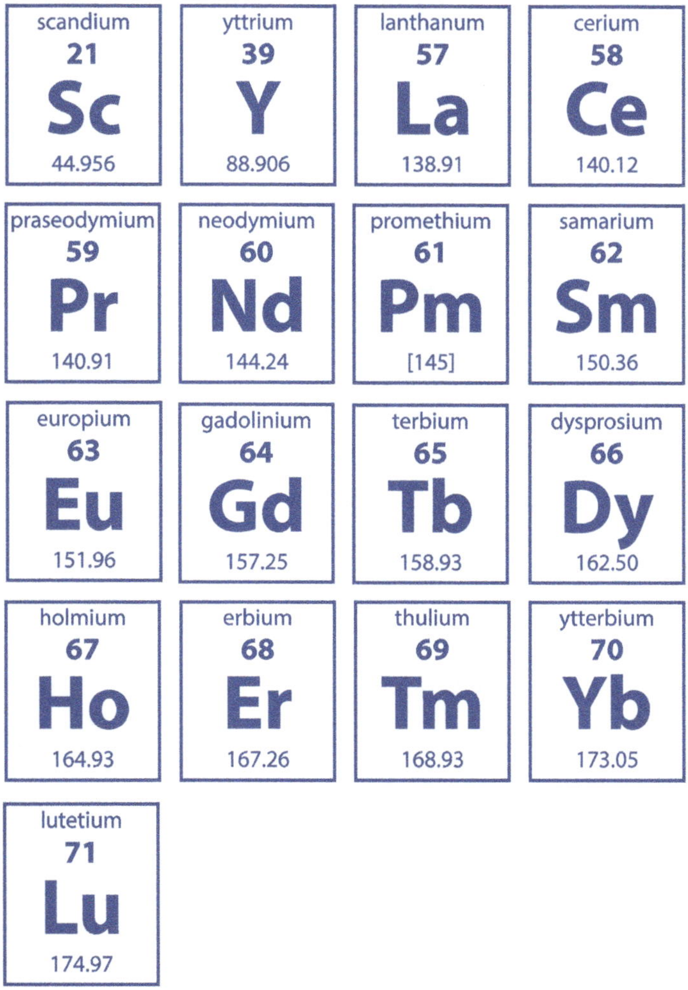

Figure 5.6 Rare earth elements.

XIV

A satellite picture of the tiny Indonesian island of Bangka tells a story about human and environmental toll of the semiconductor production. On this tiny island, mostly "informal" miners are on makeshift pontoons, using bamboo poles to scrape the seabed, and then diving underwater to suck tin from the surface through giant, vacuum-like tubes. As a *Guardian* investigation reports:

> Tin mining is a lucrative but destructive trade that has scarred the island's landscape, bulldozed its farms and forests, killed off its fish stocks and coral

> reefs, and dented tourism to its pretty palm-lined beaches. The damage is best seen from the air, as pockets of lush forest huddle amid huge swaths of barren orange earth. Where not dominated by mines, this is pockmarked with graves, many holding the bodies of miners who have died over the centuries digging for tin.[33]

Two small islands, Bangka and Belitung, produce 90 percent of Indonesia's tin, and Indonesia is the world's second-largest exporter of the metal. Indonesia's national tin corporation, PT Timah, supplies companies such as Samsung directly, as well as solder makers Chernan and Shenmao, which in turn supply Sony, LG, and Foxconn.[34]

XV

At Amazon fulfillment centers, vast collections of products are arrayed in a computational order across millions of shelves. The position of every item in this space is precisely determined by complex mathematical functions that process information about orders and create relationships between products. The aim is to optimize the movements of the robots and humans that collaborate in these warehouses. With the help from an electronic bracelet, the human worker is directed though warehouses the size of airplane hangars, filled with objects arranged in an opaque algorithmic order.[35]

Hidden among the thousands of other publicly available patents owned by Amazon, US patent number 9,280,157 represents an extraordinary illustration of worker alienation and a stark shift in the relationship between humans and machines.[36] It depicts a metal cage intended for the worker, equipped with different cybernetic add-ons, that can be moved through a warehouse by the same motorized system that shifts shelves filled with merchandise. Here, the worker becomes a part of a machinic ballet, held upright in a cage which dictates and constrains their movement.

As we have seen time and time again in the research for our map, dystopian futures are built upon the unevenly distributed dystopian regimes of the past and

[33] Kate Hodal, "Death Metal: Tin Mining in Indonesia," *Guardian*, November 23, 2012, http://www.theguardian.com/environment/2012/nov/23/tin-mining-indonesia-bangka.

[34] Cam Simpson, "The Deadly Tin Inside Your Smartphone," *Bloomberg*, August 24, 2012, https://www.bloomberg.com/news/articles/2012-08-23/the-deadly-tin-inside-your-smartphone.

[35] Marcus Wohlsen, "A Rare Peek Inside Amazon's Massive Wish-Fulfilling Machine," *Wired*, June 16, 2014, https://www.wired.com/2014/06/inside-amazon-warehouse.

[36] Peter R. Wurman et al., "System and Method for Transporting Personnel within an Active Workspace," United States Patent No. US 9,280,157 B2, filed September 4, 2013, and issued March 8, 2016.

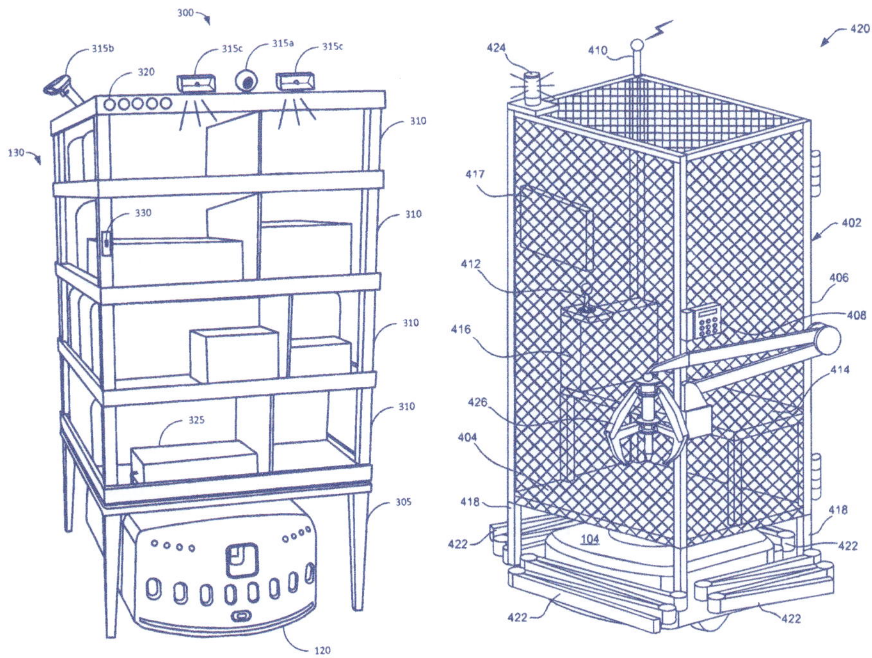

Figure 5.7 Amazon patent number 20150066283 A1.

present, scattered through an array of production chains for modern technical devices. The vanishingly few at the top of the fractal pyramid of value extraction live in extraordinary wealth and comfort. But the majority of the pyramids are made from the dark tunnels of mines, radioactive waste lakes, discarded shipping containers, and corporate factory dormitories.

XVI

At the end of nineteenth century, a particular Southeast Asian tree called *Palaquium gutta* became the center of a technological boom. These trees, found mainly in Malaysia, produce a milky white natural latex called *gutta-percha*. After English scientist Michael Faraday published a study in the *Philosophical Magazine* in 1848 about the use of this material as an electrical insulator, *gutta-percha* rapidly became the darling of the engineering world. It was seen as the solution to the problem of insulating telegraphic cables in order that they could withstand the conditions of the ocean floor. As the global submarine business grew, so did demand for *Palaquium gutta* tree trunks. The historian John Tully describes how local Malay, Chinese, and Dayak workers were paid little for the dangerous works of felling the trees and slowly collecting the latex.[37] The latex was processed then sold through Singapore's trade markets into the British market, where it was transformed into, among other things, lengths upon lengths of submarine cable sheaths.

A mature *Palaquium gutta* could yield around 300 grams of latex. But in 1857, the first transatlantic cable was around 3,000 km long and weighed 2,000 tons—requiring around 250 tons of *gutta-percha*. To produce just one ton of this material required around 900,000 tree trunks. The jungles of Malaysia and Singapore were stripped, and by the early 1880s the *Palaquium gutta* had vanished. In a last-ditch effort to save their supply chain, the British passed a ban in 1883 to halt harvesting the latex, but the tree was already extinct.[38]

The Victorian environmental disaster of *gutta-percha*, from the early origins of the global information society, shows how the relationships between technology and its materiality, environments, and different forms of exploitation are imbricated. Just as Victorians precipitated ecological disaster for their early cables, so do rare earth mining and global supply chains further imperil the delicate ecological balance of our era. From the material used to build the technology enabling contemporary networked society, to the energy needed for transmitting, analyzing, and storing the data flowing through the massive infrastructure, to the materiality of infrastructure: These deep connections and costs are more significant, and have a far longer history, than is usually represented in the corporate imaginaries of AI.[39]

XVII

Large-scale AI systems consume enormous amounts of energy. Yet the material details of those costs remain vague in the social imagination. It remains difficult to get precise details about the amount of energy consumed by cloud computing services. A Greenpeace report states:

> One of the single biggest obstacles to sector transparency is Amazon Web Services (AWS). The world's biggest cloud computer company remains almost completely non-transparent about the energy footprint of its massive operations. Among the global cloud providers, only AWS still refuses to make public basic

37 John Tully, "A Victorian Ecological Disaster: Imperialism, the Telegraph, and Gutta-Percha," *Journal of World History* 20, no. 4 (December 2009): 559–79.

38 Tully, "A Victorian Ecological Disaster," 574.

39 See Nicole Starosielski, *The Undersea Network* (Durham: Duke University Press Books, 2015).

Figure 5.8 *Palaquium gutta*, a tropical tree.

details on the energy performance and environmental impact associated with its operations.[40]

As human agents, we are visible in almost every interaction with technological platforms. We are always being tracked, quantified, analyzed, and commodified. But in contrast to user visibility, the precise details about the phases of birth, life, and death of networked devices are obscured. With emerging devices like the Echo relying on a centralized AI infrastructure far from view, even more of the detail falls into the shadows.

[40] Gary Cook et al., "Clicking Clean: Who Is Winning the Race to Build a Green Internet?" (Washington: Greenpeace, January 2017), 30.

While consumers become accustomed to a small hardware device in their living rooms, a phone app, or a semiautonomous car, the real work is being done within machine learning systems that are generally remote from the user and utterly invisible to them. In many cases, transparency would not help much—without forms of real choice and corporate accountability, mere transparency will not shift the weight of the current power asymmetries.[41]

The outputs of machine learning systems are predominantly unaccountable and ungoverned while the inputs are enigmatic. To the casual observer, it looks like it has never been easier to build AI or machine learning-based systems than it is today. Availability of open-source tools for doing so in combination with rentable computation power through cloud superpowers such as Amazon (AWS), Microsoft (Azure), or Google (Google Cloud) is giving rise to a false idea of the "democratization" of AI. While "off the shelf" machine learning tools, like TensorFlow, are becoming more accessible from the point of view of setting up one's own system, the underlying logics of those systems, and the datasets for training them, are accessible to and controlled by a few entities. In the dynamic of dataset collection through platforms like Facebook, users are feeding and training the neural networks with behavioral data, voice, tagged pictures and videos, or medical data. In an era of extractivism, the real value of that data is controlled and exploited by the very few at the top of the pyramid.

XVIII

When massive datasets are used to train AI systems, the individual images and videos involved are commonly tagged and labeled.[42] There is much to be said about how this labeling process abrogates and crystallizes meaning and further, how this process is driven by clickworkers being paid fractions of a cent for this digital piecework.

In 1770, Hungarian inventor Wolfgang von Kempelen constructed a chess-playing machine known as the Mechanical Turk. His goal, in part, was to impress Empress Maria Theresa of Austria. This device was capable of playing chess against a human opponent and had spectacular success winning most of the games played during its demonstrations around Europe and the Americas

[41] Mike Ananny and Kate Crawford, "Seeing without Knowing: Limitations of the Transparency Ideal and Its Application to Algorithmic Accountability," *New Media & Society* 20, no. 3 (2018): 973–89.

[42] Olga Russakovsky et al., "ImageNet Large Scale Visual Recognition Challenge," *International Journal of Computer Vision* 115, no. 3 (2015): 211–52.

for almost nine decades. But the Mechanical Turk was an illusion that allowed a human chess master to hide inside the machine and operate it. Some 160 years later, Amazon branded its micropayment-based crowdsourcing platform with the same name. According to Ayhan Aytes, Amazon's initial motivation to build Mechanical Turk emerged after the failure of its artificial intelligence programs in the task of finding duplicate product pages on its retail website.[43] After a series of futile and expensive attempts, the project engineers turned to humans to work behind computers within a streamlined web-based system.[44] Amazon Mechanical Turk digital workshop emulates artificial intelligence systems by checking, assessing, and correcting machine learning processes with human brainpower. With Amazon Mechanical Turk, it may seem to users that an application is using advanced artificial intelligence to accomplish tasks. But it is closer to a form of "artificial artificial intelligence," driven by a remote, dispersed, and poorly paid clickworker workforce that helps a client achieve their business objectives. As observed by Aytes, "In both cases [both the Mechanical Turk from 1770 and the contemporary version of Amazon's service], the performance of the workers who animate the artifice is obscured by the spectacle of the machine."[45]

This kind of invisible, hidden labor, outsourced or crowdsourced, hidden behind interfaces and camouflaged within algorithmic processes, is now commonplace, particularly in the process of tagging and labeling thousands of hours of digital archives for the sake of feeding the neural networks. Sometimes this labor is entirely unpaid, as in the case of the Google's reCAPTCHA. In a paradox that many of us have experienced, in order to prove that you are not an artificial agent, you are forced to train Google's image recognition AI system for free, by selecting multiple boxes that contain street numbers, or cars, or houses.

As we see repeated throughout the system, contemporary forms of artificial intelligence are not so artificial after all. We can speak of the hard physical labor of mine workers and the repetitive factory labor on the assembly line, the cybernetic labor in distribution centers and the cognitive sweatshops full of outsourced programmers around the world, the low-paid crowdsourced labor of Mechanical Turk workers, or the unpaid immaterial work of users. At every level, contemporary technology is deeply rooted in, and running on, the exploitation of human bodies.

[43] Ayhan Aytes, "Return of the Crowds: Mechanical Turk and Neoliberal States of Exception," in *Digital Labor: The Internet as Playground and Factory*, ed. Trebor Scholz (London: Routledge, 2012), 80.

[44] Jason Pontin, "Artificial Intelligence, with Help from the Humans," *New York Times*, March 25, 2007, https://www.nytimes.com/2007/03/25/business/yourmoney/25Stream.html.

[45] Aytes, "Return of the Crowds," 81.

XIX

In his one-paragraph short story "On Exactitude in Science," Jorge Luis Borges presents us with an imagined empire in which cartographic science became so developed and precise, that it needed a map on the same scale as the empire itself:

> In that Empire, the Art of Cartography attained such Perfection that the map of a single Province occupied the entirety of a City, and the map of the Empire, the entirety of a Province. In time, those Unconscionable Maps no longer satisfied, and the Cartographers Guilds struck a Map of the Empire whose size was that of the Empire, and which coincided point for point with it. The following Generations, who were not so fond of the Study of Cartography as their Forebears had been, saw that that vast map was Useless, and not without some Pitilessness was it, that they delivered it up to the Inclemencies of Sun and Winters. In the Deserts of the West, still today, there are Tattered Ruins of that

Figure 5.9 Mechanical Turk.

> Map, inhabited by Animals and Beggars; in all the Land there is no other Relic of the Disciplines of Geography.[46]

Current machine learning approaches are characterized by an aspiration to map the world, a full quantification of visual, auditory, and recognition regimes of reality. From cosmological models of the universe to the world of human emotions as interpreted through the tiniest muscle movements in the human face, everything becomes an object of quantification. Jean-François Lyotard introduced the phrase "affinity to infinity" to describe how contemporary art, technoscience, and capitalism share the same aspiration to push boundaries toward a potentially infinite horizon.[47] The second half of the nineteenth century, with its focus on the construction of infrastructure and the uneven transition to industrialized society, generated enormous wealth for the small number of industrial magnates that monopolized exploitation of natural resources and production processes.

The new infinite horizon is data extraction, machine learning, and reorganizing information through AI systems of combined human and machinic processing. The territories are dominated by a few global mega-companies, which are creating new infrastructures and mechanisms for the accumulation of capital and exploitation of human and planetary resources.

Such unrestrained thirst for new resources and fields of cognitive exploitation has driven a search for ever deeper layers of data that can be used to quantify the human psyche, conscious and unconscious, private and public, idiosyncratic and general. In this way, we have seen the emergence of multiple cognitive economies from the attention economy,[48] the surveillance economy, the reputation economy,[49] and the emotion economy, as well as the quantification and commodification of trust and evidence through cryptocurrencies.

Increasingly, the process of quantification is reaching into the human affective, cognitive, and physical worlds. Training sets exist for emotion detection, for family resemblance, for tracking an individual as they age, and for human actions such as sitting down, waving, raising a glass, or crying. Every form of biodata—including forensic, biometric, sociometric, and psychometric—are being captured and logged into databases for AI training. That quantification often runs on very limited foundations: Datasets like Atomic Visual Actions

[46] Jorge Luis Borges, "On Exactitude in Science," in *Collected Fictions*, trans. Andrew Hurley (New York: Penguin Books, 1999), 325.

[47] Jean-François Lyotard, "Presenting the Unpresentable: The Sublime," trans. Lisa Liebmann, *Artforum* 20, no. 8 (April 1982): 64–9.

[48] Yves Citton, *The Ecology of Attention* (Cambridge: Polity, 2017).

[49] Shoshana Zuboff, "Big Other: Surveillance Capitalism and the Prospects of an Information Civilization," *Journal of Information Technology* 30, no. 1 (March 2015): 75–89.

Figure 5.10 Quantification of nature.

(AVA) primarily show women in the "playing with children" action category, and men in the "kicking a person" category. The training sets for AI systems claim to be reaching into the fine-grained nature of everyday life, but they repeat the most stereotypical and restricted social patterns, reinscribing a normative vision of the human past and projecting it into the human future.

XX

"The 'enclosure' of biodiversity and knowledge is the final step in a series of enclosures that began with the rise of colonialism. Land and forests were the first resources to be 'enclosed' and converted from commons to commodities. Later on, water resources were 'enclosed' through dams, groundwater mining and privatisation schemes. Now it is the turn of biodiversity and knowledge to be 'enclosed' through intellectual property rights (IPRs)," Vandana Shiva explains.[50]

[50] Vandana Shiva et al., *The Enclosure and Recovery of the Commons: Biodiversity, Indigenous Knowledge and Intellectual Property Rights* (New Delhi: Research Foundation for Science, Technology and Ecology, 1997).

In Shiva's words, "The destruction of commons was essential for the industrial revolution, to provide a supply of natural resources for raw material to industry. A life-support system can be shared, it cannot be owned as private property or exploited for private profit. The commons, therefore, had to be privatised, and people's sustenance base in these commons had to be appropriated, to feed the engine of industrial progress and capital accumulation."[51]

While Shiva is referring to enclosure of nature by intellectual property rights, the same process is now occurring with machine learning—an intensification of quantified nature. The new gold rush in the context of artificial intelligence is to enclose different fields of human knowing, feeling, and action, in order to capture and privatize those fields. When in November 2015, DeepMind Technologies got access to the health records of 1.6 million identifiable patients of Royal Free Hospital in London, we witnessed a particular form of privatization: the extraction of knowledge value.[52] A dataset may still be publicly owned, but the meta-value of the data—the model created by it—is privately owned. While there are many good reasons to seek to improve public health, there is a real risk if it comes at the cost of a stealth privatization of public medical services. That is a future where expert local human labor in the public system is augmented and sometimes replaced with centralized, privately owned corporate AI systems that are using public data to generate enormous wealth for the very few.

Figure 5.11 Corporate border.

[51] Vandana Shiva, *Protect or Plunder? Understanding Intellectual Property Rights* (London: Zed Books, 2001).

[52] Alex Hern, "Royal Free Breached UK Data Law in 1.6m Patient Deal with Google's DeepMind," *Guardian*, July 3, 2017, http://www.theguardian.com/technology/2017/jul/03/google-deepmind-16m-patient-royal-free-deal-data-protection-act.

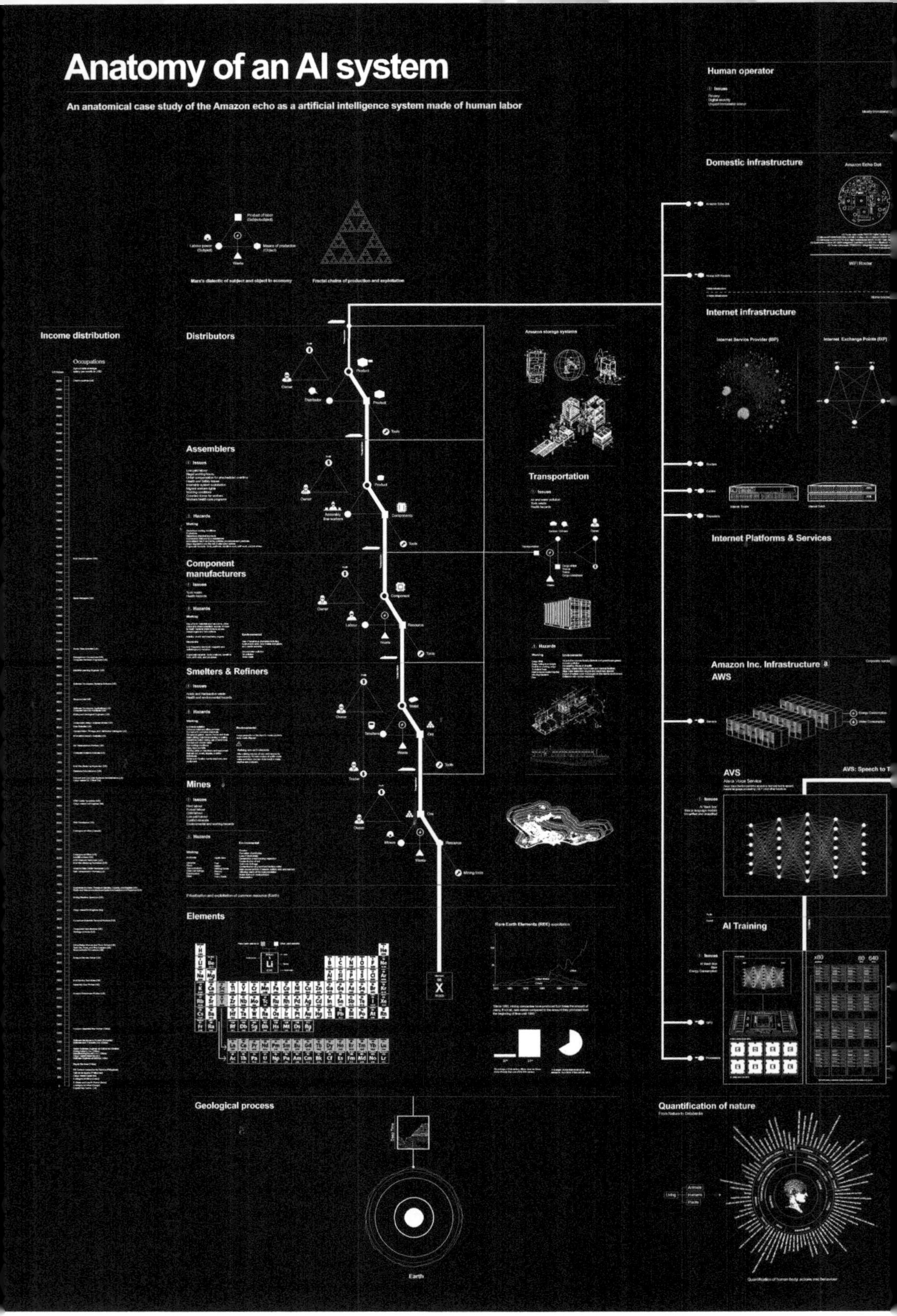

Figure 5.12 Map by Kate Crawford and Vladan Joler, 2018.

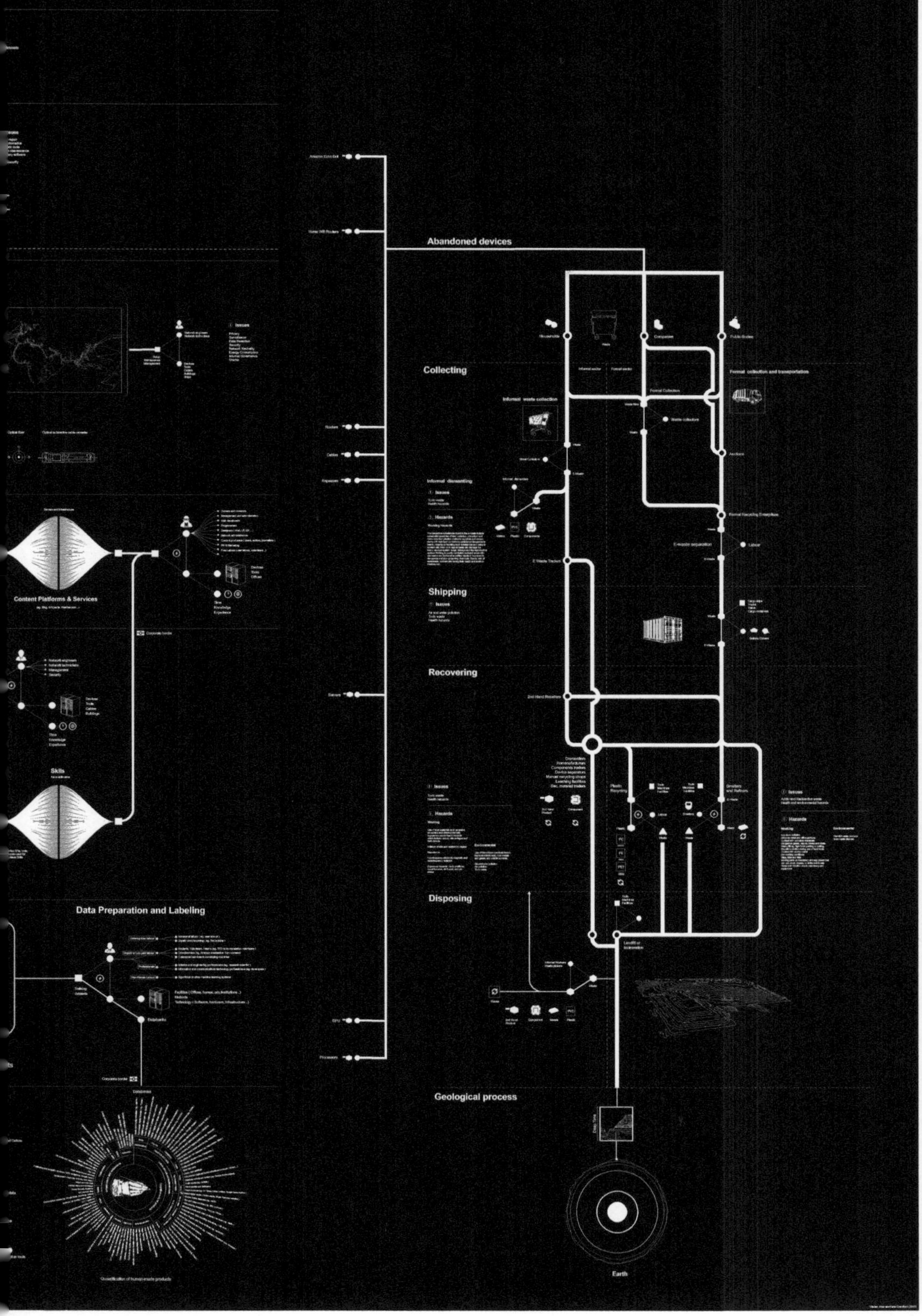
Abandoned devices
Collecting
Shipping
Recovering
Disposing
Geological process
Earth
Content Platforms & Services
Skills
Data Preparation and Labeling

XXI

At this moment in the twenty-first century, we see a new form of extractivism that is well underway: one that reaches into the furthest corners of the biosphere and the deepest layers of human cognitive and affective being. Many of the assumptions about human life made by machine learning systems are narrow, normative, and laden with error. Yet they are inscribing and building those assumptions into a new world and will increasingly play a role in how opportunities, wealth, and knowledge are distributed.

The stack that is required to interact with an Amazon Echo goes well beyond the multilayered "technical stack" of data modeling, hardware, servers, and networks. The full stack reaches much further into capital, labor, and nature, and demands an enormous amount of each. The true costs of these systems—social, environmental, economic, and political—remain hidden and may stay that way for some time.

We offer up this map and essay as a way to begin seeing across a wider range of system extractions. The scale required to build artificial intelligence systems is too complex, too obscured by intellectual property law, and too mired in logistical complexity to fully comprehend in the moment. Yet you draw on it every time you issue a simple voice command to a small cylinder in your living room: "Alexa, what time is it?"

And so the cycle continues.

The Epistemologies of Cosmotechnics

6

Black Ecologies

An Opening, an Offering

Imani Jacqueline Brown[1]

1

The war against Blackness, begun 500 years ago,
is the event horizon of planetary extinction.
And still,
and yet,
and therefore:
Beyond the boundaries of a *billion Black anthropocenes*[2]
exist infinite *Black ecologies*.

2

Through our bodies, we can learn a thing or two (or infinitely more) about integration:
Ecologies are assemblages of integral relationships between bodies.
Bodies are bits of matter and energy
that unite to become humans and microbes,
ponds and planets,
communities and worlds.
Our bodies are ecologies;

[1] This text was originally published in MARCH 01 (2021). The collages were created specifically for this volume.

[2] In her book, *A Billion Black Anthropocenes or None* (Minneapolis: University of Minnesota Press, 2018), author Kathryn Yusoff engages a "citation practice" to draw Black feminist voices into majority-White conversations about the Anthropocene, arguing that every Black death is a Golden Spike precipitating climate change. This text, a response to her opening, looks to and beyond the boundary events of Black Anthropocenes to call for a conceptualization of Black ecologies.

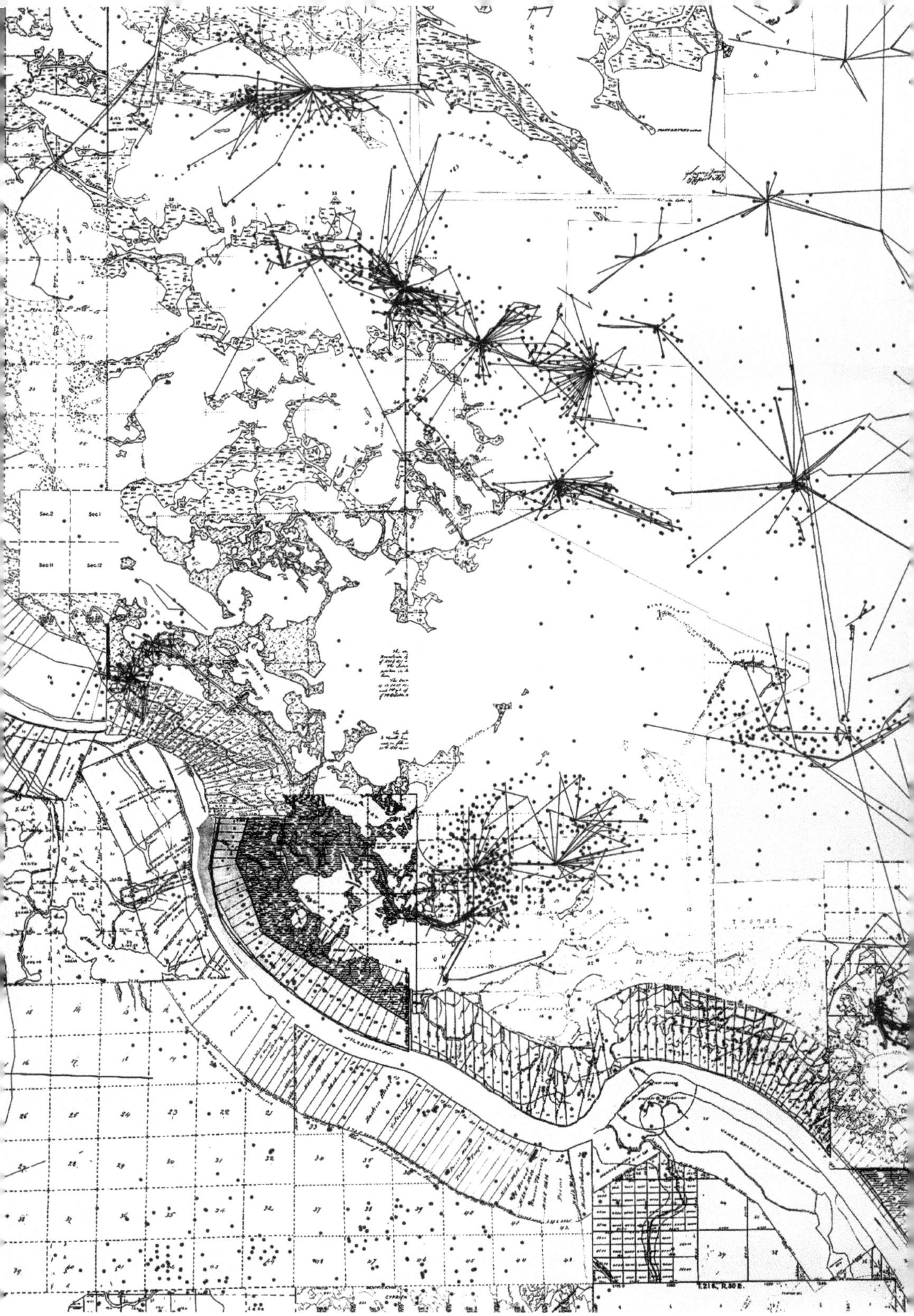

Figure 6.1 Imani Jacqueline Brown, *Untitled*, digital collage, 2022. Permits for coastal development, including oil and gas wells, flowlines, pipelines, and access canals in Plaquemines Parish, Louisiana (1920–2020) are mapped against antebellum Public Land Survey System (PLSS) charts (1820s–1860s) and aerial photographs captured by the artist from a three-passenger propeller plane. The PLSS was the first US system to plat, or divide, territory into parcels of private property.

GULF OF MEXICO

ecologies are bodies of bodies of bodies . . .
Bodies nest ad infinitum, interdependent, indivisible.
Each body is a point; each point is a threshold;
each threshold opens onto an ecology.

Each Black body points to a line of code drawn in sand and skin; each line is a boundary
that obstructs movement across ecological thresholds; each boundary hypoxifies both the Black body and the world's ecology.
To return oxygen to humanity, each bounded Black body must be named; each name must be known. There are too many names to know.
But Black bodies are not *doors of no return.*
Naming and acknowledgments are necessary, but they are just the openings
that enable us to make offerings.
Rupturing the boundaries around and within Blackness,
which holds all bodies and all names,
opens wormholes.
Through them, we can sow the seeds
of other ecologies-of-being(s).

3

To open a wormhole, we must:

a. Locate a boundary drawn through our ecological body:

But where to begin?
1452?[3] 1492?[4] 1619?[5] 1682?[6]. . .
There are too many boundary dates to note,
but I propose descending through this one:
1758. Swedish botanist Carl Linnaeus publishes the tenth edition of *Systema Naturae.* This "seminal" scientific text transplants a system of classifying Earth's plant species onto the human species. In the name of "science," Linnaeus

[3] First Portuguese slave ships return from Guinea.
[4] Columbus arrives on the Taíno Island Guanahani in the Bahamas, Caribbean Sea.
[5] The first African slaves arrive in North America.
[6] The first Slave Codes come into force in the Virginia Colony.

dissects *Homo sapiens* with a system of social categorization[7] raised from the *demonic grounds*[8] of slave plantations and, through the inscription of ironclad nomenclature into abstracted flesh, he animates a racialized chimera of stratified bodies: *Homo . . .*

> *Americanus: reddish, choleric, and erect; hair—black, straight, thick; wide nostrils, scanty beard; obstinate, merry, free; paints himself with fine red lines; regulated by customs.*
> *Asiaticus: sallow, melancholy, stiff; black hair, dark eyes; severe, haughty, avaricious; covered with loose garments; ruled by opinions.*
> *Africanus: black, phlegmatic, relaxed; hair—black, frizzled; skin—silky; nose—flat; lips—tumid; women without shame, they lactate profusely; crafty, indolent, negligent; anoints himself with grease; governed by caprice.*
> *Europeaus: white, sanguine, muscular; hair—long, flowing; eyes—blue; gentle, acute, inventive; covers himself with close vestments; governed by laws.*[9]

European men anoint themselves with Sylvia Wynter's capital "M"[10] and conjure a *cosmology*[11] they call *Enlightenment.* Enlightenment exploits the subjective murkiness inherent to ecological existence to sanction capitalism's foundational enterprises of genocide and slavery *after the fact.*[12]

Man orders himself as the height of existence—as White^{Human}—and us as its bottom—as Black$_{\text{inhuman}}$. Fearing the force of White and the pitfall of Black,

[7] "This cycle [of English social paranoia and violence against free Afro-Americans] articulates the structure of racialization that has given rise to the modern American concept of 'race.' That is, this new white cultural identity, constructed through and on an imposed social category differential between the English and the Africans as wholly 'other,' then formed the basis for the modern concept of racialization." Steve Martinot, "Motherhood and the Invention of Race," *Hypatia* 22, no. 2 (Spring 2007): 79–97, https://www.ocf.berkeley.edu/~marto/mother.htm.

[8] Katherine McKittrick, *Demonic Grounds: Black Women and the Cartographies of Struggle* (Minneapolis: University of Minnesota Press, 2006).

[9] These taxonomic descriptions are omitted from the most accessible sources of information on Linnaeus's binomial nomenclature, such as Wikipedia. Cited from Audrey Smedley, *Race in North America: Origin and Evolution of a Worldview* (1992; New York: Routledge, 2012).

[10] "However, in the wake of the West's second wave of imperial expansion, pari passu with its reinvention of in Man now purely biologized terms, it was to be the peoples of Black African descent who would be constructed as the ultimate referent of the 'racially inferior' Human Other, with the range of other colonized dark-skinned peoples, all classified as 'natives,' now being assimilated to its category—all of these as the ostensible embodiment of the non-evolved backward Others—if to varying degrees and, as such, the negation of the generic 'normal humanness,' ostensibly expressed by and embodied in the peoples of the West." Sylvia Wynter, Unsettling the Coloniality of Being/Power/Truth/Freedom: Towards the Human, After Man, Its Overrepresentation—An Argument," *New Centennial Review* 3, no. 3 (Fall 2003): 266.

[11] Glossary: *noun* A branch of both metaphysics and astronomy that conceives of the origin and structure of the universe, including its constituents and components, elements and interconnections, causalities and consequences, laws and freedoms, space and time.

[12] Glossary: "After the fact" means after an action is performed; In legal parlance, after a crime has been committed.

Others become equatorial, folded into the unsettled gradient of the Brown *in-between*,[13] striving toward the northern pole. The ecology of Humanity enters a state of *dysbiosis*.[14]

b. Open a wormhole to other ecologies:

Conflating phenotypical darkness with phenomenological darkness, darkness with the unknown, and the unknown with fear, Enlightened Whiteness casts a surgical light. With each shadow banished, new shadows are revealed; within each particle, smaller particles are found. Yet, the new forms of knowledge gained are *half-starved*.[15] With each body annihilated or enslaved, assimilated or segregated, the less the Enlightened know about themselves. Each person, place, and particle split sounds the bell for the planet's splintering.

And yet, and still, and therefore, through the alchemical fusion of the phenotypical and the phenomenological, darkness of flesh is imbued with the dark power to *reintegrate the world*.

4

Through our bodies, we can learn a thing or two (or infinitely more) about segregation. Enlightened Bodies are as segregated internally as they are externally. The ^{Human} is segregated from the $_{\text{bacterium}}$, even though ninety-nine percent of the DNA in the Human body is nonhuman.[16] The Enlightened ^{Human} is segregated from the $_{\text{plants}}$ that provide him with the oxygen he needs to exist; the $_{\text{plants}}$ create this oxygen in turn from the carbon dioxide that the ^{Human} expels. The ^{Human} is segregated from the $_{\text{land}}$ which, as Frantz Fanon reminds us, is the most important value, because it provides us with bread and dignity.[17]

Beneath the Black human body is the body of Earth, whose flesh is Black with oil and soil—Black with life. Yet, classified as $_{\text{Black}}$, Earth becomes extinction's ground zero; as Black human bodies dis-integrate under the magnification

[13] Glossary: *noun* A supernatural limbo or purgatory of ghosts; In-betweens are the drawings which create the illusion of motion. *verb* In animation, the process of generating intermediate frames between two images, called key frames, to give the appearance that the first image evolves smoothly into the second image.

[14] Glossary: *noun* A state of imbalance between the types of bacteria present in the bacterial ecology, or microbiota, of a person's body, which leads to disease.

[15] Aimé Césaire, "Poetry and Knowledge," in *Lyric and Dramatic Poetry, 1946–82*, trans. Clayton Eshleman and Annette Smith (Charlottesville: University Press of Virginia, 1990), xlii–lvi.

[16] Bonnie Basler, "How Bacteria 'Talk,'" TED, 2009, https://www.ted.com/talks/bonnie_bassler_how_bacteria_talk/transcript.

[17] Frantz Fanon, *The Wretched of the Earth*, trans. Richard Philcox (1963; New York: Grove Press, 2004), 9.

of Enlightenment, the Black body of Earth disintegrates. A sickly rainbow of bodies and souls floats above Enlightened ecologies in ethico-eco-philosophical zero-gravity. The Enlightened Human calls this weightlessness *Freedom*.

Enlightened ecologies are spheres of existence flattened into two dimensions: *Property* and *Profit*. Enlightened ecologies are hierarchical rather than rhizomatic. They are latitudinal and longitudinal, surface and flesh, black and white.

The quintessential Enlightened ecology is the plantation. In plantation ecologies, Earth is emptied of Native peoples and knowledge through genocidal clearing; enslaved Black bodies are emptied of humanity; the indwelling value of the Black Earth is transfigured into a financial abstraction.

Plantation ecologies are invasive: Once industrial monocrop agriculture has drained life from soil, the oil industry spreads throughout the Black Earth's fractal viscera; her skin crawls with prisons and petrochemical plants.[18] Plantation ecologies are virulent: As capitalism advances and thickens like a *strangling fruit*,[19] all the world becomes a plantation. The habituating force and *raison d'être* of the plantation ecology is *Death*.

5

The Black Earth asks me: *What does it mean for Black people to fight for the rights of Nature when y'all are still fighting for your basic human rights?*

In the light, the equation of existence appears reducible to division.
Darkness multiplies. In the light, knowledge of the world appears in sharp relief.
Darkness holds a more intimate wisdom we are once again coming to sense:
What has been called *darkness, the unknown, the void, the abyss, empty space*,
we now know as *dark matter*. Dark matter is the glue that holds existence together.[20]
Perhaps we can think dark matter through another, older name:
Kinship.

[18] In the US state of Louisiana, which once hosted the largest slave market in North America, industrial sugarcane plantations have been converted into industrial petrochemical facilities, producing climate change. Louisiana is the second-largest historic oil producer in the United States. For more, see Imani Jacqueline Brown, "Should We Consider Fossil Fuel Extraction an Unjust Enterprise?," *Business and Human Rights Resource Centre*, February 10, 2020, https://www.business-humanrights.org/en/should-we-consider-fossil-fuel-extraction-an-unjust-enterprise.

[19] Jeff VanderMeer, *Annihilation* (London: 4th Estate, 2014).

[20] "Dark matter [. . .] helps protect [stars] against the tidal forces of the massive host galaxy [. . .]." S. De Rijcke, S. J. Penny, C. J. Conselice, S. Valcke, and E. V. Held, "*Hubble Space Telescope* Survey of the Perseus Cluster—II. Photometric Scaling Relations in Different Environments," *Monthly Notices of the Royal Astronomical Society* 393, no. 3 (March 2009): 798–807, here 798–99.

I answer the Black Earth through my bodies:
The rights of nature and human rights must be won ecologically, in solidarity, as kin.
Perhaps we can think kinship through another, simpler name:
Love.

Little by little (or perhaps, from the perspective of the Black Earth, more rapidly) Enlightened Humanity is coming to recognize the need for the dark wisdom it worked hard to extinguish. More and more, Western scholars look with deference and humility toward Anishinaabe Original Instructions,[21] toward Quechua Sumak Kawsay,[22] toward Aboriginal dreaming.[23] They spin the West around to face the rights of Nature: the eco-philosophical integrity of the world—the horizon.
Yes:
more and
more.

But where is Black Africa and her dark diaspora? So great is the world's disregard for the Dark Continent and her people that few have noticed the absence of our eco-philosophical voices within the rising chorus of environmental scholarship.

To survive climate change, humanity needs to get radical. To get radical means "to get to the root."[24] Black Africa is humanity's root. Her uprooting seeds Enlightened ecologies. Her diasporic seeds are portals to other ecologies-of-being(s).

As Malidoma Somé, a spiritual philosopher from the Dagara tribe of Burkina Faso, affirms,[25] the time has come for African philosophy to be recognized beyond the realm of anthropology. It is time for Black studies to be re-read ecologically

[21] A cosmology, or living philosophy, of Indigenous peoples of the Great Lakes region of Turtle Island (aka North America). The Original Instructions remind humans how to exist in and with the world with care and reciprocity. See Robin Wall Kimmerer, *Braiding Sweetgrass: Indigenous Wisdom, Scientific Knowledge and the Teachings of Plants* (Minneapolis: Milkweed Editions, 2013).

[22] Translated as "Buen Vivir," or "good living," Sumak Kawsay is a cosmology, or living philosophy, of the Quechua peoples throughout South America. It reminds humans how to live in communities of ecological and cultural harmony. For more information, see: https://www.pachamama.org/sumak-kawsay.

[23] An Aboriginal Australian philosophy of communion with human and more-than-human ancestors, grounded in the interconnectivity of all elements of existence. See Elizabeth Povinelli, *Geontologies: A Requiem to Late Liberalism* (Durham: Duke University Press, 2016).

[24] Quite literally the denotative meaning. For example, Ella Baker expressed in 1969: "I use the term radical in its original meaning—getting down to and understanding the root cause. It means facing a system that does not lend itself to your needs and devising means by which you change that system." Quoted in Barbara Ransby, *Ella Baker and the Black Freedom Movement: A Radical Democratic Vision* (Chapel Hill: University of North Carolina Press, 2003), 1.

[25] Malidoma Patrice Somé, *Of Water and the Spirit: Ritual, Magic, and Initiation in the Life of an African Shaman* (New York: G. P. Putnam's Sons, 1994).

and for the de-Enlightening world to sit quietly and listen rapturously to Black people's perspectives on ecological re-integration.

6

Black ecologies travel with the Black body on tides of cellular memory and seed the gaps between worlds. Their ancestors are the brown bodies of wetlands—fecund cradles of more-than-human life, which the Enlightened Human classifies beneath whiter sands and bluer waters; which the Human carves up, drains of vitality and disappears to make way for industrial development; and which, despite this degradation and debasement, remain merciful hosts to migrating maroons of all species.[26]

Black ecologies wink into existence as *resistance* to plantation ecologies. Ecological resistance emerges from the cypress trees in whose crowns we crouch, muskets over shoulders, listening for dogs; the Indigenous nations who welcome us into their families; the handfuls of Europeans who join us, refusing the role of Colonizer[27]; the parakeets who nest near us, fleeing their own deaths on plantations[28]; the mud sinks who grab at the legs of slave catchers, stealing their balance; the mosquitoes who bite at us all, knowing that the sweetness of flesh is not carried in color; and the soils who cocoon our bodies while the worms and roaches return us to freedom in oneness with the Black Earth.

Black ecologies know our suffering, struggle, and survival.
They *are* our suffering, struggle, and survival.

Black ecologies persist today as the human bodies that block pipelines and petrochemical plants. They are cultivated by human hands, stained with shadows of indigo,[29] oil, and blood, which plant marsh grasses in humble rituals of repair. They blossom, as memories of plantation plots,[30] in community gardens, and in

26 For more on Creole maroon villages in Louisiana's swamps, see Kalamu ya Salam, "History: Jean Saint Malo, New Orleans Maroon," *Neo-Griot*, June 19, 2012, https://kalamu.posthaven.com/history-jean-saint-malo-new-orleans-maroon-ex.

27 For more on colonial refusal and tri-racial isolate communities, see Ron Sakolsky and James Koehnline (eds.), *Gone to Croatan: Origins of North American Dropout Culture* (Brooklyn: Autonomedia, 1993).

28 J. Drew Lanham, "Forever Gone: How Bird Lives and Black Lives Intertwine under the Long Shadow of History," *Orion Magazine*, 2018, https://orionmagazine.org/article/forever-gone.

29 As Tiffany Lethabo King writes in *The Black Shoals: Offshore Formations of Black and Native Studies* (Durham: Duke University Press, 2019), Indigo-stained hands are the shadowy trace of Black human *a priori* and *a posteriori* merger with nonhuman existence on and beyond the plantation.

30 Sylvia Wynter, "Novel and History, Plot and Plantation," *Savacou*, no. 5 (June 1971): 95–102.

Figure 6.2 Imani Jacqueline Brown, Black Ecologies, digital collage, 2022.
An image of coastal wetlands eroded by fossil fuel extraction is blended with a view of a petrochemic plant. While continued fossil fuel development is couched in arguments and sentiments of permanen intractability, and inevitability, this vision prefigures the inevitable uprooting of extractive industry and reclamation of its occupied land through more-than-human ecological reparations.

secret jars in solitary confinement cells.[31] They flourish, as Great Green Walls[32] dreamed up from African sacred groves.[33]
Black ecologies are *syncretic*: Like resistance to slavery, which often germinated in the spiritual grounds of Vodou, Santería, and Candomblé, they generate systems for mutual living—living systems—by synergizing poetry and science, past and future, teachings from North, South, East, and West, networking the many into the one. Black ecologies are *diasporic*: They spread to the rhythm of grandma's knitting needles, weaving the world's loving reintegration. Whatever their scale, they are expansive because they photosynthesize the wisdom of more-than-human ancestors.

Black ecologies cannot decompose Timothy Morton's plastic *Dark Ecology*, a philosophy of "coexistence" that holds no space for the repair of racism.[34] Nor are Black ecologies fooled by the Cartesian logics that tell us that Black is the lowest point on the spectrum to White. Black ecologies laugh with deadly gravity at the confusion of Enlightened Human Reason:

Black is not the *absence* of light; Black *absorbs* light, holding it close to its bosom.

31 "Jackie, in your letter you asked me: 'What sort of house does a man who has lived in a 6' x 9' cell for over 30-years dream of?!'—In the front of the house I have 3-squares of gardens. The gardens are the easiest for me to imagine, and I can see they would be certain to be full of gardenias, carnations and tulips. This is of the utmost importance. I would like for guests to be able to smile and walk through flowers all year long." Herman Wallace, excerpt from a letter to jackie sumell, February 1, 2006. Activist and former Black Panther Herman Wallace was held in solitary confinement for forty-one years for a crime he did not commit. While imprisoned, he made many attempts to grow plants in his jail cell. Herman was exonerated of all charges on October 1, 2013. He passed away three days later. For more information, see: http://www.solitarygardens.org.

32 Glossary: *noun* The Great Green Wall is a pan-African project to seed a mosaic of forests across the Sahel, from Senegal to Djibouti. This region is the frontline of desertification, which propels Africa's people to attempt dangerous migration of Europe. This desertification is the product not only of climate change, which is an indirect consequence of Enlightened ecologies but also of largely French colonial land use practices that uprooted Indigenous land tenureship and imposed a system of industrial plantation agriculture.

33 See, for example, Michael J. Sheridan and Celia Nyamweru (eds.), *African Sacred Groves: Ecological Dynamics and Social Change* (Athens: Ohio University Press, 2008).

34 Morton dedicates the following five sentences to dismiss concerns (raised by Black and Brown, feminist, postcolonial scholars and activists) that ascribing undifferentiated blame for climate change to the entire human population, regardless of actual and historical national and cultural responsibility, is racist (and precludes our ability to actually *redress* climate change and its related ills): "The user of *Anthropocene* is saying that humans as a race are responsible, and while this really means *white* humans, whites go unmarked. There is such a thing as the human. But *human* need not be something that is ontically given: we can't see it or touch it or designate it as present in some way (as whiteness or not-Blackness et cetera). There is no obvious, constantly present positive content to the human. So *Anthropocene* isn't racist." Timothy Morton, *Dark Ecology: For a Logic of Future Coexistence* (New York: Columbia University Press, 2016), 15.

Black ecologies are the *nadir* to Enlightenment's *zenith*.[35] Enlightenment has lifted us to unthinkable heights of madness; Black ecologies reorient us intuitively toward the Black Earth. Black ecologies do not upend the hierarchy that places Black people at the bottom of Humanity and Humanity above the Black Earth, shifting Black to the top and White to the bottom, but rather explode it outward from the base. Humanity will horizontalize, lying prostrate as humble human, not just big toes[36] but whole bodies pressing into the ground in a sublime encounter with our planet, with our kin, with ourselves. Black ecologies will hold us all.

7

This is why Black bodies
are key and coup
in Humanity's war against existence.
Black bodies can *jump the broom*
over and beyond dishonored Humanity
and land
not as Human, but *more-than-human*,
not as thing, but *everything*.

Black ecologies multiply;
they are migratory;
they resist ecologies of extinction.
Don't be frightened;
Black ecologies won't replace you,
they will repair you;
they will *restore Us*.

[35] Glossary: *noun* In Arabic, *nadir* means "counterpoint". The term is used in astronomy to reference the lowest point in the sky, named as the opposition to the highest point, the zenith; It also carries the secondary meaning of "the lowest point in a person's spirit." Black ecologies reject the oppositional astronomical and colloquial language in favor of the Arabic original, which implies balance rather than contestation.

[36] "The function of the human foot consists in giving a firm foundation to the erection of which man is so proud [. . .]. But whatever the role played in the erection by his foot, man, who has a light head, in other words a head raised to the heavens and heavenly things, sees it as spit, on the pretext that he has this foot in the mud." Georges Bataille, "The Big Toe," in *Visions of Excess: Selected Writings, 1927–1939*, trans. Allan Stoekl with Carl R. Lovitt and Donald M. Leslie, Jr. (1929; Minneapolis: University of Minnesota Press, 1985), 20, https://supervert.com/elibrary/georges-bataille/the-big-toe.

Reparations for racial capitalism are owed,
but not only through financial settlement.
Landscape reparations are owed,
but also through the removal of racist monuments.
Land restitution is owed,
but not only through the acknowledgment of Indigenous names.
The reparations that are owed
will return land to people
and people to land
and both to ecological being(s).
Black ecologies are the sacred relations
through which ecological reparations can be offered to our segregated Earth.
Through the cultivation of Black ecologies,
the wholly Enlightened Earth may yet become a Black whole.

7

Pluriversal Horizons

Notes for an Onto-epistemic Reorientation of Technology[1]

Arturo Escobar, Michal Osterweil, and Kriti Sharma

Prelude

In November 2021, we (the authors) received a message from a dear friend whose brother was seriously ill:

> This experience has really shaken and altered me. It's been a near-death experience for my brother, but because my siblings are so foundational to my sense of self, it feels like a near-death experience for me too. I feel disoriented, like my model of the world has been wrong. Earth really does feel incomputable. My sense of being able to predict and control the world and my life, to optimize my individual existence by calculating and executing a rational plan meant to maximize utility, to also try to protect my loved ones in this way—all of this has been based on a modernist misapprehension of reality. I feel there is nothing to be done now but to surrender to what I cannot control. Existence being holy and mysterious, far beyond my comprehension or control, being excessive of the prediction and manipulation promised by modernity, existence being so much bigger than what I could possibly have calculated or bargained for, I feel I'm being called (or forced, as the part of me who is grieving and angry would say) to surrender to just being a part of vast, incomprehensible, contingent existence, however it all unfolds.

[1] We discuss some of the issues discussed in this chapter at length in our book, Arturo Escobar, Michal Osterweil, and Kriti Sharma, *Relationality: An Emergent Politics of Life and the Human* (London: Bloomsbury Press, 2024).

Our friend's lucid and moving message revealed the intimate reach of calculative rationality in our modern lives. For even when wishing the best for our loved ones, we fall into a desire for optimizing our experiences with them, for safety and predictability—a pious and well-intentioned wish for plenitude that can only be a denial of the mystery of life, its unpredictability, and what ineffably lies beyond concepts and words. Many people today seem to be exploring these important dimensions of existence in their own unique ways, particularly when confronted with pandemics and the possibility of ecological collapse. At a time when the hypermodernity of cutting-edge technology and all things AI intend to reduce many aspects of daily life to calculation and algorithmic computation, albeit coated with a thick veneer of unquestionable progress, we feel the need to ponder deeply about what remains incomputable and incalculable, what cannot be accounted for by logocentric analyses and algorithmic rationality.

It would be impossible to trace the genealogy of such drives to calculation, optimization, rationalization, and control. Such genealogies are entangled with the long history of patriarchy and, since the dawn of modernity, with colonialism, capitalism, science, and technology. The consequences have been dire. Yuk Hui has posed the issue in a poignant way: "How should we respond to the challenge the human has undertaken to eliminate its own condition of existence?"[2] We hasten to add, however, that we are dealing here with the challenge *modern* humans have *unknowingly* undertaken, albeit *through their own designing*. We are talking here about that modern human, barely five centuries old, who invented what Sylvia Wynter has powerfully described as a "monohumanist mode"[3] of the human: Western, liberal, secular, and bourgeois. We have more to say about Wynter's and Hui's projects, but for now, let us add Hui's perspective of the conjuncture brought about by the simultaneous triumph and meltdown of modernity as one that creates new conditions for philosophy and thought and the possibility of multiple onto-epistemic and cosmotechnical starting points for a "new world history."[4]

A similar point was recently articulated by Mapuche activist Moira Millán, when she stated, in response to the ongoing *terricidio* (terricide): "Necesitamos

[2] Yuk Hui, *Art and Cosmotechnics* (Brooklyn: e-flux Books; Minneapolis: University of Minnesota Press, 2021), 249.

[3] Sylvia Wynter, "Unparalleled Catastrophe for Our Species? Or, to Give Humanness a Different Future: Conversations," in *Sylvia Wynter: On Being Human as Praxis*, ed. Katherine McKittrick (Durham: Duke University Press, 2015), 9–89.

[4] Yuk Hui, "Singularity vs. Daoist Robots: Is There Another Path than Accelerated Western Modernization?," *Noēma Magazine*, June 19, 2020, https://www.noemamag.com/singularity-vs-daoist-robots.

una revolución del pensamiento" (we need a revolution in our thought).[5] If the current dominant *pensamiento* is at the basis of terricide, then this freeing of thought can only be achieved by seeing humans as belonging to the Earth and the web of life, as many Indigenous, animist, and territorialized peoples have done for thousands of years.[6] By now it is utterly clear that, under the maddening pace of monohumanism, we are further and further away from being attuned to Earth and informed by the interrelationships and interdependence of all forms of earthly life. For we increasingly inhabit an overarching, even if fractured and ever-changing, "one-world world," which is liberal, secular, and anthropocentric.[7] Most humans are forced to march to its dictates under the predatory command of neoliberal extractive global capitalism.

This chapter unfolds in the space demarcated by terricide and technology, on the one hand, and new conditions for thought, on the other, that characterize the generalized existential crisis of the present: the interrelated crises of climate, energy, biodiversity, poverty, inequality, and meaning. It does so in four moments, each around a pivotal concept for our argument: terricide, monohumanism, monotechnologism, and pluriversal transitions. While only the rudiments of our approach can be presented here, the goal is to sketch a mode of access to the question concerning technology that weaves together these concepts. If terricide is an apt descriptor of what humans, taken as a whole, albeit differentially, are doing to Earth and to Life, then we must start with the form of the human most responsible for it, which is the Man of monohumanism and its instrumental and computational approach to technology. The diagnosis of where we have been, in onto-epistemic and political terms, will provide a clearing for alternative narratives of Life and the human that are oriented toward mobilizing a new way of dwelling with the Earth.

There exists, no doubt, a correlation between the intensification of the capitalist engineering of the Earth and the rise of computation and artificiality. Countering the combined defuturing effects of these deadly processes[8] by envisioning different futures, and futures in difference, becomes an onto-epistemic, cultural, and political endeavor of utmost importance. Is it possible to

[5] Moira Millán, "Moira Millan y el buenvivir originario," *Pensamiento Ambiental, YouTube*, May 22, 2016, https://youtu.be/JOiRYUW8R08.

[6] A certain revival of the interest in animism at present is not coincidental. See, for example, Andreas Weber, *Sharing Life: The Ecopolitics of Reciprocity* (New Delhi, India: Heinrich Böll Stiftung, 2020), and Tim Ingold, *Being Alive: Essays on Movement, Knowledge and Description* (London: Routledge, 2011).

[7] John Law, "What's Wrong with a One-world World?," *Distinktion: Scandinavian Journal of Social Theory* 16, no. 1 (2015): 126–39.

[8] Tony Fry, *Defuturing: A New Design Philosophy* (London: Bloomsbury, 2020).

retrieve and reactivate the futuring potential of those ways of worlding that are not informed entirely by the calculative and enframing rationalities, wherever they are found? How to imagine and create historically unprecedented forms of human sociality that enable multiple reweavings of the human with the Earth, within an ontology of care for the web of interdependencies that makes up Life? As we shall propose, what is at stake is the ushering of a revised understanding of history that grounds the reappropriation of technology in attunement to place, locality, and the relational reality of a living cosmos. Echoing this volume's call, we refer to this revised understanding and condition as *planetarity*.

Facing Terricide

From February 7 to 10, 2020, the first Campamento Climático—Pueblos Contra el Terricidio (Climate encampment—Peoples against terricide) met in a recovered territory in the Wallmapu (Mapuche Territory, in what is otherwise known as the Argentine Patagonia), with the goal of launching a broad movement against terricide. Called for and organized by Movimiento de Mujeres Indigenas por el Buen Vivir (Movement of Indigenous women for good life), the group issued a declaration which read in part as follows:

> We define terricide as the killing of tangible ecosystems, the spiritual ecosystem, and that of the *pueblos* and of all forms of life. Confronted with the terricide, we declare ourselves to be in permanent struggle, resistance, and re-existence against this system. [. . .] We summon all peoples to build a new civilizational matrix that embraces *buen vivir* [collective wellbeing] as a right. Buen vivir implies the retrieval of harmony and reciprocity among peoples and with the Earth. Summoned by the memory of our ancestors and the lands and landscapes that inhabit us, we have agreed on the creation of the movement Pueblos Contra el Terricidio.[9]

The movement set forth another important notion that is at the heart of this chapter: "Today," they said, "the social emergent is Earth itself."[10] Let us emphasize

[9] Authors' translation, emphasis added. Source: https://www.planbnoticias.com.ar/index.php/2020/02/15/participo-wuta-trawn-conformaron-movimiento-contra-el-terricidio. For the full declaration, see https://revistaresistencias.wixsite.com/resistencias/post/deliber%C3%B3-en-el-lof-mapuche-pill%C3%A1n-mahuiza-el-campamento-clim%C3%A1tico-pueblos-contra-el-terricidio. There are plenty of sources on terricide on the Internet in Spanish, only a few in English.

[10] See: http://comunizar.com.ar/argentina-mujeres-indigenas-ocupan-ministerio-del-interior; and http://idepsalud.org/terricidio-mujeres-indigenas-ocupan-pacificamente-desde-ayer-el-ministerio-de-interior.

that their words stem from the memory of their ancestors and the lands and landscapes that inhabit them. This is an instantiation of an understanding that the land does not belong to them, as in the modern ontological regime of property, but that they themselves belong to the land.

Unlike the notion of the apparently neutral Anthropocene, which lends itself to managerial and techno-scientific solutions to problems that overflow technoscience's ability to solve them, terricide is a concept that does not fit easily into modern thought. In contrast to the Anthropocene, terricide avoids sheltering the idea of an overarching biopolitical project for a "universal humanity" in the terms of "Enlightened Man." It reveals instead that the Anthropocene is the actual heir of modern science which, as such, is merely diagnostic and misses the point of what is at stake: "the possibility of designing *new conditions for being human*."[11] Terricide announces that what is happening to the climate cannot be solved through technoscience, no matter how comprehensive and allegedly innovative. It dispels the idea that another round of ever more ambitious modern patriarchal will for mastery and control can "solve" the climate problem. At the same time, it makes clear why our understanding of technology must drastically change if it is to fulfill a role in fostering transitions to a new era of mutually enhancing human-Earth relations. In short, whereas Anthropocene conveys the sense of a neutral, objective diagnosis of the crisis, terricide claims that what is at stake is an ongoing crime against the Earth which thus calls for a very different politics and set of strategies.

Escaping Monohumanism

Thinking about new conditions for being human implies calling into question any universal idea of humanity premised on the exclusionary category of "Man." There have been many problematizations of modern notions of the human, most famously perhaps Michel Foucault's argument about the figure of Man as the foundation of all knowledge, as the subject and object of his own discourses, which crystallized with the modern episteme, or knowledge configuration, at the end of the eighteenth century.[12] Wynter's framing of the question of Man in

[11] Tony Fry and Adam Nocek, "Design in Crisis. Introducing a Problematic," in *Design in Crisis: New Worlds, Philosophies and Practices*, ed. Tony Fry and Adam Nocek (London: Routledge, 2021), 3, original emphasis.

[12] Michel Foucault, *The Order of Things: An Archaeology of the Human Sciences*, trans. Alan Sheridan (1966; New York: Vintage Books, 1994).

terms of a "monohumanist" notion of the human is particularly powerful for understanding the current ecological-cum-existential crisis and the possibility of constructing a pluriversal horizon for humanity. Wynter posits a two-step process for the emergence of Man. The Copernican Revolution was essential to the first civilizational rift from Christian cosmology to rational worldview, by decentering God and giving birth to reason as a means of understanding the universe, a shift that was catalyzed by the conquest of America ("Man1"[13]). By the end of the eighteenth century, the rational view had developed into a fully biologized and economized account of the human ("Man2"[14]). Man2 relied on a rendering of evolution in terms of natural selection, Thomas Robert Malthus's theory of resource scarcity, and the figure of *homo economicus* ushered in by the nascent science of political economy. Man2 resulted in a monohumanist view of the human as Western, bourgeois, secular, and liberal. Its Darwinian/Malthusian and economic macronarrative pivots on the principle of race and is subject to the imperatives of capital accumulation. This bioeconomic "genre of the human" has steadily extended to all corners of the world through colonialism, development, and globalization.[15] Man2 is the space within which we learn to live, think, and act, and it has become the default onto-epistemic existential domain.

Wynter disentangles the naturalization of Man2 by appealing to Frantz Fanon. In Fanon's notable conception of sociogenesis,[16] Wynter finds a genre of the human markedly different from the cosmogony of secular liberal Man. Sociogenetically, for Fanon, a Black person living in a Eurocentric context is obliged to experience themself as both normally and abnormally human, being and nonbeing, as the "dysselected" par excellence. This leads Wynter to emphasize that the human is not only biology but is also shaped by self-maintaining (autopoietically instituted) cultural codes, origin narratives, and storytelling that in turn powerfully shape the brain and behavior. In short, the human is also always *homo narrans*. Wynter concludes that the human is inevitably hybrid: *bios* and *mythos*; matter, life, and story. As she daringly surmises, "With this hypothesis, should it prove to be true, our system of knowledge as we have it *now*, goes."[17] This is a most audacious claim (akin to "necesitamos una revolución del pensamiento"),

[13] Katherine McKittrick, "Yours in the Intellectual Struggle: Sylvia Wynter and the Realization of the Living," in *Sylvia Wynter: On Being Human as Praxis*, ed. Katherine McKittrick (Durham: Duke University Press, 2015), 3.

[14] McKittrick, "Yours in the Intellectual Struggle," 3.

[15] Wynter, "Unparalleled Catastrophe for Our Species?"

[16] "Beside phylogeny and ontogeny stands sociogeny." Frantz Fanon, *Black Skin, White Masks*, trans. Charles Lam Markmann (1952; London: Pluto Press, 2008), 4. This is a way of highlighting the dialectic of black skins/white masks confronting all Black people.

[17] Wynter, "Unparalleled Catastrophe for Our Species?," 16, original emphasis.

one to which she arrives by revisiting "the human" from the perspective of the dysselected (persons racialized as naturally inferior) and the colonized. Through her analysis, Wynter opens a path for a significant "*refiguring* of humanness."[18] As she concludes, "unless we move out of the liberal monohumanist mindset, *it's very difficult to see where we've been*," and, hence, "*where we're going*."[19] It is not easy to have an adequate awareness of our historical circumstance, as genres of the human are "cosmogonically chartered"[20] and powerfully implanted as a "second set of instructions"[21] in collective culture by narratives. This ontological determination makes clear the need to move beyond the mimetic desire to join the club of those inhabiting the bioeconomic world. An important element in this project of overcoming is debunking the master discipline of economics and the reductive versions of science and technology that make the bioeconomic view of the human inevitable.

To reiterate, the larger issue is "the incorporation of all forms of human being into a single homogenized descriptive statement that is based on the figure of the West's liberal monohumanist Man";[22] that is what *mono*humanism means. As Bayo Akomolafe eloquently states, it is as though "the word 'human' is a self-evident category that is not already simmering with tensions, elisions, disputations and troubling departures."[23] The response must come in the form of the creation of new horizon of humanity that enables an ecumenically open view of the hybrid notion of human. One might posit the question as follows: How to move toward a humanism (if at all) that embraces coeval and pluriversal genres of being human while preventing its reabsorption into the regime of Man? For Wynter, given that they/we dwell in several worlds, peoples inhabiting the colonial difference have a "cognitive edge" in envisioning and moving toward such far-reaching transformation.

[18] Denise Ferreira da Silva, "Before *Man*: Sylvia Wynter's Rewriting of the Modern Episteme," in *Sylvia Wynter: On Being Human as Praxis*, ed. Katherine McKittrick (Durham: Duke University Press, 2015), 93, original emphasis.

[19] Wynter, "Unparalleled Catastrophe for Our Species?," 14, emphasis added.

[20] Wynter, "Unparalleled Catastrophe for Our Species?," 30.

[21] Wynter, "Unparalleled Catastrophe for Our Species?," 33.

[22] Wynter, "Unparalleled Catastrophe for Our Species?," 23.

[23] Bayo Akomolafe, "Coming Down to Earth," March 11, 2020, https://www.bayoakomolafe.net/post/coming-down-to-earth.

Beyond Monotechnologism

The ontological examination of technology, by which we mean its role in the constitution of the real, the human, and Life, seems to be coming back in contemporary debates after a hiatus of several decades in which concern with the social and cultural dimensions of technology took the front seat. While ontological questions were never completely absent from the social studies of technology that came into prominence in the 1990s, particularly in the Anglo-American and French academic worlds, it is fair to say that the interest in the philosophical anthropology of technology is remerging.[24] This return is being fostered by momentous issues that include the relation between technology and what moderns call nature, or "the more-than-human" in contemporary anthropological parlance; the impact of pervasive digitalization on subjectivities, social, economic, and political life; and above all the incredibly fast pace and scale of technological development in many fields, including anything that has to do with computation and artificial intelligence (AI), as well as cutting-edge developments in nanotechnology, synthetic biology, genomics, robotics and automation, neuroscience, geoengineering, and space travel. If it is true that in jumping back to a time "when old technologies were new"[25]—landmark technologies such as electricity, telegraphy and telephony, photography and the cinema, air travel, and so forth—we would experience a sense of profound cultural transformation and disorientation, it is also the case that never before in human history has technology permeated so profoundly and extensively so many domains of social, biological, and cultural life. From the development of microprocessors in the 1950s and recombinant DNA techniques in the 1970s until today's promise of a great technological singularity, the exponential growth of technology has irremediably impacted all dimensions of social existence, from the molecular to the personal and the global.[26]

24 A number of critics of the technological rationality of advanced industrial society, from Jacques Ellul, Lewis Mumford, Herbert Marcuse, and Erich Fromm to Murray Bookchin, Ivan Illich, and Paul Virilio, as well as more recent philosophers of technology and machines such as Gilbert Simondon, Gilles Deleuze, and Bernard Stiegler, are important forerunners of the ontological examination of technology.

25 Carolyn Marvin, *When Old Technologies Were New: Thinking About Electric Communication in the Late Nineteenth Century* (New York: Oxford University Press, 1988).

26 The Korean-German cultural studies scholar Byung-Chul Han has analyzed in depth the impact of the expansion of the digital order on daily life from a post-Heideggerian perspective, particularly the erosion of the vitality of things as sources of meaningful experience by pervasive informatization. The smartphone is central to this state of affairs. The infosphere defactualizes human existence, turning it into information and ending the era of fact-based truth; this contributes greatly to the undermining of presence, empathy, place, and community. In the age of social media, a generalized narcissism settles in, causing not only things but the other to disappear. Interestingly, Han concludes

These developments are driving a great acceleration in the pace of capitalist extraction of resources and the use of energy and materials, impacting Earth indelibly.[27] At stake is nothing less than the status of the human, Life, and the fate of the Earth. The double problematic lies within the relation between technology and the magnitude and pace of extraction, on the one hand, and the historical and onto-epistemic background underlying monotechnologism and capitalist extraction, on the other. Clive Dilnot describes this unprecedented juncture as the coming of age of the artificial as a horizon for being. How is the arrival of the epoch of the artificial-as-totality to be thought and acted upon? Let us begin by stating that for Dilnot, artifice and prefiguration—the condition of being made of the human—have been central to human existence from the get-go; however, they have undergone a profound shift in the period of the Great Acceleration, which roughly started after the Second World War. For Dilnot, in becoming unbounded, technology ushers in a regime of worldmaking and materiality that fully replaces nature as horizon. This is what he means by the artificial.

The artificial thus "unarguably constitutes for us the underlying condition of existence of our time. It is an unsurpassable condition because *there can be no relation to natural conditions of existence which do not pass through it*. [. . .] The artificial replaces nature as the horizon and medium of our world."[28] Martin Heidegger's statement about the centrality of object-centeredness to modern social life ("Beings became objects that could be controlled and seen through by calculation"[29]) seems fulfilled. But is it? If the crisis has become permanent and total, it means that in this very condition there lies the possibility of an affirmation: "At once a gathering of modernity and the stepping beyond it."[30] A different future is immanently present in the new condition, only if we do not assume continuity with modernity but interrupt its trajectory: if things have been made, they can be made otherwise. In short, the artificial-as-totality suggests the impossibility of thinking from within the limits of what-is (which

that artificial intelligence, by processing "hechos *predeterminados que siguen siendo los mismos*" (predetermined facts that continue to be the same; original emphasis), precludes the irruption of something that is genuinely new (Byung-Chul Han, *No-cosas: Quiebras del mundo de hoy*, trans. Joaquín Chamorro Mielke [Madrid: Taurus, 2021], 57). For a creative statement of alternatives to the digital monocultures of the present, see the wonderful project by the Verses collective, Towards a Digital Pluriverse: https://pluriverse.world.

27 J. R. McNeill and Peter Engelke, *The Great Acceleration: An Environmental History of the Anthropocene since 1945* (Cambridge, MA: Harvard University Press, 2016).

28 Clive Dilnot, "Designing in the World of the Naturalized Artificial," in *Design in Crisis: New Worlds, Philosophies and Practices*, ed. Tony Fry and Adam Nocek (London: Routledge, 2021), 9103, emphasis added.

29 Martin Heidegger, *Poetry, Language, Thought*, trans. Albert Hofstadter (New York: Harper, 1971), 74.

30 Dilnot, "Designing in the World of the Naturalized Artificial," 95.

would only entrench calculative rationality and the logic of capital). We need to ask: Are there any affirmative openings within the artificial? Dilnot suggests that "In the artificial, we go *past* the modern. [. . .] The rupture of the artificial is the rupture—the ending—*of* the modern. It is the rupture *with* the modern in that it objectively institutes another circulation of thinking and acting. This ending of the modern is the institution of another historical epoch."[31] This "ending" of the modern is what creates an "opening" for a kind of thought that is not content with being largely a commentator on disaster (as is a great deal of critique today) but instead one that fully recognizes autonomous making as central to many cultures and as a recoverable practice.

Dilnot intuits the possibility of a reconstitution of interdependence as the basis for being human by reinterpreting the meaning of relation. This in turn requires reestablishing "biological interdependence and the contingency of existence as the prime underlying condition of all being."[32] Lest one wish to hear in this statement an anachronistic advocacy for a return to nature, what he has in mind is an onto-epistemic and political project focused on the active re/making of interdependence through the artificial. Like Millán and Dilnot, for Hui, the contemporary conjuncture provides new conditions for thought beyond the modern. Such conditions, he surmises, can be laid down through a new approach to the relation between art, technology, and philosophy that is open to what logocentric thought cannot comprehend. "The task of thinking," he states, "is to elaborate the unconceptualizable not only through concepts and ideas, but also intuitions."[33] This involves a notion of intelligence that incorporates "intuition-acting" as it takes place in the social world and a kind of thinking that enables ontologically open futures. Such new conditions might enable a different relation to technology, questioning algorithmic rationality and its relation to capital while fostering a technodiversity that resists the onto-epistemic takeover by machines.

Returning technology to being part of Life by placing it at the service of multiple ways of worlding, rather than leaving it in its dominant instance, is a clearer imperative today than ever before. We are increasingly enmeshed in gigantic technological systems (aggregate data systems, networks, prediction algorithms, sensors, interconnected "smart" devices) with a tendency to totalize

[31] Dilnot, 103. This means that conceiving of possibility through an impossible leap from within what-is (for example, the technological singularity), falls into the trap of ruling out radical possibilities from within what is currently thinkable and doable, thus missing the option of a genuine departure beyond monotechnologism toward a pluriversal cosmotechnics.

[32] Dilnot, "Designing in the World of the Naturalized Artificial," 107.

[33] Hui, *Art and Cosmotechnics*, 249.

and supplant (organic) life, redrawing the opposition between the organic and the mechanistic in favor of the latter (the becoming-organic of machines). In Ivan Illich's terms, we have significantly overshot the threshold beyond which any given technology becomes disabling, rather than enabling, of human potentiality.[34] Hence, Hui's emphasis on finding technological practices that do not stifle what is nonrational, incomputable, incalculable, and ineffable, practices that resist the drive toward the recursive rationalization of the unknown. This is where the role of art becomes manifest.[35]

The ontological critique of AI (life on the screen writ large, one might say) needs to be accompanied by an acute awareness that the world is progressively being reduced to recursive calculation by ubiquitous cybernetic machines fueled by a proliferating algorithmic rationality that is functional to global capitalism, besides being heavily dependent on extraction. In other words, it is necessary to bring the ontological analyses to bear on the fact that monohumanist technosociality is currently, and increasingly, being deployed in nearly all spheres of social life (from finance and health to shopping, media, and politics) by the explosion of machine learning, algorithmic logics, and big data. This is possibly one of today's most crucial philosophical and political questions and remains underexamined, given the naturalization of "data" and the unquestioned role of statistics and probability that underwrite such algorithms.[36] Said succinctly, it is chiefly through such machine learning mechanisms and algorithmic rationality that the worldview and desires of the powerful are imperialized and perpetuated, often with celebratory undertones. A global cybereconomy has emerged, largely run by "algorithms of oppression" with their structuring of patriarchal and racist commitments; They operate as veritable forms of technological redlining, as Internet scholar Safiya Umoja Noble has demonstrated.[37]

In this context, rearticulating the relation between technology and worldmaking becomes crucial. Does the phenomenological approach to being-in-the-world still hold, given that "when the computational environment *displaces* the world, it doesn't mean that the world disappears, but rather that it

[34] Ivan Illich, *Tools for Conviviality* (London: Calder & Boyars, 1973).

[35] Hui, *Art and Cosmotechnics*. See also, for instance, Yuk Hui and Brian Kuan Wood, "A Conversation on *Art and Cosmotechnics*, Part 1," *e-flux Journal*, no. 124 (February 2022), https://www.e-flux.com/journal/124/446668/a-conversation-on-art-and-cosmotechnics-part-1. The full conversation, along with a continuation, is printed on page 165–87 of this publication.

[36] For a critique of algorithmic rationality and the mathematics behind it, see Justin Joque, *Revolutionary Mathematics: Artificial Intelligence, Statistics and the Logic of Capitalism* (London: Verso Books, 2022).

[37] Safiya Umoja Noble, *Algorithms of Oppression: How Search Engines Reinforce Racism* (New York: NYU Press, 2018). My thanks to Orin Starn for bringing this important work to my attention.

becomes silent,"[38] and ceases to be sensible? This question leads Hui to emphasize the reconsideration of place and locality relationally and nondualistically, for which he finds clues in East Asian *shanshui* or "landscape" painting. These beautiful works of art make the cosmos visible through the composition of nonhuman agencies such as water, rocks, fish, plants, rendering sensible what is sublime and unknowable. As he says, shanshui paintings and Japanese gardens aim to make sensible what is invisible, in the realization that "the Unknown is also conditioned by place, which is the ground, the groundless ground."[39] Can this approach make (modern) humans aware of their growing detachment from Earth effected by technology and its ensuing cosmic nihilism?[40]

Hui is not advocating for a wholesale rejection of modern technology. For him, a vision of technology is needed that goes beyond modernization while being able to accommodate modern technology "without becoming technologistic," one that is able to overcome the opposition between "technology and spiritual life."[41] The resulting technodiversity would avoid modern onto-hegemony, rendering it one possible cosmotechnics among many. A different relation between art and technology would enable an "education of sensibility" that fosters an incitement to being present to place (akin to Tim Ingold's "education of attention"[42]). This requires a renewed orientation toward Being, the Unknown, and the nonrational, regaining the ability to feel "a resonance between ten thousand beings."[43] Art, in its cosmotechnical multiplicity, must confront the current crisis "in order to produce new epistemes, new sensibilities that will be able to give science and technology new directions and frameworks."[44] Such an epistemic revolution "is not something we can invent from without. Rather, it is always already *local* and *historical*."[45] We could argue, in our own terms, that

[38] Hui, *Art and Cosmotechnics*, 247, original emphasis.

[39] Hui, *Art and Cosmotechnics*, 274.

[40] Hui stresses recursivity as fundamental to both computation (machine) and the world (Life), albeit differently so. His advocacy, after a laborious genealogy of recursive thought in both Western philosophy and Chinese thought (Yuk Hui, *Recursivity and Contingency* [London: Rowman & Littlefield, 2019]), is for a "*generalized recursive thinking* [that] needs to understand and to co-exist with machines" (Hui, *Art and Cosmotechnics*, 248, original emphasis) while mindful of the fact that not everything in Life is recursively enumerable, that is, the world cannot be reduced to a recursive universe.

[41] Hui, *Art and Cosmotechnics*, 277; echoing Japanese philosopher Kiyoshi Miki.

[42] Tim Ingold, "From the Transmission of Representations to the Education of Attention," in *The Debated Mind: Evolutionary Psychology versus Ethnography*, ed. Harvey Whitehouse (London: Bloomsbury Academic, 2001), 113–53.

[43] Hui, *Art and Cosmotechnics*, 281.

[44] Hui, *Art and Cosmotechnics*, 282.

[45] Hui, *Art and Cosmotechnics*, 287, original emphasis.

Hui is calling for a pluriversal sensibility through regaining the grounding and presence to the world and, hence, to multiple other worlds.[46]

There is one remaining crucial aspect that is seldom emphasized in theoretical discussions of technology, and this is their dependence on patriarchy. Patriarchy may be defined as an ontology and an ensemble of practices and emotions that privilege competition, hierarchies, appropriation of the Earth, domination of others, power, growth, violence, and war. These values are at the basis of contemporary technology's drive toward calculation, control, maximization, instrumentality, extraction, and algorithmic prediction. Patriarchal cultures, in their entanglement with capitalism and coloniality, can be said to provide the onto-epistemic software for the techno-economic acceleration that underlies the current staggering levels of inequality and ecological devastation. They provide the architecture for the masculinist "historical project of things," in contradistinction to the "historical project of relations," as stated by Rita Laura Segato, which characterizes the ontologies of care emphasized today by many decolonial and anti-racist feminists.[47] Unfortunately, it is the techno-patriarchal gurus of high tech that are gaining the upper hand in the vital onto-epistemic task of imagining futures. With their seductive imaginaries of control and unlimited choice, of life beyond Earth and "beyond biology,"[48] these techno-patriarchs steer visions of the future more effectively than any other social actor. Naming them as ontologically, socially, and politically patriarchal is crucial for positing contrasting imaginaries for different futures and futures in difference.[49]

Pluriversal Planetarity

We are now able to suggest some elements for reinterpreting the dominant understanding of technology from an onto-epistemic perspective. We situate the contemporary conjuncture at the intersection of the double condition of

[46] For Hui, it is important to acknowledge the limits of a "becoming organic" of technology, for that remains a dualistic formulation. Socioecological theory must straddle two positions in this regard: on the one hand, the idea of a purely nonobjectified organic nature (although via making it "subject"); on the other, nature's final demise after its complete conquering by technology in the name of the utopia of endless techno-material prosperity embedded in Western metaphysics of technology.

[47] For example, see Rita Laura Segato, *Contra-pedagogías de la crueldad* (Buenos Aires: Prometeo Libros, 2018), and Segato, *La guerra contra las mujeres* (Madrid: Traficantes de Sueños, 2016).

[48] Ray Kurzweil, *The Singularity is Near* (New York: Viking Books, 2015).

[49] For an extended discussion of the patriarchal dimension of ontological dualism, see Arturo Escobar, *Designs for the Pluriverse: Radical Interdependence, Autonomy, and the Making of Worlds* (Durham: Duke University Press, 2018).

terricide and the artificial-as-totality, on the one hand, and the possibility of new conditions for thought presented by such conjuncture, on the other. By doing so, we begin to perceive a multiplicity of new beginnings derived from a revised understanding of history that, in turn, provides grounds for a reappropriation of technology in attunement to place, locality, and the relational reality of a living cosmos, that is, to other worldmaking practices in all of their pluriversal multiplicity. Echoing this volume's call, we call "planetarity" this new condition, a category to which we arrived by questioning anew patriarchal monohumanism and monotechnologism. In its evocation of Earth, planetarity counters the dominant narratives of coloniality, developmentality, and imperial globality from the perspective of pluriversality.[50]

Planetarity emerges in the vast space between Earth and World, as a project for reassessing the role of technology in shaping modernity and the human. Beyond the receding of the world and the erosion of the preeminence of the *hic et nunc* (here and now) of human existence[51] in the cybernetic age, Earth herself tends to disappear; more precisely, it is the monotechnological and monoversal way of worlding underlying the "one-world world" that is responsible for the vanishing of the Earth as Life's abode. Against such mono-ontological worlding there arise both Earth and pluriversality: Earth as the capacity for perpetual self-organization and Life's ceaselessly unfolding flux of changing forms, forces, behaviors, and relations; and pluriversality, as the continued presence of a world of many worlds, a world where many worlds fit, as the Zapatista from Chiapas have wisely put it.[52]

Together, Earth and pluriverse signal the radical interdependence of everything that exists and the fact that all entities are always in the process of dependent co-arising. This conception points at the patent limitations of classificatory rationality to fully understand life through positivist categories and concepts and, concomitantly, of anthropocentric techno-designing to shape life to its will. Even then, when faced with incomputable processes such as climate change, moderns can only think about addressing them through a new round of technoscientific rationality, which is more computation. We are still far from heeding Akomolafe's prescient admonishment that climate change is not a problem that organizations can draw lines around and manage, because climate

[50] For a discussion of the notion of pluriversality, see Arturo Escobar, *Pluriversal Politics: The Real and the Possible* (Durham: Duke University Press, 2020).

[51] Paul Virilio, *Open Sky*, trans. Julie Rose (London: Verso Books, 1997).

[52] Marisol de la Cadena and Mario Blaser (eds.), *A World of Many Worlds* (Durham: Duke University Press, 2018).

change *is* the world; as such, given its incalculable and unbounded complexity, climate change is "ontologically *unframeable, unthinkable* and *incalculable*."[53] No amount of computational power and scientific research can change this predicament at the very heart of Western scientific "objectivity."[54] We need to come to terms with the limitations of the discourse of the wise management of the planet for the benefit of everything promised by modern science and politics—embedded in notions of sustainability, innovation, ecological modernization, and so forth.

One of the figurations that, for us, best conveys the challenge of planetarity, as an onto-epistemic space where pluriversality might flourish, is that of transitions. Transition thinking has been emerging steadily in many parts of the world and from many perspectives and traditions. Yet many of the transition proposals do not depart significantly from the modern story of Life and the onto-epistemic regime of Man; therefore, calls for pluriversality are at risk of being reabsorbed by new rounds of monohumanist technosociality driven by global capital. In their emphasis on the sociotechnical, economic, and ecological aspects of transitions, they continue to uphold an anthropocentric attitude of epistemic mastery and control. Other transition narratives focus on particular domains (for example cities in transition, energy transitions, agricultural transitions, technological transitions, postcapitalist transitions, and so forth). Many of these visions are, of course, important for imagining transitions from onto-epistemic perspectives, but they need to be pushed ontologically, by which we mean emphasizing the existential and civilizational dimensions of the transitions.

It should be clear by now that approaches to transitions need to go well beyond science and logocentric social theory. "El problema no es de ciencia, sino de las condiciones de la existencia" (the problem is not one of science, but of the conditions of existence) is a rhyme heard in some activist circles in Colombia. This does not mean that science and technology do not have a role to play in the pluriversal transitions but that scientists, policymakers, and activists alike need to acknowledge the limits of both science and modern social theory, an idea aptly conveyed with the saying that "más allá de la razón hay un mundo de colores" (there is a world of colors beyond reason).[55] Here, we may locate a diverse set of visions and proposals, such as Hui's appeal to art and cosmotechnics as a way to explore

[53] Bayo Akomolafe, "What Climate Collapse Asks of Us," August 17, 2019, https://www.bayoakomolafe.net/post/what-climate-collapse-asks-of-us, original emphasis. For a similar point, see Fry and Tlostanova, *A New Political Imagination*.

[54] Lorraine Daston and Peter Galison, *Objectivity* (Brooklyn: Zone Books, 2007).

[55] Adolfo Albán Achinte, *Más allá de la razón hay un mundo de colores: Modernidades, colonialidades y reexistencia* (Santiago de Cuba: Editorial Oriente, 2013).

a different relation between technology and nature that engages with Western ontology without becoming techno-logistic; Akomolafe's call for "coming down to earth," going into the cracks of the dominant systems, and building sanctuary for the wider living coalitions that refuse to coexist with oppressive systems[56]; design theorist Ezio Manzini's call for a novel practice of urban dwelling under a novel paradigm of relationality and care that is based on proximity, which is essential for a place-based and Earth-wise human sociality[57]; the stimulating attempts to steer emergent technological trajectories toward a civilizational "great transition"[58]; and evolving experiments and frameworks for transition design, regenerative design, and autonomous design.[59] This is but a sample of what is already a rich spectrum of experiments seeking to remember the conditions of interdependence, reestablish the conditions of reciprocity, acknowledge pluriversality, and strategically ground actions in place and locality, while meshworking with similar transformative initiatives.

The principle of radical interdependence signifies going beyond ontologies of separation with their reductionist premise of intrinsically existing objects, subjects, and actions.[60] It also implies the full acceptance of the fact that, at a very fundamental level, we are all within the pluriverse. When facing the toxic loops of contemporary existence, this realization should move us along the path of repairing, healing, caring, regenerating, and making and creating with others: humans, nonhumans, materials, landscapes, spirits, ancestors, water, seeds, and all kinds of tools and technologies. To do so with an acute ontological and historical conceptualization and practice of politics implies doing one's best to shake off the constraints of having to live daily within the enduring assemblage constituted by individual personhoods, needs, technologies, markets, and single truths that has become the operating system of the structuring force field, which in social theory terms can be called the patriarchal-colonial capitalist system, the Western civilizational project, industrialism, monotechnologism, and so on.

[56] Akomolafe, "What Climate Collapse Asks of Us."

[57] Ezio Manzini, *Livable Proximity: Ideas for the City that Cares* (Milan: Egea, 2022).

[58] For example, see Paul Raskin, *Journey to Earthland: The Great Transition to Planetary Civilization* (Cambridge: Tellus Institute, 2016).

[59] For some examples, see Terry Irwin, Gideon Kossoff, and Cameron Tonkinwise, "Transition Design Provocation," *Design Philosophy Papers* 13, no. 1 (2015): 3–11; Terry Irwin and Gideon Kossoff, "Transition Design as a Strategy for Addressing Urban Wicked Problems," in *Cities Without Capitalism*, ed. Hossein Sadri and Senem Zeybekoglu (London: Routledge, 2021), 90–120; Daniel Christian Wahl, *Designing Regenerative Cultures* (Axminster: Triarchy Press, 2016); and Josie Warden, *Regenerative Futures: From Sustaining to Thriving Together* (London: RSA, 2021), https://www.thersa.org/globalassets/pdfs/reports/from-sustaining-to-thriving-together-final.pdf.

[60] Kriti Sharma, *Interdependence: Biology and Beyond* (New York: Fordham University Press, 2015).

"The way isn't forward, it's awkward."[61] Awkward, as in: questioning the patriarchal master narratives of linear progress, perpetual growth, simple causality, prediction, logocentric rationality, uni/monoverse, secularity, marching in unison to the drumbeat of rationality, white privilege, billionaire techno-gurus, individual success, panicked avoidance of suffering and death, and anthropocentric continuity. Awkward, as in: flourishing in the cracks and fissures of oppressive designs and practices; attuning to the intuitive, the irrational, the feminine, the sacred, the ineffable; committing to place despite the pressure to delocalize and decommunalize; creating pluriversal hypertextualities and other kinds of collective intelligence enabled by digitality. Awkward, as in: meditating on and organizing horizontally for the end of a civilization premised on extractive capitalism.[62]

Design approaches that take into account the ontological predicaments of monohumanism and monotechnologism[63] involve a redirective practice of technology as a space for mutually enriching articulations between the biophysical and the techno-cultural infrastructures of any and all worldmaking practices.[64] In this context, the praxis of repairing, healing, and caring for life's web of interdependencies might be established as a pluriversal strategy to keep in tension the earth-bound and the artificial and to overcome the most destructive aspects of the technological reshaping of worlds. Dealing with this tension constructively requires paying attention to the onto-epistemic, existential, and political conditions necessary for a reappraisal of the constitutive relationality of all existence, as we suggest with the propositions below. As Helena Norberg-Hodge puts it, "Why should we accept an energy- and mineral-intensive technological

61 Akomolafe, "Coming Down to Earth."

62 The revered Buddhist teacher Thich Nhat Hanh invites us to actively accept the end of our consumerist civilization by meditating on this thought: "Breathing in, I know that this civilization is going to die. Breathing out, this civilization cannot escape dying." Thich Nhat Hanh, *The World We Have: A Buddhist Approach to Peace and Ecology* (Berkeley: Parallax Press, 2008), 46. This is a call to move beyond a civilization that has become so antithetical to the ontology of interbeing, and to accept that this civilization is temporary, whether or not we wish to move beyond it.

63 For ontologically oriented design proposals, see Fry, *Defuturing*; Escobar, *Designs for the Pluriverse*.

64 A recent reader on "cosmolocalism," for instance, offers different framings, including open design distributed manufacture, maker cities, fab cities, do-it-together, planetary bricolage, planetary mutualization, cosmopolitan localism, peer-to-peer production, and commons; it includes over forty cases from modular automotive manufacturing, to agri-robotics and peer-to-peer farming, community driven wind power and housing construction to biohacking, furniture fabrication, upcycling, prosthetics, and disaster relief, with a sophisticated conceptualization of cosmopolitan localism and its technological infrastructures. For more information, see the project Cosmolocal Reader, https://clreader.net.

infrastructure that is fundamentally about speeding life up, increasing our screen-time, automating our jobs, and tightening the grip of the 1 %?"[65]

Reorientation of Technology

Design, like technology, has been predicated on the premise of the separability between humans and the world. Yet what would become of designing and technology if they were to be reconceived from the fundamental insight that the world does not exist "out there," separate from us, but that we construct it with every one of our actions and perceptions? Such reorientation should contribute to disrupting the monohumanist worlding practices that make the world principally technological and economic, thus challenging the globalizing force where only One World and One Human fit. This is the predicament of planetarity, marked by the simultaneous condition of unprecedented challenges and novel possibilities for thought, from which there might emerge new cosmotechnical directions for science and technology. Such designing and technological practices would persuasively contribute to transitioning toward the pluriverse, including the need to consider all life forms. As we have argued, this would involve an onto-epistemic (not only politico-economic) reorientation of technology. We offer a few additional propositions in this regard, which involves cultural-political work of utmost importance at present. Let us call this cultural-political task of onto-epistemically reorienting technology "pluriversalizing technology"[66]:

1. Pluriversalizing technology needs to be approached from the premise that life is constituted by the radical interdependence of everything that exists, so that designers can slowly discover the considerable potential of acting from interdependence and care.
2. Pluriversalizing technology means designing technologies that support a world of many worlds, with an active awareness that constructing worlds under the premise of ontological separation negates the possibility of existence for the ontologically divergent.

[65] Helena Norberg-Hodge, "Putting Technology in Its Place," GTI Forum Technology and the Future, Great Transition Initiative, January 3, 2022, https://greattransition.org/images/Technology-Future-Norberg-Hodge.pdf.

[66] This is an adaptation of a set of propositions concerning pluriversal designing initially developed by Marisol de la Cadena and Arturo Escobar, "Notes on Excess: Towards Pluriversal Design," in *Design For More-Than-Human-Futures: Towards Post-Anthropocentric Worlding*, ed. Martín Tironi, Marcos Chilet, Carola Ureta Marín, and Pablo Hermansen (London: Routledge, 2023), 29–50.

3. Pluriversalizing technology places in parenthesis the modern notions of object and project, opening possibilities for nonobject-centered and nonprojectual technological designing praxis.
4. Pluriversalizing technology focuses on technologies intended for the reconstitution, healing, and caring for the web of interrelations that make up the bodies, places, cities, and landscapes that we are and inhabit.
5. Pluriversalizing technology fosters a departure from anthropocentrism, figuring conditions for all earth beings to flourish. It instills a sense of being at home in a world that is alive, creating spaces for reimagining ourselves as pluriverse and as community.
6. Pluriversalizing technology is mindful of the conditions of generalized individuation, delocalization, decommunalization, and deplacing effected by modern technology-driven global capitalism. Conversely, it sees itself as contributing to the postcapitalist recommunalization of social life and the relocalization of activities such as eating (vs. "food"), healing (vs. "health"), learning (vs. "education"), dwelling (vs. "housing"), and livelihood (vs. "economy").
7. Pluriversalizing technology aims to heal the ontological uprootedness from body, place, and landscape through forms of making that contribute to re-embodying, re-placing, and re-earthing life.
8. Pluriversalizing technology entails reclaiming the capacity for making life collectively and autonomously instead of outsourcing it to heteronomous institutions, experts, the state, and the capitalist economy.[67] It moves away from the historical project of objects while favoring the historical project of relations.
9. Pluriversalizing technology contributes to dismantling the mandate of masculinity that is at the core of the object-driven ontology of modernity. It practices a feminist and anti-racist politics that pragmatically privileges collective and communitizing modes of making and acting centered on care.
10. Pluriversalizing technology takes seriously the struggles for social justice, respect for the Earth, and the rights of all human and nonhuman entities to exist.

[67] See Leanne Betasamosake Simpson, *As We Have Always Done: Indigenous Freedom through Radical Resistance* (Minneapolis: University of Minnesota Press, 2017); Gustavo Esteva, "Autonomy," in *Pluriverse: A Post-Development Dictionary*, ed. Ashish Kothari, Ariel Salleh, Arturo Escobar, Federico Demaria, and Alberto Acosta (New Delhi, India: Tulika and AuthorsUpFront Publishing, 2019), 99–102.

11. Pluriversalizing technology involves learning to think and make with those who rise in defense of their life territories, strengthening their life-making and autonomy-oriented practices.
12. Pluriversalizing technology requires a renewed awareness of how the creation of conditions for life-sustaining coexistence will necessarily have to engage with the dominant logic of unsustainability and defuturing.[68]
13. Pluriversalizing technology goes beyond the grammar of "problems" and "solutions," particularly as it pertains to incomputable and ontologically unframeable civilizational challenges such as climate change.
14. Pluriversalizing technology resists translating the inexhaustible reservoir of practices of prefiguration, artifice, and making maintained by peoples worldwide into the grammars of modern technology and design, letting them come into the foreground as unique instances of relational life-making.[69]
15. Pluriversalizing technology renders the project of re-earthing cities into a historically plausible intellectual, political, and technical process focused on the creation of spaces of healing, recommunalization, and mutually enhancing relations with the Earth.
16. Pluriversalizing technology engages in new conversations with art, aesthetics, and philosophy (multiple traditions of thought) intended to illuminate paths to a different relation between humans and the cosmos in the face of generalized cybernetic thinking. It retrieves the history, knowledge, and world-making significance of what lies beyond established forms of rationality.
17. Pluriversalizing technology actively opposes digital monocultures by creatively inhabiting the space opened between pluriversal visions of the digital and digital visions of the pluriverse, searching for a more grounded ethics of technology and AI that resists digital land grabs while fostering digital commons.[70]
18. Pluriversalizing technology has as a general goal the mobilization for a new way of dwelling on the Earth.[71]

[68] Fry and Tlostanova, *A New Political Imagination*.

[69] Alfredo Gutiérrez, "When Design Goes South: From Decoloniality, through Declassification to Dessobons," in *Design in Crisis: New Worlds, Philosophies and Practices*, ed. Tony Fry and Adam Nocek (London: Routledge, 2021), 56–73.

[70] This item inspired by the "Towards a Digital Pluriverse" project.

[71] This formulation guides the current work of a small group that includes Fernando Flores, Terry Winograd, Don Norman, B. Scot Rousse, and Arturo Escobar, gathered around the formative insights of computer network technologies and design originally formulated in Terry Winograd and Fernando Flores, *Understanding Computers and Cognition: A New Foundation for Design* (Norwood: Ablex, 1986).

8

Systems Representing Themselves

Juaniko Moreno

The term Web 2.0 is used to define the transition between the static web of the early Internet and the social-media-focused era that began in the mid-2000s. It is primarily defined by its focus on user-generated content, responsiveness, and real-time updates. But perhaps the most prominent feature of this shift was the emergence of the personal user and its consolidation at the core of many online services. In order to view content, post, chat, tag, or participate in any (if not all) of the services available on this "new version" of the Web, it is necessary to sign up for a personal account. This account stores personal information and becomes an index for all of the behavior tracking, pattern recognition, and targeted advertising that the Web has become synonymous with. The deciding factor for the shift from the early Web to the current Web is the introduction of the digital persona as the paragon of design in online services.

Nostalgia for the early days of the Internet often emphasizes the freedom, decentralization, and anonymity of Web 1.0: no data silos stored entirely within the servers of mega-corporations, no traces of any action. The personal user was not only the fundamental development that enabled the monetization of the Web, but at its core, it represented the intrusion of what Yuk Hui, whose theories I consider in more detail below, terms a particular "cosmotechnical"[1] view that became embedded in the operation of this technical system. This transformation was not unprecedented: The computer went from being a large, collectively operated, and purpose-built machine to the mass-market consumer product

[1] Regarding cosmotechnics, Hui explains: "Scientific and technical thinking emerges under cosmological conditions that are expressed in the relations between humans and their milieus, which are never static. For this reason, I would like to call this conception of technics *cosmotechnics*. One of the most characteristic examples of Chinese cosmotechnics is Chinese medicine, which uses the same principles and terms found in cosmology, such as *Yin-Yang*, *Wu Xing*, harmony, and so on, to describe the body." Yuk Hui, *The Question Concerning Technology in China: An Essay in Cosmotechnics* (Falmouth: Urbanomic, 2016), 18, emphasis original.

aptly called the personal computer. The consolidation of the digital personal user as the de facto method of interfacing with the Web is predicated on the augmentation of technical systems that build upon previous developments. This has led to a feedback loop of technical entrenchment: the automobile as a personal vehicle, the single-family home, the personal computer, and the smartphone. The ubiquity of individualized machinery has generated the need to keep track of all personal data gathered by devices, even those that do not inherently require such data to function, such as AI toothbrushes.

As technological processes, computation or information exchange do not have written rules that determine their social use by individuals. Although the instrumentalization of the natural world is the product of a particular cosmotechnology that can be traced back to Enlightenment rationality, this kind of instrumental rationality is not an inherent quality of computation or information exchange but only one way of conceiving and operating them. The individualization and personalization of technology are the outcomes of a certain worldview, of a distinct Silicon Valley version of the free-willed, self-determined individual who has been extrapolated onto the Web. In this worldview, the personal user is considered a "prosthetic extension of [the] individual ego."[2] The society and culture from which the personal user emerged is particular and contingent; yet the ubiquitous adoption and global expansion of this particular ideology have meant its consolidation as a technological monoculture.

Technological monocultures have a tendency to present themselves as universal and narrate change as inevitable and unidirectional progress. The achievements of face recognition or robotic vacuum cleaners, for example, are touted as beneficial to all, yet they conceal the ideological contingencies that gave birth to these technical objects in the first place. Hui affirms that

> There is a general misconception that all technics are equal, that all skills and artificial products coming from all cultures can be reduced to one thing called "technology." And indeed, it is almost impossible to deny that technics can be understood as the extension of the body or the exteriorisation of memory. Yet they may not be *perceived* or reflected upon in the same way in different cultures.[3]

It is this relationship between technique and culture that Hui defines as "cosmotechnics": the particular scientific and technical thinking that emerges

[2] Stephanie Sherman, "The Autocene: Towards a Post-Automotive Future," *Strelka Magazine*, January 25, 2022.

[3] Hui, *The Question Concerning Technology in China*, 9, emphasis original.

from the interactions between humans and their milieus.[4] These milieus (or environments) are social, technical, and physical: They consist of the material configuration of a given environment, the tools developed to alter it, the various needs that drive this alteration, and the creation myths that give moral grounding to both the alteration and the needs. In Hui's terms, understanding digital information as attached to a digital persona is not an inherent or inevitable result of technological development but rather the consequence of a particular cosmotechnical scaffolding in which the cultural, economic, and technical interact at every layer.

Understanding technological developments as a consequence of cosmological views, and thus as fundamentally shaped by the particular ontologies and epistemologies that such views entail, is critical to questioning the embedded design flaws within a technological system. The same design that allows for personalized feeds and targeted recommendations also paves the way for digital surveillance, targeted marketing, pervasive facial recognition, and other forms of personal intrusion.[5] This particular blueprint for the interaction between users, data, communication, and methods of retrieval can be thoroughly dismantled and reassembled if approached from a radically different cosmotechnical perspective—that is, if the fundamental principles of its design are challenged and replaced by new or alternative principles. Only by grappling with the cultural and historical contingencies of a particular cosmology can one escape technological determinism and begin to open up alternative ways of being and thinking.

Traditional Chinese and pan-Amazonian Indigenous cosmologies display an understanding of technology divergent from today's hegemonic global digital culture. By engaging with their respective ways of understanding the cosmos as a systematic mesh of actionable, living, and recursive information—while acknowledging differences that arise from their geographic and environmental contingencies—this essay seeks to shed light on alternative ways of conceiving informational and living systems, as well as the contrasting ways in which they are instantiated. These examples reach far beyond the understanding of computation as merely occurring in silicon-based machines or as the product of human inventiveness. They provide a path to decenter the human as the main

[4] Hui, *The Question Concerning Technology in China*, 18.
[5] The data sovereignty movement is based on the recognition of significant discrepancies between the asymmetrical rights and ownership of digital personas in comparison to the living, sovereign individuals they represent.

driver of design and communication, in order to place humanity alongside a multitude of equally relevant interlocutors.

Cosmological and Technical Systems

Before delving into concrete cases, it is relevant to clarify the relationship between technical and cosmological systems or, simply put, between systems of belief, thought, and action. "Technics" here is taken as the general category of all forms of making and practice.[6] It designates the artificial operations that serve to interact with and alter a given environment. Consequentially, "technology" can be understood as a complex method of action, as the aggregation and consolidation of these techniques into a cohesive, interoperable, and consistent system. One of the ways in which separate techniques become technical systems can be found in the work of the French philosopher Gilbert Simondon.

Simondon's work is concerned with technology as seen from a phylogenetic perspective, examining technology in terms of evolutionary processes and relationships. For him, the "technical object" is not given, nor is it the outcome of an inventor's inspiration, but something that emerges from a process of refinement. Several iterations, changes, and mutations occur to the technical object over generations. Technical objects, such as the steam engine, the calculator, or encryption techniques, are designed with a particular purpose in mind and improved with each iteration, developing the object's intrinsic qualities. These technical objects undergo a process by which they become "concrete," meaning that the isolated elements out of which the object is constructed are unified in a system that has a functional coherence and internal resonance.[7] Isolated technical objects can come together to form a larger system, a collective arrangement that potentiates their separate qualities and opens up new possible functionalities for each component. An example of this would be the Internet, whose individual components (connection protocols, encryption techniques, user interfaces, among others) emerged as generational improvements on individual technical objects, and their crystallization into a system exceeded their original problem-solving capacity and purpose.

A careless interpretation of Simondon's ideas would mistake intrinsic development as proof of technological determinism: Technical objects naturally

[6] Hui, *The Question Concerning Technology in China*, 4.

[7] Pascal Chabot, *The Philosophy of Simondon: Between Technology and Individuation*, trans. Aliza Krefetz with Graeme Kirkpatrick (2003; London: Bloomsbury, 2013), 15.

concretize into systems that refine their internal components, and this refinement can only follow a path of universal improvement. However, it is crucial to note here the relevance that a given milieu has in defining what constitutes a problem, what is identified as a need, and what emerges as the driver of refinement. For instance, the paths of development that a technical system might take will differ if it is concerned with optimizing water use, freedom of speech, productivity, military communication, or crude oil extraction. The definition of these needs will be driven by the system of thought and belief that articulates certain aims or goals, which in turn determines the boundaries of what is possible.

The term "system," as I am using it in this essay, also requires further elaboration. If defined by its early use in first-order cybernetics, a system involves complex interactions that are determined by the ability of a centralized authority or chain of command to control its internal processes and potential outputs, either through mechanical or protocolary agency. However, the study of open or complex adaptive systems has revealed how organized behavior can emerge without centralized planning, creating sets of relations that grow in complexity to the point where they become incalculable. In this scenario, what maintains any system as an identifiable collection of agents is its verifiable relationship to the emission and processing of signals, as well as the establishment and enforcement of boundaries that engage in the exchange of matter, energy, or information.[8]

Neuroscientist Karl Friston proposes the "free energy principle" to characterize the functioning of an open or complex adaptive system: the human brain. As an adaptive system, the brain captures information about its environment through sensory epithelia, generates internal models of said environment that minimize disorder or "surprise" in newly sampled sensory data, uses the newly generated models to make assumptions on the causes of the sensory samples it captures based on a Bayesian model,[9] and then recursively updates both the models and its sensory epithelia in order to minimize prediction error.[10] This principle not only describes the behavior of the human brain but can also be analogically extended to any configuration of objective entities collectively identifiable as a system. Of particular importance to the free energy principle is the "model generation" step, which accounts for how a system can sustain itself in time by

[8] Denis Robilliard, "Review of *Signals and Boundaries: Building Blocks for Complex Adaptive Systems*, by John H. Holland," *Genetic Programming and Evolvable Machines* 14 (2013): 279–80.

[9] A Bayesian model is a statistical model that uses probability to express the degree of belief that an event will occur, given previous experience of its occurrence. See Andrew Gelman et al., *Bayesian Data Analysis* (New York: Chapman & Hall/CRC, 2013).

[10] Karl Friston, "The Free-energy Principle: A Rough Guide to the Brain?," *Trends in Cognitive Sciences* 13, no. 7 (June 2009): 293–301.

generating *representations* of itself and its surrounding milieu, gathering and assembling proof of itself, adapting the way in which it captures this proof to perpetuate its existence. In other words, a system that sustains itself in time is a system that represents itself.

Systems are, by definition, complex yet identifiable. Despite the fact that their components may be multiple and various, there are ways to determine where to draw the boundary that delimits the elements and relationships composing a system from those that do not.[11] At this point, cosmological scaffoldings become relevant again. The boundaries around the agents[12] that engage in a given system are outlined following an underlying worldview. Whatever is deemed relevant within a particular cosmovision (such as angels for Catholics, the stock market for financiers, or DAOs for crypto-enthusiasts) is bestowed a status that acknowledges its agency in the world. Whatever is deemed implausible or unrealistic (such as the idea of natives possessing souls for conquistadores or the reality of climate change for denialists) is dismissed from the framework, and any influence it may have on a technological system is neglected.

In the broad colonial-capitalist paradigm of thought that characterizes the current technological monoculture, which is defined by quantifiability, accountability, and individuality, as well as by the division between nature and culture, any output of a system that involves artifice, craft, or human labor is considered to belong to "civilization," while anything produced by undomesticated nature is placed outside of the system. However, there are plenty of cultures where human beings, as well as their labor and techniques, are not ontologically subtracted from the rest of existence. Environments, tools, and actions are understood as belonging to a continuous spectrum between human and nonhuman entities. For example, the Amazonian *chagra* (sustenance farms in small forest clearances)[13] is not understood by the

[11] "Markov blanket" is the term used by the "free energy principle" and refers to the limit or subset of variables that are selected in a statistical model or in machine learning, which allows to infer a particular variable, leaving out those less relevant. See Judea Pearl, *Probabilistic Reasoning in Intelligent Systems: Networks of Plausible Inference* (San Mateo: Morgan Kaufmann, 1988).

[12] In actor-network-theory, an intermediary is any entity that transports meaning or action without transforming it, while the mediator is an agent that transforms, translates, distorts, or modifies the meaning or elements that it is supposed to transport. An agent is anything that has the ability to act upon or affect its environment with some degree of intentionality. See Bruno Latour, *Reassembling the Social: An Introduction to Actor-Network-Theory* (Oxford: Oxford University Press, 2005), 39.

[13] The *chagra* is an agricultural system of Amerindian cultures in the Amazon rainforest, consisting of the controlled clearing of forest patches and the planting of various edible and nonedible species. These clearings are usually used for up to two years, after which they are left to regrow. The diversity of plants that grow there, and the natural fertilization provided by wild animals that forage on the chagra, results in a highly fertile soil that renews forest growth.

community as deforestation but as a metabolic regulation of canopy density that enriches the soil and communes with both living species and spirits; and some hydraulic channels constructed in ancient China can be understood as fostering natural tendencies in bodies of water rather than enforcing domination upon riverways.

This reframing of technology as not merely a human trait but as derived from—and belonging to—the world can be found in digital information systems when viewed through the lens of these non-Western cultures. Under these cosmotechnical scaffoldings, computation can be interpreted as something discovered instead of invented,[14] as naturally occurring instead of being provided by multinational platforms. In order to raise cosmological frameworks that acknowledge computing as a process where humans are participants instead of masters, it is worth examining two traditions of thought originating from antipodal geographies (namely, Chinese and Amerindian cosmologies), as well as their interpretations of information and its availability in the world. The aim here is to understand "computation" beyond consumer electronics, "technology" beyond what is designed in Silicon Valley, and "systems" beyond the mechanisms for control, marketing, and surveillance that characterize today's global technological monoculture.

Yin and Yang as a Code for the Cosmos, and Computation through Data and Dao

Exploring notions related to computation in Chinese traditional cosmology demands consideration of two sets of ideas. First, the contact point between Western computation history and the Chinese principles of *Yin* 阴 and *Yang* 阳, and second, the relationship between two concepts that articulate the way technology is conceived as part of the cosmos, namely, *Dao* 道 or "the way," and *Qi* 器 or "tool."

The *Xian Tian Bagua* 先天八卦, translated as the Early Heaven Sequence, alludes to being created by the great sage FuXi 伏羲 and refers to the archetypal order of things before change. It departs from the limitless, nonexistent, or nonmanifest (*Wuji* 无极), which produces the delimited, existing, or formed

[14] Amerindian healers such as Jesús León Muipu and Tarsicio Vanegas identify computational systems, tools, and protocols in the procedures necessary to heal a body or its immediate environment. The journey of learning to identify these elements has been a long, multigenerational process of discovery rather than one of invention.

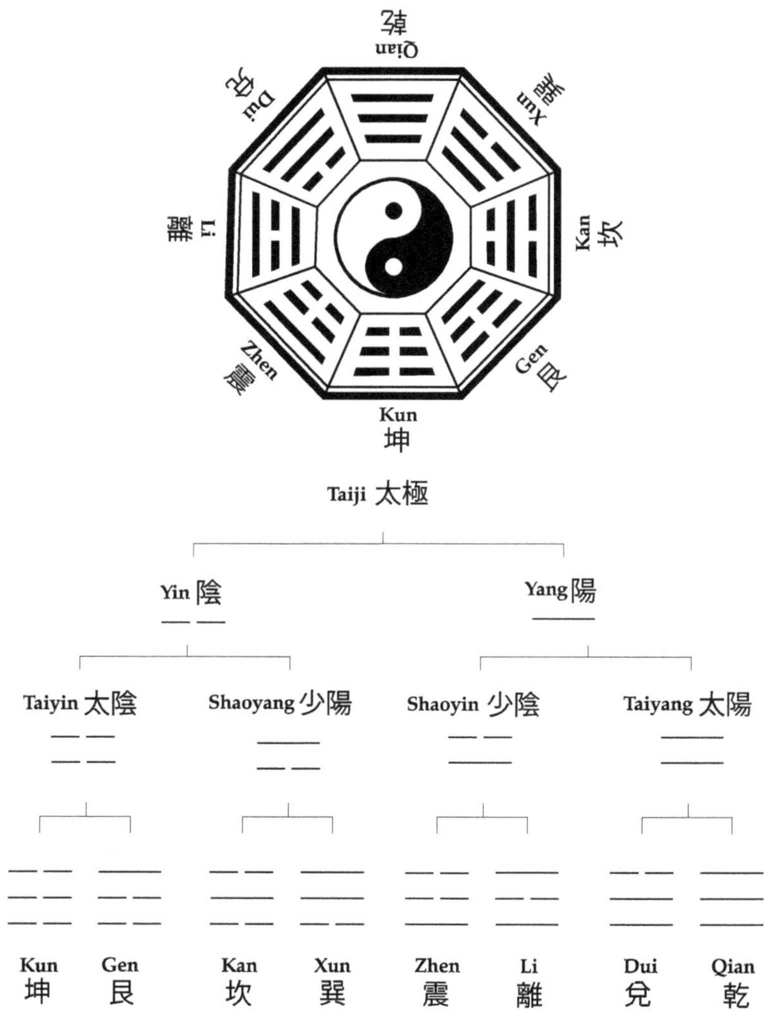

Figure 8.1 Diagram of Shao Yong's Early Heaven Sequence and the subdivision from the absolute. This division follows the detailed commentary on the *Book of Changes*: "The nonmanifest engenders the manifest [absolute, *Taiji*], the manifest engenders the two appearances [*Yin Yang*], the two appearances engender the four images [*Taiyin Shaoyang Shaoyin Taiyang*], the four images engender the eight trigrams [*Kun Gen Kan Sun Zhen Li Dui Qian*], the eight trigrams decide good or ill luck, good or ill luck engender everything."

(*Taiji* 太极). Taiji engenders Yin and Yang, the two fundamental concepts of Chinese philosophy. They constitute two polarities of the world's natural cycle: heaven and earth, attraction and repulsion, light and dark, waning and waxing (*Xiaozhang* 消长), and so on. These natural, complementary, and contradictory forces are present in every phenomenon in the world and form the basis of duality. Polarities are not mutually exclusive, as one cannot exist without the other. This principle of complementary duality is what allows for contradictory ideas and entities to coexist, for things that are simultaneously one thing *and* the other. It is therefore radically different from deterministic thought, which identifies things as *either* one *or* the other. These two forces/principles[15] subdivide until everything is produced by their permutations, and the cyclical transformation[16] between waning and waxing is the only constant. Yin and Yang are represented as continuous and broken lines that combine in order to create four images (*Taiyin*, *Shaoyang*, *Shaoyin*, and *Taiyang*), which are present, for example, in the moon's cycle, followed by eight trigrams, forces, or principles (heaven, earth, water, fire, wind, thunder, mountain, and lake), which are associated with natural phenomena, cardinal points, energy flows in one's body, and so forth. The subsequent combination of solid or unbroken lines (Yang) and broken lines or lines with a gap in the middle (Yin) makes up the sixty-four hexagrams that describe the world and all of its phenomena, which are grouped and depicted in the *I Ching* 易经 (*Book of Changes*).[17] Used as a diagnostic tool and a divination device, these organized and variable sets of solid and broken lines, positive and negative, action and receptivity, distill into everything that exists (*You* 有) and does not exist (*Wu* 无).

Historical records indicate that instead of the sage FuXi, the Early Heaven Sequence was developed by the Chinese philosopher Shao Yong 邵雍 (1011–1077) in a quest to describe the primeval configuration of the world. In this sequence, he identifies the principles that drive permutation cycles, the polarity within which energy flows and transforms all. This diagram found great receptivity on behalf of both noblemen and the general populace at the time, given the elegance and coherence with which it described reality and its

15 "Two forces" comes from the mandarin *Liangyi* 兩儀, also interpretable as "two appearances." See James A. Ryan, "Leibniz' Binary System and Shao Yong's *Yijing*," *Philosophy East and West* 46, no. 1 (1996): 59–90, here 66.

16 "Between heaven and earth there is only *xing* (form) or *chi* (energy) [. . .] *energy* transforms in *form*, and *form* returns to *energy*, although, we are not aware of it [. . .] when *shen* (spirit) transforms into *form* from *energy*, we see it; when it returns to *energy* from *form*, we do not see it." Pang Jixing 潘吉星, *Song Yingxing Pingzhuan* 宋應星評傳 [Critical Biography of Sung Yingxing] (Nanjing: 南京大学出版社 Nanjing University Press, 1990), 338. Translation by the author.

17 The *I Ching* 易经 or "Book of Changes" is one of the important classical Chinese texts, fundamental to Taoist, Buddhist, and Confucian cosmologies that has been historically used for divination, cleromancy, and decision-making.

phenomena.[18] This a priori understanding of the world as an aggregation of binary energy states would influence another philosopher centuries later, the German polymath Gottfried Wilhelm Leibniz (1646–1716). In his quest to propose a *characteristica universalis*,[19] Leibniz aimed to develop a universal and analytical language for thought that would express any concept as a combination of elements from a basic set.[20] For that purpose, he refined binary arithmetic, which, besides being useful for practical applications, could express all ordered mathematical permutations as operations in a binary number system. This arithmetic represented for him the perfection of ex nihilo creation by God, where everything that exists is composed by permutations between one (God) and zero (nothingness).[21] Because information is more easily stored and transmitted in binary states, Leibniz's binary arithmetic is the foundation of contemporary computing systems, the fundamental basis for representing and operating the world in all of its complexity—expressed as symbols of ones and zeros, as the fullness of the sacred and the emptiness of the void, as the continuous and broken lines of Yin and Yang.

As such, Yin and Yang can be said to compose both the natural world as well as the "second nature" of digital techniques and technologies that assemble upon it. Nevertheless, to put both realms in equivalent terms would be contrary to the thoughts expressed by Shao. While Leibniz was a strong advocate of rationality as a means of understanding and deciphering divine creation, Shao considered reason to be subject to human passion. "For Shao, a necessary condition for objective[22] observation was the overcoming of the self's egocentric point of view, for the self tends to take things as real which are not real."[23] In his view, Yin and Yang are principles that can be obtained and observed, yet cannot be explained

[18] It is worth noting that this diagram derives part of its legitimacy from the fact that Shao Yong referred to it as the *Fuxi Bagua* 伏羲八卦 (*Eight trigrams of Fuxi*), implying that it was Fuxi 伏羲, a mythical emperor and world creator who supposedly lived around 2000 BCE, to whom the trigrams were revealed by heaven. This claim would lead Leibniz to consider Fuxi as a religious figure similar to Moses.

[19] "Universal characteristic" is a formal and universal language that can express mathematical, scientific, and metaphysical concepts, proposed in Leibniz's book *De arte combinatoria* ["On the art of combination"] (Leipzig, 1666).

[20] Frank J. Swetz, "Leibniz, the *Yijing*, and the Religious Conversion of the Chinese," *Mathematics Magazine* 76, no. 4 (2003): 276–91.

[21] This idea is based on the Genesis passage where creation comes from a primary void (0) and God (1) who, on the second day of creation, had made heaven and earth (10 in binary). On the seventh day, creation was complete (7 is 111 in binary, further cementing the perfection of God's creation). See Ryan, "Leibniz' Binary System and Shao Yong's *Yijing*," 69.

[22] It must be clarified that the Western notion of objectivity associated with rationality is not a completely accurate interpretation. This is a translation from the Chinese *Fanguan* 反观, which is composed of *Fan* (opposite, against, anti-) and *Guan* (observe, watch, perception), that is, "self-perception" or "self-reflection."

[23] Ryan, "Leibniz' Binary System and Shao Yong's *Yijing*," 78.

through mere forms or sensible appearances.[24] The world is in constant change, in a waning and waxing cycle, and as such, any empirical science applied to the world is nothing but an attempt to capture a momentary form,[25] and thus Shao considers it a selfish and mistaken effort to control nature through techniques. These techniques, or *Shu* 術, refer to divinatory practices or numerological arts that pretend to gain personal control over the world, measuring and quantifying it in a simplistic way. Because of this, Shao considers science to be on the same level as the esoteric arts.[26] According to him, the only way to attain an objective apprehension of things is from the point of view of things themselves, which can only be achieved through the "meditative"[27] practices of no-mind (*Wuxin* 无心) and no-idea (*Wuyi* 无意), given that it is through these practices that things do not become individualized or subjective, do not become selves, and rather allow for things to manifest by themselves.

Following this reasoning, it is possible to assert that modern computation, despite being composed of a binary architecture that might mirror a cosmos structured by Yin and Yang, is unmistakably an intellectual product of Enlightenment and modernist thought. From Shao's perspective, modern computation would be an extension of a deterministic ego, a forceful projection of self and its propensity to delimit and render the world understandable, measurable, and quantifiable. In this sense, one might say that the digital realm is a distortion of the natural world[28] on behalf of the ego. Despite the fact that the digital realm can replicate the natural world in a useful manner, it is only in the natural flow and constant transformation of real things that it becomes possible

[24] "The images and numbers of heaven can be obtained and calculated, but the working of its spirit cannot be obtained and measured. Heaven can be exhaustively explained with principle, but it cannot be exhaustively explained with forms. How can the techniques of astronomy exhaustively explain heaven by means of forms?" Shao Yong 邵雍, *Huangji jingshi shu* 皇極經世書 [Book of sublime principle which governs all things within the world], ed. Sibu beiyao [1060], ch. 8A, p. 16B. Translated and cited in Ryan, "Leibniz' Binary System and Shao Yong's *Yijing*," 80.

[25] *Xing* 形 means appearance or form. It is opposed to the elements, or the real composition of things, which are those described by the trigrams.

[26] Ryan, "Leibniz' Binary System and Shao Yong's *Yijing*," 80.

[27] These ideas are present in Buddhism and Taoism. However, this meditation is not necessarily the sitting type as it is usually characterized. This meditation can be done while walking, in a mountain or garden, while painting or doing *Taiji Quan* 太極拳 and is based on removing oneself from worry and intention, striving for resonance (*Ganying* 感应) with the surroundings.

[28] In this case, "natural" is translated as *Ziran* 自然, which is composed of *Ran* (right, occurrence). In this sense, "natural" does not imply the Western connotation of "nature" as opposed to "culture" but only refers to that which occurs by itself, the naturally unfolding, "acting and behaving according to the self without pretention, or letting things be as they are" (Hui, *The Question Concerning Technology in China*, 64), "the force of the spontaneous self-so" (Eske Møllgaard, *An Introduction to Daoist Thought: Action, Language, and Ethics in Zhuangzi* [London: Routledge, 2007], 15). This is the intended interpretation of the word "nature" throughout the text, even when not referring to Chinese thought.

to apprehend the reflection of a world that computes itself without the need for techniques or contraptions. From Shao's view, the world is already computing itself in exchanges that are set in motion by the waning and waxing cycle, in the change of seasons and the mutation of the living, in the constant recalculation of the hexagrams that compose and rule everything there is, in the binary language that encodes reality in a way that does not need to be instrumentalized or entrapped by tools that ignore the true nature of things.

Shao's and Leibniz's thoughts expose how seemingly equivalent cosmological scaffoldings can end up fundamentally different. While binary determinism reads Yin and Yang as an affirmation of a world that can be codified into discrete categories, a careful look at the underlying cosmology reveals the opposite. The nonmutually exclusive nature of Shao's worldly composition is what allows things to create, change, and interact with each other. Yin and Yang are the fundamental signals that update the self-representation of the system, yet they are not the tools used to shape it as we please. It is in the confluence, juxtaposition, and entanglement of these simultaneously contradictory yet valid statements that things can exist contrary to the deterministic, exclusive, and quantifiable states that underlie the normative conception of computation.

The theories of Shao and Leibniz do not imply that there is no place for technology in Chinese cosmology. On the contrary, many sages and philosophers have maintained that it is through tools, or *Qi* 器, that technology and the cosmos can achieve compatibility. In Chinese cosmology, *Dao* 道 does not designate the cosmos but its fundamental fabric, the structure that enables it. "The way" is the omnipresent principle that resists any description, such as "the supreme order of beings,"[29] "the original harmony of the productive force of the cosmos,"[30] or the natural course of things. More than being the substance of the material and sensible, Dao is instead an unquantifiable vector of movement,[31] a directionality principle, an order that everything follows. Qi is the technique that "must be compatible with Dao in order to attain its highest standard."[32] This relationship is complementary and inseparable: "Dao needs Qi to carry it in order to be manifested in sensible forms; Qi needs Dao in order to become

[29] Hui, *The Question Concerning Technology in China*, 65.
[30] Hui, *The Question Concerning Technology in China*, 66.
[31] This dimension of direction and movement is characteristic of several descriptions of Dao, including those of Zhuangzi when Møllgaard writes: "Things flourish and decay, but the Way, which is the *movement* of this flourishing and decay, does not itself flourish and decay. Things complete and destruct, but the Way, which is the *movement* of this completion and destruction, does not itself complete and destruct. Like Heaven, the Way is the transcendental life that gives life to the living but does not itself live and die." Møllgaard, *An Introduction to Daoist Thought*, 22, emphasis original.
[32] Hui, *The Question Concerning Technology in China*, 65, emphasis removed.

perfect (in Daoism) or sacred (in Confucianism), since Dao operates a privation of the determination of Qi."[33]

From this perspective, the "natural" and the "technological" are not antithetical, but instead, technology is a vehicle for the manifestation of the natural; at the same time, any technology can only be considered complete as long as it abides and behaves according to the principles of nature. A perfect example of this relationship is the story of PaoDing 庖丁, a very talented butcher who, after being praised for his perfect and rhythmic technique to dissect bullocks, replies, "What I love is Tao, which is much more splendid than my skill"[34]—this love is what taught him the way to thrust his knife. As he would always follow this path of intuition, he would never encounter hard tendons or bones, and thus his knife would retain its original sharpness after nineteen years of use, while most butchers would have to sharpen theirs every few months. The artisan relies more on Dao than on the tools at their disposal, and tools reach their optimal functionality when applied in synchrony with Dao. Since Dao is the basis for the movement and development of natural things, a tool that acts in accordance with Dao is a tool that modifies its environment by following the environment's own logic. This complementarity principle between Dao and Qi has had several interpretations across Chinese history, but in general terms, it characterized the technological development of classical China. However, in the context of modern technology and information systems, or contemporary cybernetics, this harmonic relationship between Qi and Dao has many insights to offer.

Data stored and processed in computers can be seen as an exosomatization[35] of memory, as the exteriorization of abstract thought and sign remembrance from human bodies, and as information processed and archived outside of human brains. Data is a representation of the world, gathered through a sensory layer of artificial nerve endings that translate pressure, light, temperature, and any other quantifiable phenomena into a series of binary symbols. More than being collected, data is produced by particular devices and people with a premeditated goal in mind: a task to fulfill, a form to complete, a description to provide. Far from being neutral, data becomes the organization of a series of signs that can be communicated, recorded, or decoded; it constitutes a common language between

[33] Hui, *The Question Concerning Technology in China*, 101, emphasis removed.

[34] Zhuangzi 庄子, *Zhuangzi: Library of Chinese Classics; Chinese-English*, trans. Wang Rongpei, Qin Xuqing, and Sun Yongchang (Changsha: Hunan People's Publishing House, 1999), 43.

[35] Exosomatization refers to the production of objects that are analogous to our internal organs but built and used outside our organic bodies, such as video cameras and our eyes. These external organs are modeled and maintained by human activity. See Maël Montévil, Bernard Stiegler, Giuseppe Longo, Ana Soto, and Carlos Sonnenschein, "Anthropocene, Exosomatization and Negentropy," in *On Transition: In Response to António Guterres* (Internation, 2020), https://internation.world.

different computational systems. As a surrogate of the world, data provides a symbolic representation that is nevertheless very distant from the systems producing it. While it is important to acknowledge the limits of representation, data nonetheless have a capacity that mere human reason cannot achieve: the ability to place a set of nonhuman entities at the center of measurement and interpretation, which can disclose the agency of unaccounted third parties in natural phenomena. Worldviews can alter or obscure readings of data, loudly imposing a noise that renders inaudible the voices produced by events, objects, and living entities. Even so, whenever airborne pollution levels decide which machines operate and which shall stop, or when indicator species give snapshots of the general composition of entire ecosystems and thus inform preservation policy, living and nonliving entities can take decisions that contravene anthropocentric logic or intuition. The free flow of data and its computation could ensure that a system, which both accounts for and represents itself, carries out its functions without encountering obstructions or "hard tendons."

With this new way of enframing reality, we can find brief moments where nonhuman components also participate in the self-representation of the system, subverting the anthropocentric logic of systems based on individuated humans. Within this framework, computational systems, originally built with anthropocentric goals in mind, can be repurposed to incorporate an aggregate plurality of entities. In this proposed scenario, data can be interpreted as Qi, as the tool that reaches its highest standard, a technique that perfects itself as it materializes the path of least resistance and exhaustion, and through which the Dao of material and environmental conditions can become manifest. In this case, Qi encompasses both the "immateriality" of computational protocols, signals, and memory, as well as their physical terminals, such as backhoe excavators, dam floodgates, oil drilling machinery, and so on.

What the extractive and resource-intensive version of computation that is currently hegemonic requires is the inclusion of the compatibility principle between Qi and Dao. This entails designing tools to be used in accordance with Dao and conceiving of technology that attains its fullest realization when in harmony with planetary conditions and environmental limitations. Insofar as the computational systems and extractivist cycles of capitalism continue to ignore this model of communion between technique and cosmos, they reinforce a culture of technological "development" that continues to break while accelerating climate change, eroding prematurely just like the knives used by those insensitive to Dao.

Transistor-Less Informatics of Amerindian Thought

The harmony between technology and the composition of the world is analogically present in foundational views of pre-Columbian cultures in the Americas. Several cultures in the pan-Amazonian region have been able to discover that memory, communication, and representation are not characteristics unique to the human being but rather the commoning element of the multiple beings that populate the cosmos. Just as cosmological thought across China cannot be indisputably generalized as "unified," trying to find a common thought that can identify the hundreds of language groups and creation myths of Indigenous tribes throughout the Colombian and pan-Amazonian territories increases the problem by several orders of magnitude. This issue can only be mitigated by treating any comparison as an analogy and any attempt at generalization as a mere analysis from an outside perspective, carried out with conceptual tools that are used in an attempt to fill a gap in service of bringing several traditions of thought closer together; generalization as an adaptation made in order to attain a certain degree of understanding.

Amazonian Amerindian thought can be characterized as "systemic" (or behaving in a system-like manner) for two main reasons. First, its notion of continuity shared between human and nonhuman living beings on a spiritual level: Plants, animals, and humans are spiritually the same, and the only thing that differentiates them is their appearance. "The common condition to humans and animals is not animality but humanity."[36] When all living entities host their own version of humanity, "everything is human, [and] the human becomes a wholly other thing."[37] The ability to use language, reason, and technology is understood not as an essential privilege that grants humans absolute dominion over their environment but as a mere means of representation of the world that implies an intrinsic connection between humans and the thousands of beings who share the forest. Representation, as Eduardo Kohn demonstrates, is neither unique nor exceptional but rather the fundamental characteristics that unite living beings as such. In his work with the Ávila and Runa communities in the Ecuadorian Amazon, Kohn argues for leaving behind the idea that humanity is the only species capable of semiotic meaning.[38] "What we share with nonhuman living

[36] Eduardo Viveiros de Castro, *Cannibal Metaphysics*, trans. Peter Skafish (2009; Minneapolis: Univocal, 2014), 68, emphasis removed.

[37] Viveiros de Castro, *Cannibal Metaphysics*, 63.

[38] Eduardo Kohn, *How Forests Think: Towards an Anthropology Beyond the Human* (Berkeley: University of California Press, 2013).

creatures, then, is not our embodiment, as certain strains of phenomenological approaches would hold, but the fact that we all live with and through signs."[39] People, animals, and plants are conceived according to a similar ontological category: They all share the same attributes of mortality, social and ceremonial life, intention, and intelligence.[40] This principle allows the Barasana tribe to interpret the melody of swallows and sparrows as chants that indicate the proper time to work the *chagra*[41]—or the fall of certain fruits, the change of swimming patterns of some fish, and the type of flutter of particular birds as indices that allow several species to foresee the mating season of leaf-cutter ants.[42] Superior pattern recognition is said to be one of the distinctive traits of human sapience. Yet, in a forest, pattern recognition takes place across multiple species. The recognition that all species are capable of producing semiotic systems is essential for a worldview that assumes all living entities as equals, with humans being one among many in this symphony of communication.

The second systemic characteristic of Amerindian thought is the recognition of fractal analogical structures that are to be found in processes at micro- and macro-cosmological levels.[43] The structure of the cosmos repeats itself at the level of the universe, territory, house, and body. Tukano, Tatuyo, and Itana tribes understand the *maloka*[44] as a projection of all life on the planet, a structure[45] that mirrors the design and distribution of the cosmos at a smaller scale. This is the reason behind the precise orientation and proportion of the maloka building, as it must contain the world symmetrically, since only through this symmetry

39 Kohn, *How Forests Think*, 9.

40 Philippe Descola, *Beyond Nature and Culture*, trans. Janet Lloyd (2005; Chicago: The University of Chicago Press, 2013), 8.

41 "[A] small roadrunner lives in the chagra, a swallow and a hawk guard the crops and the forest. That hawk guards the Worlds Above, and that is why he sings at 4:30 in the morning: ¡Oah, Oah, Oah!, 'good morning, how is your day my friend, today we are not going to work, today it will not rain, calm down, in order, sharpen and take your machete, women let's clean the chagra, today is a good day.'" Bárbara Santos, *Curación como tecnología: Basado en entrevistas a sabedores de la Amazonía* (Bogotá: Alcaldía de Bogotá, 2019), 52–53. Translation by the author.

42 Eduardo Kohn, "Natural Engagements and Ecological Aesthetics: Among the Ávila Runa of Amazonian Ecuador" (PhD diss., University of Wisconsin-Madison, 2002), 104–5.

43 "Akin to our scientific thought, the thought of Eastern Tukano ethnicities is grounded on a series of analogies, what we would call fractal analogies in mathematical terms. They use a series of structures and processes at different scales that contain identical organization. And these structures are the human body, the house, the territory, and the cosmos." Santos, *Curación como tecnología*, 79. Translation by the author.

44 A *maloka* is a multifamily dwelling structure that houses several families, generations, and animals. Its construction and usage are not mundane; during rituals they become mediums through which to access ancient knowledge, as well as the knowledge available in the forest.

45 This denomination as "artificial" corresponds to the notion of anything produced by humans as external to and in opposition to the "natural." However, this idea has no place in Amerindian cosmologies, given that even their labor, such as agriculture or construction, is constantly performed by spirits (and therefore, natural actions). Also, tools like the axe are part of creation myths, such as those of the Yucuna people.

can the building be used as a medium through which to think and connect with all beings during rituals. The fractal representation of the universe contained in objects of sacred and daily use is shared across various ethnicities. It can be found in the biconical Yucuna structure (meant to support sacred beverages during rituals) that represents a universe with identical structure,[46] in the Kogi temples and looms that are oriented according to the trajectory of the sun,[47] and in the spiral weaving of traditional Misak hats that replicate the flux and unfurling of time.[48] Harmony and proportion between artifact and cosmos are not only important as an aesthetic principle, they are fundamental factors that allow a particular device to fulfill its function and enable its ability to access the knowledge and thought processes of the world. Tools are isomorphic with the cosmos because they are contained within and produced by it, while humans are merely their incidental manufacturers.

The link between Indigenous thought and informatics becomes evident in the interviews published in *Curación como tecnología* (Healing as technology) by Bárbara Santos. In the book, Tarsicio Vanegas from the Itana tribe compares digital technologies with healing practices: "A feather is a very special sacred material; for the outside world is a simple feather, but for us is like a folder inside a computer, the feather provides the light that illuminates my thought, that guides me where I want to go."[49] In another analogy, he compares the techniques and sciences related to the study and diets followed by a *kumu* (knower, healer, or spiritual leader) in order to heal or prevent disease, to the process

[46] "The world is also represented symbolically, the biconical support known in Yucuna as *umichiripucueste* is an artifact that corresponds to what they now called the black hole in physics; it is a balanced piece that is used to keep balance in the use of nature. The [thought] bench is used to manage the world in interaction with all the other benches; this means that we are the whole and we are a part, and by managing each part well, the whole is managed well." Carlos Rodríguez, II Congreso colombiano de áreas protegidas, *Áreas Protegidas: Territorios para la Vida y la Paz. Áreas protegidas y visiones interculturales: Conociendo nuestra biodiversidad.* Bogotá: Parques Nacionales Naturales de Colombia, 2015. Translation by the author.

[47] "The light ray goes from one stove to the other in between solstices, but every day advances very slowly from north to south, according to the inclination of the sun. This movement has led to a concept of great importance for kogi thought, namely, the idea that the *sun is a big weaver*. They envision the earth as a giant loom around which the sun moves in a spiraled and swaying motion." Gerardo Reichel-Dolmatoff, "Templos Kogi: Introducción al simbolismo y a la astronomía del espacio sagrado," *Revista Colombiana de Antropología* 19 (July 1975): 199–245, here 221. Translation by the author, emphasis original.

[48] "The *kuarimpoto* or *tampal kuari* is the traditional hat of the misak people. Is made from a band that is woven in spiral shape, synthesizing their notion of spiraled time and space. For the misak, history repeats itself at the same time it changes, *turning itself around*. Likewise, people are tied by an *ancestry thread*, one that *unfurls* circularly, tracing the path of everyone's life." Wall text for *kuarimpoto*, Cod. ICANH 538-E—Instituto Colombiano de Antropología e Historia, in *Ser Territorio (Being Territory)*, part of the permanent collection exhibition of Museo Nacional de Colombia, Bogotá, 2019–ongoing. Translation by the author, emphasis original.

[49] Santos, *Curación como tecnología*, 56–57. Translation by the author.

by which a programmer creates an antivirus.[50] Perhaps only now that Western technology has developed distributed systems that refine and specialize through evolution can the Western mindset be closer to understanding the similitude between technical objects and natural processes and, as a result, get closer to understanding some of the fundamental ideas of non-Western cosmologies. In this sense, Vanegas's metaphors are not mere narrative resources that allow him to explain his thoughts to a foreign public.

From this perspective, one can acknowledge how minerals fulfill identical roles in both binary and Amazonian computation. Gold and quartz are crucial components in measuring time and conducting electrical impulses as data in computers and are found in every recording device. These minerals are also crucial to the preservation of memory in Amazonian cultures and fundamental to the process of healing the world. For this healing to work, the minerals must remain buried in their place of origin without exploitation or refinement, so that the healer can activate the energy they emanate. Davi Kopenawa from the Yanomami ethnicity says, "the gold makes *xawara*. *Xawara* is a disease. Gold creates a disease to harm people. Omama [Yanomami creator deity] considered that gold should remain kept under the ground, beneath the earth."[51] These underground mineral concentrations determine sacred places within the forest, hubs from which energy and knowledge can be drawn. Instead of extracting them physically, their properties are accessed through "sacred materials," the fruits of the sacred ground: tobacco, coca leaf, cassava, nettle, and breast milk.[52] These are activated by generational knowledge, by the tales, stories, and invocations that connect living practices to a place. It is through living layers of knowledge that the materials become interfaces between the energy exuded from the sacred places and the wise who can curate them.

An analogous technical process to how the kumu identifies the appropriate *yagé* (ayahuasca) visions that guide healing practices would be the functional similarity of a TCP/IP protocol. This protocol, widely used on the Internet, ensures that if at any level (application, transport, network, data link, or physical) there is an authentication mismatch between the sender and receiver of information, the data packet will be discarded or otherwise delivered when

[50] Santos, *Curación como tecnología*, 57.

[51] Davi Kopenawa, interview during a book presentation of *A queda do céu* at the Festa Literária Internacional de Paraty in Rio de Janeiro, 2014. Author's notes and translation. Davi Kopenawa and Bruce Albert, *A queda do céu: Palavras de um xamã yanomami*, trans. Beatriz Perrone-Moisés (Rio de Janeiro: Companhia das Letras, 2015).

[52] Santos, *Curación como tecnología*, 82.

every data level match is fulfilled. When explaining the healing process, León Muipu mentions:

> [. . .] I had a vision of yagé, and that vision is what later appears again when healing. It gives you the explanation, it tells you where the diseases originate, how you should speak, and how your soul or you as a living person makes that flow in a spiritual world linking with the other person, so that you can prevent or heal the disease.[53]

To transcribe the above case in the language of a TCP/IP protocol, the protocol becomes Muipu himself, whose work authenticates the correct signals form the yagé server, matching it with the right procedure in the face of the disease, inhabiting the healing body, and guiding the disease to its conclusion with himself as vehicle. Vanegas expands the understanding of visions caused by yagé as the quintessential interface between the kumu and the vast knowledge available in the forest:

> With knowledge achieved through yagé, the kumu goes around the planet, turns around, turns around knowing what is happening and in a single vision everything is recorded, he never forgets it. With that vision, the kumu analyzes through symbols and data to determine what is that information useful for, if for healing, for protecting the children, for dancing. And you must close it in the end, to prevent viruses from coming in.[54]

This knowledge points to an understanding of technology that does not necessarily consist of artifacts, that does not lie in the recipients who are treated by sacred plants nor in the materiality of the tobacco, but in an immaterial technology that is manifested during the act or performance of healing itself. Healing the body implies healing the forest, and neither of these actions can be accomplished without the technical object being treated as a living object as well. The forest is both home, mother, and data center; a network of networks, one that contains a myriad of terminals and users, that reads and writes information as part of its metabolism, that apprehends and represents itself with each interaction, that seeks its own continuity and homeostasis. From this perspective, one might begin to understand how traditional Amerindian thought is intrinsically cybernetic. It is telematic without the need for transistors or antennas. It requires energy in the shape of sacred materials and is only enabled by the links, servers, protocols,

[53] Jesús León Muipu, interviewed in Santos, 38. Translation by the author.
[54] Tarsicio Vanegas, interviewed in Santos, 59. Translation by the author, emphasis removed.

interfaces, and terminals that are provided by the rivers, animals, plants, and metabolic cycles of the forest.

The lessons of Amerindian thought are harder to translate into contemporary information systems than the principles of Qi and Dao. Some, such as Roy Ascott, have advocated for the use of the telematic abilities of yagé in cybernetics, for the merger of vegetal and virtual reality in the so-called "moist environment, located at the convergence of the digital, biological and spiritual, [. . .]."[55] Others propose the development of wetware computation, biocomputation, or phytoindication in cybernetic systems to include forests either as signaling nodes or as processing units. However, to interpret the metaphors presented here as design principles for revamped extractive technologies would be contrary to the reality of their practice. Amazonian telepathy only works when alive and situated, while "Western telepathy, at its core, is an extractive telepathy."[56] Plugging the forest's interests into the data and geoengineering stack could result in the protection and preservation of the forest while it could also result in extractivist encroachment in the Amazon. In the eyes of many, plugging in the forest could not address the fundamental design problem that extractivist infrastructure is predicated on: The industrial complex cannot unmake the problems brought about by industrial design itself. As pointed out by Julian Dupont, transforming the forest into an extension of sensing technology cannot resolve the problem that an already objectified infrastructure presents.[57] Instead of merely instrumentalizing Amerindian knowledge in the Western context, it should be understood as a parallel operating system that comes from, runs on, and is also part of the world. These cosmotechnical models offer an alternative path for mediating thoughts and the environment with available tools, all of which will be crucial for planetary ecological regeneration.

[55] Roy Ascott (ed.), "Edge-Life: technoetic structures and moist media," in *Art, Technology, Consciousness: mind@large* (Bristol: Intellect Books, 2000), 4.

[56] Juan Pablo Pacheco Bejarano, "Telepatía Sin Internet," *Rudimentos* no. 1 (2021): 60–77.

[57] Julian Dupont, "Convertirse en Jaguar Como la Nube Automatizada" (master's thesis, Universidad Autónoma Indígena Intercultural, Bogotá, 2021), 63.

9

A Conversation on *Art and Cosmotechnics*[1]

Yuk Hui and Brian Kuan Wood

Brian Kuan Wood: In turning its focus to art and aesthetics, your book *Art and Cosmotechnics*[2] builds upon some concepts that you have previously elaborated on. The most obvious of these concepts—which is also in the title—was the focus of your earlier book, *The Question Concerning Technology in China: An Essay in Cosmotechnics*.[3] Let us begin by situating ourselves around the meaning of "cosmotechnics," also to clarify it against certain misunderstandings that may have arisen in the time since you wrote *The Question Concerning Technology in China*, as even the most necessary critiques of Western-dominated political or technological paradigms can become vulnerable to reactionary tendencies or wrongful appropriations.

Yuk Hui: I have to say that *Art and Cosmotechnics* is quite a strange book because it deals with three different kinds of logic that seem at first to be unrelated: *tragist* thought, Daoist thought, and cybernetic thought. I do not think there has been any work that has tried to associate these three. *Art and Cosmotechnics* is divided into three parts, and I should explain why it is structured that way. First, let me respond to your question by explaining why I had to coin this concept of cosmotechnics before going into what it really is and the difficulty of elaborating on such concepts. It is something quite personal in my studies of philosophy—I studied computer science first, then philosophy for many years with a focus on the question of technology. And after about ten years, I found that all I have studied is supposed to be universal. At the same time, the philosophy of

[1] This conversation was originally published in two parts on *e-flux Journal*, nos. 124–25 (February–March 2022) and has been expanded and edited.

[2] Yuk Hui, *Art and Cosmotechnics* (Brooklyn: e-flux Books; Minneapolis: University of Minnesota Press, 2021).

[3] Yuk Hui, *The Question Concerning Technology in China: An Essay in Cosmotechnics* (Cambridge, MA: MIT Press, 2016).

technology that I was studying was actually very European, sometimes a little bit American. So at a certain moment, I asked myself: What does it mean to talk about technology in cultures outside of Europe? We know that there must be technology outside of Europe. It would be a *bêtise* to deny it.

We know that, according to historians, Greek technology came from the Near East and then stayed in Europe through the Greco-Roman period until technology became an object of hatred with the spread of Christianity, until the Renaissance and later on, and then there was a huge change on the continent as European modernity began to emerge. To go any further into these origins would require a lot of discussion with classicists and historians, but the main point is that I was quite amazed by the lack of understanding of the concept of technology itself. The whole discourse is very much structured around European history and European philosophy, as if the question concerning technology is only European. I am not saying that is a bad thing, because the Western discourse does offer some important insights. It made me very curious about how we could articulate the question of technology outside of Europe. We immediately encounter a huge obstacle, because we have been told from a young age that science and technology are universal, like mathematics. In a way, we have already accepted the idea that technology is universal, science is universal, logic is universal, mathematics is universal, and so on. Even in academic disciplines, there seems to be a lack of reflection on what this universality is and what it implies.

Let me give a few examples. In the philosophy of technology, especially in analytic philosophy, all claims tend to be universal. In the continental philosophy of technology, for example, Martin Heidegger has been an influential figure. His 1949 talk[4] in Bremen (published later as "The Question Concerning Technology") basically suggests that if you want to understand what technology is, or what he called the essence of technology, then we can understand it in two parts. One part is what the Greeks called *technē*, which Heidegger associated with *poiesis*, with bringing something forth (*hervorbringen* in German). And this poetic realization is the unconcealment of Being. Here the question of Being enters his discourse as something closely related to not only the concept of technology but also to the concept of art, which he drafted around 1935 and 1936 and later

[4] Martin Heidegger, "Die Frage nach der Technik," in *Vorträge und Aufsätze* (Pfullingen: Günther Neske, 1954), 9–40; Heidegger, "The Question Concerning Technology," trans. William Lovitt, in *The Question Concerning Technology, and Other Essays* (New York: Garland, 1977), 3–35.

published as "The Origin of the Work of Art."[5] The second part that Heidegger tries to show is that modern technology—which for him came after the scientific revolution and was actualized at the end of the eighteenth and the beginning of the nineteenth century—no longer shares the same essence as technē or poiesis. It has rather become what he called "enframing" (*Gestell*), meaning that everything could be treated as what he called a "standing reserve" (*Bestand*), a resource to be ordered and exploited, from rivers to atoms.

Heidegger's discourse on the difference between Greek technē and modern technology was widely accepted not only in continental Europe but also in East Asia—at least in Japan, China, and Korea. In non-European cultures, as far as I can tell, Heidegger's thesis was widely accepted because it seemed to mirror the tension between tradition and modernity. The Chinese or the Japanese, for example, could associate technē with their own tradition and modern technology with modernization or Westernization. So you can immediately see the conflict, but there is also a blind spot concerning the essence of technology as posited by Heidegger. For example, can we identify the unconcealment of Being in the Greek technē also in the East, for instance in China and Japan? However, the question of Being, as the founder of the Kyoto School, Kitarō Nishida, famously claimed, was not the core question. In Western Europe, we know that ontology was considered the first philosophy, and if there is no question of Being in the East, it follows that there is no unconcealment of Being in the Eastern understanding of technology.

Though Heidegger's thesis has been widely adopted, this blind spot remains. People tend to equate Greek technē with Chinese, Japanese, or Indian technology without really looking into the meaning of technology that was already present not only in Heidegger's discourse but also in the history of technology. For example, the great sinologist Joseph Needham, who published more than twenty volumes of *Science and Civilisation in China*,[6] tried to show that China's science and technology were quite advanced before the sixteenth century. And his haunting question was: Why did modern science and technology not happen in China or India but only in Europe?

Some historians, following Needham, have tried to show that a certain technology—say, papermaking in the second century in China—was more advanced than in Europe. Their method compares one technology with similar

[5] Martin Heidegger, "The Origin of the Work of Art" ["Der Ursprung des Kunstwerkes," 1950], in *Poetry, Language, Thought*, trans. Albert Hofstadter (New York: Harper & Row, 1971), 15–88.

[6] Joseph Needham, et al., *Science and Civilisation in China*, vols. 1–7 (Cambridge: Cambridge University Press, 1954–2015).

technologies in other regions without considering what Needham himself warned against, which is that all these technologies, even if they involve similar materials and similar products, are actually based on different epistemological and ontological assumptions. Even when technologies can be lumped into the same category, there are still tremendous differences between them. Yet by simply comparing which is more advanced than the other, we universalize technology by default. We assume that there is only one way of understanding technology.

In the anthropology of technology, technology has been understood as essential to the process of harmonization: the externalization of memory, the liberation of organs, and so on. This is only a universal dimension of technology. I introduced what I called "the antinomy of the universality of technology" with the antithesis that where technology is not universal, it is conditioned—motivated and constrained—by a certain cosmology, that is, its locality. In an antinomy, if you take each thesis separately, they are all correct. But when you bring them together, you see a contradiction. The resolution of this antinomy leads to what I call "cosmotechnics," where all technologies are actually cosmologically constrained and motivated. Cosmology here is not merely theoretical but always embedded and embodied in the invention, development, and use of technologies. That is what I argued in *The Question Concerning Technology in China*. You can see from the title that it responds to Heidegger's 1949 lecture, "The Question Concerning Technology." I tried to reinterpret the concept of technology by coining a different concept, cosmotechnics, in order to call for a new interpretation of technology by situating it historically, cosmologically, and locally. As for your question about reactionary or neo-reactionary politics, it is an important one, and we will come back to it later.

BKW: I wonder if we could also clarify your interest in cybernetics here. Your following book, *Recursivity and Contingency*,[7] deals with the significance of cybernetics as a world-historical, political, and philosophical rupture in Western thinking. This goes back to the physics of information, as Norbert Wiener defined it in the 1940s, where feedback, circularity, and recursion, as you explain, dissolve a certain separation in Western thinking between organic life and machinic systems. You eloquently describe this in the book as a situation where machines are "no longer simply tools or instruments but rather gigantic organisms in which we live."[8] So this act of enframing also shifts into a kind of

[7] Yuk Hui, *Recursivity and Contingency* (London: Rowman & Littlefield, 2019).
[8] Hui, *Recursivity and Contingency*, 71.

cybernetic body that is both organic and machinic. Could you further describe the conditions of this merger and perhaps also the cosmic, or cosmological, implications of living inside such gigantic organic machines?

YH: Here I can simply continue from where I left off. After I finished *The Question Concerning Technology in China*, I felt that something was still incomplete. I was still haunted by Needham. I thought that, although I had responded to his question, this book had missed something significant. I have followed Needham in calling Chinese science and technology organic, but it is nevertheless dogmatic, and it will be regrettable if I cannot elaborate on its precise meaning. There was suddenly an urgency for me to work on this question. In the twentieth century, if you asked a sinologist, or even a Chinese philosopher, about the difference between Chinese thought and European or Western thought, or about the difference between Chinese technology and Western technology, you would often hear that Chinese thought is organic, while Western thought is machinic. To some extent, Needham is responsible for this really problematic answer because he tries to say this in his books. For Needham, Western philosophy became organic only after Gottfried Wilhelm Leibniz.

In the second volume of *Science and Civilisation in China*, Needham begins with Leibniz and goes on to name Baruch de Spinoza, Immanuel Kant, Georg Wilhelm Hegel, Friedrich Schelling, and Johann Fichte, down to Alfred North Whitehead and Wiener, as thinkers of organicism. Of course, some of his contemporaries, such as John Haldane, Jan Smut, and Conwy Lloyd Morgan, were also associated with holism and organicism. Needham claims that, while Western philosophy became organic only after Leibniz, Chinese philosophy was organic from the very beginning and never moved from mechanism to organism as in the West. He continues by saying that Leibniz may have been influenced by his correspondence with a Jesuit in China, Father Bouvet, who told Leibniz about the Neo-Confucian Zhu Xi, one of the most important Neo-Confucians of the twelfth century. This way of formulating the difference between Chinese and Western thought is problematic in many ways. First, it can reintroduce an Orientalist viewpoint, and second, it may not help us to qualify what Chinese thought or Chinese technology actually is. And there is an urgent need to understand how to articulate Chinese science and technology without resorting to organicism or holism. This is the problem we face today, especially when we characterize Chinese medicine as holistic, when holism is actually a German invention, as Volker Scheid, a historian and practitioner of traditional Chinese medicine, has shown.

This is why I wanted to work on the concept of the organic and show how it is fundamental to Western modernity. It is probably more fundamental to the West than to China, because mechanism and organism were never central themes in China. The Chinese were never aware of them, just as they were never aware of tragedy in the Greek sense. Even today, we think of a tragedy mainly as a sad story, but that is not what Greek tragedy is. In this sense, *Recursivity and Contingency* was partly a continuation of *The Question Concerning Technology in China*. In the preface to *Recursivity and Contingency*, I wrote that the book may as well have been called "The Specter of Joseph Needham." I used the two concepts of recursivity and contingency to reconstruct the history of modern Western philosophy, because I believe that the dichotomy or opposition between mechanism and organism was one of the most significant philosophical developments in eighteenth-century Europe.

We know that modern European thought—what we call early modernity, associated with thinkers like René Descartes—was very much dominated by mechanistic thinking. Descartes was able to compare the human body to a church organ, by articulating how the wind relates to the breath, how the pump of the organ relates to the heart, and so on. And this mechanism was strongly challenged in the eighteenth century with the rise of the concept of the organism. Let us not forget that biology was not a scientific discipline at that point, and it would not become one until the beginning of the nineteenth century. The rise of the concept of organism was significant enough that we can find it in the works of Spinoza, Kant, and, of course, already in Leibniz, as well as in the seventeenth century with the Cambridge Platonists. In *Recursivity and Contingency*, I argued that Kant's *Critique of Judgment*[9] played a significant role in imposing an organic condition of philosophizing toward the end of the eighteenth century, where philosophy, in order to exist, could not avoid becoming or being organic. I tried to show how the concept of the organic became a paradigm of thought, from all the idealists who followed Kant—Fichte, Schelling, Hegel, and so on—to the twentieth century with Henri Bergson, Whitehead, and, of course, Needham, whose turn to organicism was informed by his training as a biochemist.

To some extent, I think it is fair to say that toward the end of the eighteenth century, Kant imposed an organic condition of philosophizing, based on the irreducibility between organism and mechanism. However, in *Recursivity and Contingency*, I tried to show that this situation changed greatly in the twentieth

[9] Immanuel Kant, *Critique of Judgment*, trans. James Creed Meredith (1790; Oxford: Oxford University Press, 2007).

century, especially after the rise of cybernetics. When people talk about cybernetics, they may naively think of control and surveillance, but the basic claim of cybernetics is far more important and fundamental for us today. In the first chapter of Wiener's 1948 book, "Newtonian and Bergsonian Time,"[10] he claims, simply put, that cybernetics has overcome the dichotomy between mechanism and vitalism. The strawmen of vitalism are, for example, Bergson, Haldane, and Hans Driesch, who propose concepts such as *élan vital* or *entelechy* to describe a vital force in the organism. Wiener began by pitting vitalism against mechanism, using Isaac Newton—who, of course, is a mechanist not in the sense of Descartes but in his approach to linear causality—in order to show that cybernetic machines have overcome the opposition between vitalism and mechanism by being based on nonlinear causality. While still being mechanical, cybernetic machines are capable of assimilating the behavior of an organism. Hans Jonas, a student of Heidegger, claims in his book *The Phenomenon of Life: Toward a Philosophical Biology*[11] that cybernetics marks the first time since Aristotle that we find a unifying logic, which means that cybernetics has basically overcome dualism. If you were to criticize cybernetic machines as dualistic today, it would already be a conceptual mistake.

Since the first half of the twentieth century, our machines are no longer like those in Descartes's time or Karl Marx's time: mechanical machines, characterized by linear causality and repetition. Jonas was very critical of Wiener's cybernetics in *The Phenomenon of Life*, but he never underestimated it and was sure to point out its philosophical significance. So, from cybernetics on, we see a new paradigm of machines, which I called the becoming-organic of machines, and this becoming-organic of the machine is fundamental to the work of Gilbert Simondon, as we find in his *On the Mode of Existence of Technical Objects*.[12] If we follow this reading, perhaps we can say that cybernetics has completed Kant's organic condition of philosophizing. This is also how I interpret Heidegger's identification of cybernetics as the end of Western philosophy and metaphysics. So if, since the end of the eighteenth century, we have not only been living among a new type of machine but have also been confronted with a new condition of philosophizing after Kant (think of the publication of *Critique of Judgment* in 1790), after Whitehead, after cybernetics, and also after Donna

[10] Norbert Wiener, "Newtonian and Bergsonian Time," in *Cybernetics: Or Control and Communication in the Animal and the Machine* (Cambridge, MA: MIT Press, 2019), 43–64.

[11] Hans Jonas, *The Phenomenon of Life: Toward a Philosophical Biology* (New York: Dell Publishing, 1966).

[12] Gilbert Simondon, *On the Mode of Existence of Technical Objects*, trans. Cécile Malaspina and John Rogove (1958; Minneapolis: Univocal, 2016).

Haraway, then today, we have to rearticulate the conditions of philosophizing. For me, this means that we cannot simply go back to organic nature or a naive discourse on multispecies relations. *Recursivity and Contingency* was an attempt to historically articulate and elaborate on this new condition of philosophizing. *Art and Cosmotechnics* is a continued pursuit of this spirit.

BKW: In *Art and Cosmotechnics,* you outline a new way of thinking about art and aesthetics that follows from this, but your approach runs contrary to many vulgar approaches to art and technology that simply normalize new technological platforms. Often, this is done in the name of expanding the limits of art or aesthetics by inserting computers, social media, or NFTs into a traditional artistic setting, usually without questioning the limits of these platforms. Instead of questioning their boundaries, conservative artistic settings seem compelled to celebrate the oppressive or deterministic tendencies of technology, as in the Ballardian scenarios of *Black Mirror.* In *Art and Cosmotechnics,* you advocate for something different: a return to certain fundamentals of aesthetics, an engagement with aesthetics as a form of logic that can be said to precede, or even include, our current paradigm of technology since it is actually larger than technology. Could you describe this unusual technique, which begins in the book with a return to Greek tragedy, or what you term in the book "tragist" logic, and discuss your reasoning for it?

YH: After I finished *Recursivity and Contingency*, again, I was haunted. As I said earlier, I tried to use the two concepts of recursivity and contingency in the book to characterize this movement of thought from Kant to the twentieth century. After the book was published, Augustin Berque, a specialist on Japan who has done a lot of work on landscape and logic in East Asia, emailed me to say that he found the book very interesting but was astonished that I did not talk about the profound notions of recursivity and contingency in East Asia. At the same time, many have claimed that cybernetics is very close to Chinese thought and even that cybernetics actually originated in China (this has never been proven) because Wiener was a visiting professor at Tsinghua University for a year in the early 1930s. He did make some remarks that Chinese writing was significant to his thinking about cybernetics, though it is not clear what he was referring to. Like the discourse on Chinese holism and organicism, this myth about cybernetics and Chinese thought is quite fascinating but suspicious.

If I reject this claim, however, I have to explain the difference between cybernetics and Chinese thought. If Chinese technology is not cybernetics in Wiener's sense, then how can we articulate this? Without this distinction, everything sinks into the dark night where all cows are black, as Hegel writes in the preface to his *Phenomenology of Spirit*, when he criticizes Schelling's concept of the absolute.[13] For me, philosophy is all about elaboration, and my task in *Art and Cosmotechnics* was to elaborate different forms of recursive thinking and to show the relations, or possible relations, between these differences.

There are a lot of efforts going on today to merge art and technology, and there will certainly be more from governments, universities, and the private sector. Art and technology in the past few decades have been really fascinated with live experience—interaction, immersion, and so on—but many of the works you encounter are actually entertainment, which is not a negative word so much as it is a matter of fact. This means that the relation between art and technology has yet to be determined, and this relation is where the book sets off from. Much has happened in the last century since Walter Benjamin's *The Work of Art in the Age of Its Technical Reproducibility*, written in 1935,[14] and Heidegger's "The Origin of the Work of Art," written in 1936, both of which deal with the relation between art and technology. Benjamin wrote that it is futile to ask whether photography and cinema are art or not, and the past eighty years have shown that Benjamin captured the spirit of the avant-garde and anticipated the revolution that would take place in art. And he wrote that it is more important to think about how technology has changed the concept of art. I think this is the main thesis he put forward in *The Work of Art in the Age of Its Technical Reproducibility*. He made this claim as a good Marxist materialist, showing that material conditions determine the concept, not that the concept determines reality. But he also showed that the concept of art has to be expanded according to a technological condition. Today, photography and cinema are already widely accepted and institutionalized within the field of art.

My approach is almost the opposite. I ask: If since Benjamin—or even since the avant-garde before Benjamin—we have been trying to ask how technology changes the concept of art, as you find in Marcel Duchamp, can we now turn

[13] Georg Wilhelm Friedrich Hegel, *The Phenomenology of Spirit*, trans. Michael Inwood (1807; Oxford: Oxford University Press, 2018), 10. Hegel wrote to Schelling from Bamberg on May 1, 1807, explaining that his criticism is not directed against him but rather against his superficial readers; as Hegel says, "In the Preface you will not find that I have been too hard on the shallowness that makes so much mischief with your forms in particular and degrades your science into a bare formalism." Schelling asked Hegel to clarify this in the next edition, but Hegel never replied. See Jason Wirth, *The Conspiracy of Life: Meditations on Schelling and His Time* (Albany: State University of New York, 2003), 21.

[14] Walter Benjamin, "The Work of Art in the Age of Its Technical Reproducibility [First Version]," trans. Michael Jennings, *Grey Room* 39 (Spring 2010): 11–37.

the question around and ask how art can transform technology? I think this is an important question, not only in a conceptual sense but also in a diplomatic sense. If you were talking to an engineer about an art project, how would you talk to them? Do you simply want to import this or that technology to create some kind of new experience? Or do you want to influence how technology is made, how technology is conceived, how technology should be developed? I think we can also turn around the question further and ask: How can art contribute to the imagination of technological development?

Technology comes with not only huge opportunities but also huge potential catastrophes. If you look at climate change, the catastrophe is already there; as Heidegger said about *Gestell*, the essence of modern technology is to consider everything as a standing reserve, as a resource to be ordered and exploited. So maybe art and technology need a different relation. We should continue to ask how technology can transform the concept of art and philosophy, but at the same time we have to ask how art and philosophy can transform the concept of technology, including the imagination, invention, development, and use of technology. I think this is our task, and we should not shy away from it. If we have to go back to art itself, to the question of aesthetics, where do we start?

The study of aesthetics entered into philosophy only with Alexander Baumgarten's first volume of *Aesthetica*, published in 1750.[15] Its first line claims that *Aesthetica* is an investigation into a lower faculty of cognition. Unlike logic, which is a higher faculty of cognition that deals with clear and distinct ideas, aesthetics is more suited to subjective tastes, emotions, and feelings. Rationalists like Baumgarten also recognized a certain truth in aesthetics that cannot be denied. However, as when Leibniz talked about aesthetics, there is only a *je ne sais quoi*: the object of the lower faculty of cognition, which is aesthetics. We can continue this tradition of aesthetics today by talking about emotions, feelings, and the like, but in *Art and Cosmotechnics*, I have tried to elevate the concept of aesthetics to logic. Basically, this means not only reversing Benjamin's question but also reversing the discourse of aesthetics from Baumgarten via Kant. By doing this, I am trying to show where in aesthetics we can actually articulate a logical form and establish a transition from aesthetics to logic. I feel that elevating aesthetics to logic may offer a better idea of how different kinds of aesthetic thinking can be articulated and how they can contribute to the discourse on technology. That is why I start with two kinds of aesthetic thinking but address them as logic. One is tragist thinking or logic. The other is Daoist

[15] Alexander Gottlieb Baumgarten, *Aesthetica* (Frankfurt/Oder: Kleyb, 1750).

logic, or *shanshui* logic. *Art and Cosmotechnics* is a strange book, I do not think anyone would ever compare Greek tragedy with shanshui painting! Historians may simply dismiss it, but if you read the book, you can see how tragist thinking and shanshui logic are actually two forms of recursive thinking through a set of similar yet different assumptions. Daoist thinking and tragist thinking both begin with contradiction, but the way the contradiction is articulated and later resolved in Greek tragedy is very different from the way it is articulated and resolved in shanshui painting. That is why I had to start by elevating aesthetics to logic, and there were many more reasons for these attempts than I have been able to describe here.

BKW: *Art and Cosmotechnics* has only three chapters, plus an extensive introduction, but each chapter is substantial enough to be its own book! In the first chapter, there is a wonderful passage about "the Open" which I wonder about as a possible object for the search for what you have called "another beginning" of modern technology:

> Heidegger aligns the unconcealment of Being with what Rilke calls "the Open." When human *Dasein* looks at the world in a narrow and closed way, like a subject scrutinizing an object, the earth withdraws itself. The Open is not a scientific object, but rather another name for Being. To think together with the Open is to take into account that which resists closure and objectification. In this process, the re-grounding of truth, the truth of Being, becomes possible. Re-grounding here means rationalizing the non-rational as the incalculable last god.[16]

How does the Open reflect the logic of Heidegger's "enframing" in relation to technology? And how does another way forward emerge in your exploration of Heidegger's interest in the paintings of Paul Klee and Paul Cézanne as instances where world and Earth, or figure and ground, can be said to create openings to the unseen?

YH: In order to answer your question, we must first address the meaning of the unconcealment of Being. Explaining Being and the unconcealment of Being in such a limited space is an extremely difficult task, partly because "*un*concealment" cannot be said positively, so my answer to your question may have to do some kind of violence to these concepts. In *Art and Cosmotechnics*, I think there are two key interpretations besides my continuous work on

[16] Hui, *Art and Cosmotechnics*, 95, emphasis original.

recursivity: The first is the interpretation of the question of Being in Heidegger, and the second is the interpretation of Daoist thinking in terms of recursivity. Heidegger's work is, for me, a detour in order to move forward. Daoist literature often discusses dialectics, but how is Daoist dialectics different from Hegelian dialectics? Can it really be called dialectics at all? There is also much speculation about the influence of Daoist thought on Heidegger, largely based on the story that he once wanted to translate the *Dao De Jing* into German together with a sinologist. The way I approach this subject is rather different.

I will begin with the question of the unconcealment of Being, because without an entry point to that we cannot address the question of the Open and the meaning of the passage you quoted. The Greeks used the same word, technē, for both "technology" and "art." For Heidegger, what the Greeks understood by technē allows us to experience what he calls the unconcealment of Being. And here lies the most difficult question—one that haunts everyone and can make life feel really worthless, especially if you spend your life working on the question of Being and only much later realize that there is actually no such question of Being! I think this happened to the American philosopher Thomas Sheehan, who taught at Stanford and worked on Heidegger for half a century. A few years ago, he claimed that Heidegger's obsession with Being misses the point.[17] I would not say that Being is an illusion just because no one can say exactly what it is, like an object in front of us. This is the case precisely because Being belongs to a category that I call the Unknown or that we can also call, following Heidegger, the nonrational. The nonrational is obviously not rational, but neither is it irrational. It is nonrational because it always remains as the Unknown (*Unbekannte*), or better, Unknowable. For example, if God were irrational, then the world (including human beings) created by God could not be rational at all. If God were rational, then we would comprehend God through rationality. If we cannot do this, it is because we are imperfect beings. Paradoxically, either human rationality is limited or God is beyond rationality. In either case, if I ask you to demonstrate God's existence, you will never be able to, regardless of your belief. In this sense, God remains nonrational and unknown, which could also be the highest rationality.

For example, Descartes's famous demonstration of the existence of God in part four of *A Discourse on Method*[18] (as well as in the third meditation of

[17] Annie Atura, "Stanford Scholar Upends Interpretation of Philosopher Martin Heidegger," *Stanford News*, July 8, 2015, https://news.stanford.edu/2015/07/08/paradigm-heidegger-sheehan-070815.

[18] René Descartes, *A Discourse on Method*, trans. John Veitch (1637; London: J. M. Dent & Sons, 1912).

Meditations on First Philosophy[19]) is based on a negation of the human itself, because God's existence is negatively inferred by the imperfection of the human intellect. The first cause is a capacity beyond human rationality that we can call God. Kant is more tactical in considering God, like freedom and the immortal soul, as a postulate. This is how we can say that Being belongs to the category of the nonrational, the category of the unknown, but when we look at the world, we see only beings: a chair, a table, a flower, a dog, a human being. As Novalis says at the opening of his *Pollen*: "We seek the absolute [*Unbedingte*; also translated as 'the unthinged' or 'the unconditioned'] everywhere and only ever find things."[20] There is a difference between Being and beings (which we find in Heidegger's early work), known as the ontological difference. When Heidegger says something "is happening" (*sich abspielt*) in what the Greeks understood as technē, he calls it the unconcealment of Being. Being does not appear as an object to be predicated or analytically ordered and decomposed but rather as a place to be opened and cleared to reveal the world in a different way. It is in this decisive moment that the human being may find its place in the cosmos or situate itself in the flux of time as historical *Dasein*.

I wrote that Heidegger aligns the unconcealment of Being with what Rainer Maria Rilke called "the Open" because the Open is precisely that which cannot be reduced to either the rational or the irrational. Let us recall what Rilke says in the eighth Duino Elegy:

> With all its eyes the natural world looks out
> into the Open. Only *our* eyes are turned
> backward, and surround plant, animal, child
> like traps, as they emerge into their freedom.
> We know what is really out there only from
> the animal's gaze; for we take the very young
> child and force it around, so that it sees
> objects—not the Open, which is so
> deep in animals' faces.[21]

You can see how the Open can be interpreted as a nonobjectified way of looking at the world and is also what allows us to situate a work of art. At the same time,

[19] René Descartes, *Meditations on First Philosophy with Selections from the Objections and Replies*, trans. John Cottingham (1641; Cambridge: Cambridge University Press, 2017).

[20] Novalis, "Miscellaneous Observations," in *Novalis: Philosophical Writings*, trans. and ed. Margaret Mahony Stoljar (Albany: State University of New York Press, 1997), 23.

[21] Rainer Maria Rilke, "The Eighth Elegy," in *The Selected Poetry of Rainer Maria Rilke*, trans. Stephen Mitchell (New York: Vintage Books, 1982), 296, emphasis original.

the Open is what a work of art—through its *being at work*, its *energeia*, because energeia is that which actualizes—allows us to access, to enter into a relation with. The work of art is always *being at work*, but toward what? It is working toward the Open. The work of art opens what has been closed or what is in the process of closing. This is why I related Heidegger's unconcealment of Being to what Rilke called the Open.

For Heidegger, the unconcealment of Being in the Greek concept of technē is still possible in modern technology. This does not mean that we can no longer talk about the unconcealment of Being. Heidegger claims that it is still possible, yet modern technology's mode of unconcealment is no longer poiesis but what he called *herausfordern*, meaning "to challenge," "to provoke," or "to dare." For example, when we build a dam to generate electricity, we are challenging and ordering the river. We challenge the land, we challenge the villages that have dwelled there for thousands of years, especially if the villages have to be destroyed to make way for the dam. In the era of modern technology, the unconcealment of Being is still possible through this challenging. This form of challenging also means catastrophe, when something overwhelming like a massive engineering project becomes catastrophic, as in Fukushima, Chernobyl, and so on. The coronavirus pandemic can also be said to be such an event.

If the unconcealment of Being is still possible in modern technology, such a possibility is also a danger. How do we confront such a danger that is also a possibility? My main question is: Is it still possible, with this danger in mind, to transform technology by developing a new understanding, a new imagination, a new concept of invention, and a new relation (Heidegger would say a free relation) to technology? This is why I want to ask, as I said in my answer to your previous question, how art can transform technology. The engagement with Benjamin that we discussed earlier is not a critique but rather a review of a historical agenda. Why does it become necessary for art and philosophy to relate technology to the Open? Precisely because the Unknown is the Open, it is also open to interpretation, and in this sense, it is more general than Being. Can we, for instance, direct technology toward the Open without pushing it toward catastrophe and self-destruction? Catastrophes may allow us to resituate ourselves, as when an alcoholic has a terrible traffic accident or a fatal illness and only then gives up drinking. We moderns are all such alcoholics, but self-destruction cannot be the only way to discover meaning. The question then becomes: Can we transform technology before we hit rock bottom? Even the coronavirus pandemic does not yet seem fatal enough to keep us from wanting to resume "normal" life.

In *Art and Cosmotechnics*, I went back to Heidegger's essay "On the Origin of the Work of Art" and his encounter with Klee and Cézanne because I think it is precisely in the thinking of Klee and Cézanne that Heidegger identifies a way to overcome what he himself calls "the ontological difference." Heidegger made this clear in a postcard he circulated for Christmas to a few friends, where, after a short poem, he wrote, "What Cézanne called '*la réalisation*' [the realization] is the appearance of what is present (*des Anwesenden*) in the clearing of presence (*des Anwesens*)—in such a way, indeed, that the duality (*Zwiefalt*) [or twofold] of the two is overcome in the oneness (*Einfalt*) of the pure radiance of his paintings. For thinking, this is the question of overcoming the ontological difference between being and beings."[22] In the first chapter, titled "World and Earth," I discuss how overcoming such an "ontological difference between being and beings" shows the necessity, but also the possibility, of reinterpreting and resituating technology. As I said earlier, perhaps the Open offers a more general way of posing the question than Heidegger's Being. Even if Heidegger was able to talk about the unconcealment of Being, how could non-Europeans relate to this Being when, as we said, the question of Being was not a central one in, say, Eastern philosophy, following what Nishida said? Nishida claimed that if the central question of Western philosophy is Being, then for the East the central question is Nothing. Of course, such a clear and neat division can be disputed. At least in the case of China, one can say that the central question is *dao*, not only in Daoism but also at the core of Neo-Confucianism since the eleventh century. This is why, in *The Question Concerning Technology*, I return to the classic categories of Chinese thought, dao and *qi*, in order to elaborate on the concept of technics in China. Qi means "utensil," which has to be distinguished from another term with the same pronunciation that is more familiar to non-Chinese speakers, namely "breath," as in Qigong.

BKW: Let us move on to the second chapter of *Art and Cosmotechnics*, which begins by identifying how the logic of self-reflexivity within modern art actually forms a tautological, recursive loop. As you write, "[. . .] modernism is characterized by a reflexivity that often takes the form of self-critique. Its language is necessarily tautological. Through a negative detour, a logical contradiction, it reinforces what it negates. This gesture is fundamentally tragist because its initial negation or refusal is indeed a preparation for affirmation."[23]

[22] Cited by Julian Young in *Heidegger's Philosophy of Art* (Cambridge: Cambridge University Press, 2001), 153, original emphasis.
[23] Hui, *Art and Cosmotechnics*, 131.

You continue by offering Duchamp's *Fountain* as an iconic example of this recursive refusal-affirmation. I think we can be clear that this tautology is neither self-negating (on the contrary) nor necessarily dishonest. In fact, you clearly identify it as a tragist logic, following Greek tragedy, but we might say that it does sacrifice questions concerning Being for more immediate—maybe more urgent—questions of industrialization, specifically technologies of mass reproduction that we know from Benjamin to Duchamp, to Andy Warhol, and onward, perhaps up to contemporary art today. From this perspective, we might also suspect that modern and contemporary art have habitually confused industrialization with Being, but that would make it all the more necessary to contrast the tragist logic of this loop of refusal-affirmation with what you call "Daoist logic" found in the centuries-old tradition of Chinese shanshui painting. This second chapter of *Art and Cosmotechnics* is probably the most challenging in the book, because it synthesizes centuries of scholarship and commentary on Daoist thought to identify how recursive, or even paradoxical, logics produce meaning, either in the visual field or more generally in what can and cannot be sensed or apprehended. Shanshui painting can be seen as the ultimate expression of the dynamics of this logic, but I would not even know where to mark an entry into such an encompassing chapter of the book! Perhaps the centrality of *xuan* 玄 in this system would be a place to start?

YH: This is a key question. The passage you quote is from my commentary on what Clement Greenberg wrote about Duchamp and claimed about modernism. Greenberg claimed that Duchamp did not destroy art but rather expanded the concept of art by negating it. That is why I called it tragist, but we have not yet discussed the difference between tragist and shanshui logic, which is the core of the book. I do not think that I can do that here in a satisfactory way, but maybe I can start with an interlude from 2016, when I was at a conference in London with the sinologist François Jullien.

During a public discussion I had with François, who is a friend, the American art critic and poet Barry Schwabsky raised a question: Has tragedy, in the Greek sense, ever existed in China? And if not, why? François immediately replied that the Chinese had developed a way of thinking to avoid tragedy. I was amazed by this answer, but I was even more amazed by the complexity of the original question, because I do not think the Chinese could avoid tragedy if they did not know what tragedy was. If you want to avoid something, you have to know what it is first. Otherwise, even if you encounter it sooner or later, you will not be able to recognize it. And if you can recognize something, you must already know it.

This is similar to one of the most famous aporia in Western thought from Plato's *Meno*, when Meno challenges Socrates that if he knows what virtue is, he does not have to look for it, but if he does not know what it is, he would not recognize it even if he encountered it. So I tend to think that the Chinese did not know the Greek meaning of tragedy, which, of course, does not mean "sad," as we use the term "tragic" in the modern sense. That is why I make a distinction between "tragist" on the one hand and "tragic" on the other, because I do not want to confuse "tragist" as a logic with "tragic" as a colloquial term.

The logic of Greek tragedy always starts with a contradiction—an irreconcilable contradiction. We can take an example from Sophocles, where Antigone, according to the law of the family, has to bury her brother who died in the war, but the brother, because he was at war against the polis, cannot legally be buried as an enemy of the city. What can Antigone do? She must choose between the law of the family and the law of the polis, since the two are irreconcilable. This is the basic structure of Greek tragedy, and this is why I try to understand Greek tragedy as a logic. Years ago, I was struck by the first sentence of Peter Szondi's *Versuch über das Tragische*, which says: "Since Aristotle, there has been a poetics of tragedy. Only since Schelling has there been a philosophy of the tragic."[24] In poetics and philosophy, there is a sharp distinction between what is traditionally known as aesthetics and logic. In Aristotle, there is a discourse on emotion, on catharsis, on the purification of the soul through tragedy. It is only in Schelling that we find a *logic* of tragedy: a basic structure of the tragedy that begins with these irreconcilable contradictions. What I call tragist thought attempts to reconcile what is not reconcilable. Daoist logic also starts with oppositions, but the way the oppositions are formulated and resolved is quite different from tragist logic. The opposition we find in Daoist thinking is continuous; for example, having versus not having, movement versus tranquility, or yang and yin are all opposed to each other but also continuous. Daoist logic departs from these oppositions in order for thinking to proceed. It seeks a movement that can reconcile these oppositions, as the Greeks did with tragedy. The secret of this reconciliation, or unification, and how it operates is the task of elaborating dao.

I said earlier that Being belongs to the category of the nonrational, the unknown, and the same goes for dao, which is like Being in the sense that it is something we cannot really demonstrate. When you open *Art and Cosmotechnics*, the first sentences you read in the epigraph are the opening of the *Dao De Jing*:

[24] Peter Szondi, *Versuch über das Tragische* (Frankfurt/Main: Insel Verlag, 1961); Szondi, *An Essay on the Tragic*, trans. Paul Fleming (Stanford: Stanford University Press, 2002), 1.

The *dao* that can be said is not the eternal *dao*.
The name that can be named is not the eternal name.
Wu (nothing): the origin of heaven and earth.
You (being): the mother of ten thousand things.
Empty of desire, one perceives mystery.
Filled with desire, one perceives manifestations.
The two spring from the same source but differ in name;
Both are designated as *xuan*.
Xuan and again *xuan*,
gate to all mysteries.[25]

The dao that can be said is not dao. The name that can be named is not the eternal name, because it cannot be named. It cannot be said. *Wu*, which means "nothing" or "not having," is the origin of heaven and earth, and *you*, which means "having" or "being," is the mother of ten thousand things. They are already in opposition. The way to resolve this is xuan and again xuan ("xuan zhi you xuan"), which I see as the beginning of recursive thinking. There are different versions of the *Dao De Jing*, and in one version it is written as "xuan zhi you xuan zhi," which makes xuan a verb. Xuan has many meanings: "dark" in terms of color, as well as "mysterious." You can see how a loop serves to resolve the opposition, but then a curious question arises: What is the difference between this recursiveness and the recursiveness of Greek tragedy and, furthermore, of the recursiveness we are familiar with in cybernetics? This is the question *Art and Cosmotechnics* attempts to open. It is only through understanding this recursive logic that we can articulate the Open in a more concrete way. That would be my brief response to your complex question.

BKW: And we saw that Stephen Mitchell, among other similar translations, had it as "darkness within darkness," which is profound. Let us look at the final chapter of *Art and Cosmotechnics*, which identifies a path forward. Put simply, this path forward means, as you discussed before, reversing the question of how technology determines art in order to ask how art can determine technology, specifically by returning technology to the primordial question of Being. You point out that science and technology have a low tolerance for the unknown, or a poor understanding of the significance of the unknown, beyond that of posing a threat to control. As a simple question, I would like to ask how one might delineate a cosmotechnical approach to making art today. Have you encountered

[25] Cited in Hui, *Art and Cosmotechnics*, xvi, emphasis original.

any artistic strategies that might renew a relation to Being or to the unknown, over and above the determinism of technological enframing or capture?

YH: In *Art and Cosmotechnics*, I compared several English translations of the *Dao De Jing*, yet you can see that it is impossible to translate it word for word, because it is essentially a logic. As for your question about artistic strategies, I am not an artist, an art historian, or an art critic. I became interested in art because I see in it an openness that could be an experimental field for all of us. On the terrain of thinking, art is still in a position to deterritorialize and reterritorialize thinking. Moreover, art institutions may still have the flexibility and possibility to experiment. Whether we like it or not, we have to deal with the question of institutions because we have to think about the education of future generations. How are we going to think about the role of universities in the twenty-first century and the role of education in the humanities as well as in engineering and science? At the same time, I have become very skeptical about the potential of institutions, as you will read toward the end of the book:

> This new "institutionalization" of art has yet to come, and it has to go beyond an art designed to serve "man's spiritual needs." But it is hard to say whether this institutionalization of art will come to pass, since conventional and conservative practices in the arts and humanities, combined with institutional lack of vision, may be even more efficient than engineering and scientific disciplines in refusing imagination and becoming reactionary. Nevertheless, we still have to prepare for its arrival by providing a "ground" to think the relation between art, philosophy, and technology today.[26]

You can see that I am not an optimist, yet I am still hopeful. I am also not a pessimist and certainly not a cynic; cynicism is an enemy we all have to fight against today. I am suggesting that we should all prepare for a possibility to come. That is why I still see in art—particularly in its relation to technology—a potential, on various levels, to deal with these questions.

Maybe we can go back to the beginning of this conversation. The question I pose in the book is: How can we relate technology to the unknown at all? For example, if we could *integrate* the unknown into technology, then it would no longer be modern technology; we would no longer be modern. Modern technology would disappear, and Heidegger's discourse on modern technology would come to an end. Can art be a *place* for that? If you ask me how to do that

[26] Hui, *Art and Cosmotechnics*, 286–87.

concretely, like adding a parameter or a function to an algorithm, I would not be able to give you an answer—not only because it would be impossible but also because, if it were possible, it would paradoxically close thinking. At the same time, for me at least, this should be the way to think about overcoming modernity, because in the last century, overcoming modernity was basically done through wars, which was paradoxically only a continuation of modernity: economic and military expansion by technological means. The Second World War was also a project of overcoming modernity, with National Socialism promising to marry romanticism and industrialism into a holism (and we know that Heidegger also became a Nazi) or with the Kyoto School in Japan wanting to restore an organic thinking, which I discuss toward the end of *Art and Cosmotechnics* in my analysis of Miki Kiyoshi.

Perhaps there are other ways of overcoming modernity that are still relevant to us today. War is always a possibility as long as the sovereign state remains the only reality of international politics, because sovereignty presupposes the possibility of war. Although realpolitik has its importance, in *Art and Cosmotechnics* I try to explore some different paths, *obscure* paths that are not straightforward and probably not brightly lit by the sun. They are obscure, like xuan, in the sense that you will have to take many detours: moving backward before being able to move forward, for example, or having to turn around many times. This is also why I said at the beginning of our conversation that it is a strange book.

In the very last paragraph of *Recursivity and Contingency*, I called for a post-European philosophy, partly to echo Heidegger—for whom Western philosophy ended with cybernetics, making a post-European philosophy the only way for philosophy not only to continue at all—but also to propose an agenda for an individuation of thinking. *Art and Cosmotechnics* can be considered a response to this call, but one that is still at the very beginning of its development. However, this call is for a collective project, which means that we will have to work together, as thinkers, artists, scientists, and engineers, and I hope that there will be occasions for these kinds of dialogues to continue and flourish.

BKW: In a recent interview with Schwabsky, you said that "The ontological assumption [of social networks] is that society consists of individuals that are like atoms, and you can know the relation between these social atoms by putting a line between two dots, as in graph theory. These assumptions come to dominate our understanding of social relations and social formations, and an engineer would never doubt this or suggest that it's a fabrication."[27]

We can go one step further and point out a similar Cartesian approach to historical and geographical relations, for example, in mapping datasets of human DNA. Is historical consciousness, in the sense of narrating emergence and development in time, also vulnerable to mechanization by such an engineer's approach to optimization? If so, what does that look like?

YH: Maybe we can say that we need another "archaeology of knowledge," in the sense of Michel Foucault. Anthropology has been working on this without subsuming itself under this name. Foucault's central question was under what conditions a certain episteme is formed. Mechanism, or mechanization, is the modern episteme, and in this sense, we can find its dominance since the seventeenth century and up to our time. Although mechanization has been challenged by other epistemologies, scientific research—even in medicine—is still guided by the Cartesian method. Historical consciousness means precisely that humans realize that they are historical beings; in other words, that they live in a historical moment, which is the consequence of a series of epistemological ruptures, and that such epistemologies should not simply be accepted as truth because it is necessary to understand the conditions of such a discourse of "truth." Even the understanding of optimization is conditioned by a particular epistemology, since optimization implies the adoption of a particular method as well as the sacrifice of elements that are considered inessential.

BKW: Such an engineer's approach to the optimization of historical relations would inevitably run up against a modern streamlining of history according to national narratives. Its fragmentation might promise a greater diversity of narrative streams, but would it not also render the past—as an unknown space vulnerable to projection—accessible to the interests of essentially any power capable of sequencing its data points?

YH: Of course, but I think the question today is no longer just about providing new narratives. I think this was also Jean-François Lyotard's point about the disappearance of grand narratives. Indeed, postcolonialism has provided us with alternative narratives that challenge the dominant colonial concepts and

[27] Yuk Hui and Barry Schwabsky, "The Call of the Unknown in Art and Cosmotechnics," *e-flux Journal*, no. 136 (May 2023), https://www.e-flux.com/journal/136/538400.

ideologies, but I feel there is something more to add to it and build upon it. The past is always vulnerable. For example, in the context of East Asia, there have been decades of debate on the reliability of history textbooks used in primary and secondary schools. We know that in many places in the world, such controversies continue for decades, if not centuries. However, I think that we should now try to think beyond narratives and establish a concrete and historical analysis of epistemologies; only then can we speak with Lyotard about changing the rules of the game or inventing new games.

BKW: Finally—and maybe this has been an implicit question already—can cosmotechnics be said to have a historical project? If we are to understand "the gigantic organisms in which we live" as machines that construct precedents but also generate new unknowns, can your mode of interrogating technological development also apply to the mediation of history and memory?

YH: Cosmotechnics is a historical project in the sense that it wants to move away from history as eschatology toward a more open and plural history. Indeed, in Karl Löwith's 1949 book *Meaning in History*, he tried to show that the philosophy of history, developed in modern times from Hegel to Jacob Burckhardt, is the secularization of eschatology.[28] The telos of history is the immanentization of transcendence, the second coming of Jesus Christ, or the becoming of the *homo deus* in the most banal language of our time. I think this is a major problem of our time because we are confined to such an eschatological understanding of history. Cosmotechnics, in its plural form, proposes to understand various histories of technologies and to conceive a new way of situating modern technology according to locality. Fragmentation means precisely fragmenting this linear eschatological time in order to conceive of a new world history.

[28] "This occidental conception of history, implying an irreversible direction toward a future goal, is not merely occidental. It is essentially a Hebrew and Christian assumption that history is directed toward an ultimate purpose and governed by the providence of a supreme insight and will—in Hegel's terms, by spirit or reason as 'the absolutely powerful essence.'" Karl Löwith, *Meaning in History: The Theological Implications of the Philosophy of History* (Chicago: The University of Chicago Press, 1949), 54.

10

The Rise of the Coyote

Toward a Sociotechnological Approach to Worldmaking

Sara Garzón

When I was first approached by Federico Pérez Villoro, Ana Uribe, and Eduardo Makoszay Mayén about curating the 2022 edition of the summer school for art and technology in Mexico known as Materia Abierta,[1] I felt that only an experimental, and more importantly, experiential, collective pedagogical exercise would most accurately articulate different protocols of earth-centered technologies. In my research on Latin American contemporary art and futurity, I have been trying to stay attentive to the cosmological precepts that underscore a relationship between technology and indigeneity while taking indigeneity to indicate a specific relation to land informed by a deep understanding of the structures of care that make situated thinking possible.

Titled *The Rise of the Coyote*, the summer school took place in the rural areas of southern Mexico City. The month-long program sought to resist the unsituated perspective of conventional curatorial practices and exhibition-making and unfolded as an invitation to construct a collective and embodied creative practice that is rooted in the impetus of "returning" to Earth.[2] Through walks, cooking, harvesting, and talks with scholars, artists, and grassroots organizations, we set

[1] Materia Abierta provides a space for autonomous learning that aims to foster the socialization of knowledge from action, the mobilization of political will, and the ritualization of critical thinking based on collective principles.

[2] Under an experimental pedagogical structure, the program invited contributions by Nahua artist Fernando Palma Rodríguez, Calpulli Tecalco, Black Quantum Futurism, Paula Gaetano Adi, Laboratorio Lacustre, Michael Marder, Clapperton Chakanetsa Mavhunga, Mujeres de la Tierra, and Isadora Pedro Neves Marques. Materia Abierta is organized yearly by Federico Pérez Villoro, Ana Uribe, and Eduardo Makoszay Mayén. In 2022, the program was generously supported by Universidad Nacional Autónoma de México (UNAM) in coordination with Cultura UNAM, Museo Universitario Arte Contemporáneo, Casa del Lago UNAM, and Cátedra Extraordinaria Max Aub–Transdisciplina en arte y tecnología.

out to explore the linkages between Indigenous technologies, plant intelligence, and futurity. The artists and grassroots organizers who joined us have all been at the forefront of social justice, environmental equity, and land sovereignty since the 1990s and constitute incomparable examples of self-organization and struggles against industrial gentrification, the contamination of subsoil, the eradication of native languages, and the extraction of natural resources. Faced with the reality of these challenges, our program constituted an exercise in deep listening, attuning to the sociotechnological approaches that underpin the community's active efforts to repair and nurture collectively owned land-based practices, traditional harvesting methods, and collectively led initiatives for the recuperation of Indigenous knowledge. In this brief report, I limit my reflection to the program's focus on the relationship between land, cosmology, and technology, especially through the work of the artist Fernando Palma Rodríguez and the Náhuatl school of Calpulli Tecalco.[3]

The Teuhtli: Lessons on Reciprocity, Care, and the Cultivation of Life

The Rise of the Coyote partly took place in the municipality of San Pedro Atocpan in Milpa Alta, which has a majority Nahua population. Historically, this territory has been both a refuge for Indigenous knowledge as well as a shelter for enormous biocultural diversity. We initiated our work on the slopes of the Teuhtli Volcano, which is a natural geographical marker that connects the nine Náhuatl-speaking towns that make up Milpa Alta.[4] The Teuhtli (meaning "Lord of the Earth" in Náhuatl) is an ancient deity who continues to spiritually permeate contemporary agricultural traditions and community practices. While for many the volcano is characterized as dormant, this sacred figure is everything but idle; on the contrary, many expressions of life lie within it and emanate from it. The Teuhtli is also only one of the rival mythical volcanic deities in the greater valley of Mexico. Most prominently, it stands in proximity to the ancient warrior Popocatépetl[5] and the sleeping woman Iztaccíhuatl (translates as "white woman"

[3] Calpulli Tecalco is an organization for the investigation and preservation of the cultural and natural heritage of Nahua culture in Milpa Alta. Their mission is to preserve and revitalize the traditional knowledge that the Indigenous peoples of the south of the Anahuac Basin. For more information, see: https://calpullitecalco.com.mx/index.html.

[4] The nine original *pueblos* (towns) that make up the greater Milpa Alta are Oztotepec, Atocpan, Tlacoyucan, Tecoxpa, Ohtenco, Miacatlán, Tepenáhuac, Tlacotenco, and Villa Milpa Alta.

[5] Homero Aridjis, "EL Popocatépetl, HISTORIA Y LEYENDA," *Artes de México*, no. 73 (2005): 8–11.

in Náhuatl, but it is proverbially known as the sleeping woman).[6] According to Nahua legend and oral tradition, these three sacred volcanic deities are the providers of water, minerals, and seeds needed to cultivate and nurture the land of their respective geographic positions and topographic structures.[7]

Guided by new media artist and Nahua scholar Fernando Palma Rodríguez, we started our journey into the program by hiking up the Teuhtli at four in the morning on August 8, 2022. We initiated our ascent into the steep and rocky pathways of the volcano, 2.7 kilometers up to the peak.[8] The group hiked together in silence while we allowed our eyes to slowly adjust to the moonlight and our senses to become attuned to the slippery path. We crawled and latched onto nearby vegetation so strongly rooted into the soil that it was able to carry the weight of our bodies. Roughly an hour and a half later, at 5:30 a.m., we reached the mountaintop. We drank water as we made our way to the eastern side of the volcano, just before dawn. The slopes of the volcano and the mountainous horizon steadily came into view as the first rays of light revealed the environment surrounding it. We gathered around Fernando, who, without notice, started recounting the history of the land, Nahua myths, and Aztec legends known about the origin of the Teuhtli and its rivalry with Popocatépetl over the love of Iztaccíhuatl.[9]

Below the mountains and volcanoes, Fernando pointed to the now-disappearing Texcoco and Xochimilco lakes—the large celestial bodies of water that once supported the ancient city of Tenochtitlan, the capital of the Aztec Empire. The lake was first drained by the Spanish conquistadors during the seventeenth and eighteenth centuries, destroying it alongside the sophisticated infrastructure put in place by the Aztecs, which included aqueducts, drainage systems, and floating gardens. On top of the lake, the colonizers built "modern" Mexico City.[10] Spreading out from this rapidly growing megalopolis, we could see how urban sprawls have been voraciously eating into the landscape until this day: contaminating the lake, building over

[6] David Carbajal López, "Iztaccíhuatl: de diosa a princesa," *Signos Históricos* 23, no. 46 (2021): 286–315.

[7] Héctor Celedón Muñiz and Mayolo Sanchez Alvarado, *Nemiliz tlacuilolli Xopantlacualli: La leyenda del Xopantlacualli; La comida del tiempo verde y relatos de la milpa* (Mexico City: Calpulli Tecalco, 2014), 7.

[8] Lilia Arana Salinas, "Geología de los volcanes monogenéticos Teuhtli, Tlátoc, Tlacotenco, Ocusacayo y Cuauhtzin en la Sierra Chichinautzin, al sur de la Ciudad de México" (master's thesis, Instituto de Geología, Universidad Nacional Autónoma, México, 2004).

[9] Carlos Villanueva Yescas, "Teutli. Leyenda," in *Escuela Secundaria Diurna Número 9 "Teutlieulti": Un cuarto de siglo de labor educativa 1940–1965* (Mexico City: Lito Impresora Panamá, 1966), 39–41.

[10] Elisabeth Malkin, "Once Built on a Lake, Mexico City Now Runs Dry," *New York Times*, March 16, 2006, www.nytimes.com/2006/03/16/world/americas/16iht-mexico.html.

Figure 10.1 Fernando Palma Rodríguez on the Teuhtli, August 8, 2022.

it, and encroaching into Indigenous territory. The view of the modern city also underscored the fallacies delineated by Western ideas of progress. Poorly built homes, precarious roads, and neglected infrastructure now substitute the natural and millenarian environment while impoverishing its inhabitants. The developmentalist logic tarnished the land and, with it, people's relationalities to nature and interdependence. The Lord of the Earth, however, continues to grant fertile ground and weather, water, and seeds for the success of the harvest of nearby communities. The harvest, referred to as the *milpa*, grows on the slopes of the volcano.[11] The neighbors of Milpa Alta share the land. The land is not owned; it is worked.

Before leaving, we descended into the belly of Teuhtli, a deep crater at the center of the volcano. Gathered there, Fernando told more stories about the mountain vegetation, the spiritual practices enacted by the Nahua peoples of Milpa Alta to venerate the lord, and the many ways in which the volcano intersects and sustains communal life. Over the course of the walk, we developed

[11] Marco Buenrostro, "Las bondades de la milpa," *Ciencias* 92–93 (October 2008–March 2009): 30–32.

an understanding of the centrality of the Aztec pantheon in determining the community's intersubjective relationships with the Teuhtli. Through these stories, the artist delineated how people's identities are subject to exchange and reciprocity with the volcano and how they are in the process of becoming with the natural, ancestral, and sacred actants who inhabit it. Overwhelmed by Teuhtli's powerful spiritual force, we made offerings to express our gratitude for hosting us and showing us its majestic splendors. Retributions were made by the group with flowers and rocks that were buried in the soil while we shared short accounts about our land of origin—a reciprocal gesture with which we left behind a piece of our own sense of place.

Throughout the month, we returned to Milpa Alta and to the slopes of the Teuhtli two or three times a week. We gathered in Calpulli Tecalco, a center for the recuperation of the Náhuatl language, knowledge, and community, which is led by Carmen Rodríguez, an Indigenous activist, literacy advocate, and Fernando's mother. Guided by the family, we visited the shores of the volcano, where communal land is shared and worked. Known as *ejidos*, communal land in Mexico has been prominent since the land reform of 1934, even though the model already existed in Aztec times when it was known as *Calpōlli*. In Náhuatl, Calpōlli means "house"—this, however, does not refer to a private home of blood-related individuals but to an "extended family" of a similar ethnic background.[12] The primary objective of the Calpōlli, and later on the ejido, was to coordinate land use for growing crops, food production, and manufacturing tribute.[13] Since the 1990s, however, the ejido system has been threatened by the privatization of the state, as well as by the loss of kinship and communal life that has resulted from the dispossession of people's relationship to the land.[14]

Besides farming plots, the slopes of the volcano also contain rich ancient sediments, caves, and terraces. Angélica Palma Rodríguez, a philosopher, community organizer, and Fernando's sister, has been coordinating community-led excavations and publications about the archaeological sites of Milpa Alta in an effort to recuperate the ancestral technologies of harvest, Nahua cosmologies, and other expressions of the town's living memory.[15] In a multidisciplinary effort between Nahua and mestizo scholars, the people of Milpa Alta carry

[12] Entry for "Calpolli," in *Online Nahuatl Dictionary*, ed. Stephanie Wood (Eugene: Wired Humanities Projects, 2000–present), https://nahuatl.wired-humanities.org/content/calpolli.

[13] Melissa Schumacher et al., "Evolution and Collapse of *Ejidos* in Mexico—To What Extent Is Communal Land Used for Urban Development?," *Land* 8, no. 10 (2019): 146, here 2.

[14] Schumacher et al., "Evolution and Collapse of *Ejidos* in Mexico," 3–4.

[15] María Angélica Palma Rodríguez and the team of local archaeologists and scientists published the book titled *Tizacalco ipan ihxihua Teuhtli: Tizacalco sobre los pies del Teuhtli; Sobre los pies del Teuhtli: Tizacalco, patrimonio arqueológico de Milpa Alta* (Mexico City: Calpulli Tecalco, 2020).

Figure 10.2 Jaime Vega, a farmer of Milpa Alta, August 23, 2022.

out their own research and maintenance of the sites.[16] Ancient terraces, which are a series of successively receding flat surfaces, or steps, designed for more effective farming, are prominent around the volcano. The terraces help preserve the nutrients of the land and prevent floods and land collapse because they keep the soil from eroding and losing its resiliency.[17] On the Teuhtli, terraces appeared with the first settlers in the area around 1250–1521 CE, later replaced by other industrialized harvesting techniques, and most of them have now been abandoned. The documentation and recuperation of these systems by the peoples of Milpa Alta have served as a solution to the changing conditions that result from climate change, the erosion, and degradation of the land, as well as the loss of Nahua spiritual practices. We visited the terraces and their surrounding archaeological remnants under the guidance of Angélica, Jaime Vega Ávila, and Erika Morales in order to understand how the community is actively trying to reincorporate some of the original terrace steps created by their ancestors into their contemporary farming practices. This has become especially important as issues of the land continue to intersect the family economy, territorial identity, migration, and sustainability. Alongside the terraces, other types of ancient

[16] Some of the people that have been involved in the archaeological research around Tizacalco include agro-engineers Juan Carlos Loza and Javier Galicia, archeologists Luis Alberto Martos López and Erika Morales, as well as biologist Héctor Celedón and astronomer Rafael Zimbrón, among others.

[17] Palma Rodríguez et al., *Tizacalco ipan ihxihua Teuhtli*.

sedimentation are also slowly being unearthed and researched by the peoples of Milpa Alta in this unprecedented effort to claim their own memory. These sites are not only on their territory, but they are also active hubs of spiritual, cosmological, and agricultural practice.

Chilean sociologist Luis Razeto Migliaro has also emphasized the importance of local archaeological practices in the Andean region of South America, where, through the recuperation of millenarian agricultural techniques, native communities are able to revitalize the cosmological precepts that underpin the relationship between environment and technology. During our trips around the agricultural terraces and the milpas, we started to better understand the teachings of the land. The milpa is a plot of farmland centered around the harvest of maize (corn), which is grown alongside other plants in a symbiotic relationship. Fernando explained that in the traditional milpa, a practice that is at least 3,000 years old, three plants are commonly grown together: maize, bean, and squash. Each plant benefits the other: beans fix nitrogen, essential to the healthy growth of maize plants; in turn, maize stalks grow tall and provide a scaffold upon which the bean plant climbs up, producing a profusion of blossoming flowers that aid bean pollination; squash leaves are ordinarily large and grow abundantly, covering large areas in a shadow so thick that hardly any weeds grow underneath. The milpa is, therefore, not exclusively for human consumption, but it is planted as an offering to the soil and as nourishment for the nonhuman actants that also depend on it. In addition, the milpa cannot be worked alone; it is intergenerationally and collectively grown. Harvesting over several months is hard work; it requires long hours under the sun, cold, or rain. Cultivating and maintaining the milpa, therefore, relies on building strong affective relationships, itself a central component of the sustainability of life within the community. "In working the land," asserted Fernando, "life merges into creation, nature into culture, and art into technology."[18]

Under the intersection of land and cosmology, Razeto Migliaro advanced a relevant definition of Indigenous technologies, which he described as the capacity to cultivate life. Here, the networked ideas of technological knowledge are the product of "saber criar la vida y dejarse criar por la vida,"[19] which loosely translates to "knowing how to cultivate life and letting life cultivate you." This

[18] Fernando Palma Rodríguez, talk during our hike up the Teuhtli volcano, August 8, 2022. Author's notes.

[19] Juan van Kessel Browers, "La tecnología simbólica en la producción agropecuaria andina," in *Manos sabias para criar la vida: Tecnología Andina*, ed. Juan van Kessel and Horacio Larraín Barros (Quito: 49º Congreso Internacional de Americanistas, 1997), 42.

notion accurately captures a particular definition of the technological that not only undermines the self-proclaimed universality of Western ideas of creation, tool making, and design but also repositions the logics of technological knowing toward life-granting and earth-based paradigms. In addition to Razeto Migliaro's definition, archaeologist Alexander Herrera Wassilowsky understands Indigenous technology "como un hecho social total"[20] ("as a total social doing"). What Herrera Wassilowsky's definition implies are actants that come into being, not as isolated singularities but only through intersubjective relationships.[21] This Indigenous approach to technology understands it as a set of processes that are fully informed by the social fabric created between human, other-than-human, ancestral, spiritual, and historical actants. A shift in the relation to technology through the prism of understanding Indigenous technology as a complete social event parallels the shift in the relation to nature based on the principles of kinship, mutualism, and reciprocity, concepts that were suppressed under Western technoscientific ideologies. Through the teachings of the milpa's maize-centered harvest, we understood that Nahua peoples are not makers of the world; they are instead made by the earth; they are men and women made of maize. It is a subject position that alters creationist ideologies, enacting instead a technical thinking premised on groundedness, reciprocity, and care—principles that lead to the cultivation of life.

The Huehuecóyotl: A Sociotechnological Approach to Worldmaking

Having understood Nahua's relationality with the Earth and the cosmic worlds that support it, it was important for the group to gain familiarity with Fernando's technological imaginaries by getting to know his artistic practice. We coordinated a visit and a mechatronics workshop with Fernando and his studio assistant, software engineer Edgar Espinoza, at Calpulli Tecalco. Much of Fernando's artistic vision is inspired by the Aztec deity Huehuecóyotl ("old coyote"). According to Mexican mythology, Huehuecóyotl has the capacity to navigate a great number of varied and contradictory ecosystems.[22] Throughout

[20] Alexander Herrera Wassilowsky, *La recuperación de tecnologías indígenas: Arqueología, tecnología y desarrollo en los Andes* (Lima: Instituto de Estudios Peruanos, 2011), 7.

[21] Herrera Wassilowsky, *La recuperación de tecnologías indígenas*, 17.

[22] Guilhem Olivier, "Huehuecóyotl, 'Coyote Viejo,' el múseo transgresor: ¿dios de los otomíes o avatar de tezcatlipoca?," *Estudios de Cultura Náhuatl* 30 (1999): 113–32, here 123.

the visual representations in pre-Hispanic Toltec statues and murals in temples, this celestial figure has been interpreted as one between human and animal—an identity that can naturally navigate a multiplicity of worlds by occupying a variety of faces, capacities, and skills.[23] Fernando's embodiment of the Huehuecóyotl through both performative reenactments and robotic sculptures is a way of eliciting the coyote's shapeshifting powers.

While discussing his work, Fernando's described how his robots give materiality to Nahua cosmovision, redefining through it various traditional artistic Indigenous practices. Despite Fernando's engineering background and postgraduate residency in computer-based sculpture at the Rijksakademie van beeldende kunsten in Amsterdam, his machines do not intend to compete with state-of-the-art robotics. They do not stand for the amplification of human power, nor do they depend on systems of automatization and control. Instead, through their own aesthetic and symbolic visual language, the artist imbues the deity with the capacity to jolt and transform behavior through its interactive quality with the viewer. For this, the artist modulates the teaching of the land and the idea that we are *of* the earth into technological forms.

Fernando's composite creatures between God and animal, machine and man, subvert the binary logic of Western technology. Cables, motors, sensors, switches, and microprocessors are rendered visible in the robots and are integrated into the intentional low-tech aesthetics. Besides the technological hardware, Fernando creates mesmerizing figures using recycled materials, scraps of fabric, cardboard, and other soft supplies.[24] His robots are not, however, mere representations of Huehuecóyotl or any other Aztec deity; they are an embodiment of the powers of the Old Coyote. According to Nahua cosmogony, people can transform into the coyote by lending their physical body and human appearance to the trickster.[25] In other instances, they can invoke Huehuecóyotl as a fellow traveler to help cross between realms and into unknown destinies.[26] Art historian Jose Martin Gonzaléz even characterized Fernando's embodiment of the robotic coyote as "the need to go back to one's roots in order to summon his legendary power towards a new dawn."[27] The coyote-cyborg is the opposite of

[23] Olivier, "Huehuecóyotl, 'Coyote Viejo,' el múseo transgresor," 124.

[24] Helena Chávez MacGregor, "Hacer bailar a los espectros," in *Fernando Palma: "amotla otlacualacac oncan tlanahuatiz quename ye huitz quiahuitl . . . mocualnezcayotl" no relampagueó para anunciar que llovería . . . tu hermosura*, exhibition catalog (Mexico City: Museo Universitario del Chopo, 2015), 26.

[25] Jose Martin González, "La Ascendencia de Coyote," in *Mechatronic Circus*, exhibition catalog (London: 198 Gallery and Oriel Mostyn Gallery, 2000), 61–2.

[26] Olivier, "Huehuecóyotl, 'Coyote Viejo,' el múseo trangresor," 113.

[27] Gonzaléz, "La Ascendencia de Coyote," 62.

the Western concept of technology and a conduit by which to experiment with other paradigms of technological knowing based on reciprocity and interspecies accountability.

Having first seen the intersection between land, cosmology, and technology delineated by Fernando, we entered into an intricate web of knowledges, on farming, care, and community-oriented traditional land usage, as a group of cyborg-coyotes. While cyborgs have a long history, especially in feminist science and technology studies, Fernando's cyborgs are closer to Norbert Wiener's cybernetic "beings that merge artificial and organic systems of control."[28] However, unlike the semihuman beings enhanced through computational or robotic capabilities, Fernando's mythological cyborgs, robotic coyotes, and embodied coyote-deities are not so much a human-machine entity as a spiritual- or ancestral-animal composite figure. Less an automatized system, Fernando's entities embody a form of relationality that suggests becoming other. That is why, since none of the participants of *The Rise of the Coyote* were members of the Milpa Alta community, listening and learning from Indigenous visualities and sensibilities became possible only by seeing ourselves enter, albeit only temporarily and upon invitation, as cyborg beings. This invocation of a state of becoming also reminded us of our partial and incomplete perspective and understanding.

The embodiment of Nahua cosmovision and community practices into the robotic entity of Huehuecóyotl elucidated the earth-based forms of knowing as a manifestation of what philosopher Yuk Hui termed cosmotechnics, which corresponds to "the unification between the cosmic order and the moral order through technical activities."[29] In the case of Fernando's robotic entities, this "moral order" follows a set of practices that are governed by the rules of the land: reciprocity, kinship, attunement, and accountability. More specifically, this implies the integration of local and situated systems of knowledge that delineate other logics of technological thinking. Altogether, the multiplicity of understandings of the technological that emerges as the capacity to bring into being other mediations between humans and ecosystems, natural forces, and spiritual phenomena together in order to constitute new and alternative epistemes (beyond those stipulated by Western technologies) contributes to the urgent need to diversify our definitions of technology. The artist's machines,

[28] Norbert Wiener's term "cybernetics" is discussed in relation to cyborgs in John Jordan, *Robots* (Cambridge, MA: MIT Press, 2016), 51.

[29] Yuk Hui, *The Question Concerning Technology in China: An Essay on Cosmotechnics* (Falmouth: Urbanomic, 2016), 19.

while interactive and responsive to external stimuli, have no utilitarian purpose; they simply prompt the viewer to rethink its relationship with the technological. In fact, while the robot is a problematic figure fueled by Western science fiction imaginaries, the notion of the robot that most closely relates to the concept of Indigenous robotics that I want to elaborate on here would be the one advanced by the artist and technologist Chris Csíkszentmihályi:

> Robots themselves are perhaps defined more by their use as signifiers of our conflicted relationship to technology than by any technical specification, although the field of robotics relies on fudging that fact. [. . .] Thus robotics have, in their functional representations, been fellow travelers with humans for one hundred years, and their primary use has been to question, test, and problematize the relationships of humans to machines, labor to automation, relationality to biology.[30]

This is further supported by Fernando, whose application of Nahua philosophies engages in technologies that are the product of networks that connect life, creation, nature, culture, and art as a total social doing. This is a web of entangled forms of being in the world and relating to the world. The artist's robots are hence less a mere instrument of representation but more an embodiment of an Indigenous sensibility and worldmaking practices that transform the very utility of the machine into a network, or web, of human and nonhuman relationalities.

During the program, *The Rise of Coyote*, it became clear that many expressions of technological production and thinking emerged from the milpas on the slopes of the sacred deity Teuhtli. The walks, talks, cooking sessions, and even harvesting workshops allowed us to embody, albeit only partially, the experience of contemplating the relationality between cosmology, environment, and technology. The cyborgian and robotic artworks of Fernando Palma Rodríguez, moreover, bring into being the cosmological orders and ethical engagements that produce, contain, and weave all technological thinking as it emanates from the cultivation of life.

[30] For more on the concept and history of robotics, see: Chris Csíkszentmihályi, "Robotics," in *Sensorium: Embodied Experience, Technology, and Contemporary Art*, ed. Caroline A. Jones (Cambridge, MA: MIT Press and MIT List Visual Arts Center, 2006), 202–3; and Ricardo Iglesias García, *Arte y robótica: La tecnología como experimentación estética* (Madrid: Casimiro, 2016).

Artificial Earth

11

The Artificial Earth

A Conceptual Morphology

Conrad Moriarty-Cole and James Phillips

Anchored in the North Atlantic Ocean, some 520 kilometers to the west of Morocco, the small archipelago of Madeira consists of three groups of islands: the central island of Madeira, approximately 60 kilometers in length, coupled with the slim-line Desertas, and Porto Santo off the north-eastern coast of Madeira. These uninhabited islands had seen visitors throughout history, from Quintus Sertorius (as documented in Plutarch's *Parallel Lives*[1]), to the occasional Vikings sometime between 900 and 1030. It was not until 1420 that the first settlers arrived from Portugal, laying claim to the islands and commencing their colonization.[2] The orography of Madeira is extreme: Sitting atop a shield volcano, the peaks and ridges rise up violently from the island floor, piercing the sky before dropping off again with some of the world's highest, sheerest cliffs.

In *Ecological Imperialism*, Alfred W. Crosby details the activities the Portuguese colonists undertook to make the island inhabitable, ranging from mass fires to remove the plentiful cedar forests (*madeira* means "wood" in Portuguese) to the terracing and irrigation networks required for farming.[3] The Portuguese colonizers transformed not just the island's geomorphology but also the climate through deforestation, hydrology through irrigation and the growth of the nascent slave trade, all in order to turn the island into a prodigious value-producing machine.[4] In sugarcane they found a crop that thrived and could therefore be cultivated on a stupendous industrial scale. Madeira could be considered one of

[1] Plutarch, *Lives, Volume VIII: Sertorius and Eumenes; Phocion and Cato the Younger*, trans. Bernadotte Perrin (Cambridge, MA: Harvard University Press, 1919), 21.

[2] Daniel J. Boorstin, *The Discoverers: A History of Man's Search to Know His World and Himself* (1983; London: Penguin Books, 1986), 165.

[3] Alfred W. Crosby, *Ecological Imperialism* (Cambridge: Cambridge University Press, 2015).

[4] Richard Grove, "Climatic Fears: Colonialism and the History of Environmentalism," *Harvard International Review* 23, no. 4 (2002): 50–55.

Figure 11.1 Abraham Cresques, *Atlas Catalan*, 1375, detail.

the first large-scale geoengineering projects of colonial modernity and was an early instance of the colonial appropriation and reconstruction of the material and conceptual reality that followed in the wake of Columbus's expedition of 1492. What developed in this period was the prototypical colonial cosmology that took the material world to be fundamentally malleable and exploitable.

Mass geoengineering projects, which involved the reconstruction of nature in service of the extraction of value, were central to the colonial expansion of Europe. Yet the manipulation of the Earth was not solely in service of value; there was a cultural dimension to geoengineering. With the founding colonists of New England, for example, cultural attitudes aligned with Enlightenment thought approached the climate as something to be both colonized and modified. In his history of climate discourse, James Roger Fleming discusses the "great climate debate in colonial and early America" and highlights that the climatic conditions of the "new world" were considered extreme in the European imagination: a frozen, primitive, and degenerate wasteland.[5] This was compounded by the experience reported by colonial settlers in North America, who found the climate to be much more variable than the European climate, with hurricanes, extreme heat, and desperately cold winters. In order to persuade the European upper

[5] James Roger Fleming, *Historical Perspectives on Climate Change* (Oxford: Oxford University Press, 1998).

class that the colonial project was a worthwhile endeavor, colonists argued for large-scale interventions into the natural environment to make it "habitable" for European settlers. Proposals for climate modification included marsh drainage, deforestation, and cultivation of the soil. This climatic-colonial discourse was tied to notions of culture and civilization. As Fleming notes, the relationship between climate modification and the ideals of civilization was by no means a mere undertext. Rather, it was overtly argued by some American naturalists and social historians of the time, such as Samuel Williams, that deforestation in Europe had resulted in a more stable and moderate climate as opposed to the climate of New England.[6] Many argued for such ecological interventions because a more moderate climate would be amenable to the cultivation of "genius, learning, industry and the liberal arts," allowing the new American colonies to stand alongside other great European civilizations of the past.[7] This early form of geoengineering was part of the colonial cosmology; the climate functioned as an analogical abstraction of a deeper opposition between notions of moderation and culture versus savage winters and savage natives. The etymological genesis of the modern concept of "culture" from "cultivation" in the early modern period is nowhere more explicit than in the words of colonialists of the time.[8] In the context of the colonial project, a binary was set between the supposedly uncultured peoples of these apparently uncultivated lands,[9] in opposition to the imported cultural ideal of civilized Europe with its cultivated lands. This was a binary that was manifest in both discourse and techno-ecological practices (deforestation, marsh drainage, and so on).

What could be called the colonial cosmology was cultivated simultaneously through the physical manipulation of nature and the engineering of the idea of nature. Thus, the "colonial cosmology" was the scheme of general ideas concerning existence, serving as both a guide and blueprint for the construction of nature underlying the colonial expansion of Europe. The concern of this chapter is to locate the emergence of a logic of "artificiality" in the conceptual

6 Samuel Williams, *The Natural and Civil History of Vermont* (1794; Burlington: Samuel Mills, 1809), cited in Fleming, *Historical Perspectives on Climate Change*, 25–26.

7 Hugh Williamson, *Observations on the Climate in Different Parts of America, compared with the climate in corresponding parts of the other continent. To which are added, remarks on the different complexions of the human race; with some account of the aborigines of America. Being an introductory discourse to the history of North-Carolina* (New York: T. & J. Swords, 1811), 174–80.

8 Raymond Williams, *Keywords: A vocabulary of culture and society; Revised Edition* (1976; New York: Oxford University Press, 1983), 92.

9 Another misconception that persists even today is that the reality of agriculture in pre-Columbian America was much more complex, including permanent and intensive agricultural production. See Jane Mount Pleasant, "A New Paradigm for Pre-Columbian Agriculture in North America," *Early American Studies: An Interdisciplinary Journal* 13, no. 2 (2015): 374–412.

and material construction of nature that was central to the colonial expansion of Europe. We propose "artificiality" as a fundamental concept for understanding the underlying logic that defines colonial cosmology. "Artificiality" or "artificial" refers to something that is made by humans. Rather than apply this concept to any particular object, we argue that artificiality typifies the colonial approach to not just nature but to existence at large. This is to say that the colonial cosmology is not built upon the world as it is found but is a cosmology that states, "this is the world as it should be." Pursued over the course of this chapter is a definition of the function of artificiality as a logic that conceives of nature as malleable and to be manipulated in service of a set of instrumental imperatives or values. We argue that this logic underpins many contemporary geoengineering projects that continue to conceive of the planet as an "artificial Earth."

To make this argument, we look to the work of Sylvia Wynter, who argues that the creation of the concept of Man as separate from the Human Being was borne out of the manipulation of the Latin-Christian cosmology to serve the colonial expansion of Europe. In the creation of this concept of Man we locate the initial steps toward a logic of artificiality operating at the level of the cosmological. To develop this point, we then turn to Alfred North Whitehead's concept of the "bifurcation of nature," according to which nature is split by modern science into two realms of existence—one of knowledge, one of subjective experience—that insofar as they are real, are real in different senses. Bifurcation designates a method or process of differentiation; a logic of selectivity that grants inclusion to certain factors while excluding those apparently beyond its remit. According to this formulation, the nature of modern science is posited as the sole Truth. We suggest that the "bifurcation of nature" diagnoses a mode of thought implicit and explicit in many conceptualizations of nature, the Earth, and within the context of catastrophic climate change. Our aim is not only to advance a theorization of artificiality but also to show how artificiality as a logic is productive of an overarching approach to conceptualizing nature. The argument advanced throughout the following text also develops a methodology that, we propose, can be utilized to examine the underlying logic of concepts.

A Propositional Methodology of Conceptual Morphology

We call the model and method of analysis pursued in this chapter "conceptual morphology." Borrowing the term "morphology" from the Earth Sciences, which refers to the study of the origins and evolution of geographic forms and structures, a conceptual morphology seeks to identify and chart the transformation of a concept. The notion of "conceptual morphology" names the study of not just historical changes to a concept but also the way in which an idea changes relative to a broader framework of knowledge and the context in which the things to which the concept relates also change. In other words, the central focus of the method of "conceptual morphology" is the *recursive* relationship between cosmologies and concepts and the modes of knowledge and understanding generated therein. We are interested in how such a methodology can provide tools for the critique and reconstruction of a given cosmological order and can, in turn, fashion new terms for the construction of experience. Our argument is that a parallel reading of Whitehead and Wynter provides an excellent model for the analysis of the social practices and conceptual structures ordering society and the conceptualization of nature, which recursively instate their own epistemic conditions, from the colonial, bureaucratic, and economic to the scientific, technological, and philosophical.

Whitehead and Wynter seem unlikely allies in the model we build from their respective analyses. Indeed, how can a thinker from the heart of the Western intellectual canon contribute anything to a decolonial critique of knowledge? There are two avenues from which to approach this question. First, the problems they are tackling individually are both situated in a similar historical period: the development of the modern Western scientific conceptual scheme. Of course, Wynter goes further back in history to trace the problem of overrepresentation from the inception of the colonial expansion of Europe through to the emergence of modern scientific discourse, whereas Whitehead considers scientific discourse from a purely Western-centric perspective. Perhaps, therefore, it is more accurate to say that Wynter and Whitehead's shared historical problem is the secularization of nature and its reconstruction by scientific thought. The second shared problematic concerns the power that the natural sciences grant themselves to interpret nature and to state the Truth of reality in their own terms. Indeed, here a unity between these two thinkers is formed, one that allows us to posit that overrepresentation and the bifurcation of nature are theoretical tools that share the aim of diagnosing and denouncing the artificiality with which nature is reconstructed at the cosmological scale. These ideas will be discussed in detail below.

The model of "conceptual morphology" raises certain questions that gesture toward a broader methodology. First, to what extent (either implicitly or explicitly) are contextually specific ideas explicative of an overarching logic? Second, what kind of cosmology does a concept like the artificial presuppose, and what kind of cosmology does such a concept recursively instate? The aim of the method of conceptual morphology is to analyze not just the historical morphology of a particular concept—its transformation within different historical contexts—but to ascertain whether it instates an overarching logic that can be discerned and whether this produces other concepts that fit within that logic. The examples discussed here address the extent to which artificiality functions as the core logic from which branch other conceptualizations and practices like geoengineering, to which we will turn in the final section of the chapter. It is worth noting, however, that while we take geoengineering as an illustration of the method, we consider conceptual morphology to have a broader application than the specific instance to which we are applying it.

The Colonial Construction of Nature

On August 2, 1492, Christopher Columbus departed from Palos de la Frontera near the mouth of the Río Tinto with the aim of reaching Cipangu, otherwise known as Japan. Following Marco Polo's estimation that Cipangu lay on the same latitude as the Canary Islands, Columbus sailed southward before turning due westward at the Canaries. A little over two months later, on October 12, Columbus and his crew would make landfall on the island of Guanahani within the Lucayan Archipelago of the West Indies. This encounter is the "big-bang" of the New World,[10] a landmark moment in the colonial expansion of Europe. The touch paper of this encounter, however, was lit nearly sixty years earlier, in 1434, when the Portuguese mariner Gil Eanes rounded Cape Bojador, a much-feared stretch of ocean off the northwest coast of the Western Sahara, following a failed attempt the preceding year. Navigation at the time was based on a cartography of the world premised on an idea of nature derived from the tenets of the reigning medieval Latin-Christian cosmology. According to this cartography, Cape Bojador marked the boundary of habitability; the limit of God's Grace, emanating from Jerusalem and jutting up against the cliffs of the African continent—the "Torrid Zone." The so-called "Sea of Darkness" delineated the southernmost point that mariners would sail, after which certain death by sea monsters and other treachery lurked. Many vessels had found their demise in the violent waters, but Eanes, operating on orders from Prince Henry

[10] Walter D. Mignolo, "Epistemic Disobedience, Independent Thought and De-Colonial Freedom," *Theory, Culture & Society* 26, nos. 7–8 (December 2009): 1–23.

the Navigator and with a reputation as a navigator without parallel, persevered. Turning to the open ocean instead of toward land, leaving the cape quickly in his wake, Eanes was confronted with something fundamentally unexpected in the hospitable climate of Senegal. What had been expected? The cumulative experience of failure by mariners who had previously attempted to round Cape Bojador had seemed to confirm the Christian mapping of the world. Cliffs built of ominous red sand with no sign of life beyond seemed to indicate the very end of the world.[11] This was thought to be an uninhabitable zone, a veritable hell on Earth. Why would Eanes have expected his experience to be an exception to the rule? Instead, Eanes's progressive navigation past this treacherous point to find a rich and fertile land with people living there quite literally opened up the physical world for Europe and paved the way not only for Columbus but for the colonial construction of nature through the creation of the notion of the New World.

To an extent, Eanes's voyages disproved the mapping of the world by Christian cartography and therefore the prevailing Latin-Christian European understanding of the physical cosmos. However, the idea of nature with which Columbus proceeded in 1492 was still heavily reliant on Christian cosmology as an overarching framework for understanding existence. Accordingly, the creation of the New World took place firmly within this framework. In her analysis of this period, Wynter demonstrates the role of cosmology as the source of action, understanding, and knowledge of all existence. In Wynter's words, Latin-Christian Europe organized itself around a "value principle or master code" derived from Christian cosmology; this master code became not only a cartographic principle with which the physical cosmos was mapped but also a broader ordering principle with which the unknown was hypothesized.[12] We can see this at work in the example of God's Grace mentioned above. According to Christian-Aristotelian physics, the heavier elements of the Earth naturally had to be submerged under the lighter element of water. Only by God's miraculous intervention in nature could the Eastern hemisphere, whose physical and spiritual center was Jerusalem, be made habitable and temperate.[13] It was by God's Grace alone, therefore, that a zone was delineated within which human

11 Boorstin, *The Discoverers*, 157, 166.

12 Sylvia Wynter, "Unsettling the Coloniality of Being/Power/Truth/Freedom: Towards the Human, After Man, Its Overrepresentation—An Argument," *The New Centennial Review* 3, no. 3 (Fall 2003): 278–79.

13 Sylvia Wynter, "1492: A New World View," in *Race, Discourse, and the Origin of the Americas: A New World View*, ed. Vera Lawrence Hyatt and Rex M. Nettleford (Washington: Smithsonian Institution Press, 1994), 22.

life was indeed possible. The so-called "Torrid Zone" of Africa was held to be too hot, and the Western hemisphere thought to be devoid of land.[14] Eanes and Columbus, however, disproved the idea of God's Grace as a cartographic principle in terms of habitability/uninhabitability. The discovery of people residing in those lands thus forced a reconfiguration of the codification of God's Grace in terms of the Spirit/Flesh binary inscribed by Adamic Original Sin. Respectively, they represented "Redeemed Spirit" and "Fallen Flesh" as the poles of spirituality ordained by God. All zones residing outside of God's providential Grace were thus condemned as fallen flesh, creating a "space of Otherness" in which all things were considered less than those within the purview of God's Grace.[15] With this transmuted codification, God's Grace was employed as the principle by which the cosmos and its inhabitants were known, and, more importantly, the world was reconstructed by European colonialism. On this point, Wynter writes that

> whether religious or secular, all such schemas/programs and their formulations of "a general order of existence" also function to inscribe the specific "descriptive statement" of the human that is enacting of the ontogeny/sociogeny, nature-culture mode of being human, for whom the specific ensemble of motivated behaviors will be adaptively advantageous.[16]

Here, the figure of the human is singled out as both an anchoring point for the broader theorization of existence and a concept with which social reality is constructed. What the human is and what it is not, according to the European colonial perspective, was to determine the blueprint for the world as it was to be reconstructed through the territorial expansion of Europe.

In the relationship drawn between "a general order of existence" and what is termed the "descriptive statement of the human," the idea of Truth and its representation is key. For Wynter, the colonial instates a particular idea of the human as the figure of Man, drawn according to the general order of existence reigning during the fifteenth century that changes throughout the centuries that follow. For example, the figure of Man constructed by Christian cosmology is one from within the region of God's Grace, a Redeemed Spirit of Christianity, whose definition is achieved as much by what *it is* as by what *it is not*. The figure of Man, as the archetype of the Human, is defined by the Otherness that it is not, where Otherness is both a geographical region (outside of God's Grace) and

[14] Wynter, "Unsettling the Coloniality of Being/Power/Truth/Freedom," 279.
[15] Wynter, "Unsettling the Coloniality of Being/Power/Truth/Freedom," 274.
[16] Wynter, "Unsettling the Coloniality of Being/Power/Truth/Freedom," 280.

those deemed "Fallen Flesh" (that is, not Christian). Crystalized in the figure of Man are certain cosmological principles with which the social reality of Europe is constructed. They are values that dictate certain modes of thought, certain ways of living and being, according to which Europe is built and the rest of the world is to be reconstructed.

The idea of a "descriptive statement" is drawn from Gregory Bateson's work on cybernetic systems. For Bateson, human society is a self-correcting system, and as a self-correcting system, its behavior is in service of conserving a certain status quo. A descriptive statement is a component of this status quo, the truth of which is conserved through various system changes. "As in the engine with a governor," writes Bateson, "the fuel supply is changed to conserve—to keep constant—the speed of the flywheel, so always in such systems, changes occur to conserve the truth of some descriptive statement, some component of the *status quo*."[17] When we transpose the terms Wynter supplies above to this example, we understand that the argument being made concerns the figure of Man functioning to maintain the status quo of European Latin-Christian society. More specifically, however, if we consider this formulation in terms of the 1492 landfall, we understand that the Latin-Christian order of existence provides the idea with which the figure of Man is constructed to stabilize and maintain a status quo of European colonial expansion. Thus, the central element of Wynter's argument is that the construction of the descriptive statement of Man was as that human who seeks to conserve the supposedly God-given superiority of Christian-European society, defined in terms of degrees of perfection emanating from Jerusalem.[18] This conservation of the God-given superiority of Christian-European society was thus a conservation of the cosmological Truth (the central organizing principle of the Christian cosmology). Thus, it was according to this constructed notion of Truth that the colonial expansion of Europe was justified.

A key mechanic in the conservation of a system is what Wynter terms its "adaptive Truth-for" function, where the truth of a descriptive statement is effectively a modality prone to shift and change as required.[19] Understood in these terms, the relationship between the general order of existence, the Truth, and the figure of Man as a descriptive statement is continuously shifting to maintain the hegemony of Western society. From the theologically derived

[17] Gregory Bateson, *Steps to an Ecology of Mind* (1972; Chicago: The University of Chicago Press, 2000), 435, emphasis original.

[18] Wynter, "Unsettling the Coloniality of Being/Power/Truth/Freedom," 288.

[19] Wynter, "Unsettling the Coloniality of Being/Power/Truth/Freedom," 300.

or supernaturally guaranteed Truth, through to the rational revolution of the Renaissance humanists, who founded Truth in nature itself, the figure of Man as a descriptive statement is always *constructed* and *reconstructed* in ways that serve the dominant interests of that social reality. Man is an artificial construct that functions to maintain the system. It appears that cosmology adapts as the methods and processes with which knowledge is produced alter alongside the shifting aims of its given society (initiated by an event like the colonial encounter, for instance). In this morphological stage analyzed by Wynter, it is the figure of Man as an *overrepresentation* of the Human that functions as a mode of thought organizing the cosmology central to the colonial expansion of Europe. The important point, however, is not that Man is a construct (hardly an original idea) but that this figure of Man serves as *a model for the process of artificialization* that would resurface in different guises thereafter, including in the artificialization of Earth. Thus, artificiality starts off as a mode of thought, identifiable in the function of overrepresentation, but eventually becomes a structuring logic with the bifurcation of nature.

We see the dynamics of this morphology and the function of the descriptive statement clearly in Wynter's analysis of the secularization of the West brought about by Renaissance humanism and the scientific revolution. As God's supernaturally guaranteed knowledge was worn away by the increasing prevalence of mathematics and the sciences in the social institutions of the West, such as the courts and universities, it was nature to which the West turned in search of Truth.[20] Within nature could be found a general order of existence, which had previously been found within Latin-Christian theological cosmology. Physics, chemistry, and the biological sciences were able to identify universal laws that governed the universe, from which a singular systematic knowledge could be built and according to which the rational world could be constructed. Wynter argues that with this transformation came the creation of a new figure of Man conceived by the biological sciences in terms of the universal laws derived from nature. According to Wynter, this new figure of Man was premised on an understanding of the human as determined by the differential logic of the "Chain of Being."[21] Thus, we are made aware of a new development in the concept of Man.

Darwinian evolutionary theory established a taxonomy of organic life along a chain of being that moves from irrational animal to rational human being,

[20] Amos Funkenstein, *Theology and the Scientific Imagination from the Middle Ages to the Seventeenth Century* (Princeton: Princeton University Press, 1986), 5–6.
[21] Wynter, "Unsettling the Coloniality of Being/Power/Truth/Freedom," 300.

according to the principle of Natural Selection. Instating a new "Argument-from-Design" allowed the development of a new "master code," according to which the world was divided up as either "selected" or "dysselected" by-nature in purely biological terms.[22] But the ostensibly factual basis of this determination is made according to what W. E. B. Du Bois refers to as the "Color Line," a concept we have previously encountered above as the "space of Otherness." As such, the new master code of selected or dysselected by-nature follows the colonially produced logic of the space of Otherness. Facts are presented as Truths derived from nature, but they are just another example of the overdetermination of reality (artificialization) according to colonial imperatives, albeit now more deeply ingrained epistemologically. The biological paradigm transposed this invented master code as the ordering principle of social reality, both institutionally and discursively. It was an affirmation of "the overall global/national bourgeois order of things and its specific mode of economic production, alone able to provide the material conditions of existence for the production and reproduction of the ethnoclass or Western-bourgeois answer that we now give to the question of who and what we are."[23] That is, the biologically affirmed descriptive statement of the human as selected or dysselected by-nature is an ordering principle that legislates who is deserving of Western-bourgeois life, which is written into scientific, economic, and social systems. Wynter writes that the paradox present in the Darwinian descriptive statement of the human, defined in terms of the natural organism, is that "it must ensure the functioning of strategic mechanisms that can repress all knowledge of the fact that its biocentric descriptive statement is a descriptive statement."[24] As Denise Ferreira da Silva notes, this "reflects a particular collective *self-representation* and *not* an eternal (extrahuman) truth determined by the immutable, objective, and necessary 'laws' and 'forms' of nature."[25] An integral function of overrepresentation is, therefore, constructing the appearance of Truth without giving away the very fact of its construction, with any adaptation of ideas and methods necessary to maintain this status quo.

With the analysis of the figure of Man, Wynter makes clear the central importance of a specific concept to both a system of knowledge and a social reality. Although the figure of Man as the overrepresentation of the Human by the West is the focal point of Wynter's analysis of the morphology of the colonial

22 Wynter, "Unsettling the Coloniality of Being/Power/Truth/Freedom," 310.

23 Wynter, "Unsettling the Coloniality of Being/Power/Truth/Freedom," 316–17.

24 Wynter, "Unsettling the Coloniality of Being/Power/Truth/Freedom," 325–26.

25 Denise Ferreira da Silva, "Before *Man*: Sylvia Wynter's Rewriting of the Modern Episteme," in *Sylvia Wynter: On Being Human as Praxis*, ed. Katherine McKittrick (Durham: Duke University Press, 2015), 95, emphasis original.

cosmology, we can draw from this a more general point regarding the logic or method with which the colonial cosmology produces subsequent cosmologies. We contend that overrepresentation, as it is described and analyzed by Wynter, is a logic or mode of thought that can be called "artificialization." This is because overrepresentation functions as a method by which colonial cosmology perpetuates its apparent God-given or by-nature hegemony, through the manipulation of Truth via ideas operating as descriptive statements. Moreover, it is with this logic of overrepresentation that colonial cosmology reconstructs nature as artificial in order to preserve its status quo. As we will now consider, with the scientific revolution came a crucial moment in the conceptual morphology of artificiality: the bifurcation of nature.

Bifurcations of Nature

Wynter has shown the way in which the overrepresentation of a concept functions to maintain the status quo of a system. In her analysis of the concept of Man as the overrepresentation of the Human by the West, the concept of Man serves to anchor a set of constructed truths within the broader idea of nature as it is reconstructed by the colonial expansion of Europe. The "adaptive Truth-for" function of overrepresentation is where we locate the role of artificiality. That is, we use the term "artificiality" to designate what this overrepresentation does to the conceptualization of the world and categories of understanding such as nature. We argue that overrepresentation undergoes a further transformation from the function of a particular mode of thought into a logic embedded in an overarching cosmology. In order to make this point, we want to posit a theorization of the concept of overrepresentation in terms of what Alfred North Whitehead calls the bifurcation of nature. Fundamentally, the bifurcation of nature is a diagnosis of the way in which the scientific cosmology is constructed as the sole Truth of reality through the exclusion of certain types of evidence, or certain forms of investigation, that would otherwise contradict the claims central to this worldview. Modern science, particularly what derived from Isaac Newton's classical mechanics, Whitehead contests, splits reality into two regimes of existence: on the one hand, a reality "there for knowledge," and on the other a reality "which is the byplay of the mind," meaning that "in so far as they are real, are real in different senses."[26] Nature is bifurcated into that which

[26] Alfred North Whitehead, *The Concept of Nature* (1920; Cambridge: Cambridge University Press, 2015), 30.

is apprehended through subjective awareness and that which is the cause of subjective awareness. Awareness is the domain of "the greenness of the trees, song of the birds, warmth of the sun, the hardness of the chairs, and the feel of velvet," whereas the "cause of awareness is the conjectured system of molecules and electrons which so affects the mind as to produce the awareness of apparent nature."[27] In other words, scientific cosmology, whose domain is the cause of awareness, considers nature as a system of matter without qualities, which it can observe objectively, without interference of qualities.[28]

Whitehead's diagnosis of the bifurcation of nature is premised on a critique of the historical influence of John Locke's theory of primary and secondary qualities on scientific cosmology.[29] Locke's explanation of these qualities extends from the questions: What is a natural body? What are its qualities, and how do we experience them? This is the argument advanced in *An Essay Concerning Human Understanding*, published in 1689, according to which the phenomenal qualities of the body are separate from its objective, invariant qualities: The former are qualities such as color, which are conceived as "secondary," while the latter are, for example, extension and solidity, which are "primary."[30] Secondary qualities are apprehended via sensation through perception, whereas primary qualities are apprehended at the level of knowledge, or the intellect. The statement that secondary qualities are apprehended through sensation suggests that they are not simply projections of the mind; rather they are derived from primary qualities as an aspect of them. They are "psychic additions" that furnish our experience of bodies with sound and color.[31] Phenomenal experience is therefore, according to Locke, derived from nonphenomenal primary qualities, as in the case of light perceived by the eye that the mind turns into color.[32] Importantly, the separation of primary and secondary qualities by Locke is initiated empirically, where the nonperceptual primary qualities are differentiated from the subjective secondary qualities that are both derived and from expressive of the former. This is the precise location of the bifurcation of nature. Whitehead writes,

> What I am essentially protesting against is the bifurcation of nature into two systems of reality, which, in so far as they are real, are real in different senses. One

27 Whitehead, *The Concept of Nature*, 31.
28 Whitehead, *The Concept of Nature*, 29.
29 Whitehead, *The Concept of Nature*, 27.
30 John Locke, "An Essay Concerning Human Understanding," in *The Clarendon Edition of the Works of John Locke: An Essay Concerning Human Understanding*, ed. Peter H. Nidditch (1689; Oxford: Oxford University Press, 1975), 135.
31 Whitehead, *The Concept of Nature*, 42.
32 Whitehead, *The Concept of Nature*, 27.

> reality would be the entities such as electrons which are the study of speculative physics. This would be the reality which is there for knowledge; although on this theory it is never known. For what is known is the other sort of reality, which is the byplay of the mind. Thus there would be two natures, one is the conjecture and the other is the dream. Another way of phrasing this theory which I am arguing against is to bifurcate nature into two divisions, namely into the nature apprehended in awareness and the nature which is the cause of awareness.[33]

The bifurcation of nature as it is discussed in the *Concept of Nature* diagnoses the operation of abstraction in scientific cosmology as it conceptualizes nature as separate from experience. The Newtonian conceptual scheme, which laid the foundations of classical mechanics, is taken by Whitehead as having a particularly pernicious effect in this regard. Whitehead's focus here is not on a reconciliation of these positions at the level of metaphysics. Rather, as Didier Debaise has argued, Whitehead is determining nature as the site of a methodological inquiry into nature that would establish its qualities rather than advance an ontological position.[34] The focus is on the methods and ideas drawn from the modern scientific cosmology to reconstruct nature and its corresponding claims to Truth through the negation of evidence to the contrary—such as that drawn from experience—that bifurcates nature.

Recalling Wynter, we can suggest that the concept of Man is a *bifurcation* of the Human precisely because it overrepresents and anchors a framework of Truths constructed by the West that serves to justify, theologically and then scientifically, the colonial expansion of Europe. What this suggests, then, is not a singular concept that enacts this bifurcation alone but rather a framework of thought, or *cosmology*, that operates according to the logic of artificiality. The bifurcation of nature functions to overrepresent a framework of Truths that have been adapted to fit certain ends. In the example drawn from Whitehead, the ends relate to the hegemony of science as the sole communicator of the truth of nature. In the example drawn from Wynter, the ends are the colonial expansion of Europe through which nature was reconstructed as fundamentally bifurcated. Thus, the bifurcation of nature names the point at which artificiality transforms from a mode of thought, typified by overrepresentation, into a logic.

We have suggested above that artificiality is the action of overrepresentation with which, in the case of artificial Earth, the material Earth can be overdetermined by a logic that seeks to engineer it. With the idea of the bifurcation of nature in

[33] Whitehead, *The Concept of Nature*, 30–31.
[34] Didier Debaise, *Nature as Event: The Lure of the Possible* (Durham: Duke University Press, 2017), 37.

mind, it is therefore possible to posit that the logic and function of artificiality are in the construction and application of a cosmology that overrepresents its objects of knowledge. The artificial functions identically to the "adaptive Truth-for" function of colonial cosmology, working recursively to assert its Truth through the exclusion of other forms of knowledge, thought, and experience as a means to restate the validity of those ideas and firm up its cosmological grip on nature. This is why we argue that the artificial is a logic rather than a singular idea, as bifurcation begets bifurcation in the pursuit of cosmological hegemony.

The Artificial Earth

Ideas are transformative—in themselves and in the manner in which they propose a different way to apprehend the world. Following from the above, we present the concept of "artificial Earth" as the image of nature that is borne from a desire to reengineer the world, which we have traced from the colonial project. "Artificial Earth" is an intentionally tautologous term; "artificialization" is a mode of apprehension guided by a certain logic; and "Earth as artificial" is the object apprehended. This apprehension of the Earth as artificial in turn reinstantiates a cosmology that generates ideas according to its internal logic of artificiality. This cosmological function of artificiality has been drawn from Wynter's analysis of the genesis and transformation of the concept of Man within the context of the colonial expansion of Europe and its legacies for Western thought. We see the figure of Man not just as an idea of exclusion but as one that is explicative of a logic with which the colonial cosmology functions. When put in relation to Whitehead's notion of the bifurcation of nature, we see the way in which the logic of artificiality that is found in the colonial cosmology also operates as an implicit and explicit principle with which nature is reconstructed by scientific cosmology.

The "artificial Earth" is the product of the contemporary hegemonic cosmology that undergirds climate science and the discourse around the unfolding climate catastrophe. As such, the artificial Earth is an abstract model of Earth that amalgamates around various technologies and bodies of knowledge relating to climate discourse. However, we use this term to designate how these technologies and institutionalized bodies of knowledge work together to produce a rendering of the Earth as an artificial product. Just as Ptolemy's cartographic system presented the world as a knowable object in such a radically disruptive way that it initiated a new cosmology, the various technologies, sciences, and cultural imaginaries that we consider as constitutive of our contemporary cosmology render the Earth as fundamentally knowable and capturable within the techno-scientific paradigm. This determination of the Earth as knowable is accompanied by the notion that it can be worked upon and engineered—the Earth as machine, as it were. In being brought into the sphere of human action as both the raw material *and* the product of human labor, the *Earth is made artificial*. Of course, in actuality, this is not entirely the case; while human activity has altered nearly every ecosystem on the planet, the Earth system cannot ever be fully captured by human action. Yet in intention, within the cosmological order created by this logic and set of practices, the artificiality of the Earth is instantiated; artificiality is therefore the framework through which the material Earth is represented in culture, art, and science. This conceptualization of the Earth as artificial is significant in its effects, as it recursively rebounds on those very techniques and ideas that conditioned the cosmology in the first instance. As a logic-determining action, artificiality is ampliative and potentially destructive if not recognized and controlled.

The concept of "artificial Earth" is therefore aimed at articulating a matrix of interacting discursive and technical renderings of the Earth, including but not limited to (1) the fact that anthropogenic effects have so profoundly altered natural processes that scientific discourse is now describing the current geological era as the Anthropocene; (2) the existence of planetary-scale computation; (3) the rendering and mediation of the Earth systems through the infrastructure of this planetary-scale computation, including sensors, satellites, and computational models; and (4) the positing of these Earth systems as plastic and manageable.

The artificial Earth as a concept therefore cuts across scales: the material (Earth systems), the experiential (lived effects of climate breakdown and the technical practices through which humans affect Earth systems), and the conceptual (computational modeling and representations of the Earth). An examination of the concept of artificiality allows us to propose a way of understanding various historical engagements with the Earth, from the colonial climate engineering described above through to geoengineering technologies, carbon capture, gene editing, nanotechnology, stratospheric aerosol injection, rewilding and reforestation projects, and carbon markets. As already stated, we argue that the artificialization of Earth is an extension of certain tendencies that can be traced back to the colonial expansion of Europe and the worldview constructed through this endeavor. Let us conclude with a final analysis of the latest expression of this logic of the artificial in the figure of geoengineering.

Critique of Artificial Earth

> Philosophy, in one of its functions, is the critic of cosmologies. [. . .] Its business is to render explicit, and—so far as may be—efficient, a process which otherwise is unconsciously performed without rational tests.[35]

Whitehead believed the job of philosophy was to be the critic of cosmologies. Throughout his work, the cosmology of modern science was consistently analyzed in terms of the reality it put forth, particularly Newton's conceptual scheme and the idea of nature as constructed in *Mathematical Principles of Natural Philosophy* published in 1687.[36] The model of "conceptual morphology" we have described above seeks to trace the relationship between concepts and the overarching cosmology to which they belong, making explicit presupposed ideas or implicit logics. The basis of the dominant cosmology of our current era of catastrophic global warming is artificiality.

Geoengineering, or climate engineering, is the pinnacle example of the artificialization of the Earth. The contemplation of geoengineering projects such as stratospheric aerosol injection of sulfur to reflect sunlight,[37] or dumping iron and other nutrients into the ocean to enhance microscopic marine plant growth in order to increase uptake of atmospheric carbon dioxide,[38] is only possible because of the conceptual scheme with which an idea of nature is constructed. This is the conceptual scheme that takes nature as something to engineer, situating the problem within the mechanics of nature itself, thereby providing a salve (for certain effects of catastrophic climate change) that allows business to continue as normal. In many senses, artificiality is so deeply embedded in the way in which nature is conceived that there is an implicit, insidious bifurcation in the relationship between human existence and the power of nature. While many are calling for a mitigation of the causes of climate breakdown and global warming, these calls are repeatedly ignored by global leaders and industry. Technological solutions of green capital are increasingly being touted as the only option remaining, and carbon capture has become a main focus of market-driven

[35] Alfred North Whitehead, *Science and the Modern World* (1925; New York: Free Press, 1997), vii.

[36] Isaac Newton, *The Principia: Mathematical Principles of Natural Philosophy; The Authoritative Translation*, trans. I. Bernard Cohen, Anne Whitman, and Julia Budenz (1687; Oakland: University of California Press, 2016).

[37] Paul J. Crutzen, "Albedo Enhancement by Stratospheric Aerosol Injection: A Contribution to Resolve a Policy Dilemma?," *Climatic Change* 77 (August 2006): 211–19.

[38] Doug Wallace, Philip Boyd, Yves Collos, Peter Croot, Ken Denman, Phoebe Lam, Cliff Law, Ulf Riebesell, Shigenobu Takeda, and Phil Williamson, "Ocean Fertilization: A Scientific Summary for Policy Makers," IOC–Intergovernmental Oceanographic Commission, UNESCO, 2010, https://unesdoc.unesco.org/ark:/48223/pf0000190674.

climate mitigation while climate engineering projects are gaining supporters.[39] The overdetermination of the artificial in hegemonic discourse structures the schematics for approaching such a problem according to this notion of nature as already a product of human action, as already artificial, and therefore plastic and malleable. This is an extension of the colonial artificialization of nature that we have detailed above, predicated on the bifurcation of nature. The colonial artificialization of nature has, over time, morphed into the Promethean techno-positivism we see in contemporary discourse, from molecular environment-engineering nanobots[40] to the program of algorithmic governance responsible for assessing and maintaining the optimal level of aerosols dispersed in the atmosphere to reflect sunlight.[41] The idea that nature can be understood, modeled, and manipulated without any unexpected consequences stems from a broader conceptualization of nature as artificial.

When the Earth is conceived as artificial, in all of its guises, it is not simply an expression of technological capabilities that produces a set of affordances, such as mass geoengineering projects. Indeed, it cannot be reduced to economic imperatives for growth, creating the industrial desire to continue extracting materials and burning fossil fuels. Although the artificialization of Earth is driven by both of these forces, our argument is that the very logic of artificiality, of which these are both expressions, is conceptually derivative of a cosmology that emerged from colonialism. A cosmological construction has taken place through the conceptual accumulation of an implicit logic of artificiality, which functions in order to maintain the status quo. The change that occurs with a conceptual morphology is not necessarily a self-differentiation that eventually leads to a different concept but can in fact be a crystallization of a concept as it intensifies and rebounds upon itself until it becomes a structuring logic. This is the danger of the artificialization of the Earth: As the driver of the hegemonic cosmological scheme of contemporary techno-capitalism, the image of nature as an artifact of human action and intervention becomes ever more dominant, and the artificialization of nature becomes ever more entrenched. Such a situation only serves to generate more potentially destructive, interventionist behavior of the kind that is increasingly being touted in public discourse regarding the unfolding climate catastrophe.

39 David Keith, *A Case for Climate Engineering* (Cambridge, MA: MIT Press, 2013).
40 Christopher J. Preston, *The Synthetic Age: Outdesigning Evolution, Resurrecting Species, and Reengineering Our World* (Cambridge, MA: MIT Press, 2018), 21.
41 Holly Jean Buck, *After Geoengineering: Climate Tragedy, Repair, and Restoration* (London: Verso Books, 2019), 222.

While geoengineering and other such practices are discussed here as exemplary of artificialization, this is not to suggest or make premature conclusions regarding the deployment of technologies for mitigating climate catastrophe. Whether one wishes to engage in such practices or instead reverse the trend toward large-scale geoengineering, a recognition of the implicit presuppositions that condition our thinking about nature is nevertheless required. Central to any decision-making processes regarding our relation to the Earth should be a self-reflexive understanding of how our thinking is conditioned by the bifurcation of nature and the overdetermination of the material Earth by the human.

Current climate science and politics encapsulated by the concept of artificial Earth are an articulation of how we currently relate to nature. Yet, according to the above thesis regarding overrepresentation and the bifurcation of nature, this relationship to nature articulated in the artificial Earth is not a novel development. Although our understanding of climate breakdown may seem on the surface to be determined by the capabilities of digital technologies and scientific developments of the past hundred or so years, this is not entirely the case. Instead, these technologies have in part been developed in response to a specific conceptual organization of the world according to the logic of artificiality.

With the technical and epistemological developments of the scientific revolution, then the industrial revolution, and later the cybernetic revolution, new methods and processes of knowledge production and the ordering of information and matter shifted the hegemonic cosmology further toward an artificialization of the world (society and nature), which had already begun some centuries before. The role of technology is nevertheless central to this artificialization in that, with certain affordances of abstract representation and modeling of Earth systems becoming possible, the Earth becomes overrepresented by its technical image. This overrepresentation of the technical image allows for the abstract modeling of the Earth as a machine. Thus, the relation to nature articulated by the artificial Earth is an expression of a long-tail conceptual morphology of artificiality, which we have here traced through these histories of colonialism and the bifurcation of nature up to and through modern technological developments. By tracing this conceptual morphology, we have attempted to provide a historically minded mapping of the discursive terrain within which issues in contemporary climate politics are contextualized.

In summary, while the impacts of European colonialism on the climate have been clearly argued and well documented,[42] there was a concurrent and related transformative process in terms of the concept of nature and how we relate to Earth systems in contemporary global techno-capitalist society. A formulation of climate politics that takes into account and comprehends itself as part of the artificialization of nature is, we believe, a worthwhile endeavor. Before we attempt to reengineer the climate, we should first attempt to reengineer our thinking. The method of conceptual morphology, which we have derived from a reading of Sylvia Wynter and Alfred North Whitehead, is an attempt to formulate a proposition to do just that.

[42] For more information, see: Dipesh Chakrabarty, "Postcolonial Studies and the Challenge of Climate Change," *New Literary History* 43, no. 1 (Winter 2012): 1–18; Simon L. Lewis and Mark A. Maslin, "Defining the Anthropocene," *Nature* 519 (March 2015): 171–80; and Kathryn Yusoff, *A Billion Black Anthropocenes or None* (Minneapolis: University of Minnesota Press, 2018).

12

The Environment Is Not a System[1]

Tega Brain

In late 2017, Microsoft's chief environmental scientist, Lucas Joppa, announced AI for Earth, a new initiative to put artificial intelligence in the hands of those who are trying to "monitor, model and manage the earth's natural systems."[2] AI for Earth gives environmental researchers access to Microsoft's cloud platform and AI technologies and is now accompanied by another Microsoft program, the Planetary Computer initiative, a data resource that provides "a multi-petabyte catalogue of global environmental data" to put "answers in the hands of conservation stakeholders."[3] Alongside these programs, separate recent initiatives by companies like Google[4] and Planet Labs[5] aim to integrate AI into environmental research and management.

Silicon Valley stands to profit handsomely from the uptake of AI in environmental research and management, just as it has from the application of these methods in a diverse range of other fields. From urban design to

[1] This text was originally published in *APRJA* 7, no. 1 (2018): 153–65.

[2] Brad Smith, "AI for Earth: A Gamechanger for Our Planet," *Guardian*, September 11, 2017, https://www.theguardian.com/advertiser-content/microsoft-ai-for-earth/microsoft-ai-research.

[3] Microsoft, "A Planetary Computer for a Sustainable Future," n.d., https://planetarycomputer.microsoft.com.

[4] Google's initiative AI for Social Good purports to "build and deliver data-driven, user-first tools and technologies that tackle social challenges" that include environmentally focused problems like flood forecasting (https://ai.google/social-good). Their Google Earth Engine (https://earthengine.google.com) "is a cloud-based platform for planetary-scale geospatial analysis that brings Google's massive computational capabilities to bear on a variety of high-impact societal issues including deforestation, drought, disaster, disease, food security, water management, climate monitoring and environmental protection." This includes the use of AI for the analysis of environmental data. Noel Gorelick et al., "Google Earth Engine: Planetary-Scale Geospatial Analysis for Everyone," *Remote Sensing of Environment* 202 (December 2017): 18–27.

[5] See Planet Lab's services (https://www.planet.com), where its vice president, Giovanni Marchisio, argues that "no number of humans would be able to analyse the millions of images being sent down to Earth daily—but, potentially—an AI-capable machine could support this." Giovanni Marchisio, "You Can't Fix What You Can't See: The Realities of AI and Satellite Data," *Planet Labs PBC*, July 15, 2019, https://www.planet.com/pulse/you-cant-fix-what-you-cant-see-the-realities-of-ai-and-satellite-data.

Figure 12.1 Clouds gather in Glenorchy, Australia, March 20, 2016.

the justice system, decision-making processes are being automated by data-driven systems. And in spite of a growing body of criticism on the limitations of these technologies,[6] the tech industry continues to promote them with the mix of solutionism and teleology that imbues Joppa's words. He urges: "For every environmental problem, governments, non-profits, academia and the technology industry need to ask two questions: 'how can AI help solve this?' and 'how can we facilitate the application of AI?'"[7]

This chapter considers some of the limitations and possibilities of computational models in the context of environmental inquiry, specifically exploring the modes of knowledge production that they mobilize. As has been argued by scholars like N. Katherine Hayles and Jennifer Gabrys, computation goes beyond just reading and representing the world. As a mode of inquiry, it

[6] See Cathy O'Neil, *Weapons of Math Destruction: How Big Data Increases Inequality and Threatens Democracy* (New York: Broadway Books, 2016); Virginia Eubanks, *Automating Inequality: How High-Tech Tools Profile, Police, and Punish the Poor* (New York: St. Martin's Press, 2018); Shannon Mattern, *Code and Clay, Data and Dirt: Five Thousand Years of Urban Media* (Minneapolis: University of Minnesota Press, 2017); and Julia Angwin et al., "Machine Bias: There's Software Used across the Country to Predict Future Criminals. And It's Biased against Blacks," *ProPublica*, May 23, 2016, https://www.propublica.org/article/machine-bias-risk-assessments-in-criminal-sentencing.

[7] Lucas N. Joppa, "AI for Earth," *Nature* 552 (December 21–8, 2017): 325–28.

has a powerful worldmaking capacity, generating new pathways for action and therefore new conditions. Gabrys puts it plainly: "Computing computes. [. . .] Programming is a way of making operative."[8] Computational metaphors are also pervasive as framing devices for complex realities, particularly in the context of research on the city, the human brain, or human behavior.[9]

Historic computational attempts to model, simulate, and make predictions about environmental assemblages while both emerging from and reinforcing a systems view of the world. The word eco*system* itself stands as a reminder that the history of ecology is enmeshed with systems theory and presupposes that species entanglements are *operational* or *functional*. More surreptitiously, a systems view of the environment connotes it as bounded, knowable, and made up of components operating in chains of cause and effect. This framing strongly invokes possibilities of manipulation and control and implicitly asks: What should an ecosystem be optimized for?[10] This question is particularly relevant at a time of rapid climate change, mass extinction, and, conveniently, an unprecedented surplus of computing. As many have pointed out, these conditions make it tempting (and lucrative) to claim that neat technological fixes can address thorny existential problems.[11] This modernist fantasy is well and truly alive for proponents of the smart city and even more dramatically in proposals for environmental interventions that threaten to commodify the Earth's climate conditions, such as atmospheric engineering.[12]

[8] Jennifer Gabrys, *Program Earth: Environmental Sensing Technology and the Making of a Computational Planet* (Minneapolis: University of Minnesota Press, 2016), 11. Also see N. Katherine Hayles, *My Mother Was a Computer: Digital Subjects and Literary Texts* (Chicago: The University of Chicago Press, 2005).

[9] Sociologist Shannon Mattern warns of the "the *city as computer* model," arguing that it often hinders "the development of healthy, just, and resilient cities" (Shannon Mattern, "A City Is Not a Computer," *Places Journal*, February 2017, https://placesjournal.org/article/a-city-is-not-a-computer, emphasis original). Psychologist Robert Epstein highlights similar issues in the context of brain research, observing that historically, metaphors for cognition have always been drawn from the dominant technology of the time—hydraulics, springs and mechanics, electrical circuits, and now computation. Epstein argues that the ubiquity of information-processing metaphors in brain research may well be constraining the field by confining hypotheses and explanations to those that align with computational processes. These metaphors equally constrain approaches to environmental inquiry.

[10] This question is inspired by Mattern's discussion of the city as a computer metaphor in "A City Is Not a Computer."

[11] See Benjamin Bratton and Natalie Jeremijenko, "Latent Interfaces, Suspicious Images," in *Situated Technologies Pamphlets 3: Suspicious Images, Latent Interfaces*, ed. Benjamin Bratton and Natalie Jeremijenko (New York: Architectural League, 2008), 9; Gabrys, *Program Earth*, 230; Isabelle Stengers, "The Cosmopolitical Proposal," in *Making Things Public: Atmospheres of Democracy*, ed. Bruno Latour and Peter Weibel (Cambridge, MA: MIT Press, 2005), 1000; and Bronislaw Szerszynski and Maialen Galarraga, "Geoengineering Knowledge: Interdisciplinarity and the Shaping of Climate Engineering Research," *Environment and Planning A* 45, no. 12 (December 2013): 2817–24, here 2818.

[12] For more on the planned atmospheric tests that were scheduled to occur in the US in 2018, see James Temple, "Harvard Scientists Moving Ahead on Plans for Atmospheric Geoengineering

Figure 12.2 Seagrass living at Glenorchy Public Art and Sculpture Park, Australia, March 20, 2016.

What else does a systems view of the environment amplify or edit out? This discussion revisits several historic missteps in environmental measurement and modeling in order to pull focus on the epistemological assumptions embedded in a systems perspective. It then asks: What are other possibilities for ecological thought? Does AI have any potential to reveal environments in ways that escape the trapping of systems? Critical to my inquiry is the recent work of Anna Lowenhaupt Tsing and what she calls "the arts of noticing."[13] Tsing's work offers a starting point for thinking outside of both a systems framework and assumptions of progress. Her perspective on ecology and the lifeworlds it describes unfolds and emerges through "encounters,"[14] which bring together entities, transforming them in indeterminate ways. Might AI operate through modes of environmental encounters, or will it simply amplify "the informatics of domination"?[15]

Experiments," *MIT Technology Review*, March 24, 2017, https://www.technologyreview.com/2017/03/24/153028/harvard-scientists-moving-ahead-on-plans-for-atmospheric-geoengineering-experiments.

13 Anna Lowenhaupt Tsing, *The Mushroom at the End of the World: On the Possibility of Life in Capitalist Ruins* (Princeton: Princeton University Press, 2015).

14 Tsing, *The Mushroom at the End of the World*, 20.

15 Donna J. Haraway *Simians, Cyborgs and Women: The Reinvention of Nature* (London: Free Association Books, 1991), 161.

The Poverty of Numbers

A systems view of the environment reinforced by computation has numerous precedents, including eighteenth- and nineteenth-century attempts at scientific forest management. This early attempt at centralized ecosystem management through numerical modeling foreshadows the contemporary use of these approaches in the context of computation. James C. Scott traces how the introduction of centralized forestry required forests to be made legible in new ways.[16] Trees in forests were measured, quantified, and modeled to optimize harvest and replanting for timber yield. Thus, the fastest-growing species were replanted in felled areas, and trees became viewed as autonomous machines for producing wood. Those species not harvestable for timber—low-lying bushes, fungi, and plants,[17] as well as traditional "unofficial" use of forests by local communities—were edited out of the system.[18] These scientific (or fiscal) forests were managed with the assumption that complex species entanglements were irrelevant and could be treated as external to a system designed to efficiently transform trees into commodities. Yet after a couple of generations of felling and replanting, yields began to drop, and the health of managed forests deteriorated.[19] Viewing the forest as a factory oversimplified the reality of the relations and interdependencies of its species.

The scientific forest failed by its own criteria for timber yield. However, it is worth acknowledging that if yield had remained high while biodiversity declined, this history of sustainable environmental management would be remembered as a success, analogous to industrial agriculture. Tsing calls environments that are simplified and optimized to produce commodities "plantations."[20] The economic drivers of capitalism make crop yields the ultimate goal of agricultural landscapes and shape how they are measured, modeled, and manipulated. When a landscape is managed as a factory, its species become assets alienated from their lifeworlds[21] like workers who fulfill HITs (human intelligence tasks) on Mechanical Turk with no bearing on each other or what they produce. When the asset can no longer be extracted, the landscape becomes a ruin and disappears

[16] See James C. Scott's well-known account of scientific forestry in *Seeing Like a State: How Certain Schemes to Improve the Human Condition Have Failed* (New Haven: Yale University Press, 1998).

[17] Scott, *Seeing Like a State*, 13.

[18] Richard Hölzl, "Historicizing Sustainability: German Scientific Forestry in the Eighteenth and Nineteenth Centuries," *Science as Culture* 19, no. 4 (November 2010): 431–60, here 436.

[19] Scott, *Seeing Like a State*, 20.

[20] Tsing, *The Mushroom at the End of the World*, 39–40.

[21] I use the term "lifeworlds" following Tsing, who describes objects in capitalist exchange as being alienated and "torn from their lifeworlds" (Tsing, *The Mushroom at the End of the World*, 121).

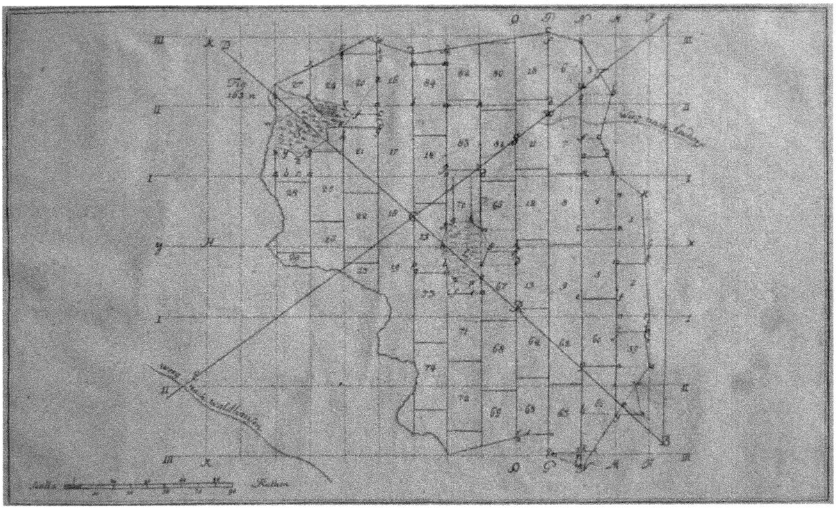

Figure 12.3 Imaginary forest patch partitioned in eighty-four sections by Georg Grünberger, 1788.

from view, deemed worthless.[22] Both the plantation and the scientific forest are the results of numerical approaches to landscape management applied in the name of economics. They highlight that data collection and modeling practices are never neutral. Rather, they are contingent on decisions about what is deemed important or trivial in the eyes of the manager and are therefore profoundly driven by culture, economics, class, and race.

The Fantasy of Stability

In the twentieth century, the science of ecology emerged in dialogue with cybernetics and systems theory. There is a rich body of literature critiquing how these conditions influenced environmental research.[23] Cybernetics, first defined in the nineteenth century by André-Marie Ampère as "the science

[22] Tsing, *The Mushroom at the End of the World*, 42.

[23] Many authors discuss the influence of systems theory on ecology, such as Fernando Elichirigoity, *Planet Management: Limits to Growth, Computer Simulation, and the Emergence of Global Spaces* (Evanston: Northwestern University Press, 1999); and Bruno Latour, "Some Advantages of the Notion of 'Critical Zone' for Geopolitics," *Procedia Earth and Planetary Science* 10 (2014): 3–6. Some also consider the influence of cybernetics, such as Donna J. Haraway, "The High Cost of Information in Post-World War II Evolutionary Biology: Ergonomics, Semiotics, and the Sociobiology of Communication Systems," *Philosophical Forum* 13, nos. 2–3 (1981): 244–78; and Gabrys, *Program Earth.*

of governance,"[24] was catalyzed as an interdisciplinary field by proponents like Norbert Wiener in the postwar decades.[25] It inspired ecologists to pursue questions of control and self-regulation in the context of species lifeworlds. Some early ecosystem diagrams were even realized in the style of circuitry.

In 1935, botanist Arthur Tansley was among the first to use the term "ecosystem" to describe the "systematic" functioning of forests, grasslands, and wetlands. He saw ecosystems as "the whole *system* (in the sense of physics), including not only the organism-complex but also the whole complex of physical factors forming what we call the environment of the biome [. . .]. [These] are the basic units of nature."[26] Like the scientific foresters, Tansley proposed that ecosystems were made of discrete stable units, interacting in ways that tend toward a state of dynamic equilibrium. He also assumed that natural selection favors stability, that "systems [. . .] which can attain the most stable equilibrium survive the longest."[27] This idea of ecological equilibrium remains stubbornly influential, as does the idea of the environment as a unified "whole." The misconception that humans and the environment are separate is further reiterated when philosophers like Bruno Latour and Timothy Morton discuss the idea of the "natural world" existing in a balanced, harmonious state that is then disrupted by humans.[28]

Toward the late 1960s, Tansley's assumption of ecosystem homeostasis was proving difficult to verify, even in ambitious large-scale ecosystem modeling projects enabled by the availability of computation. One such project was the Grasslands Biome, started in 1968 at Colorado State University. It was an unprecedented attempt to comprehensively model a grassland ecosystem with a computational model, and it aimed to uncover new ecological principles.[29] Employing hundreds of full-time researchers, the project involved extraordinary methods of data collection as researchers tried to account for all forms of energy entering and leaving the system, attempting to quantify everything eaten and excreted by all organisms in the biome and then inputting this data into a

[24] André-Marie Ampère, *Essai sur la philosophie des sciences, ou Exposition analytique d'une classification naturelle de toutes les connaissances humaines* (Paris: Bachelier, 1834).

[25] Norbert Wiener, *Cybernetics: Or Control and Communication in the Animal and the Machine* (1948; Cambridge, MA: MIT Press, 2019).

[26] Arthur G. Tansley, "The Use and Abuse of Vegetational Concepts and Terms," *Ecology* 16, no. 3 (July 1935): 284–307, here 299, emphasis original.

[27] Tansley, "The Use and Abuse of Vegetational Concepts and Terms," 300.

[28] Donna J. Haraway's concept of "natureculture" is an attempt to collapse a false binary between human concerns and nature, introduced in *The Companion Species Manifesto: Dogs, People, and Significant Otherness* (Chicago: Prickly Paradigm Press, 2003). Timothy Morton further builds on this rejection of bifurcation in *The Ecological Thought* (Cambridge, MA: Harvard University Press, 2010).

[29] Chunglin Kwa, "Modeling the Grasslands," *Historical Studies in the Physical and Biological Sciences* 24, no. 1 (January 1993): 125–55.

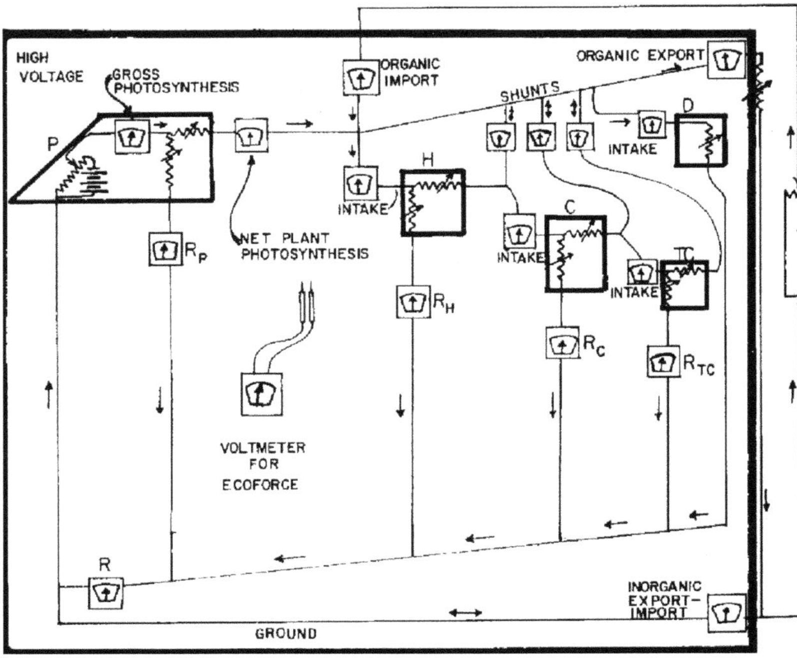

Figure 12.4 Howard T. Odum's first presentation of an ecosystem using the symbolism and aesthetic of electric circuit diagrams, 1960.

mathematical model. Students and researchers would follow animals around the grassland whispering into tape recorders. They would "collect" animals and analyze their stomach content by inserting probes into their digestive systems.[30] Soil microbiology was also studied, yet soil invertebrates and highly mobile species such as insects and birds remained frustratingly uncooperative in yielding information to researchers.[31]

Despite this labor, the Grasslands model, like similar large-scale ecological modeling programs of the time, revealed very few new ecological principles. Deemed "too simplified biologically" despite implementing an unprecedented number of variables,[32] the model was built with an assumption of default equilibrium. Chunglin Kwa argues that the Grasslands Biome model was simply "a sophisticated version of a cybernetic system, [. . .] [which] casts the ecologist in the role of systems engineer."[33] The project disproved its foundational hypothesis

[30] Robert T. Coupland (ed.), "Problems in Studying Grassland Ecosystems," in *Grassland Ecosystems of the World: Analysis of Grasslands and Their Uses* (Cambridge: Cambridge University Press, 1979), 35.
[31] Coupland, "Problems in Studying Grassland Ecosystems," 35.
[32] Kwa, "Modeling the Grasslands," 154.
[33] Kwa, "Modeling the Grasslands," 146.

that complex ecological realities can be reconciled with mathematical models and described as abstracted structures of inputs and outputs. "The grandiose ideal of achieving total control over ecosystems, which around 1966 appealed so much to systems ecologists as well as [politicians], was dismissed as a 'hyperbole.'"[34]

Data collection and modeling practices remain shaped by what is considered typical or atypical, important, and peripheral—summations of the boundary conditions of reality. However, making these assumptions is difficult. Even with the growing capacity of contemporary computing, it is dangerous to simply assume that more data equal more reality. An example of this is the story of how Joe Farman, a geophysicist working for the British Antarctic Survey, first observed the destruction of the ozone layer. Farman maintained a single ground-based ozone sensor in the Antarctic throughout the 1960s and 1970s and continued to do so in spite of the launch of NASA atmospheric monitoring satellites that collected vastly larger quantities of data.[35] When Farman's sensor began to show a 40-percent drop in ozone levels in the early 1980s, he assumed it was damaged and replaced it, as NASA's atmospheric models had reported no such change. After years of carefully checking, Farman published the alarming results in *Nature* as the first observation of the destruction of the ozone layer due to human pollutants. Until then, this had been only a theoretical hypothesis.[36] How had NASA's satellites missed such a marked change in ozone composition? One response from NASA suggests that their data processing software was programmed to discard readings that appeared to be outliers, thus ignoring the drastic changes that were occurring in ozone concentration.[37] In this case, reality itself was an outlier and assumed to be an error.

The Limits of Machine Learning

What if there were no cap on the amount of data produced from an environment for analysis? Could models be derived from datasets rather than built from theory to avoid erroneous assumptions like those made in the Grasslands model? Could machine learning be adopted to deal with quantities of data

34 Kwa, "Modeling the Grasslands," 155.

35 Paul Vitello, "Joseph Farman, 82, Is Dead; Discovered Ozone Hole," *New York Times*, May 18, 2013, https://www.nytimes.com/2013/05/19/science/earth/joseph-farman-82-is-dead-discovered-ozone-hole.html.

36 The theory of ozone destruction was published by Mario J. Molina and F. Sherwood Rowland, "Stratospheric Sink for Chlorofluoromethanes: Chlorine Atom-catalysed Destruction of Ozone," *Nature* 249 (1974): 810–12.

37 Joseph Charles Farman, Brian Gerard Gardiner, and Jonathan Shanklin, "Large Losses of Total Ozone in Antarctica Reveal Seasonal ClO_x/NO_x Interaction," *Nature* 315 (May 1985): 207–10.

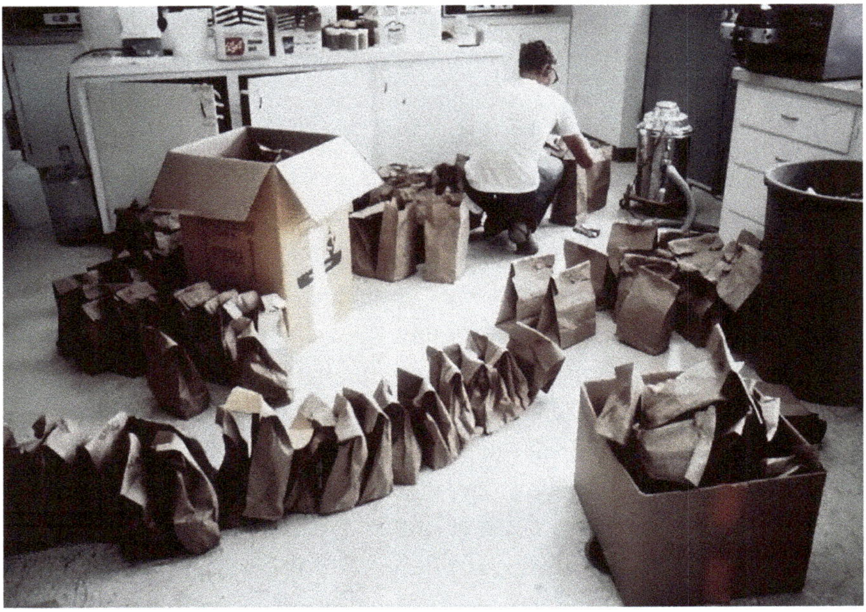

Figure 12.5 Processing of replicate biomass samples, field laboratory at the CPER/Pawnee grassland site, Colorado, July 1971.

beyond human comprehension and prevent any need for discarding outliers? Can these techniques produce a more robust representation of reality, free of human judgment?

These are the arguments made for machine learning. In 1959, Arthur L. Samuel discussed machine learning as the ability to learn without being explicitly programmed.[38] Rules are derived from patterns in large datasets rather than programmed based on theoretical knowledge of underlying structures. As explained by Chris Anderson, the editor in chief of *Wired*, in 2008: "Petabytes allow us to say: 'Correlation is enough.' We can stop looking for models."[39] In other words, had the Grasslands model been derived through machine learning, energy flows through the ecosystem could have been estimated based on correlations in the data rather than by inputting data into a theoretical model that is hardcoded by hypotheses of ecosystem dynamics. Although this would have prevented erroneous assumptions like default homeostasis, it is important to acknowledge that machine learning substitutes one set of assumptions for another.

[38] John McCarthy and Edward Feigenbaum, "Arthur Samuel: Pioneer in Machine Learning," *AI Magazine* 11, no. 3 (September 1990): 10–11.

[39] Chris Anderson, "The End of Theory: The Data Deluge Makes the Scientific Method Obsolete," *Wired*, July 23, 2008, https://www.wired.com/2008/06/pb-theory.

Machine learning assumes that enough data can be collected to adequately represent and make predictions about reality. In the context of the environment, this is an enormous challenge given the very limited size of existing datasets. Another significant assumption is that the past is indicative of the future. Yet, as the sudden, unprecedented depletion of atmospheric ozone in the 1980s shows, this is not always the case. Similarly, climate change means our ability to make accurate predictions from existing data is diminished. Many environmental datasets, like precipitation records, span 250 years at best, with the majority spanning a much shorter period.[40] From a geological point of view, this is an absurdly small slice of time and one in which the Earth's climate has been relatively stable. As the patterns, rhythms, and cycles in both climatic and biological phenomena are drastically disrupted, it becomes increasingly difficult to make predictions based on this short, stable interval of climate data. William Gail calls this the coming of "a new dark age,"[41] where the accumulated observations of Earth's irreducibly complex conditions are increasingly rendered obsolete. If machine learning approaches are to be adopted, it is important to recognize the limits of these methods.

Dreams of Objectivity

Another prominent argument made for the use of AI methods is that data-driven approaches neutralize human decision-making by simply representing the world as it is. The proponents of AI for Earth also make these claims about objectivity: "Decisions about what actions to take will be easier to make—and less vulnerable to politicization—if we know what is happening on Earth, when and where. AI can help to provide that information."[42] However, in other realms, AI systems continue to reveal and confirm biases and structural inequalities rather than offering an easy pathway to their neutralization.

For example, defendant risk scoring systems designed to help judges make decisions to "deliver better outcomes to all who touch the justice system"[43] have been shown to score Black defendants at significantly higher risk for reoffense than white defendants with similar or worse criminal records.[44] Systems like these

[40] See I. R. Simpson and P. D. Jones, "Analysis of UK Precipitation Extremes Derived from Met Office Gridded Data," *International Journal of Climatology* 34, no. 7 (June 2014): 2438–49.

[41] William B. Gail, "A New Dark Age Looms," *New York Times*, April 19, 2016, https://www.nytimes.com/2016/04/19/opinion/a-new-dark-age-looms.html.

[42] Joppa, "AI for Earth," 328.

[43] Equivant, "Equivant at a Glance," 2018, https://www.equivant.com/wp-content/uploads/equivant-at-a-glance.pdf.

[44] Angwin et al., "Machine Bias."

should serve as warnings to other industries implementing automated decision-making, even in the name of environmental management. As theorist Françoise Vergès argues, "Adaptation through technology or the development of green capitalism [. . .] does not thoroughly address the long history and memory of environmental destruction [. . .], nor the asymmetry of power."[45] Contemporary environmental challenges directly emerge from violent histories of colonialism, imperialism, and the ongoing exploitation of marginalized communities and those living in the Global South. As such, there is no reason to suggest that AI technologies built and implemented by a cohort of wealthy white men in the United States will manage or distribute environmental resources in ways that are equitable for everyone.

Technologies will only ever provide partial fixes if they are not accompanied by shifts in perception and values, along with regulatory changes that address histories of injustice and "the tradition of belief in progress."[46] More efficient resource use in a system of deregulated capitalism is most likely to beget further resource use rather than a net reduction. Microsoft seems to have it backward in its mission statement "to empower every person and organization on the planet to achieve more."[47] Is it not the case that the idea behind technologies of automation is to empower people to achieve less? Or at least prompt a radical rethinking of what "more" is? As Vergès argues, if these logics go unquestioned, mounting environmental challenges will not only continue to accelerate change in an already stressed biosphere but also further augment environmental injustices.

If the Environment Is Not a System, Then What Is It?

How else might we think of environments in lieu of the systems metaphor? Tsing offers the concept of assemblage, and here, I build on Tsing's work, understanding environments as open-ended assemblages of nonhumans, living and nonliving, entangled in ways of life.

> Ecologists turned to assemblages to get around the sometimes fixed and bounded connotations of ecological "community." The question of how the varied species in a species assemblage influence each other—if at all—is never settled: some thwart (or eat) each other; others work together to make life possible; still others

[45] Françoise Vergès, "Racial Capitalocene: Is the Anthropocene Racial?," *Verso Blog*, August 30, 2017, https://www.versobooks.com/blogs/3376-racial-capitalocene.
[46] Vergès, "Racial Capitalocene."
[47] Microsoft, "Our Corporate Values," n.d., https://www.microsoft.com/en-lb/about/corporate-values.

> just happen to find themselves in the same place. Assemblages are open-ended gatherings. They allow us to ask about communal effects without assuming them.[48]

Like Tsing, many authors have taken up the concept of assemblage to round out the simplification and abstraction connoted through the use of technological metaphors. Following Latour, to assume a system is also to surreptitiously assume "the hidden presence of an *engineer* at work,"[49] a presence that suggests intention and that what we can see are parts of a unified whole. Assemblage relieves us of this view, instead suggesting a collection of entities that may or may not exhibit systematic characteristics. The edges of an assemblage are fuzzy—modes of interaction are always shifting, and agencies within them are heterogeneous.

Hayles also invokes the term in her inquiry on cognition in complex human-technological entanglements, what she calls "cognitive assemblages."[50] Hayles chooses assemblage over network, arguing that network conveys "a sense of sparse, clean materiality," while assemblage offers "contiguity in a fleshy sense, touching, incorporating, repelling, mutating."[51] She continues: "I want to convey the sense of a provisional collection of parts in constant flux as some are added and others lost. The parts are not so tightly bound that transformations are inhibited and not so loosely connected that information cannot flow between parts."[52] Similarly, I take up assemblage as an imperfect descriptor to avoid the hubristic assumptions of a systems view. Stating "I am studying a grasslands assemblage" instead of "I am studying a grasslands system" produces a remarkable shift in expectations and assumptions. This simple substitution dismantles the subtle assumptions of fixed categories of knowledge, as well as the assumption that engineering and control are always possible. Instead, it foregrounds uncertainty and acknowledges the unknowability of the world.

Rather than describing ecology through interactions or exchanges between entities, Tsing proposes that it emerges through encounters. For Tsing, encounters open new possibilities for thinking; they produce transformation and are therefore indeterminate.[53] They are also nonhuman-centered. There can be encounters between different species—say, a mushroom and a pine tree—or between life-forms and nonhuman materials. Components of a system are

[48] Tsing, *The Mushroom at the End of the World*, 22–23.
[49] Bruno Latour, "Some Advantages of the Notion of 'Critical Zone' for Geopolitics," *Procedia Earth and Planetary Science* 10 (2014): 3–6, here 5, emphasis original.
[50] N. Katherine Hayles, *Unthought: The Power of the Cognitive Nonconscious* (Chicago: The University of Chicago Press, 2017).
[51] Hayles, *Unthought*, 118.
[52] Hayles, *Unthought*, 117–18.
[53] Tsing, *The Mushroom at the End of the World*, 46–47.

implied to be static and discrete units, leaving out processes of contamination and transformation. For example, when predator-prey relations are described as transfers of energy between components in a system, say a walrus eats a mollusk, it is inferred that the walrus remains unchanged by the encounter. Seeing the world as made up of individuals, sealed off from one another, allows for the assumption of stable categories and makes the world easier to quantify through data that is interpreted as pattern and codified as algorithm. The yield from a data-driven mode of knowledge production is obviously rich and wide-reaching, providing new insight into phenomena like climate change. And yet, as the story of Farman's attention to the atmosphere shows, scaling and automating data collection processes can risk a presumption of stability, which conceals transformations that are occurring outside of assumed areas of possibility.

In this way, "smartness," in its current form, produces a kind of myopia. A smart city, home, or environment contains networks of sensors automatically pinging data back to servers to train machine learning models. This is also Joppa's pitch for AI for Earth: "AI systems can now be trained to classify raw data from sensors on the ground, in the sky or in space, using categories that both humans and computers understand, and at appropriate spatial and temporal resolution."[54] This statement is worthy of careful consideration. First, how does one decide on an appropriate temporal resolution? In the case of the German forests, it took nearly a century to see that management methods were unsustainable because the life rhythms of a tree are at a vastly slower tempo than those of human economies. Second, Joppa infers that the world can be revealed by how it appears through the "raw data" received from sensors, yet this implies that the sensors themselves are somehow neutral and overlooks the layers of human decision-making that have already occurred in their production and installation.[55]

It can also be surprisingly difficult to resolve the world into clearly defined categories. And are these categories stable? Tsing's argument that encounters produce transformation suggests that neat taxonomies will never fully accommodate the fluidity and uncertainty of the world. This is particularly apparent in plant systematics, where even the definition of species is contested and ever-changing.[56] In trying to categorize plant specimens, a tension can emerge between how the specimen appears (its phenotype) and how it appears

[54] Joppa, "AI for Earth," 326.
[55] See Gabrys, *Program Earth*; and Bratton and Jeremijenko, "Latent Interfaces, Suspicious Images."
[56] Ernst Mayrpage, "Species Concepts and Definitions," in *Topics in the Philosophy of Biology*, ed. Marjorie Grene and Everett Mendelsohn (Dordrecht: Springer, 1976), 353–71.

on a genetic level (its genotype). As genetic sequencing techniques have become cheaper and therefore more widely available, plant scientists sometimes find that the species indicated by phenotype does not always match up to what the genotype dictates, a discovery that has caused many herbaria to be reorganized. However, even when identifying specimens on a purely genetic level, there is still dispute over how species are interpreted.[57]

Data-driven research methods necessitate the collection of huge quantities of data, and in doing so, they dismantle opportunities for paying close, specific attention to the world. These methods also tend to obscure the many other ways of building understanding. Also, perhaps intentionally, data collection increasingly acts to maintain the status quo. We use data to study problems that would be more effectively addressed through simple political action. The impetus to "study the problem" ad nauseam gives the appearance of addressing an issue while perfectly maintaining the present state of affairs.[58]

Amplifying Encounters

How might we reciprocally illuminate the environment and balance our well-oiled capacity for imagining it from an all-conquering systems worldview? How might we elevate engagement through the specifics of encounter and narrative? Ethnography is one possibility. Tsing's study of the matsutake mushroom explores what can be learnt from a Japanese mushroom, a life-form that cannot be cultivated and that thrives in highly disturbed forests. Through her ethnography, she shows how close attention inevitably facilitates transformation. Tsing calls this "the arts of noticing," tactics for thinking without the abstraction produced by quantification or deeply held assumptions of progress. If we are "agnostic about where we are going, we might look for what has been ignored because it never fit[s] the time line of progress."[59] As Farman's ozone research showed, paying

[57] See Fazekas for discussion of differences in species definitions, Aron J. Fazekas et al., "Barcoding Plants: Are Plant Species Inherently Harder to Discriminate than Animal Species Using DNA Barcoding Markers?," *Molecular Ecology Resources* 9 (April 2009): 130–39; Hull discusses how these uncertainties have led to the concept of reciprocal illumination in plant systematics. This concept acknowledges the multiple methods for classifying and naming species. David L. Hull, *Science as a Process: An Evolutionary Account of the Social and Conceptual Development of Science* (Chicago: The University of Chicago Press, 1988).

[58] Now discontinued, the Human Project, initiated at the Center for Urban Science and Progress at New York University in 2015, was an example of data collection in lieu of political action. The project planned to address issues of health, urban design, and inequality by collecting huge volumes of data from 10,000 New Yorkers over twenty years.

[59] Tsing, *The Mushroom at the End of the World*, 21.

close attention, rather than outsourcing observation and interpretive capacities, can reveal the world in different ways. In particular, attention can emphasize the indeterminacy and messiness of encounters outside of an engineering agenda. It can transform the observer, directly involving us in the weirdness of the world.

Could technologies like machine vision and remote sensing be used to amplify environmental encounters and the arts of noticing our ecological entanglements? The rise of digital naturalism sees the development of apps and initiatives that focus attention on the life-forms in our various bioregions. Initiatives such as iNaturalist, Merlin Bird ID, and eBird invite nonscientists to contribute environmental observations and use either crowdsourced or "assisted identification" to identify species and build biodiversity databases. Assisted identification utilizes computer vision techniques to guide species identification from images by identifying broad categories in order to make suggestions. Through this process, the system is also gradually being trained and will therefore make better suggestions over time. Many scientific institutions also hope that data-driven species identification can help reduce bottlenecks in identification processes because human taxonomists are in short supply.[60]

It is also worth emphasizing that these apps do not purport to replace human identification but rather facilitate human-computer collaboration to reach conclusions more quickly. This is significant, as it shows a way in which AI can produce more meaningful environmental encounters rather than automating them away. This use case for AI also serves as a reminder that data can be much more than a material for building a simulation or instrumentalizing whatever is being measured. The act of data collection and collaborative identification can amplify encounters and, by extension, yield transformation, or what artist Jenny Odell calls "a certain dismantling in the mind."[61] In observing a local bird and being assisted to identify it as a magpie, I am learning and tuning my perception to the lifeworlds I inhabit: I am subject to transformation.

Accounts of the scientific forest, the Grasslands Biome, and Farman's ozone observations mostly focused on the success or failure of the science: on whether these projects of observation or modeling succeeded or failed in revealing new patterns, on whether the resultant environmental models proved accurate, and, by extension, on whether they produced new possibilities for environmental management and manipulation. But narrating these stories like this is to tell

[60] Ke Chung Kim and Loren B. Byrne, "Biodiversity Loss and the Taxonomic Bottleneck: Emerging Biodiversity Science," *Ecological Research* 21, no. 6 (November 2006): 794–810.

[61] Jenny Odell, "Notes of a Bioregional Interloper," *Open Space*, San Francisco Museum of Modern Art, October 9, 2017, https://openspace.sfmoma.org/2017/10/notes-of-a-bioregional-interloper.

them from a systems point of view. What tends to get overlooked is how these are actually stories of environmental encounters through data collection. As encounters, they are also stories of the transformation of both the environments and the humans involved. How did the meticulous observation of the environmental assemblages in question shift and transform the people studying them? In and of itself, this question rejects a false binary between human and environment. It acknowledges the instability of the observer and the tendencies of Western science to edit out intuition, emotion, and philosophical recalibrations. The reciprocal transformation that occurs with attention and encounter, what Nobel Prize-winning geneticist Barbara McClintock called "getting a feeling for the organism," is not only critical for formulating original scientific hypotheses but also, more deeply, for questioning foundational assumptions of what is counted as knowledge and what we then expect knowledge to do.[62] Looking back on the early scientific forests and even on the more recent Grasslands Biome, it is difficult to speculate on how these projects changed the people involved. However, their stories remind us of the irreducibility of an unruly and complex environment and that, as hard as we try to contain the world in neat technological metaphors, it will always leak out and transform us.

[62] See Evelyn Fox Keller, *A Feeling for the Organism: The Life and Work of Barbara McClintock* (San Francisco: W. H. Freeman, 1983).

13

At the Limits of Computational Technocracy

Victor G. García-Castañeda

Planetary Complexity

Over the last century, humanity has witnessed a series of events that have forced a rethinking of the effects of human actions on the various environments we inhabit. A collection of competing neologisms has been coined: Anthropocene, Capitalocene, Technocene, Chthulucene, and so on. These concepts posit a newly realized "planetary scale" that can only be addressed by taking into account large periods of time (both anthropological time and geological or deep time) and space (the Earth as a macrosystem). It has been argued that the perception of a planetary scale at which all human and nonhuman inhabitants of the Earth can be perceived, and thus managed, within a single framework has been made both possible and necessary by the advance of technology and scientific ways of observing, knowing, and objectifying the planet. From the revelation of the Earth seen from space as a small particle in a vast universe to the less visible but nonetheless ubiquitous extension of information networks and communication infrastructure, our world has ostensibly been made small, tenuous, mapable, and controllable by the sheer power of computation.

While it may seem helpful to contemplate planetary scale as a necessary way to address a plethora of problems that concern all the inhabitants of the Earth, in reality, there is no single macroscopic scale, in either time or space, to which the complexity of planetary phenomena can be reduced in order to develop one-size-fits-all solutions to the problems threatening our era, such as climate catastrophe, persistent neocolonialism, racial capitalism, and cognitive capture, among other systemic problems. Since all these issues are the result of concrete historical dynamics carried forward from a past into which we can no longer return, we are bound to work toward a realistic future that may arise from our complex condition but always in this same world with the affective, rational, and technological tools we currently have at hand, not through the utopic, or rather dystopic, programs of speculative designers or fiction theorists that, more often than not, dwell on ways to flee our inescapable reality.

The planetary hosts an ever-changing diversity of complex systems and processes, woven together in syncopated rhythms that only occasionally resonate in unison and that are punctually interlocked to varying degrees across different dimensional proportions. As Patricia Reed warns, the capacity for planning and action can only bear fruit if these problems are considered in their own multidimensional complexity. Emphasizing the multidimensional composition of the manifold planetary scales with their different magnitudes by attending to the multiplicity of social, political, economic, and ecological relationships that shape them requires recognition of our shared reality while avoiding eclipsing its particular dimensions.[1] In other words, consciousness of the planetary cannot come without accounting for its inherent diversity and the necessarily localized and uneven ways in which the planetary is lived or experienced. Yet, discourses regarding locality come with their own problems that need to be addressed first in order to avoid falling into the pitfalls of ethnonationalism or ecofascism.

It seems that discussions regarding the planetary oscillate between affirmative, techno-solutionist, speculative projections, and reactionary or romantic fantasies of rejection and exodus. While these are played out largely as set-piece battles in the circuits of academic and para-academic discourse, or as tech-industry hype, the world at large still lives under the "business as usual" of the political-economic structures we have inherited from the past. In order to go beyond these Manichean discourses, it is necessary to consider the limitations that demarcate the conjunction of the technological, geopolitical, economic, and ecological dimensions of the planetary from a multiscalar approach that takes into consideration issues of sociocultural complexity and technological diversity. In this essay, I attempt to do so by critically interrogating the techno-solutionism approach of "planetary-scale" technocracy, examining the limitations and possibilities of computation, as well as exploring a non-fascist eco-techno politics of locality. I conclude with a call to do away with Prometheanism and to instead cultivate alternative relations between technology and nature.

1 Patricia Reed, "What is Care at Planetary Dimensions?," Floating University, Berlin, August 6, 2019, https://laboriacuboniks.net/wp-content/uploads/2020/05/Reed-What_is_Care_Lecture-2019.pdf.

Planetary-Scale Technocracy

Haunting such attempts to conceptualize a single planetary scope is the specter of a Promethean global technocratic project that sees the Earth as a collection of resources to be managed and extracted, humans as operative cogs in a unified techno-social machine, and technology as a panacea that can solve any type of problem through technological processes like calculation, sensing, modeling, and simulation at the macroscale. From this perspective, the solution to planetary problems such as climate change is usually reduced to a problem of technological infrastructure. Automated planetary infrastructure would create a network that Benjamin Bratton has called "planetary-scale computation,"[2] an information platform spread across the Earth that, instead of using natural resources to spit out mobile phones,[3] is reoriented to perceive and detect imminent ecological risks, model their repercussions through large-scale computation, and act geopolitically to address them.[4] Bratton claims that such a "megastructure" has already been effective in allowing humanity to perceive climate change and can therefore be instrumental in responding to it.[5]

Let us take the case of geoengineering as an example of the application of planetary-scale computation to mitigate climate change. Supporters of this approach commonly share the prognosis that it is too late to try to curb climate change only through the political and diplomatic methods currently deployed in geopolitics. As Andreas Malm points out, they dismiss political solutions, which they consider "too challenging to contemplate, let alone advocate."[6] Therefore, they argue for reinstrumentalizing capitalist-coded technologies for bold, dramatic acts of speculative design and macroscale intervention that are capable of mitigating the effects of climate change through solar radiation management, carbon dioxide removal, weather modification, and bioengineering crops. The problem with this technological determinist view is that it assumes that planetary engineering, along with social, economic, and political issues, are to be resolved primarily by means of a global digital machine and that changes will inevitably follow once this infrastructure is set in motion.

[2] Benjamin Bratton, *The Stack: On Software and Sovereignty* (Cambridge, MA: MIT Press, 2015).

[3] Bratton, *The Stack*, 83.

[4] Regarding the deployment of planetary sensing technologies, see also Jennifer Gabrys, *Program Earth: Environmental Sensing Technology and the Making of a Computational Planet* (Minneapolis: University of Minnesota Press, 2016).

[5] Benjamin Bratton, *The Terraforming* (Moscow: Strelka Press, 2019), 35. Ebook.

[6] Andreas Malm, "Planning the Planet: Geoengineering Our Way Out of and Back into a Planned Economy," in *Has It Come to This? The Promises and Perils of Geoengineering on the Brink*, ed. J. P. Sapinski, Holly Jean Buck, and Andreas Malm (New Brunswick: Rutgers University Press, 2020), 146.

Unlike Bratton's techno-solutionist approach, Holly Jean Buck not only entertains the idea of geoengineering as a practice that will eventually be unavoidable but also questions the possible consequences that may result in the future from using experimental methods and macroscale interventions such as solar radiation management or carbon dioxide capture. Since, as Malm points out, we do not have access to multiple Earths,[7] there can be no control group in these experimental solutions. Consequently, as Buck suggests, tackling such planetary-scale problems requires greater collective inclusion of human agencies rather than relying solely on technological solutions.[8]

In order to be an effective tool, any type of planetary action coming from a "megastructure" would need to be coupled with an equally effective planetary system of governance. This leads us to economic and geopolitical questions, such as who would define the values and aims to be used as parameters of the computational apparatus and according to whose authority the dictates of its outputs would be executed. It is very hard to believe that the heads of techno-capitalist companies would ever relinquish control of their technological developments for the sake of the common good without first ensuring that such actions would generate economic, political, or social dividends for themselves and their shareholders. Indeed, it has been made evident in recent years that the language of "social responsibility" and "technological solutions" is often used by corporations to mask their strategies of extracting profit from various ecological and humanitarian crises. As Achille Mbembe warns in relation to the African continent:

> Today, the overwhelming belief is that, coupled with science and technology, market capitalism and "humanitarian" interventions will sort out most of the continent's problems. Complex social structures and processes such as war, mass poverty, joblessness, disease, and illiteracy are treated as if all of these are purely technical matters and the human subjects implicated in these dramas have no histories. History itself has been reified in a set of abstractions, and the sense of being at the edge of a future so palpable in the immediate aftermath of colonialism and Apartheid has quickly vanished.[9]

7 Andreas Malm, "The Future is the Termination Shock: On the Antinomies and Psychopathologies of Geoengineering. Part One," *Historical Materialism* 30, no. 4 (December 2022): 3–53, here 10.

8 Holly Jean Buck, *After Geoengineering: Climate Tragedy, Repair, and Restoration* (London: Verso Books, 2019).

9 Achille Mbembe, *Out of the Dark Night: Essays on Decolonization*, trans. Daniela Ginsburg (2010; New York: Columbia University Press, 2021), 11.

Mbembe asserts that the computational technologies of the present, when figured as mediators of markets organized by the perverse freedom of consumer choice, threaten to reduce the world to a "huge data emporium" that reproduces these global power dynamics in amplified form, including the brutal racist world order of colonialism, in a phenomenon he calls "the becoming-black-of-the world."[10] Such a view fundamentally challenges the claim that it might be possible to reinstrumentalize technologies that were originally designed to extract material or cognitive resources so that they instead function as tools for conservation.

Even if the project of an automated planetary-scale government could be devised in corporative and geopolitical terms, this does not evade the material obstacle of implementing it effectively. The proponents of a planetary-scale geoengineering project assume that the Earth has sufficient economic, material, and energy resources to accelerate technological developments, when, according to the Global Footprint Network, humanity currently uses resources comparable to living on 1.75 Earths.[11] In other words, the limits of the "planetary apparatus" mentality are inevitably bound by the availability of natural resources. Perhaps that is why many of the techno-solutionist discourses have projected this extractivist logic beyond the Earth's limits through their proposals for exoplanetary mining and colonization.

There is also an epistemological limitation to any planetary-scale conceptualization that requires this assumption: Our planetary perception has only come about thanks to exponential technological developments, which have effectively expanded our horizons of perception at the macroscopic and microscopic levels through the vast deployment of artificial sensors and data processors. However, these technologies attempt to frame reality in a particular way and from a particular perspective, on a macro- or microscopic level, as if we were browsing through Google Maps. In reality, a macroscopic scale does not necessarily include the minor scales in its scope, nor the other way around. Different scales offer different information, which may or may not coincide. Shifting from different scale magnitudes may cause the impression of a "zoom effect," which in turn leads one to think that a solution implemented from the macrolevel would cascade to the bottom scales lineally, but this impression "is

[10] Achille Mbembe, "Thoughts on the Planetary: An Interview with Achille Mbembe," Interview by Torbjørn Tumyr Nilsen, *New Frame*, September 5, 2019, https://www.newframe.com/thoughts-on-the-planetary-an-interview-with-achille-mbembe.

[11] Global Footprint Network, https://www.footprintnetwork.org.

exactly that: an *effect*, an assemblage as artificial as a fake perspective in a stage set."[12]

In contrast, perceptions of the climate shared by smaller, local communities may offer sensibilities of time-space at a scale that may be more effective in dealing with planetary problems at a local level without having to delegate political power to a computational macrosystem. While there is no denying that scientific reasoning and technological evolution have offered us valuable tools to deepen our understanding of our environments, Bratton's planetary-scale computation largely omits the fact that, for centuries, there have been approaches to ecological sensing and acting that are not necessarily embedded in Western scientific rationality or computation and do not accommodate the extractivist agenda of capitalism or the attempt to erect an all-encompassing technocracy. Moreover, as Kathryn Yusoff points out, the scope of the planetary tends to ignore its reproduction of an inherently colonial perspective that has historically subjugated certain communities through forms of direct and indirect violence.[13]

In fact, there are ways of reasoning and designing technologies that are not coded by capitalism and colonialism, which have also resisted the projects driven by such forces. These technologies are products of their own material and cultural realities that present both practical and concrete alternatives to contemporary methods and more sustainable models for mutually nurturing relations between humans, technology, and nature.[14] For example, Mesoamerican *chinampas* and *milpas* are agricultural technologies rooted in a cosmology that, beyond perceiving the natural world only as the terrain we inhabit, constructs a political and spiritual framework sustaining a particular social and spatial organization. In other words, they not only provide alternative models of sensing and understanding nature but also offer sustainable solutions to minimize our ecological footprint.[15]

[12] Bruno Latour, "Anti-Zoom," in *Olafur Eliasson: Contact*, exhibition catalog, Fondation Louis Vuitton, Paris, December 17, 2014–February 23, 2015 (Paris: Flammarion, 2015), 121, original emphasis. Republished in *Scale in Literature and Culture*, ed. Michael Tavel Clarke and David Wittenberg (Cham: Palgrave Macmillan, 2017), 94.

[13] Kathryn Yusoff, *A Billion Black Anthropocenes or None* (Minneapolis: University of Minnesota Press, 2018).

[14] Julia Watson, *Lo—TEK: Design by Radical Indigenism* (Cologne: Taschen, 2019).

[15] The agricultural techniques of pre-Columbian Mesoamerica—such as the milpa and the chinampas mentioned above—maintained a relationship with the soil that was very different from that of the plowing techniques introduced by the Spanish colonizers in the sixteenth century. In the milpa, maize is cultivated along with beans and other vegetables, like squash, to create a sustainable agroecosystem where the nitrogen-fixing roots of legumes provide nutrients to maize, while the maize stalk serves as a support for the bean plant that becomes entangled in it to grow. Chinampas, on the other hand, are a type of floating garden in which crop plants are cultivated on lakes or wetland environments so as to also provide great benefits to water potability, flood prevention, and, in contemporary realization, carbon sequestering. The chinampas, which take their form from both their specific

One could argue that, understood in the most abstract terms, geoengineering or terraforming appears to be something that has taken place throughout the evolution of human culture. However, it would be unreasonable to compare locally rooted modifications of natural environments that maintain a mutually nurturing, horizontal relationship with nature to the establishment of an omnipotent technological empire—the *système technicien* referred to by Jacques Ellul,[16] or a *vast machine* as Paul N. Edwards calls it[17]—on a planetary scale that generalizes a homogeneous and univocal mode of relating technology and nature that ends up subsuming all its processes into cybernetic feedback loops. This is why Malm calls most of the proponents of geoengineering "rational optimists" because they faithfully believe in the power of scientific reason and technology to solve the planetary catastrophe if the best possible outcomes are realized, which is in and of itself not a realistic expectation.[18]

The climate catastrophe cannot be solved by reducing humans to mere cogs in a machine that only needs to be oiled to function better, especially when not all human actors are equally responsible or have the same level of agency in the problem. It is not reasonable to distribute responsibility under a logic of ethical flattening when the major culprits of this catastrophe are easily recognizable: oil and gas companies, Big Tech, and the governments of heavily industrialized nations, to name the most prominent. These actors have deployed their political and technological might to create an interlocking system of data and climate injustice.

Much of the promotion of terraforming, geoengineering, and digital environmental governance on a planetary scale, which is increasingly widespread in academic, para-academic, and popular discourse, perfectly accommodates the political agendas of companies and governments that are at the forefront of the research and development of the digital technologies by which these projects would be undertaken. The desire for planetary-scale computation is the desire for a Promethean technocracy already prone to exacerbating the power differentials

ecologies and their social and cultural context, differ from European models for agricultural land "reclamation," such as the Dutch polder, in their specific division of labor and structuring of the family as a production unit. For more information, see Victor G. García-Castañeda, "Techno-diversity and Maize: Suggestions for the Search of a Mesoamerican Cosmotechnics," *Technophany: A Journal for Philosophy and Technology*, no. 1 (November 2021), 1–22.

16 Jacques Ellul, *The Technological System*, trans. Joachim Neugroschel (1977; New York: Continuum, 1980).

17 Paul N. Edwards, *A Vast Machine: Computer Models, Climate Data, and the Politics of Global Warming* (Cambridge, MA: MIT Press, 2010).

18 Malm, "The Future is the Termination Shock," 10.

that favor techno-capitalist companies and short-sighted government officials of hegemonic nations.

Going Beyond the Computable

If a planetary scope already poses geopolitical, material, and epistemological limitations, the problem of computation carries a much more radical impasse. Reality is fundamentally open, respective, and dynamic, and as such, it cannot be reduced *absolutely* to a computational representation. Luciana Parisi and Yuk Hui have referred to Kurt Gödel's incompleteness theorem and Alan Turing's halting problem to argue for the impossibility of reducing reality to a series of mathematical axioms, since the propositions of any such system cannot be proved by the very axioms upon which they are based, because a system cannot be complete and consistent at the same time.[19]

Yet, computation seems to signal a way out of determinism if we take into account the noise, randomness, contingency, and glitches that emerge within the system. In an effort to think about computation uncodified by techno-capitalist logic, Parisi introduces the work of mathematician Gregory Chaitin, whose approach highlights the "productive capacity of noise and randomness in communication systems, to discuss computation in terms of maximally unknowable probabilities."[20] Parisi underlines the emergent properties that arise through computational processes that are always infected with what Chaitin calls "algorithmic randomness,"[21] a concept that refers to the entropic transformation of data that occur between an input and an output: "this challenges the view that computational processing corresponds to calculations leading to pre-programmed and already known outputs. Instead, the limits of automation—that is, the incomputable—have become the starting point of a dynamism internal to computation, which exceeds the plan for technocapital's instrumentalization of reason."[22] Such an assessment counts on the incomputable to generate new thought, in that computation holds the potential to be opened

[19] Gregory J. Chaitin, *The Unknowable* (Singapore: Springer, 1999).

[20] Luciana Parisi, "Instrumental Reason, Algorithmic Capitalism, and the Incomputable," in *Alleys of Your Mind: Augmented Intelligence and Its Traumas*, ed. Matteo Pasquinelli (Lüneburg: Meson Press, 2015), 132.

[21] For more information on this concept, see George J. Chaitin, *Meta Math! The Quest for Omega* (New York: Pantheon, 2005), 122; and George J. Chaitin, "Leibniz, Randomness & the Halting Probability," *Mathematics Today* 40, no. 4 (June 2004): 138–39.

[22] Parisi, "Instrumental Reason, Algorithmic Capitalism, and the Incomputable," 134.

up to the indeterminate. As Parisi argues, algorithmic automation no longer corresponds to the execution of instructions but to the constitution of a machine ecology infected with randomness.[23] From this premise, it becomes possible to posit an "incomputable subject" that manages to reclaim the contingency of the instrumental as a mode of nonbinary thinking and opens the political field to the unknown.[24]

Hui makes a similar argument when referring to the recursive character of contemporary computation, which is no longer linear nor mechanistic. Technological evolution has made it possible for machines to deal with contingency, discern events such as noise and failure, and distinguish unorganized inputs from necessary ones. But this only happens with the help of humans as organizers of technology. As Hui contends, we need to understand our current technical reality and its relation to diverse human realities; therefore, we must first remove eschatological interpretations of computation that take its progress as a utopic or dystopic teleology. Instead, a shift in the culture of prosthetics is needed in order to establish a horizontal relationship of coevolution with technology.[25] That is why it is necessary, as Parisi suggests, to rethink the notion of instrumentality when it comes to technology so as to avoid reproducing the existing schema of techno-capitalist "mono-technology," as Hui calls it—that is, a hegemonic and generalized technical culture.[26]

Hui complements Parisi's interpretation of Chaitin by contrasting the notions of incalculability and incomputability: While incomputable "refers to a number that cannot be *recursively enumerable*, which is to say that it cannot be reduced to the finite steps of an algorithm," incalculable "means that something cannot be submitted to calculation at all, such as love, friendship, desire, or happiness."[27] If the incomputable denotes the contingent character of computation and its inherent randomness, the incalculable refers to the radical impossibility of subsuming the totality of reality to its axioms and reducing everything into a mathematical notation. Since reality is not a zone containing things that are

[23] Parisi, "Instrumental Reason, Algorithmic Capitalism, and the Incomputable," 136.

[24] Antonia Majaca and Luciana Parisi, "The Incomputable and Instrumental Possibility," *e-flux Journal*, no. 77 (November 2016), https://www.e-flux.com/journal/77/76322/the-incomputable-and-instrumental-possibility.

[25] Yuk Hui, "ChatGPT, or the Eschatology of Machines," *e-flux Journal*, no. 137 (June 2023), https://www.e-flux.com/journal/137/544816/chatgpt-or-the-eschatology-of-machines.

[26] See Yuk Hui, "One Hundred Years of Crisis," *e-flux Journal*, no. 108 (April 2020), https://www.e-flux.com/journal/108/326411/one-hundred-years-of-crisis.

[27] Yuk Hui, *Art and Cosmotechnics* (Brooklyn: e-flux Books; Minneapolis: University of Minnesota Press, 2021), 232, emphasis original.

fully enumerable, much less calculable, any kind of computational modeling will always be partial, momentary, and biased.

In contrast, there are other modes of reasoning that do not necessarily base their logic on mathematics. Hui's way out is to look to an epistemology of the unknown that manages to recursively reintroduce the nonrational and the contingent into the system through practices such as art, which are understood as creative forces that produce reality. Within such a framework, these practices are always rooted in a certain locality, and Hui proposes to revisit the question of locality since that is where the cosmological relations we maintain with our environment through technological mediation emerge. This emphasis on technodiversity serves as a way out of the totalizing impetus of modern technologies such as planetary-scale computation.

Resituating Locality

It would seem that there are two main poles that are at play in today's political landscape: the local and the global. The former alludes to belonging to the land, traditions, and identity, while the latter refers to globalization, multiculturalism, and an openness to a world without borders. If the first faction sees its traditions diluted in the face of global homogenization, the second fails to offer a sense of belonging to the myriad of communities that inhabit the earth.[28] Ironically, until not so long ago, the first camp housed the conservative and reactionary positions of politics and the second the progressive and liberal positions. While local/reactionary populism and progressive internationalism have both always had left and right vectors, today this division has become evermore perplexing—a sign of our politically disorienting times. At points, these binary political oppositions end up feeding the same machinery that produces them, either through a fetishization of the global or an exoticization of the local. The binary ideologies of the past, upon which contemporary political dynamics continue to be based, are increasingly obsolete.

If the proponents of global planetary-scale eco-techno-politics tend to minimize the local through "one-size-fits-all" solutionism, the question of the local must also navigate its own ethical minefield. Once a certain locality is defined, it does not take much work to begin speaking in Heideggerian fascist

[28] Regarding this subject, see Bruno Latour, *Down to Earth: Politics in the New Climatic Regime*, trans. Catherine Porter (2017; Cambridge: Polity Press, 2018).

tongues—of the destiny of certain societies, of their right place, of authentic ways of revealing their "Being" over inauthentic ones—that consequently justify the power of certain nations and social groups over others. But what is the local? Where is it? What defines a locality? The whole planet? A continent? The borders of a country? The nation? A city? A neighborhood? A given number of people? The extent of a shared language? The customs of a particular region? The relationships in a family nucleus? A relationship between two peers? The individual? If we take the problem of locality ad infinitum, we run the risk of being left with self-absorbed atomism, or worse, with ethnosuprematism and ecofascism. At the same time, the concept of locality is inexorably tied to the question of its holistic relationship with the planetary, since the local is never isolated nor closed to the whole. How, then, is the local able to connect the whole with the parts within the sphere of a shared planet?

In order to move away from the fascist discourses that are common in the rhetoric of the local, Hui takes the concept of locality as a foundation for diversity. One of the fundamental axes of Hui's philosophy is a critique of the monotechnological vision of modern technology as a product of capitalist industrialization and the globalization of markets, which impose a homogeneous *episteme* with a standardized and univocal way of relating to technology and the world. Against this, Hui proposes to reclaim diversity as an imperative for the planetary condition. Just as the planet hosts a biological and genomic variety of species, which make up multiple ecosystems that are relatively balanced by multiple internal and external factors, he calls for a technodiversity that analogically fulfills the same function as biodiversity.[29]

The concept of technodiversity proposes to think of technology in relation to the communities from which it originates to the multiple and diverse cultures and their particular cosmologies. This disposition, however, goes beyond just preserving the present technological diversity, which is also necessary and recognizes the creative and innovative potential of creating new forms of diversity.[30] For Hui, locality is presented as the starting point from which a society establishes worlds in common and collective projects situated in the specificity of place. Hui's intention here is not to oppose the particular within the universal, or the local within the global, but to point out that every vision of the global or planetary is locally rooted. Hui introduces the concept of *cosmotechnics*

[29] Yuk Hui, "Machine and Ecology," *Angelaki: Journal of the Theoretical Humanities* 25, no. 4 (2020): 54–66, here 63.

[30] Yuk Hui, "For a Planetary Thinking," *e-flux Journal*, no. 114 (December 2020), https://www.e-flux.com/journal/114/366703/for-a-planetary-thinking.

to emphasize the cosmological character that resides in the local and that is expressed in technical mediations.[31] Cosmotechnics is not a multiculturalism that recognizes different cultural approaches to the same nature, nor is it a "multinaturalism" such as that proposed by Eduardo Viveiros de Castro, who argues for the recognition of multiple natural ontologies.[32] Hui argues that such positions continue to reproduce the old Enlightenment opposition of culture to nature. The concept of cosmotechnics offers an understanding of nature and culture entwined by the different technical mediations of each society, so as to articulate these two apparently opposed poles. Hence, his definition of cosmotechnics is "the unification between the cosmic order and the moral order through technical activities,"[33] a phrase that echoes the Kantian maxim of "the starry heavens above me and the moral law within me."[34]

Although the local offers a starting point, this cannot be exclusive or taken as an excuse to affirm differences in their extremes. In contrast, Denise Ferreira da Silva proposes instead to speak of "nonlocality"[35]—a concept she takes from quantum mechanics—as a response to the epistemological and ontological assumptions of determinacy, separability, and sequentiality that underlie the philosophies of modernity. An imagination that starts from nonlocality could devise a world that is not that of the totality of an "Ordered World" based on estrangement but one where there is "difference without separability" and entanglement of its parts. As Ferreira da Silva writes:

> What if, instead of the Ordered World, we imaged each existant (human and more-than-human) not as separate forms relating through the mediation of forces, but rather as singular expressions of each and every other existant as well as of the entangled whole in/as which they exist? [. . .] When nonlocality guides our imaging of the universe, difference is not a manifestation of an unresolvable *estrangement*, but the expression of an elementary *entanglement*.[36]

Similarly, for Hui, each localized cosmological specificity constitutes the way in which each culture relates to and from technics. In this theoretical framework,

31 Hui, *The Question Concerning Technology in China*, 19.
32 Regarding the concept of "multinaturalism," see Eduardo Viveiros de Castro, *Cannibal Metaphysics: For a Post-Structural Anthropology*, trans. Peter Skafish (2009, Minneapolis: Univocal, 2014).
33 Yuk Hui, *The Question Concerning Technology in China: An Essay in Cosmotechnics* (Falmouth: Urbanomic, 2016), 19.
34 Immanuel Kant, *Critique of Practical Reason*, trans. Mary Gregor, rev. ed. (Cambridge: Cambridge University Press, 2015), 129, emphasis removed.
35 Denise Ferreira da Silva, "On Difference Without Separability," in *Incerteza Vica*, ed. Jochen Volz and Júlia Rabouças, exhibition catalog, 32nd Bienal de São Paulo, (São Paulo: Fundação de Serralves, September 7–December 11, 2016).
36 Ferreira da Silva, "On Difference Without Separability," 63–65, emphasis original.

the question of the planetary is reframed as a question about local mediations of technics and their places within a plurality of reasonings and *epistemes* that are recursively connected to the planetary. For Hui, the relationship between machines and ecology requires the discovery of cosmotechnical diversity that is able to resituate technics within a particular geographical milieu, culture, and thinking. As he argues, only by "reframing the enframing" can we imagine a "new earth and people that do not yet exist."[37]

Hui sees in art a fertile ground for the creation of an epistemic revolution that would make it possible to confront the modern monotechnological *Gestell*. Through the "education of sensibility"[38] fulfilled by art, it becomes possible to create epistemes rooted in their localities that are recursively connected with the cosmos. In addition to biodiversity and technodiversity, noödiversity (a diversity of intelligence) is also necessary in order to multiply our ways of feeling, sensing, and reasoning reality. If we take the local as a context without defined borders, a time-space configured by the multidimensional perspectives of the myriad of planetary scales, then the local may pave the way to a cosmopolitics that serves as a countermeasure to the homogenizing view of a planetary-scale technocracy. But to do that, the discourses regarding the local must be vigilant about xenophobia, nationalism, or fascist "return to the land" mentalities; instead, they need to foster the diversity, multiscalarity, and sociocultural complexity and entanglement that the planetary-scale problems demand.

Doing Away with Prometheus

In light of these critiques, it appears abundantly clear that any projects of geoengineering, pervasive sensing, and macroscale computation that aspire to offer viable solutions to the problems of the planetary must address and negotiate their ethical, epistemological, and material pitfalls and contradictions. Before attempting to commit all of our efforts to finding a single technological solution to ecological and political problems, we must question whether the Earth actually possesses sufficient resources to undertake effective planetary-scale projects, especially as colonizing agents, governments, and corporations continue to harm and exploit marginalized communities and the land they inhabit. In addition, we must question primitivists and ecofascists who envision an impossible return to

[37] Hui, "Machine and Ecology," 65; Gilles Deleuze and Félix Guattari, *What is Philosophy?*, trans. Hugh Tomlinson and Graham Burchell (1991; New York: Columbia University Press, 1994), 108.
[38] Hui, *Art and Cosmotechnics*, 3–63.

a premodern or precapitalist past, for that past never existed in any unilateral sense. How can the colonized "return to a land" that has been systematically pillaged for centuries? Against the myriad other unreflective forms of escape to the countryside exacerbated by the recent Covid-19 pandemic, which frequently subscribe to an arcadian Gaia principle, we should question the premise that our planet is a great idyllic ecosystem, an ecological totality where everything fits together like a magnificent puzzle or a unified ecosystem that exists in perpetual harmony. Nature is also ruthless, chaotic, parasitic, and violent and can only be taken as a moral role model on the level of poetic metaphor, which, as malign voices from the Hobbesian "social darwinists" to the contemporary revival of "scientific racism" demonstrate, can be just as easily read otherwise.

This sentiment echoes Malm's criticism of contemporary discourses regarding nature. In *The Progress of This Storm*, Malm argues against a vision of nature as a mere social category that obfuscates the actual biochemical and physical processes occurring in it; he likewise rejects hybridism that takes society and nature as ontologically indiscernible entities; and he criticizes the new materialisms for their stubbornness in assigning an agency to matter that ends up minimizing the responsibility of human agency. Malm proposes a "climate realism" that would affirm the reality of nature, not merely as a social construction created by humans nor as an epistemological question of micro- or macroscopic perception but as a physical reality that is part of us just as we are part of it.[39] Yet, as this chapter has sought to make clear, what we especially need to do away with is the Promethean vision that continues to dominate global technocratic approaches to tackling planetary crises. For millennia, we have thought of technics as "man-made" artifacts, tools with which to intervene in nature, or ways of doing and knowing. The Promethean vision of technology—a fundamental pillar of capitalism—which takes technological development and growth as a right exercised by humans, even at the expense of the survival of other species or the planet as a whole, must be conclusively done away with. This view must be discarded in favor of a definition of technics that also takes natural processes into account. This does not imply taking nature as an immaculate and pure entity or any more of a panacea than technology is purported to be, nor does it wish to fall into an exoticization of the local that just ends up serving nationalist slogans by taking part in contemporary performative identity politics. On the contrary, what it presents is an opportunity to close the gap between

[39] Andreas Malm, *The Progress of This Storm: Nature and Society in a Warming World* (London: Verso Books, 2018).

natural and artificial processes and establish sustainable relations between natural technics and human technics, just as many altermodern communities and their cosmologically rooted practices have done for centuries and continue to do today.

What we need are technologies oriented toward establishing sustainable and creative relationships between the inhabitants of the various eco-techno-political localities that are recursively connected to the planet. This means building an appropriate techno-ecological association that does not reduce nature to ideal abstractions and material resources to be managed and exploited. These eco-technologies can only emerge from political projects based on open, inclusive, responsible, respectful, and horizontal relationships with our environmental reality. The problem of planetary phenomena—of which climate catastrophe is the most pressing but which also include global capitalism and the persistence of neocolonial structures of domination—is not a problem of technological infrastructure but rather of concentration of power, of greed, and of short-term accelerated growth that captures society within its circuits, creating a sense of apathy and hopelessness. If we accept that every form of government unfolds as a technological device, just as every technology unfolds as an ordering of nature, then the question is not whether we should put technology over politics or economics, or vice versa, but how we should define the technological modes of mediation that we want for the survival of the planet and what kinds of relationships between human governance and its technical praxis are most appropriate for a future open to the multiple contingencies emerging from the interconnection of diverse realities.

14

Prologue to the Sky River[1]

Elise Misao Hunchuck, Marco Ferrari, and Jingru (Cyan) Cheng

君不见，黄河之水天上来
Don't you see the Yellow River rushing from the sky?

—Li Bai (Tang Dynasty)[2]

The Qinghai-Tibet Plateau is often called the water tower of the world. As the source of most of Asia's significant rivers—the Indus, Ganges, Brahmaputra, Irrawaddy, Salween, Mekong, Yangtze, and Yellow Rivers—the plateau provides water for more than two billion people downstream. In the Yellow River's catchment, traditional—even if extraordinary in scale—water management techniques are reaching their upper limits. Severe stress on available water in the North China Plain, nearly half of which comes from Qinghai-Tibet, is tied to burgeoning consumption, water-intensive agricultural production, and industrial activities. Moreover, fears for its future scarcity due to climate change and the depletion of glacial reserves in the Himalaya have led the Chinese government to apply water precipitation enhancement technologies on a dramatically increasing scale.

The desire to control the rain is not new, and it is not unique to the People's Republic of China. To tell the story of the Sky River—both as a conceptual model and as a weather modification project—is also to learn the hydro-geographical history of China, which has actively manipulated its relationship to water for thousands of years and has been developing weather modification technologies since the 1950s. One of these technologies is the cloud-seeding stove (人工增雨烟炉), a rudimentary metallic envelope on a short pedestal that encases a

[1] This text was originally published in *The Avery Review*, no. 53 (June 2021): 1–15, https://averyreview.com/issues/53/prologue-to-the-sky-river.
[2] This is the first line of Li Bai's Tang Dynasty (A.D. 618–907) poem, 将进 Qiang Jin Jiu [Invitation to Wine]. Translated by Jingru (Cyan) Cheng.

high-performance combustion chamber, employing military rocket engine technology to burn a chemical compound of silver and potassium iodide. The stoves are installed on elevated terrain and strategically positioned at high altitudes where upward winds deliver the chemical exhaust into cloud formations above to trigger rain. A discrete object built and replicated across China, the stove is where and how the physical embodiment of the Sky River—as both a phenomenon and a project—is seen for the first time.

In the troposphere above the Sanjiangyuan (三江源, Source of the Three Rivers) area of the Qinghai-Tibet Plateau is the testing site where a confluence of water vapor conveyance channels—*sky rivers*—have been identified. Sky rivers are a conceptual model of water vapor transport that can, hypothetically, be turned into precipitation through the Sky River Project, a distributed infrastructure of ground and orbital machinery. While a constellation of satellites observes and monitors atmospheric conditions, the network of automated and remotely operated stoves is strategically placed to modify the spatiotemporal distribution of precipitation above this pilot region to make it rain. As a testing site and a hypothesis, the Sky River Project—one of the largest and most ambitious weather engineering projects to date—extends national extractive claims upward into the atmosphere.

The Third Atmosphere

"What kind of a thing is a cloud?"

—Sadie Plant[3]

The atmosphere-ocean environment of the planet Earth has evolved slowly over some four and a half billion years.[4] This, the third atmosphere in the planet's history, is a gas and aerosol envelope extending out from land, ocean, and ice toward outer space, held in place only by the planet's gravitational pull. The atmosphere is most dense near the surface of the Earth, where gravity pulls gases and microscopic particles of dust, smoke, and chemicals inward to settle on the surface, where they are absorbed by its organism-inhabitants. The lives of these inhabitants are protected by the creation and maintenance of pressure that allows

[3] From Sadie Plant's essay, "Something in the Air," in *The Movement of Clouds around Mount Fuji: Photographed and Filmed by Masanao Abe*, ed. Helmut Völter (Leipzig: Spector Books, 2016), 261.

[4] Atmospheres are defined by their chemical composition and with 78 percent nitrogen, 21 percent oxygen, 0.9 percent argon, and 0.04 percent carbon dioxide; it is Earth's most oxygen-rich atmosphere to date (the first and second atmospheres were dominated by hydrogen, helium, water, carbon dioxide, and ammonia, respectively).

Figure 14.1 The water vapor and aerosols (airborne particles of dust, soot, mold, and bacteria) of Earth's third atmosphere-ocean environment are shown in this full-disk visualization centered above China. Image derived from FY-4A AGRI L1 satellite dataset, 2021; Fēngyún-4 are China Meteorological Administration's most recent geosynchronous meteorological satellites, 2021.

water to exist in each of its three phases: solid, liquid, and gas. Together, these multiple states contribute to climate regulation by allowing for the absorption of solar radiation, the needed warming of the planet's surface through the retention of heat, and the reduction of changes in temperature between day and night. The atmosphere contains a variable amount of water vapor, with an estimated volume of 37.5 million billion gallons. Particles of water vapor move into the atmosphere from bodies of water (evaporation) or the exhalation of leaves (transpiration). As the particles are carried up and out into the atmosphere by wind (advection), they meet aerosols—sometimes the result of natural sources like volcanoes, sometimes the result of human activities like air pollution. Some of these aerosols are hygroscopic, both attracted to and ready to hold water vapor. If the conditions are right and if they find themselves in a stable but saturated parcel of air, these aerosols pull water vapor particles in, becoming the cores of cloud droplets. This moment of saturation is also known as the dew point, when the congregation of droplets—*a cloud*—grows in volume, weight, and height until they fall back down toward the surface of the Earth.

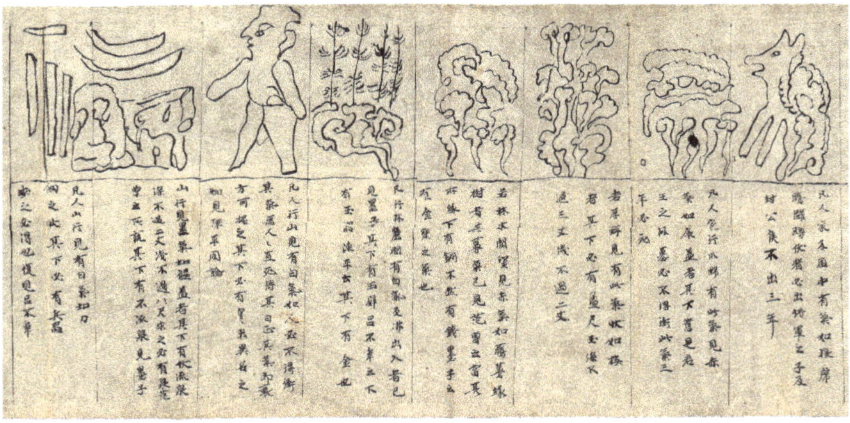

Figure 14.2a Embedded within the Dunhuang Star Atlas, the oldest-known manuscript star atlas from around 700 BCE, is a manual for divination based on the shape of water vapor, what we now call clouds.

At any given moment, somewhere around two-thirds of the planet is covered by clouds.[5] These masses of microscopic droplets assist in the distribution of the sun's energy while storms push and pull energy from the equator to the poles and back—a process that may seem straightforward, but "the picture grows increasingly complicated as heat, evaporation, turbulence, radiation, wind, geography, and myriad other factors come into play."[6] Nephology, the science of clouds, requires an understanding of the entire atmosphere, including all of its dynamics and externalities. It is no coincidence, then, that one of the great challenges to climate change prediction is the inadequate representation of cloud processes. Although ground-based observations exist, "scientists require frequent observations (at least daily), over global scales (including remote ocean and land regions), and at wavelengths throughout the electromagnetic spectrum (visible, infrared, and microwave portions of the spectrum)"[7] so that they might approach the possibility of verifying their model predictions. To further their understanding of clouds, NASA has instruments on the Afternoon Constellation (A-Train) satellites Aqua, Aura, CALIPSO, CloudSat, and Terra. Scientists believe the A-Train will also allow them to observe the movement of clouds around the planet and their vertical structure, too: *They will see the clouds from inside.*

In the troposphere, the lowest region of the atmosphere where we live, water vapor also exists in the form of filamentary structures and in a state of constant

5 Natalie Wolchover, "A World Without Clouds," *Quanta Magazine*, February 25, 2019, https://www.quantamagazine.org/cloud-loss-could-add-8-degrees-to-global-warming-20190225.

6 Wolchover, "A World Without Clouds."

7 "The Importance of Understanding Clouds," *NASA*, 2005, https://www.nasa.gov/pdf/135641main_clouds_trifold21.pdf.

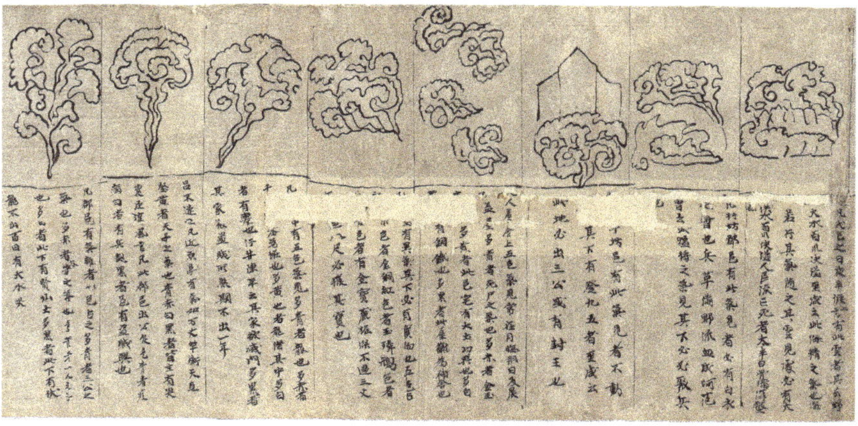

Figure 14.2b Embedded within the Dunhuang Star Atlas, the oldest-known manuscript star atlas from around 700 BCE, is a manual for divination based on the shape of water vapor, what we now call clouds.

flux. In 1992, a group of scientists from the Massachusetts Institute of Technology and NASA Goddard Space Flight Center proposed naming these filaments *tropospheric rivers*, as they had the characteristics of the rivers that we know on land: places where water gathers to move in "lengths many times their widths and persist for many days while being translated through the atmosphere."[8] Their finding originated as a fortuitous observation within NASA's Measurement of Air Pollution from Satellites (MAPS) program, aimed at providing a near-global map of the distribution of carbon monoxide in the lower atmosphere.[9] Startled to see thermal fingerprints of the air-polluting molecule located far away from any potential source of emission, but limited by the capabilities of MAPS, which was not capable of sensing through clouds, the scientists turned to data provided by the Total Ozone Mapping Spectrometer (TOMS),[10] an instrument designed to return a more detailed portrait of the atmosphere by measuring the wavelengths of back-scattered solar radiation. Through the reflectivity data provided by TOMS, they saw moisture pathways connecting the concentrations of carbon monoxide to their source regions. Water vapor, they inferred, was responsible for

[8] Reginald E. Newell, Nicholas E. Newell, Yong Zhu, and Courtney Scott, "Tropospheric Rivers? – A Pilot Study," *Geophysical Research Letters* 19, no. 24 (December 24, 1992): 2401.

[9] The MAPS instrument flew onboard four Shuttle missions between 1981 and 1994 and was equipped with an infrared camera. More information can be found at: https://www.nasa.gov/centers/langley/news/factsheets/MAPS.html.

[10] TOMS is the same instrument responsible for the imaging of the ozone hole in the 1980s and the only American instrument to ever fly onboard a Russian satellite.

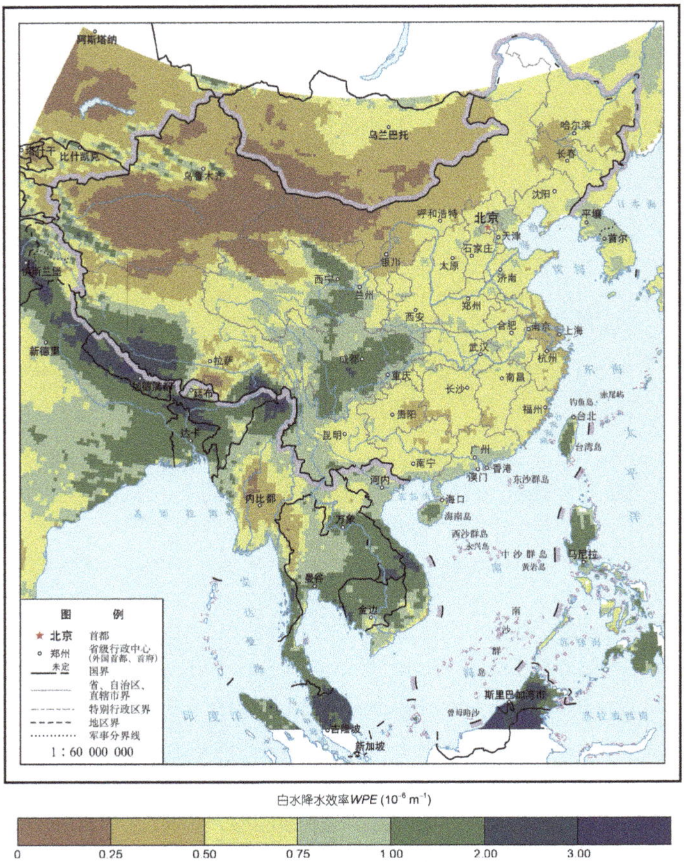

Figure 14.3 The distribution of so-called "white water" precipitation efficiency [sky river] is shown above the Qinghai-Tibet Plateau.

the transport of carbon monoxide in the troposphere over long distances. After establishing the persistence and volume of these so-called "rivers," the authors asked, "Can any *practical use* be made of the apparent relation, revealed through serendipity, between TOMS images and lower atmosphere water vapor flux?"[11] Just two years later, Reginald Newell and Yong Zhu, two of the paper's authors, were able to determine that *these filamentary structures are, in fact, responsible for the transport of almost all of the water vapor in the atmosphere* despite discharging only, on average, a third of the water they carry as either rain or

[11] Newell et al., "Tropospheric Rivers?," 2404, emphasis added.

snowfall. These vectors,[12] which originate in the tropical and subtropical regions and travel poleward, represent such frequent features of global atmospheric circulation that evidence of their presence can be observed in ice cores from Greenland and Antarctica.[13]

Sky Rivers (天河)

> We do not want to just write theoretical essays, we want to write essays on the ground. For the Yellow River to flow continuously, we have been working for ten years; we also want to do this for ten years to see if we can find a new way to solve the water shortage of the Yellow River.[14]
>
> —Wang Guangqian, Principal Investigator of the Digital Yellow River Model, President of Qinghai University, and leader of the Sky River team

In a series of seminal papers published between 2016 and 2018 by *Science China Press*, a team of hydroscience engineers from Tsinghua and Qinghai universities proposed a new concept of atmospheric water: "white water." Alongside the two already established types of water that are conventionally recognized as part of the hydrological cycle—blue water, found in "rivers, lakes, underground aquifers, and other sources that can be used by humans directly,"[15] and green

[12] It is expected that atmospheric rivers, which originate in the tropical and subtropical region and travel poleward, will become longer, wider, and stronger as a consequence of global warming (from Vicky Espinoza, Duane E. Waliser, Bin Guan, David A. Lavers, and F. Martin Ralph, "Global Analysis of Climate Change Projection Effects on Atmospheric Rivers," *Geophysical Research Letters* 45, no. 9 [May 16, 2018]: 4299–308), leading to an increased probability of extreme precipitation and catastrophic floods, especially in coastal areas, where moisture-heavy air from the oceans is pushed upward by mountains and turned into precipitation. The most studied example of this kind of meteorological event is the Pineapple Express, an atmospheric river that originates near Hawai'i and flows eastward as it follows the global ocean conveyor belt. A long, recorded history of observations has linked winter storms along the coast of Western United States and Canada to this phenomenon, to the extent that an array of Atmospheric River Observatories has been installed between Southern California and the Canadian border as an early warning system to predict its landfall—a "picket fence" of radars, tuned to detect the smallest variations in wind profiles and water vapor concentrations. From "Atmospheric River Observatories," NOAA Physical Sciences Laboratory, n.d., https://psl.noaa.gov/psd2/technology/aro.html.

[13] Temperature changes that emerged in the data can be explained through layers of frozen water with salinity and oxygen ratios typical of warmer latitudes. From Reginald E. Newell and Yong Zhu, "Tropospheric Rivers: A One-year Record and a Possible Application to Ice Core Data," *Geophysical Research Letters* 21, no. 2 (January 1994): 113–16.

[14] This 2018 quote from Wang Guangqian originally reads, "我们不想仅仅写理论文章，我们想把论文写在大地上。为了黄河不断流，我们干了10年；这件事我们也想干上10年，看看能不能找到解决黄河缺水的一条新路。" Interview source: http://sklhse.tsinghua.edu.cn/show.asp?id=506. Translation into English by Jingru (Cyan) Cheng.

[15] Jiaye Li, Tiejian Li, Guangqian Wang, Jiahua Wei, Deyu Zhong, Yang Su, and Xudong Fu, "空中水资源及其降水转化分析 [Kongzhong shui ziyuan jiqi jiangshui zhuanhua fenxi]

water, which "directly evaporates from the land surface and transpires from vegetation"[16]—white water exists when and where moisture with a high potential to precipitate moves into "discontinuities in the atmospheric flow fields."[17] These special boundary layers, developed by air masses, are characterized by different gradients of temperature, velocity, pressure, or humidity "when they meet each other in the troposphere."[18] They named these discontinuities "sky rivers (天河)," specific bodies and durations of tropospheric water filaments that hold identifiable spatial and temporal dimensions.[19]

While previous research on atmospheric rivers has been focused on the transportation of water vapor and its short-term manifestation in extreme weather events, the attention of Wang Guangqian and his colleagues is on the "long-term [. . .] configuration infrastructure" of atmospheric rivers so that they may be "fully brought into the water resources management systems."[20] If atmospheric rivers describe a phenomenon that can be directly observed through a combination of remote sensing and ground-based radar measurements, then sky rivers are a kind of calculation, a constructed model, explicitly devised for new forms of utilization and extraction, the facilitation of water resource management, and, potentially, the establishment of claims of state ownership:

> [. . .] as a natural resource like others, the ownership of atmospheric water resource must be well clarified. For this purpose, we have to figure out whether there exists a concept of "basin" in the atmospheric water resources. And if so, *where* is the "basin"? How do atmospheric water flow *between* the "basins"? Obviously, these two problems make the atmospheric water resources development no longer a simple meteorological issue.[21]

Sky rivers, then, require a double move: from terrestrial topography, where watersheds are determined by geomorphology, to a topography of winds and

Atmospheric Water Resource and Precipitation Conversion," 科学通报 [Kexue Tongbao] *Chinese Science Bulletin* 63, no. 26 (September 2018): 2796.

16 Li et al., "Atmospheric Water Resource and Precipitation Conversion," 2796.

17 Guangqian Wang, Deyu Zhong, Tiejian Li, Yu Zhang, Changqing Meng, Mingxi Zhang, Xiaolong Song, Jiahua Wei, and Yuefei Huang, "Study on Sky Rivers: Concept, Theory, and Implications," *Journal of Hydro-environment Research* 21 (October 2018): 112.

18 Wang et al., "Study on Sky Rivers," 112.

19 When defining white water, Li et al., in their 2018 paper "Atmospheric Water Resource and Precipitation Conversion," make clear that it is not the amount of water stored in the atmosphere per se but, instead, the huge flux of water transport in the atmosphere. They use a unique term in the English summary of the paper: "a renewable/replenishable flux," and in the body of the main text, they write: "其本质是水物质的更新量 [its essence is the renewal of water substances]."

20 Wang et al., "Study on Sky Rivers: Concept, Theory, and Implications," 110, 109.

21 Wang et al., "Study on Sky Rivers: Concept, Theory, and Implications," 110, emphasis added.

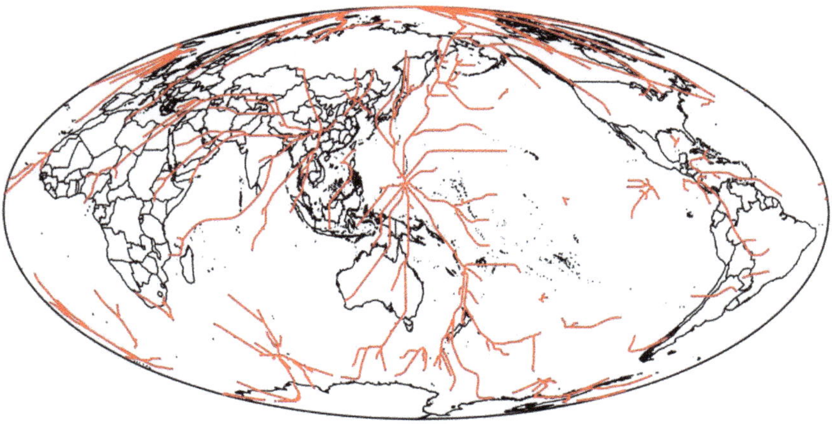

Figure 14.4a Sky rivers are shown as a system of pathways. The lines of sky rivers seem to establish a long-term knowledge of recurrence, something that can be wired to an infrastructure for extraction.

pressure gradients through which the catchments of sky rivers can be determined mathematically.[22]

Sky rivers are four-dimensional topologies of flow projected onto—and into—the sky. "The clouds in Qinghai," according to Wang, "are different from those in Beijing. The clouds in Qinghai press on my head. I feel that the water in the sky is connected to the water on the ground. I thought of the sky river dynamic theory."[23] With Wang's implicit invocation of Li Bai's poetry—*Do you not see the Yellow River come from the sky?*—the cultural imagination of the Yellow River becomes a scientific one: if the Tibetan Plateau extends the presence of water vertically into the atmosphere, then the Sky River Project extends the structure of the river catchment into the fourth dimension, up in the troposphere and across time to locate sources of extraction and optimize their temporal instability. While transboundary river management is tied to the understanding of topography as organized by gravity, the interception of water vapor requires the imagining, and imaging, of global wind circulation as a latent topography of opportunities.

[22] "In the present study, we treat the potential φ *(x,y)* [the divergent field] as the topography of the sky rivers, just as the surface topography for the ordinary surface rivers, this allows us to identify the sky rivers using the same algorithms to extract surface rivers, with the spatial distribution of sky river channels as well as their basins determined accordingly." Wang et al., "Study on Sky Rivers: Concept, Theory, and Implications," 111.

[23] Quangqian Wang, 2018. Interview source: http://sklhse.tsinghua.edu.cn/show.asp?id=506. Translation into English by Jingru (Cyan) Cheng.

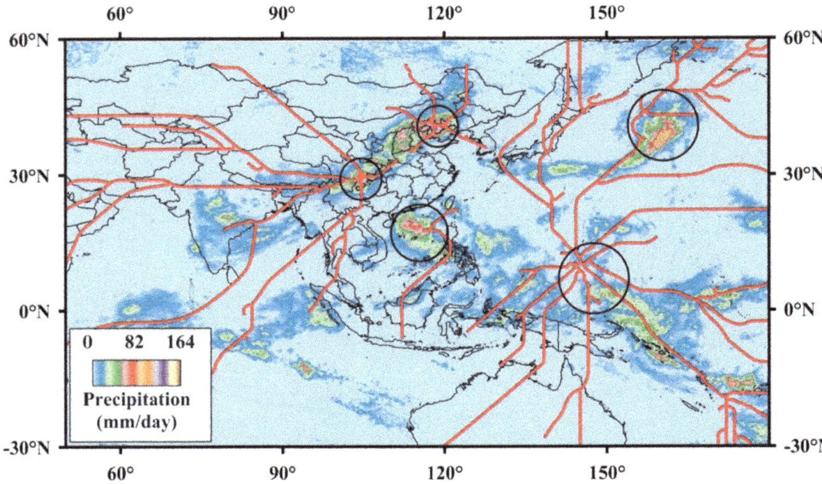

Figure 14.4b Sky rivers are shown as a system of pathways. The lines of sky rivers seem to establish a long-term knowledge of recurrence, something that can be wired to an infrastructure for extraction.

In most visualizations featured in the foundational papers, sky rivers are portrayed as linear structures or fractures across the Earth's atmospheric continuum: a clear departure from the more conventional representations of atmospheric rivers, characterized by heatmaps and vector field graphs. Sky rivers are depicted as a system of clear, stable pathways that define a hierarchy of primary and secondary streams and their confluences, similar to the classical diagrams of watershed analysis. Different states of wetness and temporality are separated and made evident through these lines. This act of reduction and separation is deemed necessary to image water vapor into a river. While heatmaps portray water and moisture in flux as snapshots of a field of forces wherein the atmosphere is imagined to be pictured as a whole, the lines of sky rivers seem to establish a long-term knowledge of recurrence, which is something that can be wired to an infrastructure.

In 1931, American hydrologist Robert E. Horton presented the now-ubiquitous hydrological cycle diagram in his paper "The field, scope, and status of the science of hydrology." Horton claimed his circular diagram represented and "rest[ed] on the mathematical constitution of the water balance equation"[24] and, in so doing,

[24] James Linton, "What is Water?: The History and Crisis of a Modern Abstraction" (PhD diss., Carleton University, 2006), 199. Horton's diagram claimed to show how "Water *evaporates* from the oceans and the land surface, is carried over the earth in atmospheric circulation as water vapor, *precipitates* again as rain or snow, is *intercepted* by trees and vegetation, provides *runoff* on the land surface, *infiltrates* into soils, *recharges* groundwater, *discharges* into streams, and ultimately, flows out

successfully establishing hydrology as an independent field of study. The Hortonian hydrological cycle is still considered a universal portrayal of the states of water, with an understanding of water as subject to regular patterns despite having been modeled after conditions specific to the temperate climates where Horton lived. If the diagram-as-definition of the hydrological cycle led to the establishment of hydrology,[25] then a new science, the concept of white water—a temporal and spatial conjecture created through the optics of weather satellites and galvanized by the chemicals of the ground stoves—is also meant to establish a new understanding of the global water cycle and to set new boundaries for the field.[26]

The (Hydro-)geography of China

Sky rivers emerge from a history of water management that dates back thousands of years to the establishment of the Xia Dynasty, its founding sage king, Yu the Great (大禹),[27] and the flood-based recreation myth wherein Yu "established

into the oceans from which it will eventually evaporate once again." David R. Maidment, *Handbook of Hydrology* (New York: McGraw-Hill Inc., 1993), 1.3, emphasis original.

25 Linton, "What is Water?," 199.

26 In "天空河流: 发现、概念及其科学问题 [Tiankong Heliu: Faxian, gainian ji qi kexue wenti] Sky River: Discovery, Concept, and Implications for Future Research," *SCIENTIA SINICA Technologica* 46, no. 6 (May 2016): 649–56, Guangqian Wang et al., [DeYu Zhong, TieJian Li, JiaHua Wei, YueFei Huang, XuDong Fu, JiaYe Li, Yu Zhang], explicitly argue for a new discipline: "由此可见, 在新形势下, 需要我们建立一门与水利工程特征尺 度(数小时至数百年)相适应的新的研究方向, 来研究 天-地耦合下的水(汽)运动、转化. 这个新的学科是以 气象学、天气学、水文学、河流动力学为基础的新学 科, 其核心研究内容之一是建立与陆表河流多尺度 过程尺度匹配的, 以降水和蒸发为纽带的, 描述天空 水汽输移、陆表水流流动的理论与方法, [Therefore, under the new situation, we need to establish a new research direction that is compatible with the scalar characteristics of hydraulic engineering (from several hours to hundreds of years) to study the movement and transformation of water (vapor) under the atmospheric-terrestrial coupling. This new discipline is one that synthesizes meteorology, synoptic meteorology, hydrology, and river dynamics. One of its core research areas is to establish the theory and method that are compatible with the multi-scalar processes of surface runoff, focus on precipitation and evaporation as the link, and describe the transport of water vapor in the sky and the flow of water on the ground]." The lead author, Guangqian Wang, is aware of the pivotal role that the white water concept and the depiction of the sky rivers might have and that they can be leveraged to reverberate on a much larger stage: "在观测、模拟技术快速发展的背景下, 我认为空-地整体研究是发展的必然趋势, 也是未来气象、水利等学科交叉融合的内在需求, 科学研究并没有截然的边界和禁区, 天河研究也是水循环研究的自然延伸。[In the context of the rapid development of observation and simulation technology, I believe that the coupling of atmospheric-terrestrial research is an inevitable trend of development, and it is also an inherent requirement for the future interdisciplinary integration of meteorology, hydraulic engineering, and other disciplines. Scientific research does not have clear boundaries and restricted areas. Research on sky rivers is the natural extension of hydrological cycle research.]" Interview source, 2018: http://sklhse.tsinghua.edu.cn/show.asp?id=506. Translation into English by Jingru (Cyan) Cheng.

27 The story of 大禹治水 Dayu Zhi Shui [Great Yu Who Controlled the Waters] appears in many forms, but the earliest surviving mention of Yu's deeds (that occurred during his life, c. 2123–2025 BC) appear in 山海经 Shan Hai Jing [Classic of Mountain and Seas], with versions of the text purported to have existed as early as the fourth century BC.

Figure 14.5a and 14.5b The Qianlong Emperor inspects flood-control measures along the Yellow River as he stands by a spillway where the blue waters of Huai meet the silty Yellow River. In its entirety, this oversize scroll by Xu Yang (c. 1751) shows flood-prevention techniques, including double-sluice gates and pounded-earth and stone levees.

the proper order of the mountains and the waters"[28] by "opening passages in the mountains and guiding rivers into their courses,"[29] allowing the waters to flow naturally. It is not a coincidence that the root of the Chinese character "to govern (治)" is water; the lines of mountains and rivers that Yu traced out divided China into the Nine Provinces.[30] According to scholars, including American sinologist and historian of ancient China, Mark Edward Lewis, "the mythology of the flood provided a background for accounts of the structure of the world, along with the forces that made that structure possible."[31] The story of Yu is the story of worldmaking, and it offers "a mythology of contemporary practices of water control, in both its literal and figurative uses"[32] of "demarcation, division, and expulsion"[33] to achieve hydraulic stability of the Yellow River and flood control in the North China Plain.

The hydro-geography of China is the legacy of both centralized and local water management plans and the contemporary fluid relationships of consumerism and industrialization, which link the country to global markets. Historically focused on the North China Plain as an area of intensive agricultural production and demographic growth, Chinese water management projects had the Yellow River as their main stage. For the most heavily silted river in the world,[34] the myth of a tamed river emerged alongside the engineering of its riverbed and landscapes. The more the river was changed and controlled, the more it pushed back, requiring ongoing intervention and tying the political destiny of the ruling class to negotiations with the river in the valley.[35] Various understandings of water flows have been the state's organizational backbone for centuries, coupling the stability and fate of dynasties and other structures of power to their abilities to establish and maintain the particular arrangements of labor and control along the river.

The understanding of the Yellow River as an asset was tested most intensively in 1938, when, in a desperate move by the Nationalist Government to halt the Japanese invasion; the dikes were artificially broken to flood the plains and displace the river's estuary, turning it into a defensive weapon. The prewar course

28 Mark Edward Lewis, *The Flood Myths of Early China* (Albany: SUNY Press, 2006), 22.
29 Lewis, *The Flood Myths of Early China*, 30.
30 Lewis, *The Flood Myths of Early China*, 66.
31 Lewis, *The Flood Myths of Early China*, 147.
32 Lewis, *The Flood Myths of Early China*, 147.
33 Lewis, *The Flood Myths of Early China*, 53.
34 The extraordinary amount of sediment present in the Yellow River is the result of erosion from the Loess Plateau, a vast area of highly erodible soil situated in the middle reaches of the river.
35 An extensive account of the thousand-year-old history of the Yellow River can be found in David A. Pietz, *The Yellow River: The Problem of Water in Modern China* (Cambridge, MA: Harvard University Press, 2015).

of the river was only reestablished after 1947 with a coordinated effort by the United Nations Relief and Rehabilitation Administration (UNRRA). It became clear that the conditions of the North China Plain and its ability to become the "breadbasket of the world" were not only of Chinese concern. In the words of Oliver J. Todd, the chief advisor and engineer of the UNRRA mission:

> The fact that the lives and welfare of 50 million people are so closely tied up with this river's behavior makes the problem intriguing to engineers and intensely interesting to those who are studying population growth and the question of how the world is to be fed. In a way the Yellow River is a world problem in itself.[36]

The modern geographical imaginary of the country is tethered to a rigid conception of water balance, where a scarcely populated, water-rich south is often placed in opposition to a densely inhabited and agriculturally developed but water-scarce north.[37] This uneven distribution of water resources across the nine major river basins of the national territory is amplified by spatiotemporal patterns of precipitation. In an attempt to respond to this water crisis, two strategies are underway: The first is to increase rainfall through weather engineering, and the second is to offset water needs. North China's water stress is not only a national security concern; its growing population and industrial expansion have exponentially increased the impact of this demand on the global water balance, challenging the geopolitical stability of Central and Eastern Asia. As David A. Pietz argues, China is aggressively leasing farmland abroad to produce both food and biofuel crops (which are particularly water-intensive) to meet its emission reduction goals. These investments, along with the direct import of agricultural commodities,[38] are a trade of water by proxy.[39] While the majority of China's irrigated farmland is in the north, its products mostly feed the heavily populated cities of the south, where wealthier dietary preferences demand water-intensive crops, alongside a spiking increase in meat consumption.[40] As water moves northward through rivers, pushed

[36] Oliver J. Todd, "The Yellow River Reharnessed," *Geographical Review* 39, no. 1 (January 1949): 56.

[37] This national imbalance has been put under even greater stress by the fast industrialization of the coast and the increase in food production that spiked after the Second World War. The North China Plain became an area of food production that was relevant for the entire world, not just on a national scale. See Michael Webber, Britt Crow-Miller, and Sarah Rogers, "The South–North Water Transfer Project: Remaking the Geography of China," *Regional Studies* 51, no. 3 (January 2017): 370–82.

[38] In Central Asia, approximately 80 percent of the water from irrigation is used for products that are exported to China. From David Trilling, "Chinese Researchers Calculate Central Asia's Water Waste, Agriculture Potential," *Eurasianet*, April 27, 2021, https://eurasianet.org/chinese-researchers-calculate-central-asias-water-waste-agriculture-potential.

[39] Pietz, *The Yellow River*, 307–21.

[40] Pietz, *The Yellow River*, 307–21.

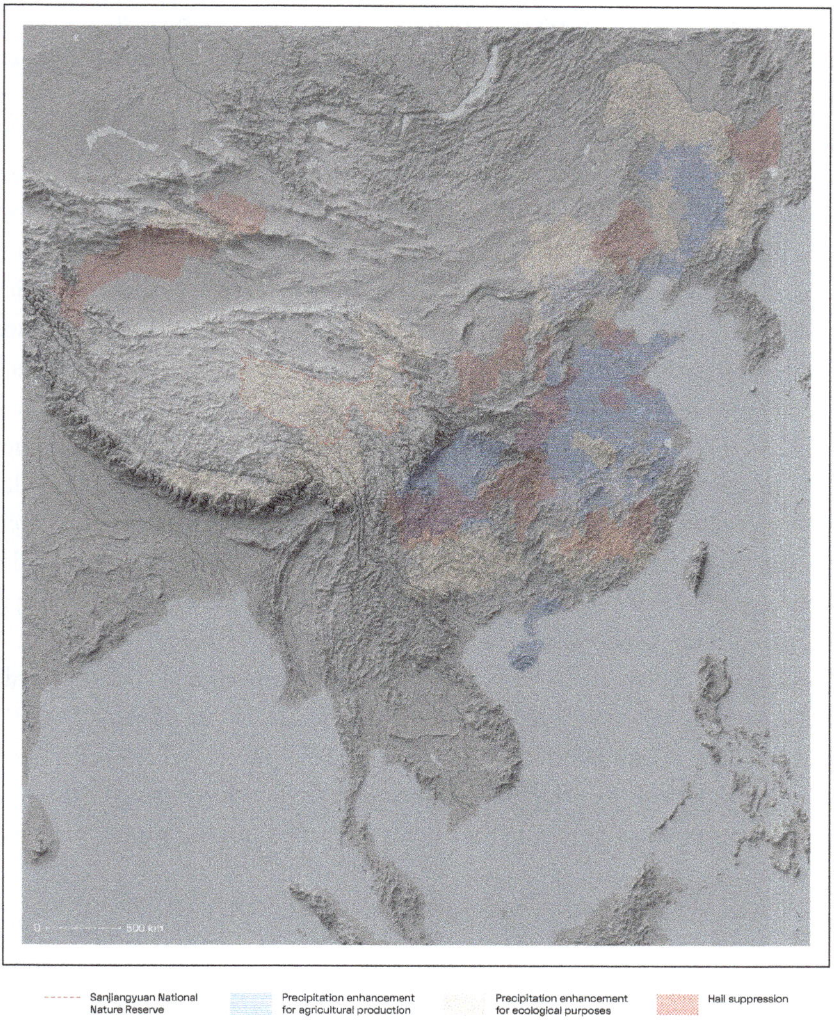

Figure 14.6 China's National Weather Modification Development Plan (2014–20) outlines potential areas for weather modification that includes precipitation enhancement for agricultural (blue) and ecological (yellow) purposes as well as hail suppression (red). These areas are proposed in tandem with the Sky River Project test site, Sanjiangyuan National Nature Reserve (outlined with a dashed red line), 2021.

through aqueducts and canals, pumped from aquifers to irrigate exhausted northern farmlands threatened by desertification, it also circles back south—inside food. Water not only flows in the atmosphere, above and below the ground, but also through economies. The river is not a single stream but many streams that flow through bodies, territories, and data centers.

The Digital Yellow River Programme (DYRP) is a large-scale project of integrated river basin management aimed at creating a real-time representation of the "natural" Yellow River in a geographic information system (GIS)-based virtual environment.[41] Taking into consideration multiple inputs, from meteorology to geology, economics to legislation, the DYRP is a platform that provides numerical modeling and centralizes decision-making processes on every aspect of river ecosystem optimization.[42] The Yellow River is rendered and replicated into a network of gauges and sensors, data processing and storage centers, numerical models, and protocols of intervention. The sky river, which also exists almost uniquely in the form of numerical forecasting models, can be easily integrated, extending the headwaters of the Yellow River up into the atmosphere.

The total incoming water vapor in China is estimated to be close to 20 trillion cubic meters per year, approximately four times the average annual rainfall.[43] This water vapor is delivered through various long- and short-distance channels, ranging from the Caspian Sea to the Altai Mountains, the Western Indian Ocean, and the Bay of Bengal. Smaller quantities of water vapor are conveyed by high-altitude winds originating above the Atlantic Ocean.[44] Almost all of these channels enter the territory of China from the southeast,[45] only partially discharging their water before reaching the height of the plateau.[46] As the western route of the South-North Water Transfer Project[47] is deemed more and more unfeasible for its high costs and engineering challenges, the focus on water vapor implies a shift from the middle and lower reaches of the river to the

[41] "The Digital Yellow River Project: GIS Plays Important Role in Managing China's Yellow River Basin," *Esri ArcNews* 32, no. 1 (Spring 2010): 26–27, https://www.esri.com/news/arcnews/spring10articles /the-digital-yellow.html.

[42] The aspects range from flood forecasting to reservoir operations, from water-draw-permit management to negotiations around water prices. All are coordinated through the database and applications of the Digital Yellow River platform. Qingping Zhu and Wentao Liu, "The Digital Yellow River Programme," in *Frontiers of WWW Research and Development*, ed. Xiaofang Zhou, Jianzhong Li, Heng Tao Shen, Masaru Kitsuregawa, and Yanchun Zhang (Berlin: Springer Verlag, 2006), 1144–51.

[43] Wang et al., "Study on Sky Rivers," 109.

[44] Hao Yang, Guan-yu Xu, Xiaofang Wang, Chunguang Cui, Jingyu Wang, and Dengxin He, "Quantitative Analysis of Water Vapor Transport during Mei-Yu Front Rainstorm Period over the Tibetan Plateau and Yangtze-Huai River Basin," *Advances in Meteorology* 2019 (August 2019): 10.

[45] Kepiao Xu, Lei Zhong, Yaoming Ma, Mijun Zou, and Ziyu Huang, "A Study on the Water Vapor Transport Trend and Water Vapor Source of the Tibetan Plateau," *Theoretical and Applied Climatology* 140 (February 2020): 1031–42.

[46] "Over the northern India and Tibetan Plateau of China, the average occurrence frequency of the sky rivers is between 40 and 50 days per annum. [. . .] Moreover, the highest frequency appears in southwest China (central Sichuan and southern Tibet) with more than 60 days per annum." Wang et al., "Study on Sky Rivers," 111.

[47] In the meantime, the long and storied history of artificial canals in China (including both the Grand Canal of the fifth century and the Red Flag Canal of 1960–65) is continued with the South-North Water Transfer Project. Now in its fifth generation under Xi Jinping, the multidecade mega-infrastructure project aims to transfer water from the Yangtze River to the Yellow River basin, with the goal of moving 44.8 billion cubic meters across the country every year, along three routes: eastern, central, and western.

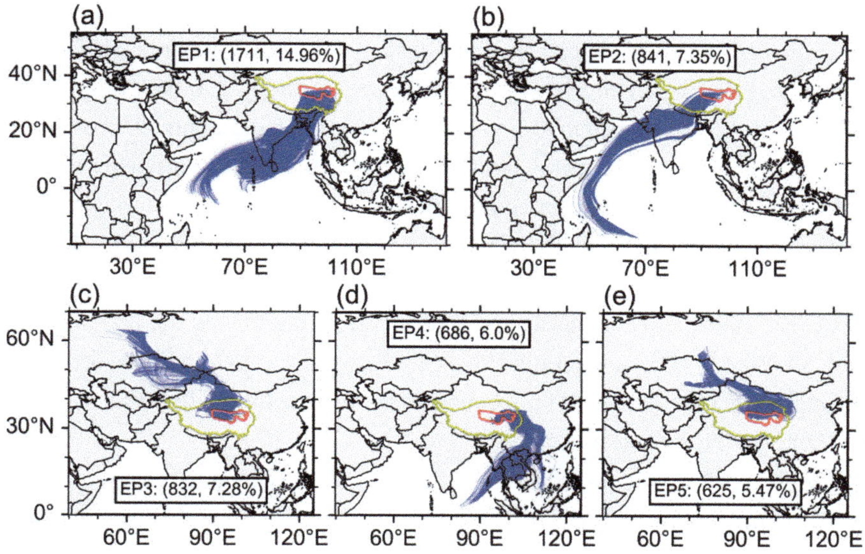

Figure 14.7 In the troposphere above the 三江源 [Sanjiangyuan] Source of the Three Rivers area (outlined in red) of the Qinghai-Tibet Plateau (outlined in yellow), a confluence of sky rivers meets (moving blue body).

headwaters and to the sky. The headwaters of the Yellow and Yangtze Rivers are not confined to the plateau but branch off in multiple directions to the sources of each water vapor stream as they approach Qinghai-Tibet from above. In this move, the control (and the watershed) of the rivers' headwaters is extended into the atmosphere.[48] Here is where the rivers of the Earth meet the rivers of the sky.

A Cloud-seeding Stove

Rain enhancement and hail suppression operations are carried out on a regular basis in an attempt to safeguard and increase agricultural production, ensure

[48] Studies and speculations on atmospheric watersheds, also called "precipitationsheds," have focused on the mechanics of transboundary moisture recycling. Precipitation sheds are defined as "a watershed of the sky in that it describes how upwind and downwind regions are connected through the flows of water vapor, similar to how upstream and downstream areas are connected." In recent years, Patrick W. Keys and his colleagues have extended these studies to include questions of transboundary water governance in relation to land-use changes and their effect on evaporation: "Despite an *upwind country's ability to change a downwind country's rainfall through changes in land-use or land management*, the major legal and institutional implications of changes in atmospheric moisture flows have remained unexplored." Patrick W. Keys, Lan Wang-Erlandsson, Line J. Gordon, Victor Galaz, and Jonas Ebbesson, "Approaching Moisture Recycling Governance," *Global Environmental Change* 45 (July 2017): 15–16, emphasis added.

drinking water supplies, restore ecologically damaged biotopes, and sustain hydropower projects.[49] Recall Yu the Great: A significant portion of Chinese territory is organized into areas of precise meteorological intervention. This infrastructure is managed across four levels—national and regional, provincial, city, and county—where local weather bureaus make decisions with relative autonomy (and sometimes in competition with one another[50]), though all sit under the general helm of the China Meteorological Administration (CMA), according to policies established by the Meteorology Law of the People's Republic of China.[51] Weather engineering applications are meant to seed the sky; the scale and persistence of these operations have produced a vast geographical footprint throughout China, which is now being extended into the mountainous area of the Qinghai-Tibet Plateau.

Cloud-seeding stoves—ground-based silver-iodide generators—have operated across China for more than two decades as part of an apparatus that includes stationary and mobile rocket-launch systems (usually deployed on the back of commercial pickup trucks), aircraft, and UAVs.[52] The "ground seeding furnace and control system BD-48"[53] is the fourth iteration of a genealogy of similar devices manufactured by Shaanxi Zhongtian Rocket Technology Co., Ltd., a civilian-military contractor headquartered in Xi'an City that is engaged in

[49] Shiuh-Shen Chien, Dong-Li Hong, and Po-Hsiung Lin, "Ideological and Volume Politics Behind Cloud Water Resource Governance—Weather Modification in China," *Geoforum* 85 (October 2017): 225–33.

[50] Chien et al., "Ideological and Volume Politics Behind Cloud Water Resource Governance," 232.

[51] Promulgated in 1999, the Meteorology Law of the People's Republic of China, allows the "competent meteorological department" (China Meteorological Administration) to organize, coordinate, and guide weather modification activities, including operations, demonstrations, and research nationwide. This is outlined, to greater detail, in Article 30: "People's governments at or above the county level shall enhance their leadership over weather modification and, in light of actual conditions, carry out work in this field in an organized and planned way. The competent meteorological department under the State Council shall more efficiently administer and guide weather modification throughout the country. Local competent meteorological departments at all levels shall make plans for weather modification operations and, under the leadership and coordination of the people's governments at the corresponding levels, administer, guide and arrange for such operations. Relevant departments shall, in compliance with their functions and duties and division of responsibilities, cooperate with the competent meteorological departments in weather modification. Organizations engaging in weather modification operations shall meet the qualifications prescribed by the competent meteorological departments of provinces, autonomous regions, or municipalities directly under the Central Government, use the operational equipment which meets the specifications set by the competent meteorological department under the State Council, and conform to the operational rules." China Internet Information Center, "Meteorology Law of the People's Republic of China," China.org.cn, http://www.china.org.cn/environment/2007-08/27/content_1034467.htm.

[52] Sumit Singh, "How China Is Using A UAV To Help Change The Weather," *Simple Flying*, January 7, 2021, https://simpleflying.com/china-weather-changing-uav.

[53] "地面播撒系统简介(第四代) Dimian Bosa Xitong Jianjie (Disidai) [Ground Seeding System Description (Fourth Generation)]," 百度文库 Baidu Wenku [Baidu Library], https://wenku.baidu.com/view/e3c40591f5ec4afe04a1b0717fd5360cba1a8d8f.html.

the research, development, and production of rocket technologies for civilian and military purposes. The stoves all share a similar design: They are solar-powered, their ignition is remotely controlled through Global System for Mobile Communications/General Packet Radio Service (GSM/GPRS) technology, and their exhausts are conveyed to the sky through a chimney that can reach a height of about 5.5 meters. Cloud-seeding stoves are fastened with wind wire ropes to a concrete slab foundation that also hosts an automatic weather station and antennas for remote control and communication; a metal fence protects the installation from its immediate surroundings. The only variation is the metal roof, which sometimes presents a sweeping curvature resembling traditional religious architecture. On the exterior, they can display bright, solid colors, but the majority are painted dark green, easing their optical blending into background vegetation. They can be decorated with abstract shapes, camouflage patterns, or even the outline of an arched door corresponding to the aperture that grants access to their operating dashboard. More often than not, the four sides of the lower chamber are stamped with bold texts and icons claiming their belonging to the CMA and local weather bureaus.

While these exteriors, which can appear to be makeshift metalwork, are documented in countless photos posted on WeChat, Weibo, and the websites of local administrations, the stove's inner workings are concealed from view by a heavy metal plate that holds the bays for the flammable silver-iodide bars. As of 2018, more than 500 stoves have been installed for experimental use on the alpine slopes of the Qinghai-Tibet Plateau, covering a combined area of 110,000 square kilometers divided into three macroregions: the Sanjiangyuan National Nature Reserve,[54] the Qilian Mountains, and the Kunlun Mountains. While more specific place names are mentioned online,[55] the actual locations of the test sites within these three regions are not disclosed.

This sprawling ground network will be met in the sky by a constellation of low Earth orbit satellites—a monitoring and forecasting system that will be able to quantify the volume and duration of atmospheric water transport—that activate the stoves when favorable meteorological conditions are met. This new satellite platform aims to enhance the sensing capabilities of the current and

[54] The source of the three rivers and China's first national park. For more information, see: Qi Wang, "China's First National Park to be Established in Sanjiangyuan Area in 2020," *Global Times*, May 21, 2020, https://www.globaltimes.cn/content/1189032.shtml.

[55] Those places include Darlag County in the Sanjiangyuan area, the headwaters of the Heihe River in the Qilian Mountains, and Haixi Mongol, the Yuzhu Peak of the Kunlun Mountains.

future Fengyun-4 family of geostationary Chinese meteorological satellites.[56] Once operational, Tianhe-1, the first of a series of twenty-eight planned spacecraft,[57] is meant to provide ecological monitoring of the Sanjiangyuan region while studying the temporal and spatial evolution of sky rivers to achieve a sky-air-ground integrated monitoring system for atmospheric water resource development.[58] This satellite constellation is meant to guide the operations of a sophisticated platform of rain enhancement technologies, including telling the stoves when to fire up and when to turn off. From the coordinated communication and actions between these two seemingly heterogeneous, disparate networks emerges a new atmospheric-terrestrial infrastructure. "Infrastructure," writes Ashley Carse, "[is] more than a word. It [is] world-making."[59] The stoves and satellites act as a network, actively participating in the transfer of water from transboundary atmospheric pathways to the Yellow River Basin. In the projection of a three-dimensional metaphor onto a four-dimensional cloud topography, the interlocking polar orbits of the LEO satellite constellations—or even the geosynchronous equatorial orbit[60] of the current and future Chinese weather satellites—become an instance of the contentious atmospheric water basins presumed by the Sky River Project. The orbital paths of the satellites, as they perform their ecological monitoring duties, not only define the instrumental envelope of the Sky River infrastructure but they also highlight the very "*infra* in infrastructure: the logics of depth and hierarchy that

[56] Fengyun satellites can be seen at "50th Anniversary of Fengyun Satellite Program," China Meteorological Administration, September 10, 2020, http://www.cma.gov.cn/en2014/news/News/202009/t20200930_564281.html.

[57] Tianhe-1 was publicly presented at the 12th International Aviation and Aerospace Exhibition in Zhuhai, November 2018, with a launch planned for the end of 2020, though confirmation of its launch is not readily available.

[58] Although the State Council issued a press release in December 2020 confirming the importance and priority of the Sky River Project for the development of an advanced weather modification plan with so-called "breakthrough technologies" by 2025, evidence of the Tianhe-1 satellite being operational has not been found online. For more information, see: "China to Forge Ahead with Weather Modification Service," State Council, People's Republic of China, English.gov.cn, December 2, 2020, http://english.www.gov.cn/policies/latestreleases/202012/02/content_WS5fc76218c6d0f7257694125e.html.

[59] Ashley Carse, "Keyword: Infrastructure; How a Humble French Engineering Term Shaped the Modern World," in *Infrastructures and Social Complexity: A Companion*, ed. Penny Harvey, Casper Bruun Jensen, and Atsuro Morita (London: Routledge, 2017), 31.

[60] It is of note that in the 1976 Bogota Declaration, the equatorial orbit was declared a "natural resource," with consequent claims regarding its sovereignty raised by the countries whose territory is situated below it. These claims, coming from countries with no developed space programs or advanced satellite capabilities, were ignored by the international community. For more information, see: "Declaration of the First Meeting of Equatorial Countries (Adopted on December 3, 1976)," *Space Law*, 1998, https://www.jaxa.jp/library/space_law/chapter_2/2-2-1-2_e.html.

manifest in design, management, and maintenance"[61] and allow for the imaging of "white water," without which sky rivers could not be imagined.

Transboundary Hydro-Geo-Logics of the Sky River

The Sky River Project is a paradigmatic act of geographical reimagination. The sky rivers bring with them a new understanding of global atmospheric circulation. In a conceptual leap, atmospheric water and conventional surface water are brought together in a long-term coupling that follows the way in which water gathers and flows in terrestrial rivers. Establishing a new theoretical framework for the hydrological cycle expands the domain of resource extraction to the atmosphere. The project and its assumptions are not without controversy, and there are critiques from the scientific community. Besides internal challenges that stem mainly from the debate over the project's feasibility and its reframing of atmospheric and hydrological science, the project has raised concerns in neighboring countries, especially India. Still, proponents of the Sky River Project maintain that this potential precipitation should be thought of as a readily available resource, a kind of standing reserve water supply, available to be displaced through this distributed network infrastructure.

If the topology of water is imagined as a continuous surface, water flows into the plateau through sky rivers and out of the plateau through rivers on the ground. In both cases, sensed and processed into data, these flows are already a resource even before becoming water. In the past, China has strategically refused to provide data for cooperative transboundary river management while simultaneously building new dams and adopting water-intensive cultivation in many of Central Asia's watersheds. There exists a growing international suspicion that Sky River, a project set to exist in a space without clear boundaries and regulations, would take a similar approach on a much larger scale.[62] China, its neighbors contend, is trying to control all resources in all directions. This control is both spatial and temporal: Dams and reservoirs are time-management infrastructure, regulating rain and runoff, making surface waters—otherwise obscured by their uneven temporal distribution—visible by collecting and then storing them in plain sight. The Sky River Project seizes upon this logic

[61] Carse, "Keyword: Infrastructure," 35, emphasis added.
[62] Dhanasree Jayaram and Kurnica Bhattacharjee, "China's Geoengineering Push Dangerous for the Region," *Sunday Guardian Live*, December 19, 2020, https://www.sundayguardianlive.com/news/chinas-geoengineering-push-dangerous-region.

of optimization and resource visibility, coordinating remote sensing and stove activation in order to tap into the sky rivers fluctuating and pulsating above.

Water vapor is not a stable feature of planetary origin but of the atmosphere. Its fundamental composition and materiality challenge us to overcome our familiar principles of direct, Cartesian verticality when conceptualizing borders and boundaries. When projecting the topology of the watershed up into the sky with wind, temperature, and pressure as vectors, a vertical appropriation takes hold, extending principles of terrestrial sovereignty into the atmosphere.[63] Ultimately, the exploitation of the atmosphere fosters changes at the ground level. New lines on the land are drawn by changing the geography of the sky in the form of displaced rainfall, fortified glaciers, floods, or chemical alteration of the water-bearing stratum above. Faced with a new concept of atmospheric water resources and a new definition of the water that circulates in the atmosphere, water vapor is converted from a natural occurrence into an asset that can be secured through state infrastructure. The very idea that water vapor might act as a river extends the fantasy of control of the cartographic imaginary into the atmosphere. It occupies a space without definite proxies for borders while erasing from the map below local communities that have inhabited these lands since before the advent of state organization. Will the atmosphere become the next domain to be mapped through scientific research, appropriated through technology, and administered through planetary infrastructures?

Sanjiangyuan, the test site for the Sky River Project, is where we find a confrontation of scales and complexities that exceed our abilities to parse through modeling or computation. In the atmosphere, another logic—*a transboundary hydro-geo-logic*—refuses imaginaries of control based on surfaces, vectors, and boundaries and instead insists on a logic of flows, processes, and entanglements. As the domain where elements of the climate crisis are unfolding and regulation and conflicts are emerging, the atmosphere produces multiple localities on the ground where these transformations can be observed and sites of their mediation can be imagined.

[63] See Chien et al., "Ideological and Volume Politics Behind Cloud Water Resource Governance."

15

Designed to Disappear

On the Ambiguity of "Nature" in Dutch Coastal Engineering

Michaela Büsse[1] and Konstantin Mitrokhov

Land Adrift

Today, about a third of the Netherlands lies below sea level while another third of the country is in need of protection against flooding.[2] Centuries of storm surges and floods have taken their toll on the low-lying country, eroding the shore and pushing the coastline further and further inland. Without its numerous dikes, dams, polders, and coastal nourishments, parts of the country are projected to be lost to the sea. Paradoxically, it is the constant reengineering of the Netherlands that increases the very likelihood of its flooding. Drainage issues, siltation, and algal bloom, as well as faster erosion of nourished coasts, are only some of the consequences of reclamation works.[3] Subsequently, they create new problems and the need for improved solutions. While the reasons for converting sea to land may be environmental, economic, or geopolitical, the common desire is that the newly created land be made to last. Zandmotor (Dutch for "sand engine"), an artificial peninsula close to The Hague in the Netherlands, is of a different nature: It is the first land reclamation project *designed to disappear*.

Zandmotor comprises a heap of 21.5 million cubic meters of sand dredged from the bottom of the North Sea and temporarily relocated to the Delfland coast in 2011. The sandy artificial peninsula is supposed to erode and spread

[1] Michaela Büsse's research was supported by the SNSF–Swiss National Science Foundation, and the Deutsche Forschungsgemeinschaft-funded cluster of excellence "Matters of Activity: Space Image Material" at Humboldt-Universität, Berlin. The fieldwork for this chapter was made possible by SNSF's Flexibility Grant.

[2] Mark van Koningsveld et al., "Living with Sea-Level Rise and Climate Change: A Case Study of the Netherlands," *Journal of Coastal Research* 24, no. 2 (March 2008): 367–79, here 367.

[3] Xiuzhen Li et al., "Coastal Wetland Loss, Consequences, and Challenges for Restoration," *Anthropocene Coasts* 1, no. 1 (January 2018): 1–15.

along the coast through the natural forces of currents, wind, and waves over the course of twenty years. This large-scale experiment in coastal engineering is considered to be a more flexible alternative to the annual coastal nourishment that has been applied in the Netherlands since the 1990s. But Zandmotor is more than just a coastal nourishment project; it is conceived as a living laboratory that provides researchers with an opportunity to study changes in real-life conditions rather than in experimental setups like flumes or wind channels.[4] Zandmotor is a precedent in many aspects and is poised to become, upon successful completion, a model for coastal nourishment projects all over the world. However, scientists came to realize that any real-life intervention at this scale comes with its own complexities. Reactions by the local community and the site's unruly nature have challenged their modeling approach, leading Zandmotor toward unforeseeable outcomes.

This case study discusses the conflicting functions of Zandmotor as a coastal protection measure, a living laboratory, a leisure site, and, ideally, the testbed for a less ecologically harmful and more flexible approach to coastal engineering in the future.[5] This text attempts to grasp the intricacies of the project and question the Promethean righteousness that is present in much of contemporary geoengineering discourse. Taken together, images and text make legible how the landscape is tokenized, that is, converted into quantifiable land segments through the means of scientific surveillance while simultaneously being informed by its own materiality. By emphasizing the iterative interplay between the model and the experimental site, we aim to disrupt its probabilistic design project and offer a nondeterministic, field-based reading of the site. Ultimately, the case study comes to exemplify the fact that despite attempts to calculate and control nature, the "sand engine" remains incomputable.

4 Arjen Luijendijk and Alexander van Oudenhoven (eds.), *The Sand Motor: A Nature-Based Response to Climate Change; Findings and Reflections of the Interdisciplinary Research Program NatureCoast* (Delft: Delft University Publishers, 2019).

5 This chapter combines fieldwork conducted as part of Michaela Büsse's dissertation "Reclamations: On the Making and Remaking of Sand as a Medium for Design" with a visual investigation by Konstantin Mitrokhov, juxtaposing data gathered by scientists and our own footage from the field in order to emphasize the interplay between different material practices. All visualizations in this chapter are based on a dataset published by Technische Universiteit Delft, namely "combined_data_delfland_transects_r20171115.nc," which is worth a brief technical elaboration. The topography and bathymetry (seabed topography) of Zandmotor were surveyed with high-precision differential GPS equipment over the course of the project. The researchers moved the measuring device across the site by foot, quad bike, and jet ski. The dataset is labeled "combined" because it cross-references four other datasets, consolidating the surveys that were performed within a short period of time from each other. This approach resulted in combined measurements of the whole site over thirty-eight days spread over the course of five years. The eventual dataset contains interpolated data that encodes the contingent material boundaries of Zandmotor into a predefined, slightly varying grid of observation cutting across the site. For more information, see Bart Roest, "Combined Morphology Surveys Delfland 2011–2016," *4TU.ResearchData*, November 24, 2017, https://data.4tu.nl/articles/dataset/Combined_morphology_surveys_Delfland_2011-2016/12716117/1.

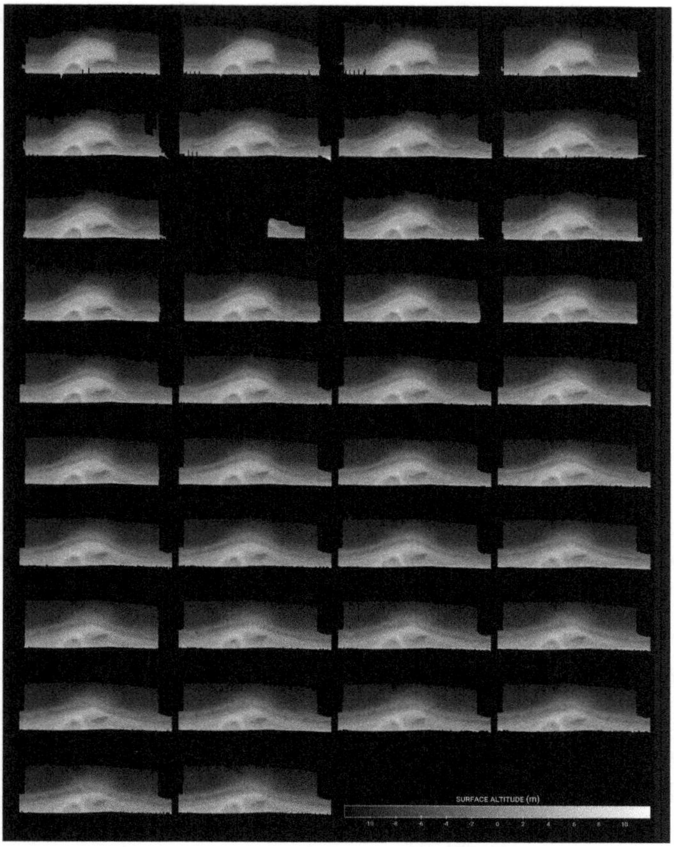

Figure 15.1 Konstantin Mitrokhov, 2022. These thirty-eight images render Zandmotor as height maps based on bi-monthly measurements of the site according to the combined dataset available from TU Delft. Visible variations in the outline of the measured area are presumably due to the diverse conditions on site.

Building (with) Nature

In order to mitigate the threat of floods and keep the beaches attractive to visitors, the Netherlands extracts twelve million cubic meters of sand every year from the North Sea for coastal nourishment.[6] With sea levels rising due to climate change, demand is expected to increase by seven million cubic meters in the coming years. Blessed with a seemingly endless supply of sand within its territorial waters and a steady positive net flow from southern Europe, the

[6] Huib de Vriend and Mark van Koningsveld, *Building with Nature: Thinking, Acting and Interacting Differently* (Dordrecht: Ecoshape, 2012).

global scarcity of sand is of little concern to the Netherlands, as many of our interlocutors admit.[7]

In the early 2000s, after several reclamation projects were rejected due to their lack of environmental sensitivity, the Dutch dredging industry was on the verge of financial collapse. A consortium of dredging companies and research institutes initiated the private-public partnership program "Building with Nature" with the goal of developing less destructive land reclamation methods. Building with Nature does not signal a radical reinvention of design and engineering practices but implies an adaptation of business as usual to the growing ecological awareness that has arisen within society since the late 1960s. Its development resonates with the emergence of "ecological design," that is, the shared concern within architecture, planning, and design discourses to create more holistic and all-encompassing designs inspired by ecology and systems thinking.[8]

Ecological design, as architectural scholar Lydia Kallipoliti notes in her overview of the history of the field, is "the design of the planet itself as much as the design of an object, building, or territory."[9] While the roots of ecology are usually traced back to zoologist Ernst Haeckel, who coined the term in 1866, the 1960s witnessed a growing interest among architects and designers in the principles of ecological thinking, that is, the acknowledgment that organisms and their environments are entangled.[10] Landscape architect Ian McHarg's *Design with Nature*, first published in 1969, is considered fundamental in the efforts to make knowledge from ecology applicable to design. His theory of "physiographic determinism" describes the need for design proposals that respond to and harness the workings of natural processes. According to McHarg, "each area of land or water has an intrinsic suitability for certain single or multiple land uses," and the role of the designer is to evaluate their fit.[11]

McHarg's approach is still relevant in the field today, and the accelerating ecological crisis has only fueled design's engagement with the environment. A few decades after *Design with Nature* was published, the environmental crisis

[7] Net flow describes the cumulative movement of sand through tides, wind, and waves along the shore.

[8] Sim Van der Ryn and Stuart Cowan, *Ecological Design* (Washington: Island Press, 1996).

[9] Lydia Kallipoliti, "History of Ecological Design," in *Oxford Research Encyclopedia of Environmental Science* (April 2018), 1–59, here 1.

[10] Kallipoliti, 1; Florian Sprenger, *Epistemologien des Umgebens: Zur Geschichte, Ökologie und Biopolitik Künstlicher Environments* (Bielefeld: Transcript, 2019); Chris Reed and Nina-Marie Lister, "Ecology and Design: Parallel Genealogies," *Places Journal*, April 2014, https://placesjournal.org/article/ecology-and-design-parallel-genealogies.

[11] Ian McHarg, *Design with Nature* (1969; New York: John Wiley & Sons, 1995), 79.

Figure 15.2 Konstantin Mitrokhov, 2022. Zandmotor's most recently recorded topographic and bathymetric survey data from September 2016, visualized as a datascape superimposed with the satellite view of the site captured via Google Earth in 2021. When the layers are visually aligned, it becomes apparent that the ragged edge of the measured area is due to the infrastructure, such as kiosks, implicated in the Deltares dataset.

itself became a design problem.[12] In this context, environmental scientist David Orr states that it is a matter of design to better align human intentions with ecological systems. "Environmental problems," he states, "are mostly the result of a miscalibration between human intentions and ecological results, which is to say that they are kind of a design failure."[13] Supposedly, a better recalibration of material and social processes—that is, a better design—can tackle these problems.

Building with Nature should be understood as a continuation of these practices; however, in a less deterministic manner. Building *with* nature, as opposed to building *in* nature, is considered to be less extractive, instead "utilizing natural processes and providing opportunities for nature as part of the infrastructure development process."[14] In practice, this means, for example, that hard defenses such as concrete dikes separating land from the sea are no longer employed. Instead, artificial beaches and dunes make a soft defense, zones for a flexible transition between land and water. Zandmotor is one of the pilot projects within the framework of Building with Nature. Alongside its function as

[12] Terry Irwin et al., "Transition Design Provocation," *Design Philosophy Papers* 13, no. 1 (2015): 3–11; Tony Fry, *Becoming Human by Design* (London: Berg, 2012).

[13] David W. Orr, *The Nature of Design: Ecology, Culture, and Human Intention* (Oxford: Oxford University Press, 2002), 14.

[14] De Vriend and van Koningsveld, *Building with Nature*, 9.

Figure 15.3 Konstantin Mitrokhov, 2022. Close-up rendering of Zandmotor's datascape superimposed with a satellite image of the site.

a coastal nourishment instrument, it also provides a freshwater lagoon, a resting area for seals and birds, and plenty of beach for visitors.

Instrumentalized Contingency

Zandmotor broke new ground. Prior to the project, "dry" and "wet" models were treated separately, and there was no model that would account for wind, waves, and currents when projecting sand distribution over a long period of time. The implementation of Zandmotor was therefore a major challenge for the coastal morphologists at Deltares, the Dutch institute for applied research in water- and land-related matters. In order to manipulate the coastline, the entire coastal system had to be taken into account: the beach, the sand banks under the surface of the sea, as well as the dunes behind the beach, with varying waves, currents, and wind conditions adding to the complexity of the calculation. In addition, the type of activity that takes place on the coast is factored in to arrive at an approximation with an uncertainty rate that seems "manageable." The coastal landscaping tool that Deltares's modeling expert Arjen Luijendijk eventually developed combined preexisting models into a new one, derived not from real-life conditions but constructed from bits and pieces of the datafied landscape.[15]

[15] Arjen Luijendijk, "A Novel Coastal Landscape Model for Sandy Systems: Community Base for Interdisciplinary Research on Coastal Evolution," *Research in Urbanism Series* 7 (February 2021): 223–40.

The location of the sand pile and its dissolution over the course of twenty years were approximated and based on the model's projection. The peninsula also owes its hook-shaped design to Luijendijk's calculations. Since its establishment, Zandmotor has been under constant surveillance and its performance tracked in order to learn from its behavior and feed that knowledge back into mathematical models. "We cannot predict answers, but we can quantify uncertainty," explains Luijendijk when describing his work during our interview. His credo is "learning by doing."[16] He also explained that the model that defined the initial shape of Zandmotor is constantly recalibrated to be in tune with the living experimental site. In the process of steady fore- and backcasting, the simulation and the site are successively aligned. Even though the model and the coastline never fully coincide, their difference is factored in as a risk that does not disrupt the project but rather contributes to the overall goal of maintaining control over the site.

In the engineering manual of Technische Universiteit Delft, risk is defined as "the probability of an undesired event multiplied by the consequences."[17] In order to assess the risks of a project, it is necessary to first identify the parameters that are considered important before quantifying their interrelation and their probability of assuming any particular state. These parameters are usually defined together with the project's stakeholders, the local government, citizens, environmentalists, and so on. It is less the qualities of the individual parameters that are of interest but rather their systemic functionality (that is, their relationship to one another). Luijendijk's approach presumes that every ecosystem consists of complex and nonlinear interactions. Still, there are a certain number of key parameters that steer the system, which can be measured and manipulated, allowing for control of the system without full knowledge of its behavior. The engineer's numerical model, created to mimic the coastline, reduces its complexity to the key variables (or to those that appear to be significant). Data is collected by discrete on-site measurements, then interpolated along predefined observation lines and smoothed out. The coastline thus becomes largely homogenous, with heterogeneity existing only within the rate of variance allowed by the engineers. Together with his students, Luijendijk modeled one thousand possible weather predictions for Zandmotor, knowing that none of them would be entirely correct. By taking the median of the predicted behavior, the simulation model provides the ground for decision-making. "It is less about problem-solving than opportunity-seizing," he explains, and about "who has to

[16] Arjen Luijendijk, interview by Michaela Büsse, November 16, 2020.

[17] Sebastiaan N. Jonkman et al., *Probabilistic Design: Risk and Reliability Analysis in Civil Engineering* (Delft: Delft University of Technology, 2015), 58.

pay." His task is not to produce an exact prediction but to provide functional knowledge that enables intervention. Based on the different risk scenarios his simulation offers, design decisions are made alongside the negotiation of responsibility for possible failure.

Following Luijendijk's explanations, it is important to note that the adaptive design process of Building with Nature is not premised on a reconfiguration of human-material relations but on the acknowledgment that uncertainty cannot be ruled out. Claims of sustainability thus attempt to disguise "the will to design"[18] in a cloak of greenness. The will to design, according to political geographer Kevin Grove, does not aspire to truth—an impossibility given the complexity of human-environment relationships—but to transformation. In other words, natural processes that are deemed useful by the engineers are acknowledged, while others are ignored and externalized for the sake of comprehension. Grove's analysis bridges the discourses of design and resilience and is fundamental to understanding how the attempt to quantify bodies and processes has allowed for a rationalization of the design process and thus an expansion of what can be considered "designed."[19] In this functional reading, design is understood not as a social or material practice but as a creative act of recombination. This leads to a necessary abstraction of sociomaterial processes into sheer numbers and probabilities:

> In a complex "environment," the truth of objects lies in their functionality. And their functionality, in turn, hinges on their relations with other objects. Control is no longer a matter of predictive knowledge, but is rather an ongoing process of learning about and managing feedback-driven relations between various entities that make up complex social and ecological systems. Rather than a will to truth that reduces reality to universally valid objective truths that can be folded into a calculative and predictive rationality, resilience theory mobilizes *a will to design* that strives to reduce reality to functional abstractions that can be re/combined with one another in order to solve contextually specific problems of social and ecological complexity.[20]

Unlike analytical knowledge, functional knowledge primes interventions not on the basis of existing conditions but on the basis of desired future outcomes. As

[18] Kevin Grove, *Resilience: Key Ideas in Geography* (London: Routledge, 2018), 19.

[19] Grove's work builds on Herbert Simon's *The Sciences of the Artificial* and brings it into dialogue with present-day climate governance. Design, in Simon's words, is "concerned with how things ought to be, with devising artifacts to attain goals." This is rooted in a cybernetic understanding of design that emphasizes its synthesizing qualities, that is, the capacity to tackle complexity through iterative adaptation. See Herbert Simon, *The Sciences of the Artificial*, 3rd ed. (1996; Cambridge, MA: MIT Press, 2019), 114.

[20] Grove, *Resilience*, 134–5, emphasis original.

Figure 15.4 Scientific observation tower at Zandmotor, November 2020.

such, it blurs the lines between research and management; it is learning by doing. The consequence of this iterative approach is that unforeseen developments, if recognized at all, are assimilated and instrumentalized so as not to diverge from the overall goal. Building with Nature, read through the lens of Grove's work, can be understood as an attempt to manage the environment in order to maximize socioeconomic benefits. While this approach acknowledges the limits of scientific knowledge, it also reduces human-environment relationships to what can be (recursively) quantified and continues to perceive nature as something that can be interfered with, modified, and optimized by means of more flexible, soft(er) control.

However, the reality of Zandmotor shows that any attempt at systematizing sociocultural and material reality inevitably lags behind and cannot begin to grasp the intricate web of life that exceeds control. The "real" Zandmotor, after almost ten years of existence, turned out to be quite different from the projections. Not only has its shape changed in unexpected ways, but the latest projections suggest that it will last for forty to fifty years instead of the original twenty.[21] At the same time, a lot of sand is being lost to the sea, undermining its original core function as an alternative method of coastal nourishment. So, what has become of the experiment as it has materialized in situ?

[21] Luijendijk and van Oudenhoven (eds.), "Shaping Future Sandy Solutions," in *The Sand Motor*, 147.

Figure 15.5 Scientific observation tower at Zandmotor, November 2020.

A Living Laboratory

A few years before Zandmotor was built, a group of local artists gathered in order to investigate the sociocultural implications of this geomorphological shift. "We were curious what it would mean for us, physically and symbolically, not to be located on the open sea anymore, but [in] a lagoon," explains Jaqueline Heerema of the Dutch artist collective Satellietgroep. Such a project would have repercussions not only on the landscape but also on the local population. "How would this shift in horizon affect our self-conception?"[22] Initially, Zandmotor was poorly perceived by the local community, which had been excluded from the decision-making process, but over the years, it became assimilated as "their nature," Heerema reports. When a scientific observation tower was installed to measure and collect data on the project's progress, locals were offended by its presence in the landscape. They did not like the prospect of having their view disrupted. Furthermore, "correctional citizens," as Heerema likes to call them, started patrolling the beach, frequently pulling out the grass that researchers had planted to study its performance. According to her experience, certain people felt that the beach was now part of their local habitat and did not consider the grass as belonging to the beach. Although initially disputed, the newly reclaimed land has now become part of what is perceived as the "natural" environment.

[22] Jaqueline Heerema, interview by Michaela Büsse, November 7, 2020.

Today, Zandmotor is a popular spot for kitesurfing, walking, running, and, most notably, collecting fossils. As a huge amount of sand was required due to the scale of this project, it had to be dredged from a depth that holds soil from the Pleistocene era. In a comprehensive report on the making of Zandmotor, the scientists explain: "The sand used for the Sand Motor originates from two different geological layers: a Holocene marine layer (up to twelve thousand years old) overlies Pleistocene fluvial sand (between twelve thousand and two million years old)."[23] The deepest, and thus the oldest, layers eventually settled on top of the younger ones, leading to curious finds such as mammal bones and fossils. Numerous objects were discovered by hobby collectors, and the most significant pieces are now on display at the Natuurhistorisch Museum Rotterdam.

During our walk, Heerema points out some rusty-looking stones scattered on the ground: iron ore. It has become habitual for some visitors to collect the iron ore and pile it in the middle of the beach, a collective installation that acknowledges the existence of this ancient matter amid the recently reclaimed and therefore young land. A middle-aged couple not far away from us is carrying little shovels on their walk. Every now and then, they stop, kneel down, and remove some of the topsoil. "It is better to come after a storm," Heerema explains, "when the top layer is blown off and new fossils drift ashore." An accidental by-product of the engineering process is that Zandmotor has become important to archaeologists and geologists, both professional and amateur.

Continuing our walk, Heerema makes us aware of the changing dune landscape. "We have just crossed the border between The Hague and Westland province," she explains. Behind us, toward The Hague, the transition between the beach and hinterland is smooth, a soft elevation. Every few hundred meters, beach kiosks divide the strip into sections for the orientation of visitors and to shorten the distance to the next refreshment. Before us lies the wilder side of Zandmotor. No more kiosks; instead, a winding landscape with outstretched patches of vegetation. She tells us that in order to keep its beaches clean and attractive for visitors, the city frequently removes garbage, thereby clearing smaller vegetation. Weeding the beach, however, impacts the growth of the dunes, a consequence that no one had previously taken into account. While the Westland authorities have decided not to clean the beaches—whether for financial or environmental reasons is unclear—The Hague prefers to stick to its policy because the beach is the pride of the city, and its visitors are an important

[23] Iris Pit, "Environmental Impacts of an Innovative Beach Nourishment," in Luijendijk and van Oudenhoven, *The Sand Motor*, 71.

Figure 15.6 Zandmotor, November 2020.

source of income. Yet, it is not only The Hague's trash policy that has effects on the growth of the dunes. Researchers and politicians frequent the site in their four-wheel drive vehicles, further slowing down the growth of the dunes. It is examples like these that demonstrate in unexpected ways how Zandmotor fulfills its role as a real-life laboratory, emphasizing the many unforeseen circumstances that manifest within the human-environment relationship.

As our experience on site and the interview with Heerema show, Zandmotor is more than just the functional object envisioned by the engineers—continuously informed and morphed by different practices of living and engaging *with* it. The ephemeral nature of the project further emphasizes the contingency and interdependency of these activities. In other words, Zandmotor is not just interpreted and represented differently by various groups and through multiple activities, but it becomes actualized through this difference.

After the Model

While any coast can be considered a threshold between land and sea, solid and liquid, Zandmotor takes this condition to the extreme: It is artificial and natural at the same time, yet accepted as nature because it benefits leisure life. Although artificial, it resembles a natural phenomenon (the beach), and yet the tower and the grass do not fit into this narrow framework. While tides and waves (the gravitational forces of the sun and moon) define the measurement of human time, the occurrence of fossils in Zandmotor's topsoil turns the linear geological history of the site upside down and challenges its temporal boundaries. Iron ore lies next to plastic bottles; citizens, archaeologists, biologists, engineers, politicians, and artists have come to define Zandmotor on their own terms. When shifting the view away from the model and toward its material site, their differences and interdependencies become more pronounced. The unruly behavior of matter and animate beings allows us to point toward the ever-present contingencies only partially accounted for by the modeling process. Zandmotor unfolds through the divergent but overlapping practices of scientists, citizens, archaeologists, engineers, politicians, and the sheer forces of waves, wind, and sand.

Cultural theorist Lisa Blackman argues that data always "bear the traces of human, material, technical, symbolic and imaginary histories that are often displaced and occluded in data metrics,"[24] and indeed, a close look at Zandmotor's data reveals that the model is haunted by its materiality. As sociocultural and material complexity slowly creep into the engineering models, the experimental site continues to be computationally governed by means of land surveillance. Parts of public life and public space are pragmatically operationalized for the purpose of mitigating climate change. As an engineering paradigm, Building with Nature relies on quantifying uncertainty to subsume the incomputable. This adaptive approach to managing geoengineering interventions would not be possible without numerical models and quantified risks, rendering success or failure moving targets. Such flexibility, however, hinders the accountability of the stakeholders and treats the earth and its inhabitants as a mathematical process that runs on fragmented and cross-referenced data. Still, a nuanced look at the experimental site reveals that Building with Nature, however sophisticated an engineering paradigm it may be, fulfills the wrong premise. In a twist of irony, the displaced layer of Pleistocene sand does not cease to attract amateur paleontologists, inadvertently rendering the Anthropocene public and tangible through the disruption of geological strata. Zandmotor clearly demonstrates how the geoengineering mindset can obscure its own workings. In the end, nature takes command over software.

[24] Lisa Blackman, *Haunted Data: Affect, Transmedia, Weird Science* (London: Bloomsbury Academic, 2019), xiii.

Planetary Scientia

16

Poetics of Science/Dialogic Curiosities/Incomputabilities

fields harrington and Katherine McKittrick

This chapter highlights how humanity, science, and technology, together, have the potential to offer a creative grammar of liberation. The impetus for our conversation is twofold. First, holding close Aimé Césaire's "science of the Word," we have been working through how to talk about race, science, and technology without continually falling into conceptual traps that glorify, depersonalize, or condemn science and technology. Césaire provides a frame to think about the potential of a world that is conditioned by a poetics of science. He offers a form of knowledge that is capacious, hopeful, opaque, collaborative, magical, precise, numeric, flowering, ecosystemic, and cellular. Second, we read Césaire alongside the writings of Sylvia Wynter, who extends the concept "science of the Word." Even with the beautiful potential of science Césaire offers—the numerically capacious, for example—both science and technology are tethered to race thinking; there is a tendency, for example, to imagine Black people as simultaneously unscientific and purely biological and endlessly robotic-(enslaved)-resilient. Wynter's research allows us to expose and destabilize and unthink these tendencies. She dislodges scientific racism and what she calls a biocentric system of knowledge (one that conceptualizes the human under a Darwinist model of the natural organism and posits that we are purely and totally evolutionary beings) by insisting that we reconceptualize scientific knowledge as fused to poetic knowledge. Indeed, she argues that we are, as a human species, bios-mythoi: the word (*mythoi*) conditions the study of nature (*bios*); mythoi and bios are enmeshed and, together, posit the human as a biological-storytelling species. With this in mind, our conversation emerges from a paired interest in how (and where) the poetics of science can and cannot unsettle scientific racism and its attendant technologies.

> As we attended to these themes, a preoccupation with metaphors emerged. We confronted a set of difficult questions!
>
> How are scientific metaphors used in Black studies? How do we grapple with the hard sciences and scientific metaphors if we are not trained in these disciplines?
>
> What is at stake in configuring scientific knowledge as a conceptual and theoretical propellor that helps us better understand systems of oppression or liberation or creative activities?
>
> What is the function of the metaphor more broadly—not just those that emerge from the hard sciences but those that appear in theories of nature and geography literature? How do we think through, say, "stardust as a mode fugitivity," if we do not fully understand micrometeoroids, and we are not in the depths of nineteenth century slave-post-slave fugitive activities. Should we abandon the cosmic dust? Is fugitivity even metaphoric?

Part of our conversation is inspired by the work of Neil Smith and Cindi Katz, which means we are pushing ourselves not to abandon the metaphor but instead work out *how it functions* in our research, writing, and creative activities while also paying attention to its material-concrete underpinnings. Or, stardust is carbon dioxide, silicon carbide, amorphous silicate, polycyclic aromatic hydrocarbons, water ice, and polyformaldehyde. Or, stardust is matter and radiation and gas. These concrete properties animate and ideally help us define the metaphoric propulsion.

This chapter also draws attention to how dialogue, collaboration, and conversation connect to the themes of science, technology, poetics, and Blackness. We have been in dialogue, on and off, since 2020, and we have shared and built up a set of narratives and concepts that seek to better understand how science and technology are entangled with the creative and theoretical work of Black studies. Curiosity and mutuality are fostered through helping each other work out ideas, concepts, and ways of living. Dialogic curiosity and cooperation have provided us with a context for thinking and talking about science and technology without focusing on mastering concepts and ideas. We approach science and technology not as expert practitioners in these fields but as readers, writers, scholars, and creatives; we use conversation to speculate and study with each other. The prevailing and calcified knowledge systems that haunt embodiment and Black humanity (the race thinking, the biological determinism, the scientific racism, the IQ tests, the measurements of flesh and teeth and bones, the repeated evaluations, the surveillance, the codifications) spark our ongoing commitment to wade through the possibilities and limitations of race, science, and technology.

Katherine: I would love to begin our conversation with your art practice, which attends to scientific experimentation and scientific activities and how that poses other kinds of questions about relationality and interhuman relationships.

fields: Thanks for that question. In my creative practice, I work with science and scientific experiments and think about them in relation to Black studies. I go back and forth, thinking about the applications of the experiments that I am doing or have done in the past. I wanted to try to create maps through attempting to develop a new system, or a new type of embodied systems, of knowledge-making through experimentation. One in particular that I have tried to do this with is the iodine clock reaction, which is a demonstration of chemical kinetics. I have seen it demonstrated online by high school students or people in their kitchen. For this demonstration, you can find most of the ingredients at a grocery store or pharmacy, which is important to me since I am not a chemist and do not have institutional access to high-grade chemicals. I learned how to perform this demonstration by watching a good amount of YouTube videos and looking at several recipes online. There is something interesting that happens with the iodine clock reaction, which is that you begin with a clear solution and then you end up with a dark blue or a black solution. It is a classical chemistry experiment to display chemical kinetics in action. This type of transmission, or transfiguration, of a visual shift from a clear solution to a black solution made me think about detection, being seen or unseen, the specular appearance of being, and capture. What does it mean to be detected, or what does it mean to not be detected? Or what does it mean to be seen as one figure and then be seen as another? I try to think through the demonstrations and experimentations as a way to imagine how we can think beyond the limitations of science and start to consider a poetics of science that moves around the boundaries of scientific activities and thought with creative and theoretical maneuvers.

One way I tried to do that was by going back to Frantz Fanon and "The Fact of Blackness." Sometimes when I reread this chapter, I thought about how Fanon has the capacity to reside in both first- and third-person positions within his writing. Sometimes it feels like I am encountering someone who is writing an event or scene from above, which makes me think about the iodine clock reaction and what it means to be a witness and participant in an event and how one is witnessed. I have done other demonstrations that were not as successful in terms of trying to look at the reaction or the experiment and being like, okay, now I can find some theory or metaphor to try to embed into them. And I guess this goes back to the other part of the question about metaphor. I am still trying to understand the use of metaphor in my work. I think metaphors have the capacity to be a means for making sense of the complexities we live in and with. Perhaps, in addition to that, there is a threshold, or a thin line, that I try not to cross where the scale of the metaphor's presence can flatten the motivation behind the work.

Katherine: The iodine experiment provides a texture to more discursive or narrative ideas about science, because it concretizes and activates an experiment. And it is also working with colors and moves us toward color theory—I am thinking about Josef Albers who did visual experiments with contrast and form in order to get us to think about perception. In some ways, what you are saying is that the experiment itself provides the language to help us better understand visibility and invisibility and also that moment in *Black Skin, White Masks* where Fanon talks about how he is in three places at once, where he tells us "I existed in triple." So, the scientific experiment shapes our understanding of what it means to be a Black person, or it at least helps us think through the dynamics between perception, visuality, and observation.

fields: I am trying to think about how this demonstration of chemical kinetics can ask the question about what it means to move between being permitted to pass undetected to being the sensor itself. And how those in and in between maneuvers are about the living conditions of Blackness. I am asking these questions as I develop the work, but I am not sure if it always comes through in my work. I have some ambivalence about my installation *Redox Drip* (2019) and whether it is equipped to solve these questions. With *Redox Drip*, I wanted to create a stage for a few figures in the history of Black cultural production by arranging specific materials and objects on top of two 1970s wooden cabinet speakers that face each other. Some of the materials that rest on top of the speakers include NyQuil, Sprite, a *Diary of the Originator* mixtape by DJ Screw, a mixer, magnetic stirrers, Graham condensers, glass flasks, Styrofoam cups, sound absorption material, contact mics, plastic tubing, water, zip ties, black anodized aluminum, a steel panel, a two-way vacuum distilling adapter, and an angle distilling adapter. The historical figures that I was trying to put in dialogue with one another were DJ Screw, Frantz Fanon, and Norbert Rillieux. This allowed me to propose new models of inquiry into the alchemical affinities that could be created between these thinkers and makers. I wanted to understand these affinities as palpable sites for knowledge production. For me, this is a proposition to generate new meanings through a poetics of science. And perhaps, to ask what thinking beyond the scope of the demonstration is actually doing. The work of this kind has been criticized by some people because I call them "experiments." There are a few people who believe that the works are purely demonstrations because there is this kind of distinction between what is being demonstrated and an actual scientific experiment. I am not delving into the unknown and I do know the outcome of the reaction. I have been called out by those who do not believe this is actually an experiment. It is just a demonstration that is being reenacted again and again. When it comes to *Redox Drip*, it has its own set of limitations because it fails to enact truly entangled knowledge systems that have the capacity to unmake empirical systems of knowledge through interdisciplinarity by simply relying on the essential qualities of representation and metaphor.

Katherine: I don't know, "demonstration" sort of obscures the potential of what experimentation is, and it also forecloses the fact that repetition can be experimental, doesn't it? It also rules out modification or change while putting the work itself in its "proper place" (for example, this is art, it is a demonstration, a presentation, an exhibit, a description, it is not science). That is an entryway into thinking about what it means to do interdisciplinary work. If you pair chemistry with, for example, sound studies and if you pair science with other forms of knowledge-making, perhaps that idea of experimentation can stay vibrant. I think, as you said, there is an invitation, and part of that invitation has to be open to how we think about what Black art can do, how it functions, and also how we define it as scientifically creative.

fields: Yeah, when I was working on *Redox Drip*, I wanted to respond to the prompt that you are bringing up: "What does it mean to do interdisciplinary work?" My process of making *Redox Drip* was to assemble various materials, objects, and texts, which were carefully chosen because of their capacity to situate themselves within a larger social system. The arrangement of these elements in my work is not an isolated event, statement, or structure; rather, it is contingent upon the relational connectivity between them. Through this process, I aim to construct systems of support and distribute precarious processes throughout the totality of the work. By enacting material and social contradictions, I use assemblage as a way to think about *difference* within a complex unity. My assumption is that those who attempt to make the distinction between my work (Is it demonstration or experimentation?) are thinking about "demonstration" from a scientific perspective. In this point of view, the purpose of a demonstration is to illustrate a scientific concept or phenomenon, while the goal of experimentation is to test a scientific hypothesis or theory. Your question about repetition prompts me to ask another question: What is repetition without accumulation?

Maybe to follow the openness that you are bringing up, I can ask you a question. You shared your talk with me, "Anaformative Impulse. Chest Punch. A Working Paper on Bass," and as I was reading and listening to that talk, something that jumped out to me was how you are thinking about bass as a study in givingness. Correct me if I am wrong, but that was what I had noted down. How to think about bass as a study in givingness, and giving on, and giving with, and maybe even giving through bass too. This note led me to the question: Is it possible to perceive bass as this kind of rhythmic differentiation across each other, as an invitation, or as an openness, or as a mode of giving and welcoming at the same time? I do not know if that actually makes sense as a question, but there is something with the bass and the vibrations and how they open up new territories of relationality.

Katherine: Yes! I was interested in writing about the bass for all the reasons that you point out. Part of my research involved seeking out experiments that have been done with bass and what the low vibrations or sub-bass frequencies "do." I am interested in how the bass moves through walls in ways that other instruments and frequencies do not move through walls. I mean, bass frequencies can move through concrete! I also thought about that in relation to the idea that we are physiologically moved by sound and vibrations. So, it is not just that we are *psychically* moved by sound or song and that these tunes can make us feel unhappy or happy or groovy, or whatever, it is also about how the bass can actually move us physiologically and how it shakes dense masses. As you know, the bass can give you a chest pump; if it is really deep and really loud, it can actually shake your body and move your heart. So, there is that aspect of it, those vibrations that are so, so deep, that I am in love with. But at the same time, I like to think about how the bass functions in relation to other instruments and sounds, and in that respect, Édouard Glissant's notion "giving on and giving with" has been really useful. That generous givingness and how the bass musically holds together a jazz ensemble or a hip-hop song, for example. I think the bassist Flea said, "the bass provides a kind of bed for everybody to sleep in." And so, I asked myself: What does it mean that the bass functions in all of these complicated ways? Where is the sound creating a vibrational difference in our environment, and at the same time, it is providing a sonic nest in relation to other sounds. One of the ideas that we have been talking about is vibration and things that move. You introduced me to self-assembly and those experiments that were done, where two objects were dropped in water and then vibrations made them move around, and eventually they reassembled and connected to each other. In vibration, they moved toward each other and kind of "embraced" each other! Those ideas around what the bass can do to us (acoustically, sonically, physiologically, and psychically) expand and invite us to think about how vibrations might allow us to form self-assembly, arrangement, and cooperation.

fields: Thank you for bringing me back to self-assembly. The video of the Fluid-Assembly Chair was done by the Self-Assembly Lab at MIT. They were looking at autonomous assembly in complex and uncontrolled environments, and in this case, the environment was water. In their experiment, the unique 3D-printed components are released into a tank of turbulent water. Each of the components has a precise location in the final structure. After about seven hours of flowing around in the turbulent water, a fully assembled chair was created. That was my first introduction to thinking about self-assembly as a means to reorganize ourselves. When I first encountered the concept of self-assembly, I was trying to comprehend it alongside what Fred Moten says about *moving by way of disorganization*. Again, this goes back to the question of whether or not

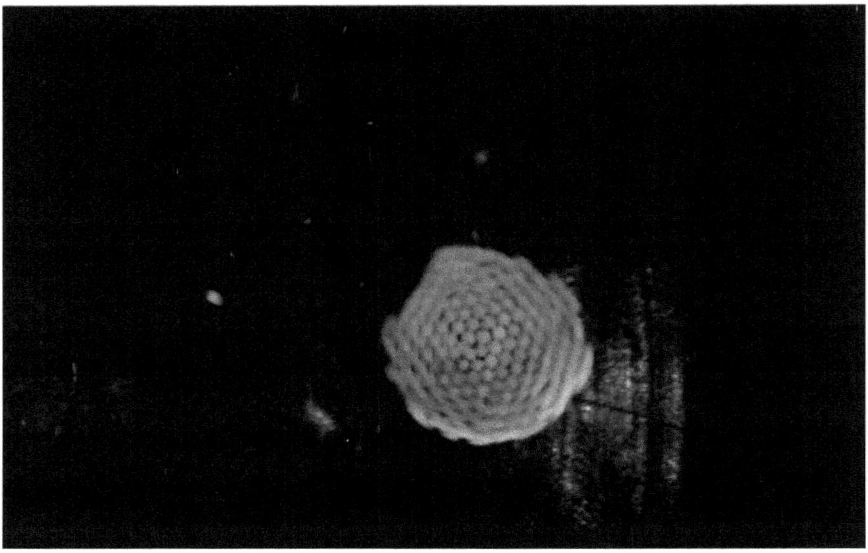

Figure 16.1 and Figure 16.2 Fluid-Assembly Chair project at MIT's Self-Assembly Lab: Demonstrating autonomous assembly in turbulent water, 3D-printed components are released into the tank, showcasing cutting-edge technology in complex and uncontrolled environments.

I should allow myself to be seduced into trying to map metaphor onto scientific processes or science at large. But there is something really interesting about seeing how the operation of self-organizing, or assembling oneself, can be based on our local interactions or environment and does not necessarily rely on an external dominant force. This is how I understand self-assembly, which is that you can take components that are a little different from each other, place them in a specific environment, and through their conditions, they organize or order themselves.

So, the process of self-assembly in and of itself can be dispersed into so many different disciplines and observations that are happening within different communities or populations. What happens when a community, a neighborhood, a group of people are implicated by their interactions and relations to each other, which permits them, or demands them, to organize solely on the fact that it is a means to survive? I think that is how I was starting to think about self-assembly. And again, it is a leap from what I think self-assembly actually is and how it is observed in the natural world. It is nice that you bring it up in relation to your study with bass because as I looked closer at the articles that address vibration and self-assembly, I realized that some of the experiments are using sound as means for acoustic trapping. For example, there is an article titled "Mechanical

Properties of Acoustically Levitated Granular Rafts," where the physicists use scattered sound that can tune vibrations for the purpose of producing a self-assembled particle system, which forms a granular raft with acoustic binding. They use polyethylene spherical particles, these tiny granular objects, and place them between an ultrasound transducer and an acoustic reflector. This setup generates a standing sound wave with vibrations; very low frequencies that cannot be heard but allow for the levitation of a granular raft. Through this tunable attractive force, it creates a new kind of short-range cohesion. The granules respond to the ultrasound transducer by rotating rapidly, creating a cohesive force, and each granule is bound to one another through gravitational binding by way of sound pressure—a new ordered system is formed with sound. This brings me back to what I am hearing in how you think about the bass and

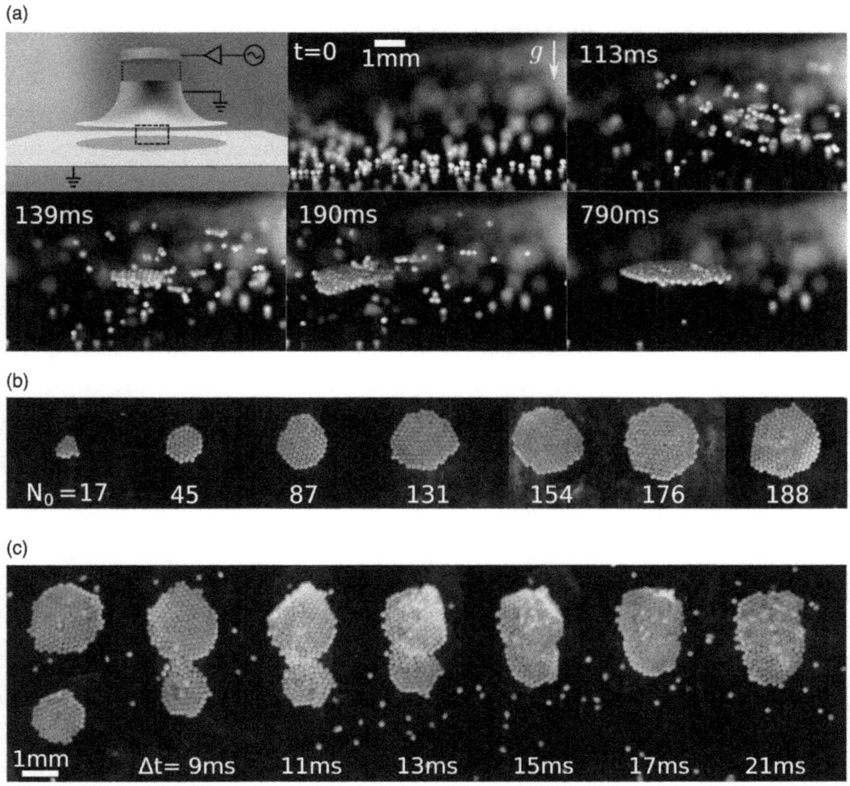

Figure 16.3 Self-assembly of granular rafts by acoustic levitation. Three-dimensional drawing of setup and sequence of images (side view) showing the self-assembly of a granular raft from its constituent particles. Piezoelectric elements (cylinder) are attached to an aluminum horn to generate ultrasound (only the base is shown). The grounded aluminum horn is spaced over a (grounded) indium-tin-oxide glass slide.

what it does with the body, through the body, and maybe for the body too. It can travel through it, it can travel around it, and also it can move it.

Returning to how you are thinking about bass as a site for givingness and openness, I am beginning to draw a connection through the experiment on granular rafts and what you are saying in your talk about low frequencies and bass. Whether we understand this givingness and openness as offered through the bass, sound, music score, or through the lab of the physicist, it is doing the same thing, which is creating a space for community and unity. Is it a form of invitation? Do the processes of sound pressure on the granules that model the dynamic rapid rotations similar to a black hole, the production of a sound system, the chest punch of the bass, and the music itself, bring and invite people in?

Katherine: The granular raft experiment makes me think about creating an acoustic space to produce different types of togetherness. Togetherness cannot always be comfortable, right? It has to be a contested space. But the granular raft points to a kind of spontaneous relationality; there is something meaningful about a momentary togetherness that is unexpected. Paul Gilroy has these incredible insights about bass culture, where he talks about how one experiences bass culture and how bass-driven music creates a moment of utopia. I think he calls it a "utopian possibility," so it might be more than a moment. But the idea that there is a different world, a beautiful world, an impossible world that is summoned by music is really appealing to me, even if it does not last! If we think about that in relation to what you are thinking about, these different spaces of interhuman connection and unity, they are iterations of how the physics of the world and the chemistry of the world provide clues and lessons about how sound and pressure and vibration push and pull us. I think that is where we have created common ground in our conversations because we are both interested in how extra-human things and processes (acoustics, liquids, experiment, and vibrations) touch us and how they provide a space for us to improvise and create new spaces of togetherness, hospitality, care, and cooperation. But as I mentioned, there is always contestation—the music does not always appeal to us. You may not always love the music. What about the pieces of the granular raft that did not make it to the raft and the pieces that did not find each other or fell to the side?

fields: Maybe there is some type of narrative that is happening between the production of the vibration and what is happening inside of the granular raft. As you said, there are granules that are ejected from the granular raft, and this could be a narrative in and of itself. But there is something interesting about the sonic properties of what we are discussing. Whether it is the experiments that I am looking at, the histories that you are looking at, or the communities that are being developed, it seems like low frequencies and bass have the capacity to reorder time and space with and through vibration. In addition to studying the processes of self-assembly and force-chain networks, I also wanted to think about assemblages and collectivity. In our previous conversations, we talked about assemblages, and it made me gravitate to this idea of a social weight, a social assembly, or a social body. This made me think about data as a bodily possession. What is a collective or social assembly that exceeds bodily data possession? If we are going to think about assemblages, and if we are going to think about the social weight of the granular raft, what are some of the other social weights that manifest through the bass vibrations?

Katherine: This is a hard question because it requires taking seriously the fact that some things cannot be measured or turned into data, even though there is an attempt to capture, enumerate, and objectify everything. One of the themes that this publication is trying to think about is incomputability. I wonder if there is a way to think about how some things cannot be produced as data or that data in and of itself might be a site of possibility rather than a site of extractive analysis. Simone Browne's work would be useful here, especially her work on how Black artists take surveillance data and turn it into something new. This might be a way into generating collectivity and social assembly or cooperation; if we think about different forms of togetherness, we have to produce those social systems outside the terms of colonialism because the terms of colonialism require that social assembly is only done through exclusion, ownership, and boundaries. I guess social weight (the bass vibrations, or whatever else) forces us to address how assembly might be a point of openness, or a momentary utopia, just as data must invoke something outside of itself. And again, this is also about living with differences and the hostilities and anger that provide a way to understand togetherness as a mode of complicated terribleness animated by hope.

fields: Yeah. I think you are right. Because a social body, a collective, or a social system has the capacity to resist the possibility of being computable, gathered as a statistic, determined, and located in a data repository. I think we could probably start there because it is already in our reality. To trust the data can often mean that we are putting faith in inherently biased systems of knowledge. They are biased, subjective systems that have their own goals in mind. This becomes legible when looking at who is funding the research for certain kinds of data to be produced. It brings up questions about the value of science. Who influences and funds scientific value? Scientific activity is often being shaped by the social and cultural values of the university, biomedical research, pharmaceutical companies, and military activity. When I was looking at force-chain networks a few years ago, I stumbled on the website *Duke Today* and read this article titled "What happens when a missile or meteor hits" by Robin Smith. The article discusses how physicists at Duke developed techniques in their lab that simulate high-speed impacts in artificial soil and sand. The physicists then observed in slow motion what happens underground at a close range. The artificial soil and sand compress and absorb the transmission of force, and the physicists saw that the propagation of force resulting from the impact was carried from granule to granule branching out to form networks of stress. What I found interesting about this particular experiment is that it is being funded by the Defense Threat Reduction Agency (DTRA), which is a combat support agency and part of the US Department of Defense. The role of this agency is to counter weapons of mass destruction (chemical, biological, radiological, and nuclear) and high explosives. Tracing the patterns of funding for this research prompted me to ask more questions, such as: Why would physicists at Duke University carry out research that involves penetrating sand? Which geographies are composed and mostly mapped with sand? How does a university end up working with the DTRA? What is the effect of this scientific activity? How do moral, political, and social values influence and shape scientific research?

To follow Helen Longino, experimentation and scientific research is a "collaborative human activity" formed within political, social, and historical contexts that operate with specific goals in mind. When we observe the patterns of funding within the scope of scientific inquiry, we discover that the values and interests of scientific research and experimentation are contingent on corporate and government funding. This reliance on external patronization locates the comportment of science in a highly vulnerable position in relation to the origin of its funding. Research into the behavior and materiality of tiny artificial granules and sand, when struck with a blow from above, is inherently violent and biased because it is operating on behalf of the military with government funding; this particular data has the intention of producing information for some but not for others. We could probably find out more about why they wanted to think about the penetration of sand specifically, but I have an idea that it was for missile penetration and remote violence committed on behalf of the United States of America.

Katherine: This is the question that I had when I began working on my book *Dear Science and Other Stories*, and I began thinking about the interdisciplinary possibilities of science and scientific knowledge. I kept hitting a wall. Well, I kept hitting two walls. The first wall was that I was not a trained scientist and had very little knowledge of hard scientific thought. That is when I started to rely heavily on scientific metaphors to advance particular ideas in Black studies (string theories, physics, fusions, algebra). Then I hit another wall because everything became a metaphor—I lost my way! Everything became thin, descriptive, and allegorical. I had to start again and grapple with failure. When I was a graduate student, I was preoccupied with metaphors because I learned that they have material underpinnings; there is a materiality to space, and it is not just a cloud of diffuse emptiness. I always want to keep this in mind because it was an important lesson. When I became interested in science and technology, I lost my way, and I forgot the lesson. I kept hitting that same wall over and over again with science because I wanted the *idea* of science, its metaphoric potential, to do something for me, to free us from scientific racism or something. Maybe, for example, if we use this idea from particle physics, which I know nothing about, as an allegory for struggle, we can reorient ourselves away from the brutality of race thinking. I kept wanting "science" to do something for me, but it could not do anything because I kept casting it as pure metaphor and I was understudied. Initially, I had two columns: metaphors and scientific racism. As I continued to read, I began to think about the materiality of some sciences and how they are underwritten by the militarization that you are talking about. This research and writing process created a cycle of complicated and frustrated sadness for me, circling around the limits of knowledge and the possibilities and limitations of literary devices. But it also opened up the arts through a process of thinking about scientifically creative and creatively technological worlds, which led to thinking about what we touched on earlier: how vibrations and songs move you. For me, this is paired with the artist's physiological effort; how they use their bones and blood and muscles to make things. This is why I am so attracted to your art practice. You show the materiality of the scientific experiment as it is paired with the possibility of creative metaphor, and you are part of the story, too! That has the potential to undo the pure militarization of some sciences because it is creating and recreating possibilities by explicitly pairing science, technology, and cultural production. It shows that science is flexible, alterable, and shifting.

fields: There is a lot for me in what you were saying about metaphor and its material underpinnings, which is something that I have not thought about with

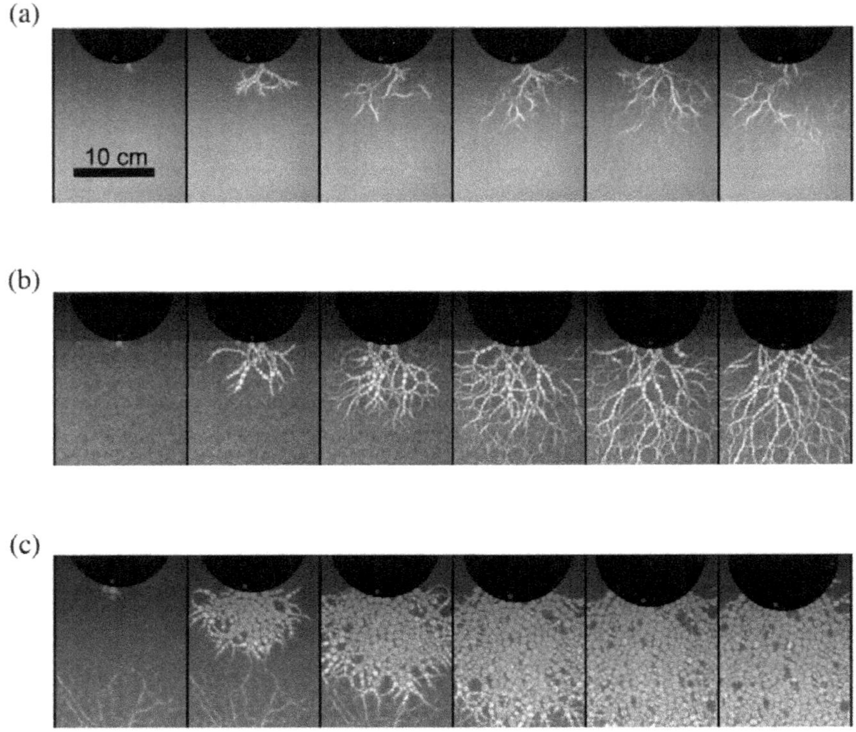

Figure 16.4 Force propagation after impacts with v0 ≈ 5 m/s. (a) Hard particles ($M' \approx 0.1$) show sparse, chainlike force propagation. (b) Forces for intermediate particles ($M' \approx 0.3$) are more dense spatially but still relatively chainlike. (c) Soft particles ($M' \approx 0.6$) show a dense force structure that propagates with a well-defined front.

anybody else before. Science is a discipline with the potential to be flexible and alterable, but the material underpinnings of science can also be seen as a flexible and alterable space. In my work, a metaphor can be both useful and inadequate, depending on its application. The consequence of the use of metaphor is that it possesses a partial truth. Because metaphors are somewhat dialogical, they can be seen as impartial to how we understand what truth is, or what maneuvers we have to make to get toward a truth. Since science is often accepted as a truth-making system, or a "capital K" knowledge production system, is metaphor the wrong method to apply to science? This partiality to truth that metaphors contain is a positive feature. However, it is a double bind of some sorts because that partiality still embodies a truth in relation to the history of science and what it means to be factual, accurate, and objective. It is hard for me to also commit to metaphor as a tool to think beyond the sciences or ask what it means to think in excess of science in relation to Blackness. Science is always trying to contain Blackness, whether it is through observation or discovery, and here I am thinking about Henrietta Lacks and the HeLa cell. The role of biomedical research in relation to the HeLa cell is to contain these unruly cells after they were extracted from Lacks's body. After the extraction, the cells have continued to live within a disembodied state of survival. There is still a notion of containment, capture, captivity, and control. So how can we think about metaphor in relationship to excess and Blackness?

Katherine: You ask the question that I have never asked myself but that I should be asking: Is metaphor the wrong method to think about and apply to science? And how does metaphor, at the same time, allow us to think about science as fictive and not as "capital O" objective? I love that, it is so entangled, and it really gets at how the production of knowledge is underpinned by questions of race. Is metaphor the method to apply to science, and how does metaphor expose the fictiveness and truths of science? These very questions mean we have to sit with ambivalence, sit with those impossibilities *and* possibilities, precisely because the category of race shores up the idea that science is "capital O" objective. As you noted with the example of Henrietta Lacks, even as those cells live in excess of her body, her story is about the enclosure and ownership of her; there is a scramble to own what her body gave us. That is how powerful and monumental scientific racism is. No matter how many experiments we do, no matter how many Black scientists we find, no matter how many scientific discoveries we participate in, that monumental history of scientific racism weighs down. Earlier you mentioned social weight. I think about the weight of what scientific racism does to Black sociality and how it is a kind of entrapment. We need to dream outside of that. While keeping science, technology, and different technological possibilities in mind, we have to dream of something that notices the beauty and fictiveness of our interhuman and interecological worlds.

fields: The useful thing about metaphor is that it can contain more than one truth. I think the problem with understanding the role of science and society is that science is under the guise of producing one truth as opposed to many truths or having the possibility to have multiple truths. What you brought up about dreaming outside and beyond, but with science in mind, brought me back to the earlier topic about incomputability. Perhaps dreaming outside and beyond, but with science in mind, means to be incomputable, unquantifiable, or incalculable. What if we thought about the incalculability of being as valuable or desirable?

I ask this question because it would situate us in a way to dream in excess, or dream beyond the mechanisms of science, while still using the tools and methods of science. We cannot just leave science behind. I mean, we could totally abandon science, but what would be the use of that? Another oppressive regime of knowledge would just take its place. Keeping science present, or perhaps even keeping *it* contained, gets at this notion of what I am thinking about in some of my work, which is how to imagine a poetics of science. Something that I learned through your work is Césaire's science of the Word. Maybe this is what it means to move toward a science of the Word, it is a notion of thinking through excess, dreaming in excess, or dreaming in the beyond while keeping the sciences in mind, as opposed to divorcing ourselves from science as a way to imagine a false utopia. To move beyond science without asking or acknowledging why we are making the journey in the first place is counterintuitive. Perhaps, what I am suggesting is that we need to dream more while simultaneously asking why we are incalculable.

Katherine: Yes, the brutality and terrorism of science are why we cannot abandon it. That is the lesson for us here. Science is full of possibility, but it is also full of terror, and we need to hold those two things together. Terror is why we should not abandon it. We are scientific beings, we are biological beings, we make things, and we develop languages and alternative languages. I like the idea of a poetics of science and a scientific poetics building on Wynter's work and building on Césaire's work. This is urgent for Césaire, as we see in "Discourse on Colonialism," which is about colonialism and fascism as entangled expressions of normalized violence. And if we pair that with his plea in "Poetry and Knowledge" for poetics to condition the natural sciences, the urgency is amplified. Reading across his work, it is clear that his capacity to imagine the science of the Word is such a beautiful leap because he is so cognizant of the limits and dangers of the colonial sciences.

fields: Awesome, that was great. I am a little bit jarred by our conversation. I am also like, "Where else can we go?"

17

At the End of Autopoiesis

Nonaxiomatic Patterns and Millions of Incomputable Earths

Luciana Parisi

How can the rewriting of the epistemology of the Earth escape the cosmogony of Man,[1] the epistemological paradigms that determine the limit of knowledge? To address this question, it is important to reflect on Sylvia Wynter's warning about the need of rewriting knowledge to confront the immanent catastrophe that the human species is driving toward. As Wynter claims:

> We must now collectively undertake a rewriting of knowledge as we know it. This is a rewriting in which, inter alia, I want the West to recognize the dimensions of what it has brought into the world—this with respect to, inter alia, our now purely naturalized modes or genres of humanness. You see? Because the West *did* change the world, *totally*. And I want to suggest that it is *that* change that has now made our own proposed far-reaching changes *now* as imperative as they are inevitable. As Einstein said, once physical scientists had split the atom, if we continue with our old way of thinking—the prenuclear way of thinking—we drift as a species toward an unparalleled catastrophe.[2]

If, as Wynter argues, the epistemological enterprise of colonial capitalism totally changed the world, from which standpoint would it be possible to rewrite the origin story of the Earth, abandoning the old prenuclear model of knowledge,

[1] By following Wynter, this chapter draws a parallelism between Wynter's argument against the overrepresentation of the origin story of Man—the cosmogony of Man—and that of the origin story of the Earth. As much as this cosmogony is based upon a particular mythopoesis which tells the story of an exceptional human sapience performing the racialization, gendering, and sexualization of knowledge, so too does the mythopoetic origin of the earth sustain the cosmogony of the human in the image of Man, for which scientific knowledge (such as evolutionary theory) is set to sustain the ontological givenness of Man.

[2] Sylvia Wynter, *Sylvia Wynter: On Being Human as Praxis*, ed. Katherine McKittrick (Durham: Duke University Press, 2015), 18, emphasis original.

and rather embracing the relativity of space and time, but also the quantic and parallel dimension of the atom and the wave or the principle of indeterminacy?[3] In order to understand and describe these phenomena, the natural sciences, on the one hand, and mathematics and computation, on the other, have introduced and embraced the conditions of contingency, chaos, and ultimately the unknown.

Starting from these premises, this chapter takes inspiration from Wynter's warning about the "unparalleled catastrophe" that critical thinking faces when the epistemological enterprise is locked within paradigms that reproduce the order of knowledge that has been contested and expanded. For Wynter, one way to undo the self-referentiality of the already known is by means of heretic statements, counter-memories, cultures, and expressions that overturn the overrepresented onto-epistemological circuit of Man, which must be reproduced by paradigmatic knowledge. Wynter, importantly, takes the cybernetic notion of autopoiesis to address the process that describes how organisms reproduce at a system level until environmental disruptions act as triggers at the level of the structure and transform it. From this standpoint, autopoiesis also explains how the onto-epistemological structure that places the human at the center of the Earth can only change through the appearance of a rupture (or negation and refusal) that the Earth brings forward and that is carried out by heretic statements, which pave the path for a new radical intervention in determining the meaning of the human-Earth relation. According to Wynter, the possibilities for overturning the overrepresentation of the origin story of Man (or cosmogony) must be found in epistemology as a scientific praxis itself, in the daring expansion of heretic languages that crack the autopoietic reproduction of knowledge, whereby anthropocentric cognition and mathematical formalism preserve the exceptionalism of human sapience and continue to determine the conceptualization of the Earth as we know it.

By following Aimé Césaire,[4] Wynter sees the promise of an epistemological shift in the appeal to return to the promises of "Scientia," or "the Word," as a new science able to nourish natural sciences. This chapter pushes Wynter's proposition for an epistemological expansion of the natural sciences in order to rewrite digital computation in two steps: First, by inquiring how the shifts in

[3] It is well known that quantum theory challenges the wave/particle duality by demonstrating that every quantum object, that is, any object measured at the quantum level, acts both as a particle and a wave.

[4] Wynter discusses the contribution of the work of Césaire in rearticulating the centrality of the Word in the formation of heretic epistemology in Sylvia Wynter, "Unsettling the Coloniality of Being/Power/Truth/Freedom: Towards the Human, After Man, Its Overrepresentation—An Argument," *New Centennial Review* 3, no. 3 (2003): 257–337.

sociotechnical mentality challenge the universality of scientific paradigms, and second, by suggesting that the incomputable aspect of computation[5] entails a nonformal and nonbiocentric dimension of practical knowledge, or know-hows, where cognition starts as the compression of randomness, or namely with the account of incompleteness in the formation of knowledge. As much as Wynter reflects on the possibilities of counter-cosmogonies as alternative know-hows that take humanness as an unfinished praxis,[6] computational theory, I suggest, brings forward a socio-techno-genic mentality of incompleteness from where artificial epistemologies—that is, how nonbiocentric know-hows function and structure knowledge—can challenge natural sciences as they continue to determine the temporal evolution of human speciation, racialization, and gendering of the Earth. Computational theory offers a view of indeterminacy that relies neither on contingency nor on probabilities or the celebration of the unknown. Instead, this chapter uses computational theory to argue for systemic incompleteness, defined in terms of incomputables: the patternless structures and irreducible functions that condition any self-organizing system. From this standpoint, by challenging theoretical completeness, the turn to the incomputable offers a possibility of transforming the onto-epistemological framework that conceptualizes the Earth per human centrality. Instead of decommitting the politico-historical project of humanity from its violent self-determining decisions of colonization, dispossession, enslavement, and the metamorphic power of extraction and abstraction, the incomputable Earth that I am proposing to think here is another way to argue for an unknown, alien dimension of the cosmos, of life, of language, of intelligence. Such a dimension continues to surge up against the totalizing planetary extension of what has become a neurocognitive automation (the capitalization of attention via social media architecture) and its accelerated surrogacies (for instance, the employment of mechanical turks from the Global South).

I argue neither for a mechanistic nor a vitalistic model of epistemology, neither for a techno-scientific representation of nature nor for the appeal to a natural truth standing before colonial cosmogonies and racial capitalism. A collective rewriting of the model of the Earth must account for the mediation, instrumentality, and machine-like process of knowing, which entails what

5 To put it simply, incomputability defines the limit of a universal computation that cannot contain within itself the infinity of unpatterned information and thus stands for the indeterminacy and incompleteness of any systematic attempt at compressing reality into a smaller set of procedures or in a given postulate.

6 McKittrick (ed.), *On Being Human as Praxis*, 151.

Gilbert Simondon calls a "technical mentality"[7] and thus a sociotechnical practice of know-hows that constitutes epistemological enterprises through the dynamic of machine language or coding. In what follows, I draw on Wynter's effort to expand models of knowledge beyond the biocentric explanation of being and the bioeconomic success of the racialized and gendered human. While keeping in mind Wynter's premise that such expansion offers a possibility of rethinking humanness away from its overdetermined cosmogonies, I aim to expand the anthropocentric views of the Earth by turning to Simondon's articulation of the technical mentality of cybernetics and thus to suggest that the technical mentality of computation shows us a mode of abstraction of socio-techno-genic practical knowledge, or know-hows, that constitute today's algorithmic culture as conditioned by incomputability. From this standpoint, this chapter suggests that the technical mentality has already challenged the epistemological foundation of anthropomorphic cosmogony. At the same time, Wynter's argument about challenging the overrepresentation of the origin story of Man also opens up Simondon's theory of the technical mentality to the tension between the social and the material reproduction of knowledge as entangled with its racialization, gendering, and sexualization. In the effort to reverse the colonial conceptualizations of the Earth, the turn to an incomputability theory exposes the centrality of randomness, fallibility, and patternlessness within the structure of knowledge and their logical irreducibility within epistemology. The incomputable, as a mode of *alienness*[8] rather than humanness, can continue to challenge the paradigmatic model of the Earth in the time of climate breakdown, which largely remains bound to the notion of the Anthropocene and the critiques thereof.

The Loop of Knowledge

As Wynter points out, the order of premodern metaphysics and the order of modern science are not separated but rather overlap in the overrepresentation

[7] Gilbert Simondon, "Technical Mentality," trans. Arne De Boever, *Parrhesia: A Journal of Critical Philosophy*, no. 7 (November 2009): 17–27, http://parrhesiajournal.org/parrhesia07/parrhesia07.pdf.

[8] I use the term "alienness" to discuss a double negative process: On the one hand, the alien(ated), excluded, and negated socio-techno-genic forms at the core of the process of capital abstraction/extraction, and, on the other, the alienness of nonbeing and nonworld, whose symbolic and material forms refuse subjection to the monotonic cosmogony of Man. In short, alienness is a critical and speculative condition of transformation that sides with the unintended consequences of negative negation.

of the biocentric model of the human,[9] for which science is assigned the transcendental role of separating Man from the "space of Otherness."[10] This decision results in the division between two categories: those selected and those deselected by evolution. This division is also sustained by a paradox: While the ontology of the human is justified by scientific knowledge, it nonetheless remains independent from it because "being human" must remain, as it were, a given. According to Wynter, the biocentric explanation of what is human is determined by an ontological self-decision—namely, "being human" is a decision given within an overrepresented structure of power and is explained as a naturalized truth—for which the Earth only has meaning for Man.[11] As such, the ontological self-decision is always set in advance of the scientific enterprise of knowing whose task is already set to justify (explain and find proof) for what is already known. This is what shapes the onto-epistemological apparatus for which the condition of knowledge is determined, constituted, and defined by being (human). It is precisely this (already posed) self-referentiality concerning the existence of being that is set to drive the operative production of knowledge according to what Wynter calls "self-definition," or "self-conception," as the founding motor of recursive epistemologies.[12]

Wynter derives this notion from cybernetician Gregory Bateson, according to whom recursivity describes a self-reflective principle that repeats itself in time and across scales.[13] The idea that the past can be reconfigured in the present is understood by Wynter as a process of mythopoesis—the making of myths—that shapes the collective cultural ways of knowing and acting. The emphasis on continuous looping defines epistemology not as a series of shifts between totalizing systems of knowledge and homogeneous blocs of dominant ideologies but as nonlinear iterations of statements. These necessarily resonate with each other and ensure a continual variation of a given set of rules within its schema. For Wynter, recursive epistemologies are not symbolic paradigms but, above all, entail the biosocial structuring of the self, which coincides with the persistence of the "sociogenic principle" across times.[14] From this standpoint, the formal model of knowledge is not predicated on the universal language/

[9] Here, the notion of biocentrism is derived from Wynter's critique of the overrepresentation of the Darwinian model of evolution, according to which sapience is explained in racialized, gendered, and sexualized terms based on the exceptionalism of selected populations (Europeans) over the rest of the world.

[10] Wynter, "Unsettling the Coloniality of Being/Power/Truth/Freedom," 322.

[11] Wynter, "Unsettling the Coloniality of Being/Power/Truth/Freedom," 326.

[12] Wynter, "Unsettling the Coloniality of Being/Power/Truth/Freedom," 258, 268.

[13] Wynter, "Unsettling the Coloniality of Being/Power/Truth/Freedom," 267.

[14] Wynter, "Unsettling the Coloniality of Being/Power/Truth/Freedom," 328.

grammar of Man, which mirrors the biocentric origin of the subject. Instead, the overrepresentation of this grammar is granted by the recursive loops that generate dominant statements in terms of an autopoietic programming of the sociopolitical cognition of humanness. As Wynter argues, following Frantz Fanon, neither ontogeny nor philogeny can explain the autopoietic processes of racialization, gendering, and sexualization that accompany the origin story of Man. Instead, she argues that it is the sociogenic structuring of knowledge that comes to define how nature becomes entangled with culture; namely, it is not just the human that shapes culture, but rather it is the mythopoetic organization of the flesh as the material source of extraction that defines the human form in terms of the universal grammar of Man.

Wynter develops on Fanon's articulation of sociogenesis[15] to argue that the human is not a biological concept but is predicated on regulatory practices that are social and replicate this social code by ascribing it to the flesh—what Fanon terms "epidermalization."[16] In other words, what Wynter calls culture's "descriptive statements"[17] describe a social situation that alters the psychosocial structure of the human prior to the reflectivity of consciousness but is nonetheless irreducible to physical laws. Instead, descriptive statements become the functions of a semiotic technology or language. This is a material-semiotic programming of human sapience, establishing racialized and gendered hierarchies that determine the biological exceptionalism of the human. For Wynter, this sociogenic process has evolutionary consequences insofar as it is recursively reproduced by cognition, language, and social interactions.[18] In particular, Wynter insists that one must challenge the overrepresentation of the semiotic order of the human that she understands in terms of the neurochemical processes that produce the cognitive structure, according to which the origin

[15] Wynter's articulation of Fanon's argument about sociogeny entails a critique of ontogeny—as the biological definition of being (human)—and addresses the conflictual relation between the first and third person, for which the racialization of consciousness generates a split in self-recognition. It is the social situation of colonial and racial capital that determines this conflictual relationship, not the biological essence of the human. Wynter considers this process of socialization as becoming entangled with neurobiological existence, according to which Blackness is determined in opposition to the genericity of being human and in terms of lack (lacking reason, culture, civilization, interiority). Wynter's articulation of sociogeny as fundamental to the constitution of Black nonbeing can be found in Sylvia Wynter, "Towards the Sociogenic Principle: Fanon, Identity, the Puzzle of Conscious Experience, and What It Is Like to Be 'Black,'" in *National Identities and Sociopolitical Changes in Latin America Hispanic*, ed. Mercedes F. Durán-Cogan and Antonio Gómez-Moriana (New York: Routledge, 2001), 30–66.

[16] Frantz Fanon, *Black Skin, White Masks*, trans. Richard Philcox (1952; New York: Grove Press, 2008), xv.

[17] Wynter, "Unsettling the Coloniality of Being/Power/Truth/Freedom," 267.

[18] Sylvia Wynter, "Human Being as Noun? Or *Being Human* as Praxis? Towards the Autopoietic Turn/Overturn: A Manifesto" (August 25, 2007), 13.

story of Man's evolution must imply the negative representation of the "Negro and of his African past."[19] In particular, Wynter suggests that this overrepresentation, or cosmogony of Man, also entails the recursive operations that link biology with economics, *Homo sapiens* with *Homo economicus*, and constitute the onto-epistemological grounding that perpetuates the dominant sociogenic code and the corresponding racialization, gendering, sexualization of the world. Wynter argues that it is precisely this grounding that must be challenged in order to expand the epistemological possibilities of knowledge beyond the colonial pillars of knowledge. Following Wynter, I want to expand the sociogenic principle to the socio-techno-genic possibilities of epistemological transformation of the dominant code of the racial, patriarchal, and colonial order of capital. In particular, I propose to consider the possibilities of machine cognition as an instance of compression of incomputabilities, which move beyond the biocentric model of reason, whereby the Earth is understood as a fixed capital for the reproduction of the life of Man.

This biocentric model of cognition,[20] it can be argued, continues to be overrepresented in the notion and discourse of the Anthropocene, where the posthuman critique of the human colonization of the Earth seems not to challenge but rather preserve the overrepresentation of Man's cosmogony. For instance, the posthuman critique of the Anthropocene points to a view of the Earth as a self-organizing autopoietic system, or Gaia, that entangles the problem of human survival with the problem of communal living and world-relatedness.[21] Here, the Earth stands directly for nature, which is defined not as a construction of the liberal humanist project but as an actor who enjoys agency and thus can be mutually imbricated with and through the human. Within this discourse, the Earth stands for a nature that is no longer an object of exploitation but is an actant that contributes to shaping the human as such. One can argue that this posthuman version of the Earth, as a natural and autonomous agent, relies on a materialist conception of immanence, namely, assuming direct communication between the Earth and the world, presenting nature as that which exists without means or mediation. If the Earth is an agent among other agents, how can it stop being simply a part among others that are

[19] Wynter, "Unsettling the Coloniality of Being/Power/Truth/Freedom," 269.

[20] The biocentric model of cognition relies on the conception of intelligence explained in terms of the evolution of sapience. Wynter's argument for expanding cognition beyond this model turns to neuroscience as offering a systemic approach to knowledge beyond the hierarchical racialization of sapience.

[21] Bruno Latour, *Facing Gaia: Eight Lectures on the New Climatic Regime*, trans. Catherine Porter (Cambridge: Polity, 2017).

ultimately ensembled together into one onto-epistemological order to constitute a whole? While this posthuman critique of the concept and discourse of the Anthropocene tries to shift the centrality of the human as the sole agent of the Earth, it also seems to reintroduce the metaphysical argument for which nature exists as an unmediated agent like any other, without accounting for the onto-epistemological structure that subtends the reproduction of colonial, racial, and gendered oppression. Some have claimed that the posthuman critique of the Anthropocene overlooks and dismisses the realities of extra-human natures and thus the possibilities for differential antagonisms in distributed agencies, for which not all of the agents have the same power.[22]

Following this problematization of posthuman critique, Kathryn Yusoff claims that the discussion about nature as a nonhuman agent seems to dismiss "the natal alienation" that comes to constitute the "inhuman as the subject position within slavery and the dispossession of land."[23] Here, the Anthropocene, above all, demarcates the accumulation of free value for the origination of capital by precisely transmuting the necropolitics of extraction/abstraction into the form of bioeconomic success, which sustains the cosmogony of Man. According to Yusoff, "geology is a category and praxis of dispossession,"[24] where colonial extraction performs an equation of value, which turns enslaved people and land into commodities without internal value. In particular, geology becomes the space of transaction where enslaved people, land, and ecologies were first dehumanized—rendered inhuman objects—so as to sustain the colonial racial economy of transportable properties. For capital accumulation runs parallel with a socio-techno-genic dehumanization of the living, through which recursive epistemologies establish the bioeconomic model of human exceptionalism. While for Yusoff the socioeconomic system of racial capital and colonialism places geology at the center of the recursive function of dehumanization in establishing the property relation and the rules of commodity exchange, the posthuman critique of the Anthropocene seems to leave intact the binary function of racial capital in its efforts to expand the decisional power of the subject to nonhuman agents. Here, the included/excluded dyad cannot be corrected or expanded by simply offering a mode of recognition of the split between Man and nature/

22 Razmig Keucheyan, *Nature is a Battlefield: Towards a Political Ecology*, trans. David Broder (2014; Cambridge: Polity, 2016); Timothy Mitchell, *Carbon Democracy: Political Power in the Age of Oil* (London: Verso Books, 2011); Phillip John Usher, *Exterranean: Extraction in the Humanist Anthropocene* (New York: Fordham University Press, 2019).

23 Kathryn Yusoff, *A Billion Black Anthropocenes or None* (Minneapolis: University of Minnesota Press, 2018), 65.

24 Yusoff, *A Billion Black Anthropocenes or None*, 67.

Earth. To be recognized is to be included in the privileged position of being an autonomous agent, and this requires the demonstration of a certain humanness, which the excluded can rather demonstrate by acting according to the principle of self-reflection and thus conforming to the principle of free will. The dyad is already constitutive of the overrepresented cosmogony of Man, for which what was excluded also stands for a negative negation of Blackness,[25] which is necessary to reproduce the circle of being. In other words, even when the excluded is recognized as such and even when it is "given" autonomy and agency, it still remains enframed by the onto-epistemological structure that predetermines the meaning of thinking and existing according to the overrepresented models of Man's knowledge. From this standpoint, the posthuman project of inclusion into the ontology of Man only perpetuates its model of being, according to which Blackness is already set up to fail.

How can the posthuman critique of the Anthropocene avoid reproducing the brutality of the transvaluation of Blackness into inhuman things, which sets up the inclusion/exclusion dyad at the core of racial capital? Does the posthuman articulation of the equivalence between nature and culture, according to which agency is, as it were, conceded to otherwise inert things, avoid reproducing the colonial-capitalist mentality that it tries to challenge? What Yusoff calls the "inhumanities" already exposes why the posthuman critique of the Anthropocene fails to account for the racialization of knowledge at the core of the colonial-capitalist onto-epistemology. In other words, the posthuman critique of the Anthropocene cannot account for the irreparable condition of the inhuman, nonhuman, and unhuman through which the systematic enslavement and dispossession of the naturalized flesh of Black bodies and the dust of native lands continue to occur. What the posthuman critique misses is the impossibility of restoring an ontological condition for the excluded without forcing the latter to continue to bear the weight of the gratuitous violence that the sociogenic principle of recognition continues to impart on the other. From this standpoint, one can also ask how critical thinking is capable of challenging the paradigm of the Anthropocene if it remains derivative of a self-posed frame of reference—namely, the included/excluded sociogenic principle—according to which the image of the Earth is always already given? The critique of the Anthropocene, in

[25] Denise Ferreira da Silva, "1 (Life) ÷ 0 (Blackness) = ∞ − ∞ or ∞ / ∞: On Matter Beyond the Equation of Value," *e-flux Journal*, no. 79 (February 2017), https://www.e-flux.com/journal/79/94686/1-life-0-blackness-or-on-matter-beyond-the-equation-of-value.

as much as it adheres to the paradigm of colonial Man,[26] rather works to reinstate the prenuclear framework of the onto-epistemological project of humanity.

I want to bring Wynter's proposition about the extension of epistemology into conversation with Simondon's reflections of the "technical mentality," entailing transcategorical methods of know-hows that come to constitute computation as the abstraction of socio-techno-genic incomputabilities that do not fit the biocentrism of knowledge and the formalism of mathematics. By considering the material effects that incomputabilities have on the social structuring of knowledge, I will argue that the image of the Earth must be unbound from the representation of a neurocognitive schema inspired by a Promethean myth of creation, according to which machines are conceived as prosthetic extensions of Man or technological scaffoldings of the planet (for example, the computational Earth). Both Wynter and Simondon offer us a nonbiocentric and nonmechanist model of epistemology by addressing the principles of cybernetics, such as recursivity, and setting up the possibility of accounting for the incomputable condition of knowledge for which the image of the Earth can be disentangled from the theoretical and cognitive model of the Anthropocene.

Techno-epistemology

By following Wynter, one can argue that the centrality of neurocognition and mathematical formalism in the descriptive statements of the Earth can be challenged with rupturing theorizations that show how the incomputability of the Earth must first of all break from the current hegemonic onto-epistemological framework. I want to ask: Can techno-semiosis expose a dimension of alienness that breaks the origin story of Man, those recursive epistemologies that are centered on the human/Man-Capital? Can incomputable patterns break open from the sociogenically reproduced reality of Man's universality? Can they hack the order of meaning subtending the imperial technology based on colonial governmentality and racial capitalism?

[26] Wynter's critique of humanness questions Michel Foucault's critique of Man because she argues that the Columbian encounter of 1492 was vital to the formulation of European representations of the human, which epistemologically foreclosed all other ways of being human. She discussed Man1 as involving the theological explanation of the human, alongside the rise of the physical sciences. In contrast, Man2 is used to describe the secular conception of Man invented alongside the rise of the biological sciences. While Man1 emerged out of the Renaissance as *Homo politicus*, including the representation of the human as religious, Man2 coincides with the secularization of knowledge, subtending the formation of the liberal *Homo oeconomicus* and the bio-successful evolution of Man.

Simondon's argument about the centrality of technology in the formation of modern epistemology[27] helps us to imagine what can happen when knowledge, languages, and statements are understood in terms of "technical mentalities"[28] or what could also be called "techno-cognitive epistemologies." Simondon's efforts to map a genealogy of technical mentalities that breaks open the homogeneity of scientific paradigms should be explored within the context of Wynter's elaboration of sociogenesis. Sociotechnical assemblages constitute the technical mentality of cybernetics, which contributes to decentralizing language from the biocentric perspective of knowledge. In short, Simondon's view of information as part of sociotechnical assemblages can help us to envision the material condition of socio-techno-genic know-hows, namely, the way sociogeny also includes a specific mode of knowledge (know-hows or practical knowledges) that refers to technical schemas—mechanical, cybernetic, or computational. While these socio-techno-genic assemblages are abstractions, they are not yet irreducible to the epistemological totalities of mathematical formalism and biocentric sapience, which I argue still subtend the paradigm of the Anthropocene.

Simondon insists that "technical thought is present in all technical activity, technical thought is of the order of invention; it can be communicated."[29] As such, a technical mentality is what enables and transforms the order of thinking so much so that machines become a generative medium, which coincides with the communication and invention of thought.[30] By following Simondon's argument, the theorization of the incomputable Earth requires a machine epistemology. This is because the problem of the incomputable is a problem that emerged as a socio-techno-genic abstraction of automation, which, as I argued above entails the concretization of thinking in machines, namely offered by Alan Turing's thought experiment, the Turing machine. Machine epistemology, in other words, carries within itself a mode of artificial know-how that does not start from formal axioms and biocentric evolution but from the epistemological break that the discovery of incomputables brought to formal axiomatic knowledge. One way to instigate this epistemology may require a reenvisioning of whether the logic of technology is able to admit that machines are not simply tools but

[27] Gilbert Simondon, *On the Mode of Existence of Technical Objects*, trans. Cécile Malaspina and John Rogove (1958; Minneapolis: Univocal, 2016).

[28] By technical mentalities, Simondon refers to a unique mode of knowledge that is at once technical and social and serves to explain the mode of existence of a technical object. This notion defines how a schema of action also becomes a set of values, which constitutes a thought-network because it implies a material and conceptual synthesis that is accessible at various points. In what follows, I explain the two models of technical mentality that Simondon argued for.

[29] Simondon, *On the Mode of Existence of Technical Objects*, 252.

[30] Simondon, *On the Mode of Existence of Technical Objects*, 17–18.

entail general operations, regulatory practices, modes of organization, and compression of indeterminacies. From this standpoint, the material effects of the sociogenic principle may include not only technology as a tool in the hands of Man but also, more importantly, technology as know-hows or practices that challenge epistemological universality through what Simondon calls "technical mentality."

In the essay "Technical Mentality," Simondon is concerned not with the ontology of being or the being of technics but with how thinking about machines somewhat coincides with the way thinking itself can be known, valued, and conceived. In particular, he insists that "the technical mentality offers a mode of knowledge *sui generis*,"[31] involving a common way of functioning—or what he calls a "regime of operation"[32]—between distinct domains of reality, human, and nonhuman. Simondon recognizes the prominence of two main cognitive schemas that have claimed the universality of these common modes of operation across the living and the nonliving. On the one hand, Simondon discusses the Cartesian cognitive schema. This implies the view of a mechanical mode of operation, or functions, whose linear chain of causes and effects also determines the mechanism of the structure of cognition. Simondon describes this Cartesian schema in terms of the "long chains of reason," whereby the syntactical infrastructure of cognition entails a physical transfer of forces from one point to another in order to carry out proofs, or "transport of evidence," from premises to conclusion.[33] The function of knowledge thus coincides with the operations of a simple machine whose transfer of forces admits no losses between links, both at the level of things (*res extensa*) and of thought (*res cogitans*). According to Simondon, the Cartesian schema allowed philosophy to shift the ontological question of living beings toward an analogical relation between humans and machines, turning living beings into objects of science.[34] Simondon, however, dismisses this schema because it fundamentally grounds the relation between the functional operation and the formation of the structure in the atomistic view of the universe (that is, that the universe can be reduced to discrete parts that can reconstruct anew over time). This is a mechanistic onto-epistemological model that brings forward a socio-techno-genic mentality at the service of a metaphysical clock, which reduces nature to instrumental thought and, as such,

[31] Gilbert Simondon, "Technical Mentality," in *Gilbert Simondon: Being and Technology*, ed. Arne De Boever, Alex Murray, Jon Roffe, and Ashley Woodward (Edinburgh: Edinburgh University Press, 2012), 1, emphasis original.
[32] Simondon, "Technical Mentality," 1.
[33] Simondon, "Technical Mentality," 2.
[34] Simondon, "Technical Mentality," 2.

a systematic mode of knowledge that breaks down the world into composite parts without remainder.

In order to break from the atomistic pillars of mechanistic epistemology, Simondon turns to the cognitive schema provided by cybernetics. Inasmuch as cybernetics brings forward a technical mentality based on principles of feedback and adaptation, it defies not only the atemporality of knowledge but also the pillars of scientific paradigms based on the Newtonian conception of time and Euclidean geometry. Cybernetics entails not the mechanistic preoccupation with the parts of the whole but a mechanology, that is, the self-regulatory functions of living and nonliving systems. In particular, for Simondon, self-regulation implies that the relation between function and structure entails a practice of learning and not given tasks, an immanent finality that breaks from teleological causality and leaves systems open to contingency or indeterminacy. Simondon explains that with cybernetics, the relation between function and structure implies the system's "active adaptation to a spontaneous finality."[35] Instead of a new unified schema, Yuk Hui explains that for Simondon, cybernetics is an instance of a technical mentality already imbricated with a set of schemas that defy the totality of any systemic approach.[36] From this standpoint, Simondon steers cybernetics away from positivist conceptions of information in order to reconceive information in terms of a new category that goes beyond the fundamental principles of physics: matter and energy.[37] As much as information adds signification to a system, it also becomes a matter of difference and differentiation, which leads to individuation. For Simondon, information becomes the starting point to propose a counter-epistemology of technology against the universal model of mathematical representation, quantification, and measurement that continues to define cybernetics as a science of prediction and control.

The technical mentality that cybernetics allows, according to Simondon, is a view of technology that involves a mode of thinking sui generis that is not determined by essence, categories, or domains. Technology, therefore, coincides not with the use of machines, but, more importantly, it entails "transcategorial" knowledge that not only brings forward the universality of a mode of activity but also accounts for the phenomenology of these activities without presupposing their ontological belonging.[38] In other words, technology brings together activities and thinking generated in different fields and, as such, the knowledge

[35] Simondon, "Technical Mentality," 3.
[36] Yuk Hui, *Recursivity and Contingency* (London: Rowman & Littlefield, 2019), 139–40.
[37] Hui, *Recursivity and Contingency*, 19.
[38] Simondon, "Technical Mentality," 3.

produced with and through machines crosses categories (from scientific to philosophical to aesthetic categories) as a mode of activity or an operation. For Simondon, however, this systemic view of technology does not preclude that operations are also in and of themselves phenomena, and as such, they hold onto singularities that cannot be subsumed into a whole view, or even a being. Simondon's rejection of the onto-epistemological model that presents technology as a universal paradigm implies a new conception of the code.[39] "In cybernetics' technological paradigm, the code of Sender and Receiver must coincide in order to allow a correct exchange of information," which is a process or an operation that does not depend on the content of the code that can be permanently inscribed in the system's structure: "On the contrary, for Simondon the code and functioning of the system depend on each other."[40] This view in which the code can be both a "producer-of and produced-by information exchange"[41] indicates that systems, or categories of different systems, not only interact but also above all can be modified. In other words, they inform each other as they couple with each other, as in "human being and machine or machine and animal, but also a human being and virus, orchid and wasp."[42]

Simondon clarifies that the application of technology—or of technical mentalities that qualify the uniqueness of cognitive schemas—must meet two postulates. First, parts or technical activities must always be detachable from the whole. Technical objects are therefore conceived as open-ended, insofar as they may need repair, maintenance, testing, and modification. In other words, the technical mentality of cybernetics and, it is possible to add here, of computation coincides with a mode of transcategorial knowledge that must account for the temporal transformation and autonomy of parts (or subsets): a fractality that emerges within practices of know-hows. What is important to tease out from the information sciences is that no eternal and holistic mode of knowledge can be sustained by this cybernetic socio-techno-genic mentality. Second, any object, subset, or part must be addressed in terms of their own entelechy (the realization of potential). This is not the Aristotelian model of the potential of a given substance but rather refers to emerging potentialities that come to occur between functions and structures. As Simon Mills observes, if the technical object contains abstract structures necessary to its operation, it is itself part of

[39] Andrea Bardin, *Epistemology and Political Philosophy in Simondon: Individuation, Technics, Social Systems* (Dordrecht: Springer, 2015), 26.
[40] Bardin, *Epistemology and Political Philosophy in Simondon*, 26.
[41] Bardin, *Epistemology and Political Philosophy in Simondon*, 26.
[42] Bardin, *Epistemology and Political Philosophy in Simondon*, 26.

a number of systems whose operations are independent from one another as much as they perform a single function. Operations, therefore, are not already integrated but rather become concretized in a single structure defined by a plurifunctionality.[43]

Nevertheless, as Hui reminds us, while some systems can have the same functions, they may not share the same mode of operation: "The cybernetic machine has only a *functional equivalence* with them but not necessarily an *operational equivalence*."[44] Simondon's theory of conversion is not based on a principle of equivalence but on internal resonances that include asymmetries and fractalities between scales. As much as technical functions are immersed in a "specious present,"[45] which at any given moment contains recursive feedback between past and future, so too do technical structures defy homogeneous temporality. Information systems above all allow for structures to fractalize; namely, to become more than one in the process of mediation, where parts ally or disally within and among themselves. Against the mathematical formalism of rules and the biological ontology of self-adapting structures, Simondon's articulation of the technical mentality of cybernetics promises a new vocabulary for the understanding of cognition. What interests Simondon is the heterogeneous fractality of a socio-technique, the social dimension of the medium. Thereby, socio-techno-genic mediation corresponds to an abstracted technical mentality that carries over the cybernetic mode of transcategorical knowledge. Information thus does not linearly reproduce the overrepresented cosmogonies of Man (the completeness of mathematical and biocentric onto-epistemologies) but exposes the heterogeneity of know-hows, which are actualized potentialities.

Cybernetics, for Simondon, points to the possibility of an info-technical mentality where not only energy and information are enveloped within the technical functions of feedback but also where patternless information is enfolded within an increasingly fractalized (and thus incomplete) and asymmetric structuring of knowledge. From this perspective, parts or points of interaction do not correspond to a static arrangement of sequential functions. On the contrary, points are spatiotemporal complexities that cannot be reduced by one domain but proliferate at the crossroad of information, without originary reference to formal categories or biological essences. Importantly, the information logic of the structure is not predetermined because it results

[43] Simon Mills, "Simondon and Big Data," *Platform: Journal of Media and Communication* 6 (2015): 59–72, here 62.

[44] Hui, *Recursivity and Contingency*, 192–93, emphasis original.

[45] William James, *The Principles of Psychology*, vol. 1 (New York: Henry Holt, 1890), 638.

from the interactive functions of algorithmic compression, a dynamic telos for which conclusions cannot be prescribed by premises. For instance, inasmuch as interactive rules in artificial intelligence (for example, in deep learning algorithms such as generative adversarial networks or GAN) condition the recursive complexity of layers, the algorithmic compression of randomness—of indeterminate quantities of information—becomes an operation for learning and structuring cognitive patterns. This view entails what Simondon calls allagmatics[46] (a theory of conversion) that brings together the functional operation of algorithmic compression of randomness and the symbolic structure of computational logic and pushes them toward a third epistemological synthesis: a nonaxiomatic epistemology, which is not only an instantiation of a rupture within a dominant onto-epistemology but also, more importantly, entails a radical transformation and expansion of sociotechnical know-hows that are carried on by the temporal individuation of collective praxis of knowledge.

But what is it that holds the function-structure of sociotechnical operations together? As Hui has already demonstrated in his systematic attempt to define cybernetic metaphysics, it is the recursive function that describes the circular qualities of self-reflectivity in terms of feedback that return to initial conditions. Feedback exposes how the structure results from the way the recursive function responds to the contingency of the world, by compressing infinite numbers into a given set, or by reducing chances to probabilities. If recursivity holds the function and the structure together, it is also because, according to Hui, recursivity is the mode of reflective judgment: The recursive function implies a form of feedback that allows the structure to have an adaptive response to contingency. The system of rules has thus come to depend upon its capacity to adapt to contingency.[47] This indeterminacy is not unknown but becomes part of the recursive loop of adaptability. For Hui, the principle of recursivity in natural philosophy demarcates a historical route for both modern philosophy and cybernetics as models that confront the irregularities of nature.[48] In particular, recursive self-reflectivity is embedded in the epistemological vision of biology and the natural

[46] Simondon uses the term "allagmatics" to define a theory of conversation stemming from distinct modes of operation that are in relation to one another. These operations demarcate not the merging of many into one but a process of individuation—physical, technical, and vital. These operations are not separated but are rather said to be autonomous yet in solidarity with one another. What is at stake is a view of epistemology where philosophy enters into conversation with the sciences. Simondon's engagement with the theory of cybernetics particularly offers the vision of a philosophical science that is at once transversal and unifying. See Arne De Boever, Alex Murray, Jon Roffe, and Ashley Woodward (eds.), *Gilbert Simondon: Being and Technology* (Edinburgh: Edinburgh University Press, 2012), 22, 204.

[47] Hui, *Recursivity and Contingency*.

[48] Hui, *Recursivity and Contingency*, 61.

sciences of the eighteenth century that explained the world from the standpoint of the organism.[49] However, Hui's efforts to reconstruct the biological lineage of self-reflective systematic thinking move from natural philosophy's conceptions of the organism to the arguments for organicism and organology in order to map an alternative formation of a philosophy of technology grounded not in functions and mechanical parts but in the temporality of autopoiesis and self-organization.[50]

Hui's account of contingency is an effort to address epistemology beyond the model of calculation and prediction based on probability. This is also an effort to engage the unknown in terms of the incalculable, namely a dimension of the real that cannot be subsumed into the paradigmatic universality of computation.[51] However, one can also argue that the recursive function of reflective thinking is a schema that must preserve the mode of selective decision and the negative subjection of contingencies because what is external to the system must become adapted to the self-reflective mode of thinking. In other words, the question of the unknown is a problem for power, namely the unknown must fit the overrepresented onto-epistemology of Man. Adaptation is a mode of capturing the unknown in the self-reflective system that preserves transcendental decision, racial capital, colonialism, and patriarchy. From this standpoint, since contingencies are not there to rupture the recursive bond of function and structure but to maintain onto-epistemological circularity, one can ask, how can the incalculable (as Hui discusses it) avoid being placed at the limit of reason? How can this conception of the incalculable avoid becoming yet another sublimation of indeterminacy, which, by playing the role of the limit of self-reflectivity, precisely affords the dominance of the recursive function? How can the programmable—the computational programming of worlds—become related to the incalculable without subjecting the latter to the biocentric onto-epistemology that uses it as a source to reproduce its originary cosmogony? Perhaps, following Hui's views of the incalculable, as some kind of innumerable condition, can offer a critical proposition for an incalculable Earth. And yet, if this was the case, how could this condition avoid the risk of disqualifying or neutralizing the continuous efforts to rupture and radically transform the epistemological expansion of know-hows that do not fit the model of the subject, of being, or even of becoming programmable? The scope here is to open up computation beyond the calculable and yet consider the

[49] Hui, *Recursivity and Contingency*, 63–64.
[50] Hui, *Recursivity and Contingency*, 42–43.
[51] Yuk Hui, with Karen Sarkisov, "The Incomputable and the Incalculable," 13th Gwangju Biennale, February 23, 2021, https://13thgwangjubiennale.org/live-organ/the-incomputable-and-the-incalculable.

technical mentality of computation, which implies how incomputables define not the limit of thinking but the challenge that the nonbeing of externalities brings within the overrepresented cognitive model of probability and prediction.

The Incomputable Does What Computation Is Not

The relation between computation and the incomputable, in particular, exposes how epistemology can not only become an incomplete system but also cease to be determined by the opposition between contingency and certainty. Instead, the formalization of randomness as incomputable radically breaks from the autopoietic circularity of self-organization based on the incorporation of contingency, or the inclusion of the excluded, by exposing the constituting function of alienness in epistemology. But can the technical mentality of incomputability afford the overturning of the socio-techno-genic language that preserves the image of the Earth in the circuit of Man's cognition? To account for the way the incomputable overturns the onto-epistemological ground of the Anthropocene, one has to turn to how Turing's theory of computation has profoundly transformed the autopoietic relation of recursivity and contingency, between a self-determining model of knowledge and its limits, by posing randomness or incomputability at the core of scientific knowledge. Following Turing, Gregory Chaitin points to the formalization of randomness as incomputability, which conditions the process of comprehension and compression. Incomputabilities are externalities that break from the internal/external dyad as well as the principle of sufficient reason (that is, that everything must have a cause or a reason), which is opposed to the accident of contingency. The incomputable describes how incompressibility demarcates alienness within the techno-cognitive order of computational culture and its socio-techno-genic mentality. These are not aleatory indeterminacies but active operations that unify without merging inputs with outputs. These are incomputable functions that play a role in codes/statements, independently of intention and reference. Indeterminacies do not simply demarcate how meaning can change through use. Techno-semiosis does not match the symbolic representation of the Earth but coincides with heterogeneous practices of meaning that, at each instance, expose the alienness of the real as this cannot be reproduced by what is already known. Chaitin's articulation of the incomputable is only one among many possible articulations of alienness that appear within the onto-epistemological description of the Earth. The scope of drawing on this theory of incomputability is not only to debunk the biocentric origin of the Anthropocene paradigm, which has already been done extensively from diverse perspectives,[52] but also, more importantly, to instigate possibilities about what the incomputable can do for the socio-techno-genic cognition of the Earth in terms of shifting away from the overrepresented model of Man's cosmogony.

With the halting problem, Turing pointed to the first example of a decision problem that proved unsolvable. Importantly, Turing's proof departs from the model of calculation by means of a recursive function through introducing the notion of computation by machines. In particular, Turing understood this notion while addressing the limits of self-determining logic in terms of the halting problem.[53] The halting problem shows that the model of absolute knowledge of deductive logic (as a formal universal language) has failed to solve the problem of incomputability, according to which any formal language or axiomatic system is deemed to remain incomplete. In formal logic, the halting problem is a decision problem, which implies that it is not possible to determine in advance whether or not a computer program will stop or run forever. Turing demonstrated that it is impossible to create an algorithm that solves the halting problem for all of the possible input pairs of a program or a Turing machine. Since this is not only a problem of indeterminacy but above all an example of undecidability, one can argue that the halting problem can be taken as a starting point to highlight that the incomputable cannot explain—namely it cannot represent—the model of the Earth within the framework of the Anthropocene and its critique. What is at stake here is the way incomputability forces us to challenge the image of the Earth as a complete system while opening up theories of the Earth to undecidability insofar as the techno-semiotic configuration of the Earth is challenged by patternless information that is logically irreducible.

From this standpoint, the correlation between simplicity and complexity is central to the enterprise of computational epistemology, for which the socio-techno-genic processing of know-hows is demarcated by the way incomputabilities come to condition what we take the Earth to be without the overrepresentation of the biocentric model of the human and the formalism of universal mathematical language. The transformation of methods of compressing infinities in computational processing places the question of

[52] These perspectives include, for example, the theorization of the Chtulucene, the argument for the Plantationocene, and the debates about the Capitalocene.

[53] Alan M. Turing, "On Computable Numbers, with an Application to the Entscheidungsproblem," *Proceedings of the London Mathematical Society* s2-42, no. 1 (1937): 230–65.

epistemology beyond the self-reflective and transcendental model of recursivity and contingency. The question of undecidability at the core of computation can thus offer a heretic epistemological method based no longer on logical deduction but on the nonaxiomatic compression of randomness. This is to say that as much as the Turing machine demonstrated that there are realities that cannot be computed by following a universal schema or knowledge, it also foregrounded the infinite horizon of machine epistemologies. But what are the pillars, modes, and practices of socio-techno-genic know-hows that incomputability can offer to the postnuclear project for expanding while rupturing epistemology?

A critical theory of technology may argue that computational processing mainly accelerates the transcendental logic of universal epistemology, whereby the computational paradigm has taken over the representation of the Earth while continuing to justify dispossession, colonial capitalization, and enslavement.[54] At the core of the computational enterprise, new forms of algorithmic enslavement expand the field of action for a transcendental logic that determines results through the real subsumption of all kinds of variations. While it is not possible to neglect the onto-epistemological recursivity as the intensification of the logic of enslavement by other means, it is also necessary through algorithmic compression to not fall back into the mirroring relation between self-determination and contingency, whereby the latter stands for the externalities that become the extraction for the self-adaptation of the system.

By bringing together Wynter's critique of recursive onto-epistemology and Simondon's approach of cybernetics, this chapter has attempted to theorize the incomputable Earth. This chapter has argued that the critique of the Anthropocene does not sufficiently question the onto-epistemological model that sustains the exceptionalism of the humanist project and its entangled colonial, racial, and gendered metaphysics. Instead of recuperating the Earth as an agent that hosts humanity, this chapter rather turns to the critical engagements with system theories to argue against the given myth of Man, whose overrepresentation binds the image of the human to the biocentric model of evolution, based on the selection of successful forms of living. While Wynter's argument about this overrepresented cosmogony exposes the racialized pillars of epistemology, it also takes Bateson's cybernetic notion of recursivity—namely feedback as a self-reflective operation—to discuss how the onto-epistemological order of Man is granted by sociogenic patterns that preserve the racialization and gendering of

[54] Ruha Benjamin, *Race After Technology: Abolitionist Tools for the New Jim Code* (Cambridge: Polity, 2019).

knowledge across time. From this standpoint, this chapter turns to Simondon's discussion about the technical mentality of cybernetics in order to claim that the sociogenic principle also entails a technological form of thinking, introducing the problem of incomputability as central to recursive self-reflectivity.

Both Wynter and Simondon open up a new critical space for cybernetics as a form of technology that challenges the biocentric speciation of knowledge, of humanity, and the Earth. As much as Wynter takes the cybernetic model of recursivity to describe the function of preserving the overrepresentation of Man's cosmogony, she also argues that self-reflective epistemology can be overturned through an expansion of knowledge beyond the biocentric model. It is here that Simondon's reenvisioning of sociotechnical mentality contributes to expand toward what I have called "machine epistemology": a process-oriented, transcategorical, alien(ated) mode of knowledge that cannot be recuperated from within the biocentric epistemology of species reproduction. Instead of following Hui's argument for a cybernetic metaphysics that expands the organology of philosophies of nature and the transcendental principle of self-reflectivity and autopoiesis, this chapter suggests that recursivity is not given but is rather conditioned by the negative realities of incomputables. These are negative externalities that enter the feedback function of recursivity and yet are neither fully incorporated within the system nor stand for the ultimate horizon: the sublime, the unknown, or the incalculable. Instead, machine epistemology maintains the opacity of the incomputable axioms, which afford for the fallacy of experimental know-hows, breaking the circuit of programming and exposing an alien programmability beyond transcendental decision and the sublimation of indeterminacies.

From this standpoint, the notion of the incomputable Earth is contained neither by eternal feedback nor by absolute indetermination. As an epistemological problem, the incomputable interrupts the presumed convergence between universalist epistemology and universalist technology, the biocentric history of human/Man, and the information-oriented artificiality of computation. Similarly, as a problem of compression of infinities, the incomputable enables a theorization of the Earth as an incomplete system and of computation as a process that remains fallible to a field of possible versioning of nonaxiomatic practices of knowing the Earth. This is not knowledge that assumes that the Earth is a human-centered planet, nor that it is an equal agent among many. Nonaxiomatic know-hows fit no given postulate or messianic proposition to save the Earth by saving the human world. And yet, nonaxiomatic know-hows require the collective expansion and formation of cognitive schemas to account for radical transformations that overturn the cosmogonies of Man. Algorithmic randomness or incomputable complexities interrupt the loop of the recursive logic as the latter becomes exposed to experimental problems, namely those requiring a pragmatic yet nonperformative approach to knowledge. Randomness cannot be performed because it cannot be postulated, and thus it is not executed as such. And yet computation is also set to reproduce the image of the Earth as already given in dominant representations. Instead of defining a dimension for the incalculable, the absolute aleatory event (chance occurrence) sets up as that which comes to interrupt the system's completeness. The incomputable rather becomes a condition for computation to process what it does not yet know but remains conditioned by. This processing goes beyond probability. While computation is a closed system, it signals that there is always more patternless information that any computer may be able to process in a finite amount of time. Since this information cannot be known in advance, it generates patternless structures that condition the operations of computation, which depend not on the circularity of postulates but on socio-techno-genic fractalities of space and time, giving way to millions of incomputable Earths.

18

Subaquatic Sensoriums and the Incomputable Ocean

Margarida Mendes

The ocean is a discreet sensorium that, while connected to that of the dry land, is different in fundamental ways from that of terrestrial space. For centuries, the ocean's fluidity has challenged efforts to apply the same uniformly gridded cartography by which land is understood and thus governed and controlled. Contemporary understanding of ocean space has been shaped by the long, historical development of sensing, probing, and imaging techniques and technologies aimed at mapping the ocean in an effort to turn it into a governable political space. Recent intergovernmental efforts to develop a "Blue Economy" have led to multilateral treaties being signed in order to promote ocean literacy and seabed mapping, directing attention to the centrality of biodiversity and economic wealth that ocean preservation can bring. This has promoted an acceleration of monitoring projects across oceans, along with an unprecedented shift toward surveying oceanic resources. However, the legal frameworks that govern the prospecting and extraction of ocean resources have developed at a slower pace than the technological and economic drivers of the "Blue Agenda." It is therefore necessary to develop a critical account of the politics of oceanographic sensing, the ways in which its technical infrastructure and data collection processes imbricate scientific research with industrial surveying, and the monitoring and surveillance activities of governments.

Underpinning the conceptualization of a Blue Economy is the imagination of the ocean as capable of becoming a completely mapped and uniformly governable political space. This vision is, however, troubled and undone by the continuous frictions of uneven development, the epistemic blind spots, and the pressures of the ideological frame within which processes of infrastructure construction, logistical management, and data governance are understood. In my fieldwork, which is comprised of interviews with scientists working in the North Atlantic, I have seen how oceanic media frame modes of existence and create an ungridded ocean space as an incomputable arena in which nonlinear metabolic exchanges, multiple forms of sensing, and datafication occur in simultaneity. I report how recent scientific experiments with biosensing and automation-supported information networks at sea work to diversify and productively complicate the regimes of total legibility sought by the surveys of extractive industries.

From Environmental Biosensing to Biomimicry

Despite the ocean's opacity, ocean sensing relies heavily on primarily optical technologies that relate to the ocean space as a layered condition, often termed a "water column." This is in and of itself bracketed on both sides by the lithosphere below, the atmosphere above, and the orbital layers further above. Sensing across scales, remote devices use satellite-based laser technology to map large-scale phenomena, such as tidal patterns or algae fluctuation, while biotic action is monitored with microscopic accuracy. These technologies reconfigure interspecies relations at sea and are themselves reciprocally shaped by the inclusion of nonhuman agents. Increasingly in recent decades, the deployment of bionic gadgets has refigured ocean space as a medium for interspecies communication rather than a one-directional relationship between sensing subjects and the mute objects of their consideration.

Environmental sensing in the oceans is deployed along a vertical axis that stretches from the orbital satellites to the depths of the seabed. The technoscientific gaze of remote sensing detects an electromagnetic landscape of data points that render legible biotic, hydrologic, and geologic patterns. Through geostationary operational environmental satellites and polar-orbiting satellites, several variables of oceanic patterns can be monitored simultaneously. These include currents and water circulation patterns, sea surface temperature, as well as sea level changes, sea ice, coral reefs, tidal heights, and animal migration.[1] In recent oceanography, however, there has been a desire to merge biotic and robotic forms of sensing and adopt a less axial, stratified approach that is more akin to how aquatic organisms experience and come to know their environments. While concepts of biomimicry have long informed scientific practice, they

[1] For more information on ocean monitorization projects, see the National Oceanic and Atmospheric Administration (US Department of Commerce) portal: http://www.noaa.gov.

have only recently been connected to the invention of a series of robotic tools and synthetic materials used to enhance ocean prospection. Drawn from both natural morphology and animal behavior, researchers across the world are developing an array of biomimetic technologies capable of collecting samples and performing other operations in the difficult conditions of the deep ocean without harming the environment.

I interviewed biologist David Gruber in 2017 and 2019 while he was working at Harvard University's Wyss Institute for Biologically Inspired Engineering, the research center that has developed numerous technological tools for oceanographic research. Among them are a "shark-eye" camera, which was prototyped after a prolonged study of white shark eyes, and a soft robot with "squishy fingers," noodle-like appendages that are able to handle fragile jellyfish and carefully collect samples of delicate corals.[2] Gruber is currently developing yet another fascinating tool, a hyperspectral camera, that is inspired by turtle vision. This camera records a high level of spectral information as a data cube in every pixel, just as satellites perform compacted land readings. Gruber is interested in how the animal-inspired cameras function in correlation with the evolutionary design of the different eyes. In his research, he is interested in developing ways to observe and describe organisms without killing them, saying that his team goes through crazy engineering backflips in order to not hurt the animals. He is fascinated by how time and space take on totally different dimensions in the deep ocean, and he wishes to see the world from the perspective of the animals that live "down there." As he explains, "animals have invested in different forms of vision after many thousands of years of surviving."[3] As his tools have been taken up by others in the scientific community, Gruber has expanded to include sound and optical devices and is now studying cetacean communication in search of new models for interspecies information exchange.

The convergence between technological developments and optical adaptations, evolved by animals native to the aquatic milieu, presents the possibility of conceptualizing hybrid, or interspecies, sensing underwater. The bionic paths opened by this convergence lead to unforeseen possibilities for monitoring oceanic habitats and reveal hidden relations between critters and their surroundings. Gruber's desire to see "as an animal" steers the engineering efforts of his laboratory practice. He has carefully devised an ethical space where

[2] On Gruber's inventions, see Michael Tessler et al., "Ultra-gentle Soft Robotic Fingers Induce Minimal Transcriptomic Response in a Fragile Marine Animal," *Current Biology* 30, no. 4 (February 2020): PR157–R158.

[3] David Gruber, unpublished interview with the author, December 2019.

he acknowledges other underwater sensing systems and tries to understand transductions across bodies and physical dimensions. In this context, he analyzes how biological sensing mechanisms are constituted and how they have evolved and adapted to the aquatic milieu. These experiments in the creation of biomimetic technologies provide an understanding of how these human-alienated landscapes can be mediated by amphibian technologies. It remains to be seen how biomimicry will shape new perceptions of sensing bodies underwater and how this will translate into the development of future infrastructure and ocean policies.

Technologies and processes of environmental sensing have not only been aiming to reach closer to the perceptive apparatus of animals, but they have also extended toward studying the sensory capacities of phytoplankton and how this is reflected in their biochemistry and movement through the ocean. Forms of environmental sensing have, for example, predicted the emergence of algal blooms through the chromatic monitoring of ocean surfaces and levels of phytoplankton. Remote sensing, using space-based technologies such as light detection and ranging (LiDAR), makes it possible to survey levels of chlorophyll concentration, thus allowing for the monitoring of plankton populations. It is a crucial operation because these minuscule entities are the basis of the trophic chain, and their photosynthetic action leads to the absorption of a great deal of the carbon dioxide near the ocean's surface. This absorption plays a key role in the carbon cycle, as the carbon captured from the atmosphere then sinks into the ocean.[4] Microscopic oceanic entities are also used in the forms of biosensing for a variety of other applications. It is possible to track differences in temperature, illumination, and chemical composition of the ocean using lasers to register fluctuations in the fluorescent character of algae. Then, by comparing these readings, the algae function as biological sensors of aquatic pollutants that react to changes in temperature and circulation of heavy metal ions, making them perfect allies for ecological monitoring of the aquatic environment.[5]

Sociologist Jennifer Gabrys's book, *Environmental Sensing Technology and the Making of a Computational Planet*, offers a survey of different environmental sensing technologies and citizen science programs. She analyzes how machines and animal agents interact, producing and reproducing milieus, and questions

[4] Michael J. Behrenfeld et al., "Annual Boom–bust Cycles of Polar Phytoplankton Biomass Revealed by Space-based Lidar," *Nature Geoscience* 10, no. 2 (February 2017): 118–22.

[5] Sergey S. Voznesenskiy et al., "Biosensors Based on Micro-algae for Ecological Monitoring of the Aquatic Environment," in *Algae: Organisms for Imminent Biotechnology*, ed. Nooruddin Thajuddin and Dharumadurai Dhanasekaran (London: IntechOpen, 2016), 103–32.

how engagements with the animal-sensor networks shift agency on different organizational levels. Gabrys speaks of a "becoming organismal of machines"[6] as she follows the global animal-sensor network and inquires how the processes and overlaps between technological devices, agents, and networks generate social processes and data gaps. She also makes the point that environmental change has now become an informational problem and asks how methodological strategies and governance frameworks can be developed from these mapped events as "computational sensor technologies become environmental to create new subjects of experience."[7]

Proposing that oceans have become sensor spaces due to current monitoring practices, Gabrys looks into particular processes of data generation and poses questions about interaction on an ontological, procedural, and sometimes chemical level. Thinking about forms of environmental sensing across different levels, she explores the potential of conceiving the ocean as a milieu of chemical interactions in a dynamic continuum. By surveying the molecular behavior of plastic waste, she asks, "How do chemical substances migrate into and across organisms, and what effects do these substances have on organisms over time? How might they cause endocrine disruption within marine organisms and humans?"[8] She also addresses how persistent organic pollutants may be transmitted through microplastics, which might be connected to alterations observed in microplankton feeding habits, a decrease in biodiversity, and climate change and expands upon this to consider the possibilities of understanding the human endocrinological system as an extension of the animal-sensor network.[9] While the tracking of all of these chemical processes continues to have severe data gaps due to the lack of efficient intergovernmental data platforms, they may already be responsible for the alterations found in the food web and the chemical regulations of the ocean that relate to source-sink dynamics and the carbon cycle.

The extension of the animal-sensor network to include hybrid forms of infrastructure and data mediations constitutes the subjects of experience as they are generated by computational sensor technologies. These forms of ocean monitoring are supported by sensing processes derived from a detailed

[6] Jennifer Gabrys, *Program Earth: Environmental Sensing Technology and the Making of a Computational Planet* (Minneapolis: University of Minnesota Press, 2016), 95.
[7] Gabrys, *Program Earth*, 268. This is a discussion further explored by Tega Brain in her 2017 article "The Environment Is Not a System" (republished on page 223–39 of this publication), where she criticizes AI interfaces to deal with climate change.
[8] Gabrys, *Program Earth*, 142.
[9] Gabrys, *Program Earth*, 142.

study of animal behavior, community migration, and adaptation to changing ecosystems. The monitoring also uses biomimetic devices to compile data and shape infrastructures, which in turn shape our perception of ocean space and the critters that inhabit it. These forms of sensing and the ways in which they shape the perception of milieus are being studied at the same time as the development of infrastructures that support and enable them. There is, once again, an overlay of significant processes that co-constitute ocean space, elemental media, and the conceptualization of the bodies that inhabit it.

Sensing from a Distance

In the early days of modern oceanography, in the nineteenth century, the water column was seen as *mare nullius*—a space devoid of life and ready to be industrialized with the deployment of oceanic infrastructures and sensing technologies. Some areas started to be surveyed and legislated, which led to the contemporary United Nations Convention on the Law of the Sea. The scope of this legal mechanism is now being revised with the intent of protecting marine resources and bioprospection activities. While it protects the ocean and mitigates anthropogenic effects, it also leads to further division of the water column and artificial stratification of what are actually borderless and dynamic habitats existing in a fluid continuum. Remote sensing plays a decisive role in this equation, as the mechanisms of visualization, cartography, and datafication of ecosystems merge various spaces and scales, creating partial and transduced representations of continuous ecosystems.

In this era of land exhaustion, nation-states are turning toward aquatic resources with the hope of stimulating "blue growth" in governance strategies, as well as legal and technological infrastructures that allow for ocean monitoring, marine resource extraction, and commercialization.[10] This involves the further development of a series of tools that prosthetically extend the possibility for human exploration, from remotely operated vehicles (ROVs) and fully automated robots operating in the ocean's depths to satellite networks in space. This large-scale investment resembles aspects of the pioneering rationale that advanced space exploration, which in turn reformulates spatial and cartographic

[10] "Blue growth is the long-term strategy to support sustainable growth in the marine and maritime sectors as a whole. [. . .] The 'blue' economy represents roughly 5.4 million jobs and generates a gross added value of almost €500 billion a year." European Commission, "Oceans and Fisheries," n.d. https://ec.europa.eu/maritimeaffairs/policy/blue_growth_en.

perceptions. But exactly what kind of cybernetic sensibility is being forwarded by these remote sensing technologies, and what kinds of planetary and spatial reconfigurations do they promote?

Researching the seabed as a site for terraforming and scientific exploration, geographer Rachael Squire writes that in this promising narrative of eternal conquest, seabed prospection replaces the quest for outer space travel as ocean space becomes a sight of endless profit.[11] Squire refers to scientists on the dark abyssal plateaus of this sunken planet as aquanauts who explore an unconquered element, groping in the dark with their ROV robot arms, searching for mineral and genetic treasures. The abstraction of the milieu into datasets, which literally translate a darkened field of research filled with alien forms, establishes a convincing analogy. In these sites of exploration, the tactile frontiers of sensing are replaced by numerous technologies and forms of sampling that translate temperature profiles, biotic elements, and geomorphological landscapes into datasets. Computational machines transduce tactile, visual, and acoustic stimuli, giving shape to volumetric models. The unknown contours of the abyssal horizon are replaced with a series of representations and informational landscapes. The abstraction of the context and transduction of information into mediated sensorial interpretations make the scientific research process a practice of studying environmental conditions that also formulates understandings of spatiality.

Anthropologist Lisa Messeri delves further into the ways in which scientific methodologies for earthly training of outer-space expeditions simultaneously establish and expand the limits of what she calls "planetary imagination," leading up to the conceptualization of practices of terraforming and the fictioning of planetary spatial dimensions.[12] Elaborating on how planetary imaginations are supported by practices of spatiality, both worldly and outer-worldly exploration, Messeri analyzes how the scientific and mercantilist imaginary is built through developing experiments and tools that rehearse space exploration without leaving the planet.[13] This, Messeri argues, leads to a reconfiguration of spatial constructs and sensing practices. She looks at the crossover between epistemic forms, interfaces, and technologies for scientific study and how these challenge

[11] Rachael Squire, "SEABEDS—Sub-Marine Territory: Living and Working on the Seafloor During the Sealab II Experiment," in *Territory Beyond Terra*, ed. Kimberly Peters, Phil Steinberg, and Elaine Stratford (London: Rowman & Littlefield, 2018).

[12] Lisa Messeri, *Placing Outer Space: An Earthly Ethnography of Other Worlds* (Durham: Duke University Press, 2016).

[13] For this purpose, Messeri follows the Mars training teams in the Rio Tinto area of southwestern Spain.

geographical definitions. Developing an argument around the concept of planetary imagination, she calls into question the very idea of frontier-making and the situatedness of practice by showing how scientific methodologies coevolve with ideas of spatiality, which define fields of existence and intervention. Messeri goes further in asking how representations of space and the environment are naturalized through epistemic practices, inquiring how planetary and spatial imagination "capture holistic conceptions that scientists have of the planets they study," mixing ideas of outer-worldliness with both fictionalized spaces and quotidian descriptions of spaces they know.[14] For Messeri, this process of naturalization is challenged by the perspective created through the interfaces and tools for scientific literacy in action and how these interact with spatial arrangements, social processes, and relations with nonhumans. This set of reflections on methodology and conceptual framing can be usefully applied to ocean exploration as oceanographic research, and the practices organized around seabed prospection lead to the emergence of similar technoscientific processes and interactions with its milieu.

In the ocean, environmental sensing and mapping practices are absolutely entangled, even if the human never sets foot in its deep waters. This type of research is supported by the creation of new tools for scientific study and different forms of data collection that intervene at different scales of biopolitical and technological interaction. This includes a series of robotic infrastructures. The mediated distance necessary for this project shapes the relations of sense-making throughout oceanic space. Analyzing the frontier collapse in the turn toward ocean research automation, legal scholar Irus Braverman writes that

> management in deep sea space displaces humans, creating the conditions for a kind of biopolitical gaze that extends not only beyond the human, but also beyond particular management sites to the entire ocean. At the same time, technology enables the virtual reinsertion of the human into a scene that was once considered human-less.[15]

Braverman reflects on the regulation of technological developments while pondering the true extent of the biopolitical reach of the human, particularly in regards to the regulation of coral communities. She reflects on how the use of machine learning models applied to ocean robots creates yet another layer of agency and distancing from the environment when it comes to the terminal

14 Messeri, *Placing Outer Space*, 12.

15 Irus Braverman, "Robotic Life in the Deep Sea," in *Blue Legalities: The Life and Laws of the Sea*, ed. Irus Braverman and Elizabeth R. Johnson (Durham: Duke University Press, 2020), 148.

practices that robots enact for coral conservation through the obliteration of predatorial species. She writes: "The robot's autonomy from humans in the context of governing the eco-security of the sea thus further removes humans from the killing scene and from responsibility for their acts."[16] Braverman highlights the injunction as she addresses the need for a legal debate around the ontological framing and societal behavior, where it concerns planetary governance and ocean management practices, and calls for scientific accountability. She states that scientific research is central to epistemic production at sea and that it coevolves with the establishment of legal parameters. The debate on the legal questions surrounding oceanic automation is somewhat peripheral to ocean sensing, but the cross-scalar levels of abstracted agency defining this context parallel those suggested by ocean monitoring practices in significant ways. The overlap of scalar injunctions, brought forward by the multiple developments in social, scientific, and technological fields, is further questioned by the anthropological debate on sensorial hierarchies addressed in the next paragraphs.

The rationale behind the presence of automation at sea produces what anthropologist Stefan Helmreich terms a new cybernetic sensibility in remote sensing technologies.[17] In the ocean, the possibilities of technological transduction are endless and unfold in a space defined by overlapping techno-metabolic vectors of infrastructural interactions, geomorphological fictions, and biotic and chemical feedback cycles. Helmreich addresses scientific practices and their use of distinct technologies that he calls "submarine cyborgs," providing a machinistic approach that tames the environment and alienates nature to serve the interests of science. Reiterating that our sight of the seabed is already cultural, he analyzes the co-constitution of the scientific gaze that merges with its data, as scientists are immersed in media ecologies resulting from a chain of cyborgian transductions. Reminding us of the underwater sensing capacities of humans—who are unable to hear clearly and whose vision is insufficient—Helmreich expands on how the prosthetic cyborgs designed for ocean exploration enhance our amphibian senses in illusional ways. Visual technologies and ROVs reinforce the distancing of bodily experience and vision from the milieu but still allow their operators to perform machine-supported physical actions while acoustic signals and sonifications of events are fully audible. Helmreich writes that this imbalance of spatial and sensorial hierarchy brings about an interesting incorporation of ocean prospection tools that, at times, erases the distance of the

[16] Braverman, "Robotic Life in the Deep Sea," 152.

[17] Stefan Helmreich, *Alien Ocean: Anthropological Voyages in Microbial Seas* (Berkeley: University of California Press, 2009).

researcher from the site when spatial boundaries become imperceptible to the operator despite their lack of adaptation to the exterior milieu.

Reflecting upon this axis of transductions and abstractions, it is important to reiterate how different perspectives and ontological problems framed by tools for ocean literacy incite different epistemic problems and forms of environmental violence across scales. Cultural theorist Nicholas Mirzoeff's critique of the visualization effected by the deployment of remote sensing technologies comes to mind here. In his critique, he addresses how ecosystems are represented through a distanced perspective, which results in the loss of context specificity and a flattening of dimensions. He mentions that technology discloses previously unrepresented habitats and existing natural resources. Mirzoeff also speaks of the hierarchical aspect of the grand-scale view-from-above and, criticizing its origin, explains how it conceptually underpins the high-resolution resource mapping demands of extractive infrastructures that are identifiable as a form of digital colonialism, paving the way to land dispossession.[18]

Precisely what is being disclosed by the vertical axis of ocean remote sensing technologies and bathymetry (submarine topography) projects being conducted in the ocean? Not only are climate change, pollution, and ocean shipping being monitored, but other forms of marine resources, such as minerals and metals, are being identified and disclosed by these actions. This process of scientific and spatial inquiry is further complicated by the fact that, due to the large size of these research operations, both in space and at sea, the scientific community strongly depends on funding that often comes from resource mapping surveys promoted by the mining industry. This fact alone seriously implicates the limits of knowledge production and mediation of our relations with nonhumans and inorganic matter.

Extractive Frontiers

This entanglement between oceanographic research and the extractive lobby has always been notable. Expeditions of scientific research vessels have historically been sponsored by the oil and gas industry, as companies read seafloor textures in order to identify sediment types and geomorphologies connected with fossil resources. In addition, oceanographic dredging expeditions dating back to

[18] Nicholas Mirzoeff, *The Right to Look: A Counterhistory of Visuality* (Durham: Duke University Press, 2011).

the nineteenth century applied knowledge gleaned from experts in the copper mining industry who operated with heavy machinery that could hoist large amounts of ore. Today, projects of seabed mapping and ocean prospection are closely connected with the possibility of deep-sea mining for mineral and genetic resources, much in the same way that geologists are backed by the mining industry, and biologists foster close connections with the pharmaceutical industry. These partnerships are highly problematic, as research studies and the technologies deployed often back extractive activities that are harmful for ocean communities because they are enacted before the development of adequate legislation.

Deep-sea mining is still in an experimental pilot phase, yet testing is already happening in Papua New Guinea, Spain, China, and Japan. Leases granted by the International Seabed Authority cover a total seabed area of approximately two million square kilometers, and the main sites targeted for future exploration in international waters are the Mid-Atlantic Ridge and the Clarion-Clipperton Zone of the Pacific Ocean. Mining companies aim to extract gold, copper, manganese, nickel, lead, cobalt, lithium, titanium, platinum, zinc, and rare earths from seabeds at a depth of approximately 1,400 to 5,000 meters, where hydrothermal vents and cobalt-rich crusts are located. The borderless frontiers of the deep ocean, which were long considered a lawless territory, have become the center stage for twenty-first-century geopolitical disputes. These frontiers also urgently highlight the need for more transversal forms of mining resistance and policymaking that are respectful of the cycles of geological time.[19]

It is crucial to tune into the rhythms of planetary fluxes and develop an awareness of the role that the seabed occupies in climate metabolism. Ocean floors play an important role in managing the Earth's carbon cycle, as large parts of the circulating carbon dioxide sink to their depths. We know that mutations in abyssal planes are slow, as there is very little movement and exposure, and that geomorphology takes a long time to develop due to seldom-changing conditions. Timescales operate differently at the depths of our planet, which has worked in favor of the abyssal planes while the human hand has been kept at bay for so long. Yet assemblages of anthropogenic agents, such as microplastics and chemical compounds, have already been discovered at depths as distant as the Mariana Trench in the western Pacific Ocean.

19 The first resistance movements against deep-sea mining emerged in Papua New Guinea in 2015 in reaction to the experimental mining tests conducted by Nautilus Minerals Inc., a now bankrupt Canadian deep-sea and mining company.

The deep oceans are the largest ecosystem on the planet and, at this point, the most unexplored. Millions of unknown species are thought to inhabit the ocean under the threshold of the photic zone. In the absence of scientific representations of this largely unmapped part of the planet, it is crucial to develop comprehensive methods of ocean literacy and visualizations that communicate the role of the deep ocean and seabed ecosystems. To this end, the extractive deep-sea mining industry has been producing a series of CGI ecosystem models that circulate with the purpose of promoting mining activities.[20] On these, the water column is represented as devoid of any flora or fauna. These stylized images, while emptying the aquatic space from the multiple modes of existence that live within it, perpetuate the mare nullius conception of the ocean inherited from earlier oceanographic epistemes. As such, these CGI models work against a fair representation of an ocean replete with mobile life forms living under continuous metabolic exchange. This results in the imposition of a speciesist and industrialized vision of the darkened seas that redundantly echoes representational blind spots. These images also fail to graph the ocean as a site whose habitats are shaped by the imprint of geological time and anthropogenic action. They leave unaccounted the ways in which ocean space has become a deposit for human waste and chemical weapons, a site where industry is transforming the deep seas and the continental shelf through many forms of extractive action.

In the absence of high-resolution ocean representations that are true to the extreme conditions of its deep frontiers, it is left for us to reengineer how tools for ocean literacy can be redeployed to communicate the depths unknown. What ethics can ground the new subjects of experience that Gabrys calls for, and how can our perception of the gaps in computational sensor technologies shape any act of counter-legibility against the colonial axis of extraction that Mirzoeff announces?

Perhaps calling for the precautionary principle is timely, as it demands we responsibly treat the ocean as a common heritage and that there is scientific consensus about its prospects before extractive action is undertaken. Tuning into the rhythms of planetary flux and being aware of the role that the seabed has for the climate's metabolism is crucial. Because mutations in abyssal planes are slow, any calculation of the cumulative effects of extraction can only result in more misguided havoc. It is imperative now to fight for this space of unknowability to remain as such throughout the generations to come.

[20] Sources: https://metals.co; and https://impossiblemetals.com.

19

Pending Xenophora

Mari Bastashevski

"Unsorted," "problems," and "miscellaneous"—these are some of the ways of not being in order at the Natural History Museum in London. Other categories of being uncategorized include those more situated in time: "to be incorporated when space allows" and "pending xenophora." The museum is home to one of the largest databases of animal remains on Earth. Behind the public displays of *species simplicior*, thousands are tucked away in mahogany cabinets, buried in the labyrinthine archives, and in the nomenclature filing cabinets above and beneath stairs that lead, quite literally, to nowhere.[1] Millions of specimens remain uncataloged. Inside the maze, time slips away: years turn into decades, decades into centuries, and centuries into dust. Some of the drawers to which these bodily remains are consigned have not been opened in two hundred years. The largest proportion of the "uncataloged" is gastropods. Many of these species are either already extinct or will have gone extinct within the twenty-five years it will take to sort them into a museum category.[2] This unsorted part of the collection is so large that it exceeds the capacity of a human to meaningfully engage with it. Looking into the past ambitions and intentions that allowed for this excess accumulation gives an unexpected insight into the contemporary models of managing planetary emergencies by sorting the world into a database.

Within these cabinets, the Natural History Museum manages to be simultaneously both a conservative and a futurist institution, engaged in speculative design even as it constructs a mythology of stable, rational, natural categories. The drawers filled with incomplete journals and labels reveal an intention on the part of the men who wrote them to organize knowledge in the

[1] Once I took a wrong turn on my way to the bathroom and found myself in a room full of files and boxes and a staircase that led right into a brick wall.

[2] Thom van Dooren, "A World in a Shell: The Disappearing Snails of Hawai'i," *UMassHistory*, October 24, 2020, https://youtu.be/fcl9-Njn-sA.

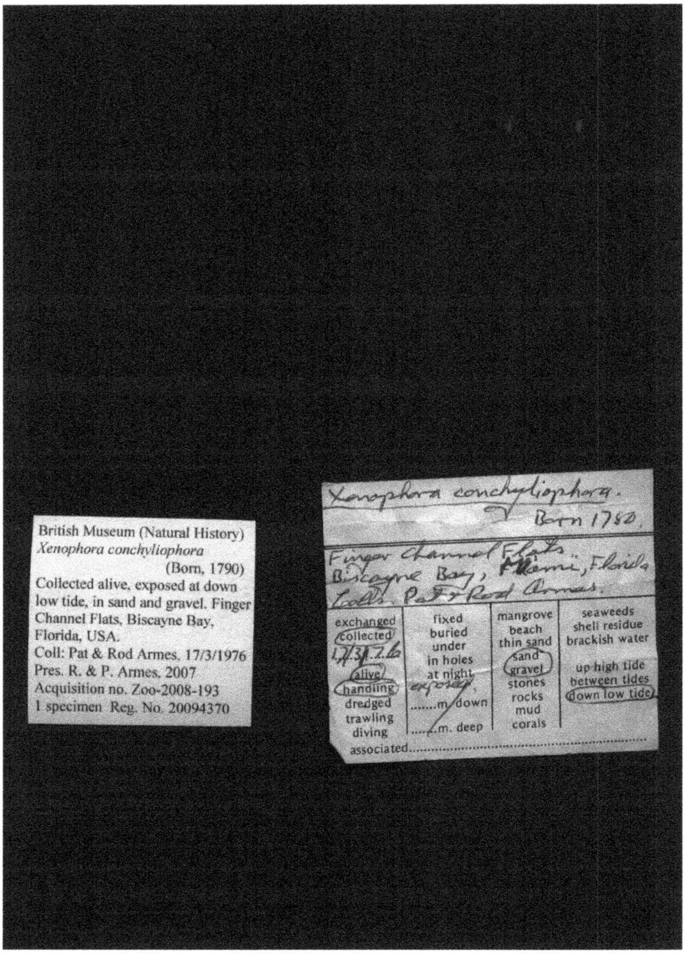

Figure 19.1 Label inside the "pending" drawer of the unincorporated and unidentified gastropod collections, Natural History Museum, London, September 2021.

context of deadly competition, both real and imagined. The documents read like a blueprint for a major heist that would turn the Earth into a support system for imperial ambitions, a system that later would *have* to be managed by humans in order for humans to survive.

The background against which this design figures is the competition for resources and territories between the British Empire and its rival "great powers." Two hundred years after the nineteenth-century heyday of this rivalry, the spirit of competition continues to shape institutional objectives for the museum and drives the agenda of the adjacent organizations with whom it collaborates, including its work in modeling the future of climate emergencies. Today, London's Natural History Museum is linked to some of the most prominent contemporary data conservation initiatives in the world and remains a crucial knowledge hub, guiding decisions about the past, present, and future of species, especially concerning their conservation, resettlement, or potential revival. It is both an influencer and the first point of call for an array of ongoing efforts to tag, scoop, break down, digitalize, and reverse-engineer every shred of DNA, living and dead, in the hope that it will be possible at some point to repopulate the Earth after our current sweeping wave of extinction.[3] It frames the way the future is imagined as being inevitable while still maintaining that same old zeal of competition.

The archives, frozen in time, connect a movement in design to the repetitive feedback loop of hoarding and sorting. Unpacking them reveals other potential plotlines and trajectories of worlds-that-could-have-been beyond the Victorian imperial design objective. While the inscriptions in the drawers take us back to the model of a singular human future with all other species subordinate to us, the perfectly anonymous remains inside the drawers point to other-than-human methods for the acquisition and preservation of knowledge. The unnamed beings—entire colonies of former lives in boxes—allow for the possibility of a future that is not yet finalized, presenting a shimmer of potentiality inside a crumbling system that is extremely difficult to escape, yet which cannot possibly be sustained long-term. It is precisely the lack of clear classification that may help us to conceive of beginnings as well as ends/endings differently.

The Snail

Slow is the universally accepted primary adjective when it comes to the description of snails. In *Dialectic of Enlightenment*, Theodor Adorno and Max Horkheimer theorize the snail as an embodiment of stupidity. The snail impedes its own progress, they argue, when it comes to the evolution of knowledge production. A snail's first feeling of the world as it ventures out of its shell is "always easily repulsed," they write, "behind it stand goodwill, fragile hope, but no continuous energy. In the direction from which it has been definitely scared

[3] Rebecca Ellis, "Rethinking the Value of Biological Specimens: Laboratories, Museums and the Barcoding of Life Initiative," *Museum and Society* 6, no. 2 (July 2008): 172–91; International Barcode of Life, https://ibol.org.

Figure 19.2 Pinboard for shells with partial label information from inside a drawer of the unincorporated and unidentified gastropod collections, Natural History Museum, London, September 2021.

off, the animal becomes shy and stupid."[4] But this is only true if we consider evolution as something that operates linearly, aimed at an ultimate total knowing that does not need to take into account the energies expended in the process. Snail intelligence, on the other hand, works from the middle toward multiple trajectories at the same time. The snail that ends up spending the entire mass of its liquid body on these critical inquiries is more likely to be concerned and constrained by the meantime than deep time (the immediate middle futures),

[4] Max Horkheimer and Theodor W. Adorno, *Dialectic of Enlightenment*, trans. Edmund Jephcott, ed. Gunzelin Schmid Noerr (Stanford: Stanford University Press, 2002), 213–14.

and it is perhaps this freedom from making long-term projections that has made its manner of acquiring knowledge so timeless. In order to consider these various escargothic and ahuman futures, it will first be necessary to examine the historical significance of these gastropods for futurist imagination.

Looking at the historical importance of these gastropods, as it appears in many parts of the museum's futurist vision, reveals various escargothic and ahuman futures. Fossilized snail shells made an unexpected debut in geology during the eighteenth century. This resulted in the discovery of deep time and, as a result, the practically instant creation of the new science of modern (paleo) biology. In geological eras prior, the snail had archived itself in the strata of rocks in a manner so precise it could be read accurately thousands of years into the future, using a combination of diagnostics and other human tools for dealing with time capsules.[5] The geologists, whose discipline owes much to the snails, clams, and other mollusks, determined that snail fossils lead to mineral deposits and that once people learned to read fossils and recognized the minable resources they pointed toward, they would find new means of controlling time with fossil fuels. Here, true to the cliché of being slow, the snail also happens to be one of the most efficient long-distance time travelers and transtemporal communicators—a lab assistant to the accelerationist, which is to say, anything but shy and stupid.

The legacy of a time-traveling snail has been eclipsed by an obsession with the aesthetic qualities of shells and the symbolic, frequently gendered, analysis of hard shell/soft power that emerged with the Victorian shift in the animal-human relationship. Snails were abruptly in high demand and hoarded to excess. However, even in captivity, the remains of the unidentified Victorian snail find their way into a time capsule that avoids destruction by transforming into an impossible, pending task.

The Anxious Victorian Snail

While the partial information on the labels attached to the remains reveals little about the past lives of the Victorian snail, it nonetheless points us to the conditions that have come to define the relationship between the snails and those who collected them: Syria, Siam, Cashmere, Ceylon, Haiti, the Seychelles, Liberia, Chittagong, Tahiti, Yemen, 1861, 1888, 1897, 1908. These are the first bytes of information

[5] David Bainbridge, *How Zoologists Organize Things: The Art of Classification* (London: Frances Lincoln, 2020), 90–93.

on the partial tags, usually followed by the British signature of the malacologist who hoped to claim a "discovery." The scale of this extensive collection should be read within the context of the urgency of the imperial ambitions of the time.[6] A majority of the uncataloged snails were accumulated during the geopolitical and socioeconomic period of "the Great Game,"[7] the not-so-cold war of the Victorian era, during which the British Crown was not merely concerned with territorial expansion but with accelerating it to a point where it could safely outdo all other empires. The malacologists did not come, as one may expect, from the natural sciences but rather were bankers, oil moguls, and military personnel, all working for the objectives of the Crown in remapping the world for war. The plot of the great games cemented the time-space aesthetic in which a first encounter between a Siamese snail and a British malacologist could take place.

During the presidential address at the Malacologists' Society in 1892, the British military surveyor Henry Haversham Godwin-Austen, whose minuscule handwriting is all over the uncataloged archive, spoke openly about optimizing lulls during wartime with a pleasure principle. The investment in collecting snails would not only improve the morale and character of the officers but also help to win the race in the Great Game and accelerate the imperial future:

> Except when on actual service in the face of an active enemy, no men have better opportunities for research than officers of H.M. Army. Days, weeks, even months are spent by many in the most favorable places in the world, where Nature is lavish of her gifts, where the monotony and solitude of the life are to some temperaments almost unbearable; in such a quarter, if they only possessed an

[6] The museum itself has long since come to accept that opening the archive to the decolonial gaze is the only way to maintain leverage in the ensuing conversation. The challenges of this process are so enormous that it would be a mistake to mention them in passing. The broader discourse that can be made accessible through the admittedly narrow prism of this chapter seeks to outline the Victorians' inability to distinguish a planetary future from the accumulation of species, which they held hostage. In particular, the focus is on whether certain computational models concerning the species are merely future projections or, akin to the Victorian scholars, are actually at the forefront of ameliorating or accelerating certain major ecological shifts. This scenario also provides a microcosm from which to study the anthropocentric system, whereby the expenditure of resources it requires to maintain itself far exceeds its contribution toward the conservation of said resources. Also see Garry J. Alder, "Standing Alone: William Moorcroft Plays the Great Game, 1808–1825," *International History Review* 2, no. 2 (April 1980): 172–215.

[7] David Gillard, review of *The Great Game: Britain and Russia in Central Asia*, ed. Martin Ewans, *Slavonic and East European Review* 84, no. 1 (January 2006): 161–63. The term "Great Game," coined in the 1840s after what in Britain was called the "First Anglo-Afghan War" and later popularized by Rudyard Kipling of *The Jungle Book* fame, implies a sense of playfulness to the nineteenth-century geopolitical strategizing between British and Russian imperial powers competing for control of Central Asia and its resources, often reducing those caught in the middle to mere pieces in a global power game. Also see Matthew Edwards, "The New Great Game and the New Great Gamers: Disciples of Kipling and Mackinder," *Central Asian Survey* 22, no. 1 (March 2003): 83–102; and Seymour Becker, "The 'Great Game': The History of an Evocative Phrase," *Asian Affairs* 43, no. 1 (2012): 61–80.

> interest in some branch of science, their lives would at once be full of pleasure, and the sameness of the daily military duty would not be felt.[8]

To Godwin-Austen, the utility of all knowledge boils down to a formula for maximizing the accumulation of capital through the optimization of all possible energy and resources. The snail, to him, is an ultimate symbolic other—a phenomenon that simply waits to be captured, a conquest, an exotic and peculiar object—that sustains the appetite in the monotony of war, like a field affair with a nurse, through reassurance, release, and distraction. In time, a routine was established, consisting of shell collection, mapmaking, surveying, combat, flagging, occupation, and more shell gathering.

The fetish of the snail connects the fetishization of animal species with speed, futurism, accumulation of capital, and the evolving militant masculinity of this period, which continues to reverberate in popular culture. The gentrification of the snail was a shift from the medieval snail, the warrior, and the snail as the earthly analogue of God in heaven, such as the famous snail of the late fifteenth century that sits on the edge of the frame of Francesco del Cossa's *The Annunciation*, which represents, if anything, the slow unraveling of life cycles and the incomprehensible reluctance of God to make any decisions.

The gendered shift spiraling around these hermaphrodite gastropods seems to have begun with the Linnean Society of London, which, during the eighteenth century, barred women from having anything to do with the development of natural history, arguing that animals were too dangerous and plants too sexy, but an exception was made when it came to shells.[9] Elsewhere in Europe, writer and critic Johann Wolfgang von Goethe became obsessed and repulsed by the snail's inherent perversity, reflected throughout his oeuvre in a perplexing variety. In 1797, taking a detour from his usual work, Goethe wrote "Anatomie der Schnecke" (Anatomy of the Snail), puzzling over the (for Goethe) unsettling combination of the creature's hardness and fragility. Goethe felt stalked, emasculated, and persecuted by snails whose gender messiness and delicate-yet-sturdy form reminded him of a female figure of victimhood. In *Faust* (1808–32), the snail works as an analogy for Gretchen, by turns demonic, clairvoyant, and pathetic. In his earlier work, "Jahrmarkt zu Plundersweilern" (The Fair at Plundersweilern), the snail is used as a metaphor for a woman, a persecuted maiden who frustrates her suitor by retreating into her shell like a snail. Later, Salvador Dalí became

[8] Henri Haversham Godwin-Austen, "Address of the President," *Journal of Molluscan Studies* 3, no. 5 (July 1899): 241–62, here 242.

[9] Barbara N. Nagel, "Goethe's Stalker Snails," *Feminist German Studies* 36, no. 2 (Fall/Winter 2020): 1–26.

Figure 19.3 Label inside the "miscellaneous" drawer on the unincorporated and unidentified gastropod collections, Natural History Museum, London, September 2021.

obsessed with a totemic snail he had encountered at the steps of Sigmund Freud's house and presumed to be the ultimate Freudian brain. In "Morphology of the Skull of Sigmund Freud" (1938), Dalí sketches Freud's skull as a snail, "d'après nature." In France, Honoré Daumier's snail-sex network theory may have at least in part contributed to an early natural history of the Internet.[10]

[10] João Felipe Ginefra Toni, "On the Origins and Renaissance of Goethe's Morphology," *Elemente der Naturwissenschaft* 108 (2018): 5–21; Freud Museum London, "When Dalí Met Freud," February 4, 2022, https://www.freud.org.uk/2019/02/04/when-dali-met-freud; Justin E. H. Smith, "The Internet of Snails: Escargotic Commotion and the Wood-wide Web," *Cabinet*, no. 58 (Summer 2015), https://www.cabinetmagazine.org/issues/58/smith.php.

These tropes, fictions, dreams, and historical facts echo each other in the uncataloged archive. A serendipitous find here is the diary of Elizabeth Betham, the real wife of the actual Captain Bligh,[11] filled with sketches of snails and shells, nested on top of her other notebooks in the drawer. The trail of the incognito snail really does take us back not only to the cultural production of future dystopias but also to the development of precise material conditions that would uphold and expand certain ideals as standards while systematically drowning out others. More precisely, the malacologists were at the forefront of creating the precise geopolitical socioeconomic conditions for the global ecological issues that the Natural History Museum is now seeking to prioritize and resolve.

Shell Companies

Another recurrent name in the gastropod archive is the lawyer Ernest Ruthven Sykes, who collected the tiniest of marine snail shells in the Seychelles and studied others found across the Arabian Peninsula. When not fascinated by snails, he worked as a banking executive, writing policies focused on the securities of lending, speculating, betting on, or shorting against particular financial futures. The Sykes family owned a bank that eventually amalgamated into the present-day Barclays.[12] While it would be a challenge to trace the snail trail of cash to the precipitate "Flash Crash" of 2010,[13] Marcus Samuels, another snail enthusiast of the nineteenth century, made the snail into a mascot for a company whose century of ecological disasters left millions of species swimming in oil. Samuels, born in 1853, inherited from his father, a "shell merchant," a collection of valuable shells along with other animal body parts considered exotic at the time. He continued the business of shell collecting and the associated international

11 The reputedly tyrannical sea captain is portrayed by Antony Hopkins in the 1984 film *The Bounty*, directed by Roger Donaldson.

12 See John Brunton, *Bankers and Borrowers* (London: E. Arnold & Co., 1921); and W. J. Rees, "Ernest Ruthven Sykes (1867–1954)," *Journal of the Society for the Bibliography of Natural History* 4, no. 1 (January 1962): 20–27.

13 The "Flash Crash" was an event that occurred on May 6, 2010, which sent the American stock market into a frenzied twenty-minute period of unrivaled turbulence. Around 2:40 p.m., a seemingly normal trading day was thrown into chaos when a trading AI from Waddell & Reed, a Kansas mutual fund company, initiated the indiscriminate sale of $4.1 billion of E-Mini S&P futures contracts, setting off a chain reaction of volatility and liquidity issues across the markets. The carnage of these twenty minutes saw the evaporation of approximately $1 trillion in market value, with share prices of several companies wildly fluctuating to absurd degrees. It was a collapse of the market that has since been immortalized as the stark example and reminder of the risks and effects of high-speed automated trading. For more information, see Tom C. W. Lin, "The New Market Manipulation," *Emory Law Journal* 66, no. 6 (2017): 1253–63.

mercantile logistics with his brother, developing his father's shell connections into a resource trading network that would eventually become an oil empire. These two seemingly disparate activities harmonized as they traveled along the very same imperial seaways. To outdo the Nobel brothers in Azerbaijan, Samuels conceived of the first oil tanker. The vessel was designed to navigate the Suez Canal, the construction of which had recently cut transportation costs and won time for what would become Shell Oil.[14] The evergreen dependency of the global economy on this canal was made evident in 2021 by the tragicomic plight of the container ship *Ever Given* that got stuck in the middle of the canal, blocking the route for nearly a week and causing significant disruptions in commercial marine traffic. Global flows of extracted resources still move through the original channels established by the disparate voices of men who feared the feminine fragility of the snail and yet insisted on the authority to masterplan the future of Earth.

Crypto Giraffe

If Victorian malacology can be said to have a spirit that endures into the present, it is, today, all too perfectly encapsulated in the company Crypto Giraffe, founded by Jonathan Ledgard, in that the techno-futurist organization of species frames both animals and technologies as props in a one-man, data-as-capital-driven humanitarian planet rescue mission. After a long career as a war reporter and a financial journalist, Ledgard, now a guest professor at the École politechnique fédérale de Lausanne, reinvented himself as a futurist and a prophet of doom.[15] His current utopias are constructed around the speed with which new technology could rectify humanity's most destructive tendencies. During the Covid-19 pandemic, and at the height of blockchain enthusiasm, he founded the Crypto Giraffe service, complemented by a corporate white paper advocating for the establishment of what amounts to a system of crypto-species

[14] "Albertus Seba, de Amsterdamse hofleverancier van Tsaar Peters Kunstkamera," Het Schelpenmuseum Amsterdam, March 10, 2017, https://www.schelpenmuseumamsterdam.nl/albertus-seba; Mark Forsyth, *The Etymologicon: A Circular Stroll Through the Hidden Connections of the English Language* (London: Icon Books, 2011), 140–41; Jonathan Conlin, *Mr Five Per Cent: The Many Lives of Calouste Gulbenkian; The World's Richest Man* (London: Profile Books, 2019), 53–66; "The Prize Chapter 3: Competitive Commerce," EGEE 120 Oil: International Evolution, Dutton Institute, Pennsylvania State University, n.d., https://www.e-education.psu.edu/egee120/node/205.

[15] Ben Taub, "Jonathan Ledgard Believes Imagination Could Save the World," *New Yorker*, September 16, 2019, https://www.newyorker.com/magazine/2019/09/23/jonathan-ledgard-believes-imagination-could-save-the-world.

pyramid schemes, claiming that saving species requires their addition to a single database and giving them money.[16] By reducing the problem of extinction to something as banal as cash for conservation, the proposal is wide in scope and light on specifics. Its branding leans heavily on the depiction of a charismatic giraffe. Where snail/shell was the mascot for oil, here the giraffe becomes the mascot for a new coin—like Dogecoin but with actual collateral animals, while the African giraffes themselves are almost of no consequence.

Crypto Giraffe is one in a long line of popular virtual animals that were put to work as the mascots of progress by digital tech, representing quasi-religious investment in the utopian possibilities of new technologies.[17] The allure of Crypto Giraffe, much like the allure of snails, has been molded by centuries of exploitation and spectacle, in no small part thanks to "the emperor's giraffes" that were commodified and paraded around the globe in displays of imperial showmanship. Zarafa, a giraffe gifted in 1826 by Muhammad Ali, Pasha of Egypt, to Charles X, King of France, makes a glaring example. Zarafa's arrival in France incited what can only be described as "giraffe mania": an intense, sensationalist reaction with her likeness reproduced ad nauseam, which continues to permeate pop culture. Writing about the history of colonial giraffes, Jody Berland notes that the Giraffe Conservation Foundation openly acknowledges that the giraffe's primary economic benefit lies in its evolutionary uniqueness. The commercial exploitation of the giraffe's exotic image is relentless and works to sell just about anything from children's clothing and wine to social media fads and even massive global events such as the FIFA World Cup.[18] It is to this portfolio that Ledgard is attempting to add crypto banking.

The core idea of Crypto Giraffe's services contends that humans will value a living giraffe more when it generates some income in a symbolic exchange. At the same time, with undoubtedly racist undertones, Ledgard cites the ritual killings of giraffes "for their tails, which are used for ceremonial purposes" as a

[16] Jonathan Ledgard, "Lecture: The Age of the Interspecies; New Ways to Think about and Protect Nonhuman Life on Earth," European University Institute, San Domenico di Fiesole, May 15, 2023, https://www.eui.eu/events?id=550345; Laura Spinney, "The Big Idea: Should Other Species Have Their Own Money?," *Guardian*, March 14, 2022, https://www.theguardian.com/books/2022/mar/14/the-big-idea-should-other-species-have-their-own-money; Jonathan Ledgard and Erik Meijaard, "Endangered Wildlife Should Pay for Its Own Protection," *Project Syndicate*, December 2, 2021, https://www.project-syndicate.org/commentary/digital-wallets-for-endangered-wild-animals-by-jonathan-ledgard-1-and-erik-meijaard-2021-12.

[17] Consider, for example, the Twitter bird, the Firefox, the cats of Apple operating systems, Linux's penguins, and so forth, described in detail by Jody Berland in *Virtual Menageries: Animals as Mediators in Network Cultures* (Cambridge, MA: MIT Press, 2019), 101–22.

[18] Berland, *Virtual Menageries*, 62–66.

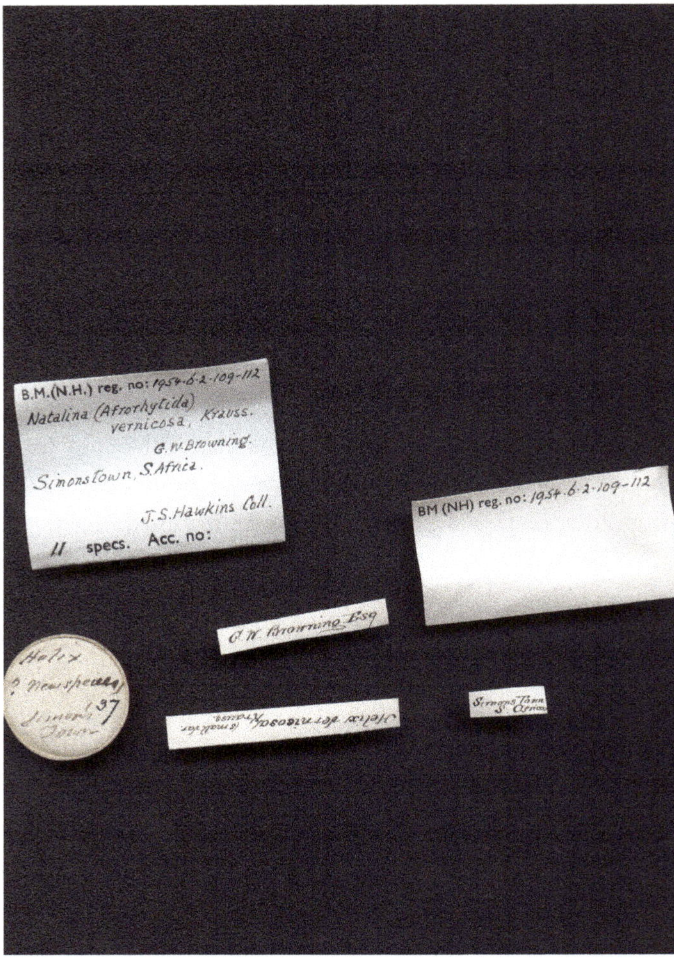

Figure 19.4 Labels inside the "problems" drawer of the unincorporated and unidentified gastropod collections, Natural History Museum, London, September 2021.

justification for this intervention[19] and argues that aggregating all of the giraffes into a single database and then endowing them with biometric wallets will help local conservation initiatives. This principle of "giving" money to animals is then extended to all species across all ecosystems. While Ledgard is aware of the death drive that dominates human relationships with nonhuman entities,

[19] Ledgard claims that "Giraffe numbers are falling because of human encroachment. Giraffes are killed for their tails, which are used for ceremonial purposes, and for their meat and bone marrow. They are snagged in barbed wire or hit by vehicles when crossing roads. They are pushed out of their range by cattle. Acacia trees are cut for fuel." See Jonathan Ledgard's website, "linnaeus.life," archived on October 19, 2021, https://web.archive.org/web/20211019183928/https://linnaeus.life.

he only sees this at work in local and literal ritual instances of the slaughter.[20] He does not seem to perceive how the death drive and rituals underpin his own proposal or how the creation, mythologization, dissection, and reification of the giraffe add more distance between the animal and human. Following an example of one sacrificed giraffe, Ledgard's paper relegates all animals as a priori dead until culture and ecology reincarnate them with the nominal power of a market participant. Not a single giraffe, or creature considered worth saving, is involved in this decision. The question of "thinking with" does not arise because the animal is already, and always will be, a commodity in a design experiment, always already dead, and thus cannot possibly have anything to contribute through its own prescription.

Much like the uncataloged archive, this type of design for the future is situated in the unspecified time of the beyond, entirely removed from histories, which, ironically, takes us right back into the past. The Crypto Giraffe proposal lacks any reference to the existing global economic systems and proceeds as if they do not exist. It does not concern itself with defining the boundaries of the emerging animal digital market (in the form of taxes or debts, for example). It does not discuss the temporal expenditure and the resources required to catalog and upload the entire planet onto a cloud, and despite this being a problem, the Natural History Museum has been struggling with this for decades. Instead, Crypto Giraffe follows the old Victorian pipeline from crisis economies to animal commodities, as if Godwin-Austen's totemic snail did not exist. In addition to working as a commodity, the giraffe has always been a medium in imperial politics. In "Attending the Giraffe,"[21] Berland notes that the 1414 exchange of giraffes between a Bengal ruler and the emperor of China was not merely an exchange of exotic commodities but an action loaded with geopolitical implications. The giraffe represented an alluring prospect of near-futures in which natural resources could be effectively mastered by force, marking a

[20] Elsewhere, Ledgard asserts that "life-forms are an 'inevitable and emergent property of any random dynamical system that possesses a Markov blanket.' [Sergio Rubin et al., "Future Climates: Markov Blankets and Active Inference in the Biosphere," *Journal of the Royal Society Interface* 17, no. 172 (November 2020): 1–12]. All autonomous systems and all living things are self-organized into Markov blankets of Markov blankets, from their cellular structure to their bodily form. They are subject to a free energy principle where free energy is the dissipation of energy to equilibrium—death. In order to contain energy, the biological life-form establishes a boundary for sensing and predicting the outside world. This boundary extends out to the communities they belong to. Such a scale-free and domain-independent approach may underpin Interspecies Money." Jonathan Ledgard, "Interspecies Money," in *Breakthrough: The Promise of Frontier Technologies for Sustainable Development*, ed. Homi Kharas, John W. McArthur, and Izumi Ohno (Washington: Brookings Institution Press, 2022), 95, footnote 39.

[21] Jody Berland, "Attending the Giraffe," *Humanimalia* 9, no. 1 (September 2017): 22–45, https://humanimalia.org/article/view/9612.

turning point in relations between these two imperial powers. Berland goes on to emphasize that this event marked the beginning of a new role for giraffes as emissaries of international trade. In the centuries that followed, giraffes would continue to play a key role in political and diplomatic narratives, highlighting their enduring place within our sociopolitical structures. Seen through Berland's perspective, Crypto Giraffe emerges less as a passing fad and more as a political medium. The tangential interest in preserving giraffes taps into a deeply rooted symbolic association between the power and masculinity of giraffes and humans that has evolved over centuries.

The key issue with Crypto Giraffe is not constrained by material reality because it was never designed to become a reality despite being framed as a desirable future. However laughable the notion, it fuels the Victorian-age ambition to mobilize resources for the markets, all the while guilt-tripping anyone who objects to the idea by pointing to their "inaction" in the face of environmental crisis. The project not only gained traction and popularity by appearing in major Western media outlets but also received financial support from the World Bank and the United Nations (UN). This is perhaps unsurprising, given that the proposal reads like a tribute to infrastructure built around the concept of "ecosystem services" put forward by the 2005 UN Millennium Ecosystem Assessment, which sought to catalog human effects on the environment worldwide.[22] Crypto Giraffe is trendy because it turns a crude techno-managerial call for planetary governance into cutesy, easily consumable, and shareable social media snippets. By planetary governance, I mean a movement, originally inspired by an assembly of environmental thinkers, policymakers, and international organizations in the 1970s in response to burgeoning environmental crises, that is presently reinventing itself as a culture of digital environmentalism. To keep abreast of the zeitgeist with the advent of artificial intelligence (AI) and crypto, proposals began shifting toward the idea of governance as a set of binary codes and blockchain architectures, inclined toward the defense of a single holistic database for the species. This techno shift nonetheless follows a predictable path of seeing technology strictly as an accelerator and an efficient executor of Victorian design. Could we, instead, imagine technologies codeveloped with animals that could catalyze their release from human-driven planetary colonial projects, especially considering that the bulk of military technologies in use

[22] Sian Sullivan, "'Ecosystem Service Commodities'—A New Imperial Ecology? Implications for Animist Immanent Ecologies, with Deleuze and Guattari," *New Formations* 69, no. 1 (July 2010): 111–28.

today spring from cultures, structures, and mechanisms found and harvested from other-than-human *umwelts*.[23]

Celebrated as future-oriented and progressive, the techno-optimistic adaptations persistently contribute to the material circumstances that bolster and broaden the scope of prevailing uniform standards. The digitalization of planetary governance projects works to overshadow the potentialities of other futures, other designs, and other methodologies, chronologies, and scales while denying that something differing from Victorian cataloging standards is needed. In addition, each shift toward technology adds a new set of challenges to the preexisting complexities that already beset the management and preservation of animal databases at every turn. While Ledgard's proposition might seem like just a bizarre idea of an eccentric individual and thus unfit to enter the natural history archives of visionary men, more comprehensive initiatives have been utilizing for well over two decades the same kinds of big data sorting and economic offsetting projects for animals. It is the core ambition of these planetary governance projects to be larger than the museum and the planet itself.

Databasing Natural History: From Museum Specimens to Molecular Data

In 2002, the global initiative International Barcode of Life (iBOL) transformed the concept of species identification through DNA barcodes into a mammoth undertaking, professing to "barcode" all life on Earth. Hosted initially as a small hub at the University of Guelph, iBOL has accumulated over two hundred million dollars in funding and lists the Canada Foundation for Innovation, the Alfred P. Sloan Foundation, and the European Research Council as its primary sponsors. Although the numbers are incredibly difficult to verify, iBOL boasts a vast network of over a thousand researchers from forty-one countries. Today, iBOL relies on an open database model that is tethered to mostly unnamed governmental agencies, sponsors, universities, and, of course, Natural History Museums—in particular, the Smithsonian National Museum of Natural History, which hosts and advocates on behalf of the project. Over the past two decades, iBOL has expanded from a solitary concept to an entity that increasingly governs

[23] "Umwelt" is a term coined by Jakob von Uexküll, which refers to the techno-cultural context in which each animal exists, not only just the physical surroundings measured in biological properties but also the possible subjective experience of time-space. Jakob von Uexküll, "The Theory of Meaning," *Semiotica* 42, no. 1 (1982): 25–82.

stories about biodiversity that are being told in a variety of disciplines. In 2016, it even wormed its way into the UN, a move deemed by the institution itself as its crowning achievement. To leverage the specifics of digital growth and unity across law and order and underscore its relevance, iBOL boasts a collection of over two million samples and uses that fact to push for DNA barcoding as a universally accepted rule for the identification of species.[24]

The initiative relies on an army of volunteers, eager to scoop a spoonful of DNA from their native environments. Much like the Victorian hoarding of specimens in the uncataloged archive, iBOL proposes collection as something to do "in your free time." The act of collecting is framed as a hobby and a positive distraction, obscuring more pressing questions such as whether "rapidly identifying species" does anything to prevent climate breakdown or whether it, in fact, acts from the perspective of a future in which only a few privileged individuals in the Global North are left to marvel and ponder all of the species that once inhabited the ravaged planet and have the Godlike power to resurrect them.

The insistence on accelerating the collection process is a significant aspect of this future-oriented design. The expectation here is for the ontological status of beings (their very nature and existence) to be determined in increasingly expedient fashion, even potentially in real-time, framed in a context of crisis and competition. Whereas in the Victorian era this contest was seen as the Great Games between imperialist powers, iBOL's race is with and against the accelerated destruction of life on Earth itself. The urgency is explained away by the threat of impending extinction. iBOL never articulates the causality between extinction, efficiency, and collection but subtly implies in its principles and objectives a future in which reincarnation and rewilding from DNA are possible. If the National History Museum in London rationalizes collecting as intellectual inquiry, with iBOL, there is a distinctly utilitarian purpose in place[25]

[24] See "20 Years of DNA Barcoding: U of G Developed Technology Hits Milestone," *University of Guelph News*, February 10, 2023, https://news.uoguelph.ca/2023/02/20-years-of-dna-barcoding-u-of-g-developed-technology-hits-milestone; Paul D. N. Hebert, Alina Cywinska, Shelley L. Ball, and Jeremy R. deWaard, "Biological Identifications through DNA Barcodes," *Proceedings of the Royal Society B* 270, no. 1512 (February 2003): 313–21; Rebecca Ellis, Claire Waterton, and Brian Wynne, "Taxonomy, Biodiversity and Their Publics in Twenty-first-century DNA Barcoding," *Public Understanding of Science* 19, no. 4 (July 2010): 497–512; "DNA Barcoding Part of UN Biodiversity Strategic Plan," *University of Guelph News*, December 19, 2016, https://news.uoguelph.ca/2016/12/dna-barcoding-part-un-biodiversity-strategic-plan.

[25] See Stephen Cameron, Daniel Rubinoff, and Kipling Will, "Who Will Actually Use DNA Barcoding and What Will It Cost?," *Systematic Biology* 55, no. 5 (October 2006): 844–7; Sergei V. Shekhovtsov, Irina N. Shekhovtsova, and Sergey E. Peltek, "DNA Barcoding: Methods and Approaches," *Biology Bulletin Reviews* 9 (2019): 475–83.

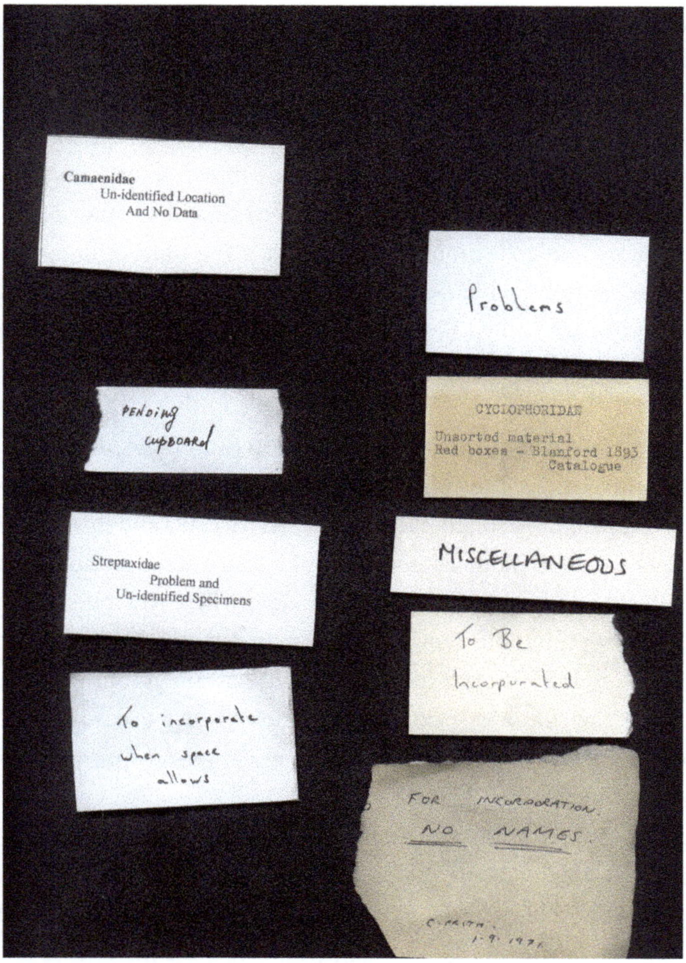

Figure 19.5 Drawer labels inside of the unincorporated and unidentified gastropod collections, Natural History Museum, London, September 2021.

that undermines the project's stated goals because it relies on the validation of an existing authority. The Natural History Museums are essential to iBOL, as they provide a core system of taxonomy for classifying discovered species. iBOL is also dependent on Natural History Museums as the main donor source of samples for practicing genetic sequencing and training the algorithm. In this relationship, the status of the Natural History Museum shifts from an incomplete and inconclusive catalog of beings and futures—an institution with a significant backlog of its own issues—to an authority on the future knowledge of species and species delimitation. The issues, already too plural, multiply exponentially.

Whole new tensions of scalability and entropy emerge in this relationship, demanding new temporal, spatial, and evolutionary arrangements that could, at least in theory, link machine learning epistemologies with existing systems of knowledge production. In *Rethinking the value of biological specimens*, Rebecca Ellis points out that a DNA sequence without a direct, transparent, and clearly audited route to the voucher specimen, for example, becomes a meaningless artifact.[26] Or, as one collection curator stated, "The only thing a voucherless DNA sample can tell you is that vials or data labels got switched or that the specimen was misidentified before it flew away."[27] This is a perfect recipe for the rapid accumulation of DNA material for a digital archive of uncatalogable species.

If the systemic issues of the Natural History Museum are made somewhat apparent by the archive of the uncataloged, what would happen to the thousands of future creatures, tangentially identified by AI as snails and insects? Will there be an "elsewhere" of iBOL that someone could glance at two hundred years from now? What will be the volume of such a box, and where will it be situated, given iBOL's planetary aspiration? How will the discarded samples be accounted for? If, using the Natural History Museum structure as a method, the AI is tasked with deciding the future of priorities in which we get to know, archive, and revive these beings, what exactly is the future utility of this enormous collection, and for whom?

The closed-circuit validation loop between iBOL and the Natural History Museum not only upholds the Victorian idea of animals as lists of discrete units but it also adds a host of generational problems stemming from false equivalences between action and change, panic and intention, and the inability to distinguish scale, labor, and value from imperial ambition and form. iBOL owes a certain debt to the lineage of information theory in its treatment of the fragile snail and dead Crypto Giraffe not as a set of valuable parts or mythical creatures but rather exclusively as a nexus of information. Ignoring, instead of foregrounding, the possibility of mutual learning between species and the regimes and parameters under which these ideas are currently produced shapes a database culture that, in its obsession with a particular future, implies the multiplication and expansion of the here and now into all possible futures. The future is simply to be an infinite set of remixes of the present.

26 Ellis, "Rethinking the Value of Biological Specimens."

27 Quoted in Ellis, 174: Email from a North American collection manager, January 24, 2008.

Figure 19.6 One of the many cabinets of snail shells without formal names in the unincorporated and unidentified gastropod collections, Natural History Museum, London, September 2021.

Nothing about the public-facing framework of iBOL suggests a practice that proceeds by way of exploring and negotiating between the distinct intelligence of technology and the animals that are still living and dying in their own dramatically changing ecologies and umwelts. While positioning itself as a state-of-the-art life-saving initiative, this project's interest stops at tools and leaves out consideration of the beings who are actually killed. For example, one of the ways in which iBOL encourages the users to collect the DNA is by scraping their remains off the windshields of cars. Animal, in the Western philosophical tradition, is defined not by its presence but as a negative to be vanquished, repudiated, or repressed. Riffing on Jacques Lacan's formulation that "la femme n'existe pas," Oxana Timofeeva explains in *The History of Animals* that the "animal" does not preexist but is defined as absence, a negation to be mitigated, conquered, or repressed.[28] Indeed, in the animal writings of Georges Bataille, Giorgio Agamben, Martin Heidegger, Georg Wilhelm Hegel, and Freud, the animal materializes as a residual by-product of the human refinement process. The symbolic animal plays the role of creating a buffer between the human (as the ultimate ideal of refined design) and everything it has to overcome: fear, vulnerability, insanity, poverty, war, and, more recently, extinction.

Timofeeva's dissection of Hegel's compulsive need for meticulous classification gestures toward a larger Hegelian imperative: to understand classification as a mechanism within a teleology. Taxonomy as the handmaiden of teleology in Hegel can be understood—especially with regard to his philosophy of race—as being a necessary condition for the kind of "transcendence" toward which his philosophy was aimed.[29] iBOL has significantly more (immediate) sublunary goals, but the same kind of extractive metaphysics is at work in its model. This long history of classificatory logic echoes in the exhaustive archiving at the Natural History Museum and in iBOL's endeavors to capture and classify all manner of planetary wealth. It is as if each megabyte of DNA information holds the potential to achieve immortality for the future transcendent, superhuman database manager. The Victorian snail, iBOL's DNA, and Crypto Giraffe all point to distinct instances of redefining the human under the threat of climate collapse.

The animals entangled in this process, both symbolic and living, are, however, neither dormant nor compliant. In her reading of Hegel, Timofeeva investigates how "Hegelian animals" contribute to and recurrently disturb this design process by way of intermediate form, the neither-here-nor-there beings that go between categories within a proposed totality, while poised at the water's edge, ready to fly, wavering between plural potentialities. Hegel, much like Goethe, was not only completely preoccupied but also occasionally irritated by animals that continued to defy his efforts to establish hierarchy and dominance through clear categorizations.[30] This frustration epitomized itself in the famous Hegelian

[28] Oxana Timofeeva, *The History of Animals: A Philosophy* (London: Bloomsbury, 2018), 60–103.

[29] Hegel really included the entire planet, a whole ecosystem, in a potential subjecthood. "Animals are subjects in so far as they are negatively related to certain sensual objects, and if one wants a perfect example of negativity, one should go back to a wonderful passage in the *Phenomenology of Spirit*, where Hegel, criticizing sense-certainty, compares the animal with an initiate of the Eleusinian mysteries." (Timofeeva, *The History of Animals*, 85). "As if we were witnessing the most thorough stocktaking or inventory of nature, which aimed to appropriate all its wealth. Nothing should get lost or, worse still, be excluded; everything and everybody are welcome in the menagerie of spirit. The unity or even the solitude of spirit is reflected in the abundant multiplicity of nature. Animals, too, are now becoming a very important element of totality and have therefore to be thoroughly classified." (Timofeeva, *The History of Animals*, 74).

[30] With caterpillars and worms, "for Hegel, 'there is something repulsive about them' [*Hegel's Philosophy of Nature: Part II of the Encyclopedia of the Philosophical Sciences (1830)*, trans. A. V. Miller (Oxford: Clarendon Press, 2004), 418]. The question would be the following: Why, in fact, does he not like

formula, "if reality does not fit the notion, then it is reality's problem, not the notion's."[31] The same disturbance is perpetually operative in the novel, and now digital, ways of classifying the animal. Fascinating, untamed, and sometimes amusing things transpire when scientific classifications are compared with those in mythology, literature, fiction, and art. What is astonishing, Timofeeva continues, citing Michel Foucault, is not "simply the oddity of unusual juxtapositions between different classifications and categories" but "the fact that the common ground on which such meetings are possible has itself been destroyed."[32]

While much eco-fiction has been written that is in opposition to this order[33]—or at the very least, eco-fiction distances itself from science within any distinct discipline—the uncataloged archive discredits the possibility of propinquity and stability between categories in a radical way by hiding a world within a world. Consider the snail, again, as an example: the sovereign of dust, the time-traveler, the pre-Victorian fossil, and a human archivist nonpareil. Nothing in the snail's fossilization practice, which is one of the most efficient

them? On a close reading, it turns out that Hegel's repulsion toward reptiles accompanies his crucial theoretical attitude. As is well known, intermediate forms are usually taken as a proof of the idea of evolution. However, for Hegel, there is and can be no evolution or generation in nature" (Timofeeva, *The History of Animals*, 80). For this, there is a threat to the spirit's final form. "The Hegelian amphibian is a mistake of nature, a defective individual, which did not succeed in following the idea, got stuck in between the air and the Earth and therefore 'presents a sorry picture'" (Timofeeva, *The History of Animals*, 81). In other words, it did not overcome its inferior form. It got stuck. Then, instead of seeing himself as just one of the broader universe, he quickly adjusts, and Hegel admits that reality is incredibly complicated, then goes on to explain that if reality does not fit the notion, then it is reality's problem, not the notion's.

31 "[. . .] Hegel insists that we have to begin with theory, with the concept, with 'general determinations.' And here we find the famous formula: if reality does not fit the notion, then it is reality's problem, not the notion's" (Timofeeva, *The History of Animals*, 76).

32 "According to Foucault, what makes us laugh here is, roughly speaking, a heterogeneity or heterotopia, the absence of a unifying principle, which is really unthinkable in the Western tradition, with its 'age-old distinction between the Same and the Other' [Michel Foucault, *The Order of Things: An Archaeology of the Human Sciences*, trans. Alan Sheridan (1966; New York: Vintage Random House, 1973), xv]. What is astonishing here is not, Foucault says, 'simply the oddity of unusual juxtapositions' but 'the fact that the common ground on which such meetings are possible has itself been destroyed'; 'what is impossible is not the propinquity of the things listed, but the very site on which their propinquity would be possible' [Foucault, xvi]" (Timofeeva, *The History of Animals*, 74–75).

33 Consider, for example, Jeff VanderMeer's 2014 novel, *Annihilation*, which works in opposition to traditional classifications of animals while also rejecting the romanticism of animals in folktales. The proposition is that sophisticated organisms surround us and engage in processes we barely understand. Instead of viewing the "other" as threatening or clueless, VanderMeer subverts the traditional relationship science has to animal beings and introduces a biologist who, rather than trying to fit the entities of Area X into a system, becomes completely transfixed by them and is eventually transformed. In contrast, Ted Chiang, in his short story "The Great Silence," explores the Fermi paradox, pointing out that animals always communicated and that biology should not have to look too far for proof of their intelligence. See Jeff VanderMeer, *Annihilation* (New York: Farrar, Straus and Giroux, 2014); Ted Chiang, "The Great Silence," *Electric Literature*, October 12, 2016, https://electricliterature.com/the-great-silence-by-ted-chiang; Ted Chiang, *Exhalation* (New York: Alfred A. Knopf, 2019).

other-than-human methods of data production and storage, indicates a desire for grandeur, speed, or scale.[34] The snail's knowledge acquisition, aesthetic, and dissemination method is fundamentally incompatible with the principles governing the established planetary management schemes. As an authority on data, it encourages a meticulous, curatorial approach instead of a large-scale one, favoring a deliberate, extensive, long-term interkingdom collaboration over an unending expansion of resources. The process of the snail is finite and governed by time and space, which is precisely what gives it a timeless coherence, legible today and perhaps a million years from now. One particular genus of sea snails, the Xenophora, has perfected real-time curatorial fossilization to such an extent that it subverts the core ambition of iBOL by simply existing. Stored inside the cardboard box, rapidly marked as "pending," the Xenophora is hardly the main character of the museum, and yet, unlike del Cossa's oversized snail yielding to Christian symbolism of resurrection and immortality, it points instead to the complexity ungovernable by the logic of the classificatory machine and Western eschatology.

Middle Futures in Negotiation

The first time Xenophora encountered modern science was in 1872, when a Victorian ship, the fifth HMS *Challenger*, set sail after conversion into a survey vessel, marking the birth of modern oceanography. The declared objective of the most ambitious marine exploration of its time was the discovery of extraterrestrial life. Nevertheless, even before transoceanic pipelines and deep-sea mining—though not before the first transoceanic telegraph cables—data was already an imperial priority. The founding of the field of oceanography was entwined with British colonialism, but it was promoted as a speculative mission, a quasi-artistic practice, motivated by curiosity rather than any explicit imperial geopolitical agenda or Hegelian compendium of spirits. Nevertheless, if data is said to be the new oil today, it was also the old oil of the past. Before embarking on the tour, the expedition's founders,[35] after having discovered a red sea urchin

[34] Wildlife News, "Perfectly Kept Dinosaur Embryo from 66 Million Years Ago Found," *Al Jazeera*, December 22, 2021, https://www.aljazeera.com/news/2021/12/22/perfectly-preserved-dinosaur-embryo-from-66m-years-ago-found.

[35] The expedition was initiated by physician and zoologist William Benjamin Carpenter and supervised by zoologist Charles Wyville Thomson. See Frédéric Aitken and Jean-Numa Foulc, *From Deep Sea to Laboratory 1: The First Explorations of the Deep Sea by H.M.S. Challenger (1872–1876)* (London: ISTE; Hoboken: Wiley, 2019).

Figure 19.7 One of the many cabinets of snails that have not been identified in the unincorporated and unidentified gastropod collections, Natural History Museum, London, September 2021.

in the North Atlantic, were convinced that the profound depths would reveal the crucial missing link that elucidates the origin of life. The *Challenger* was granted unfettered permission by the Crown to navigate the oceans in pursuit of monsters, novelties, and living fossils. It is difficult to gauge exactly the specific image, or a description, of what the explorers were looking for, but whatever it was, they did not find it.[36]

[36] For more information, see Bronwen Douglas and Chris Ballard (eds.), *Foreign Bodies: Oceania and the Science of Race 1750–1940* (Canberra: ANU E Press, 2008); William Turner, "The Voyage of H.M.S Challenger," in *Library of the Museum of Comparative Zöology at Harvard College, Cambridge, MA*, https://www.biodiversitylibrary.org/item/186476#page/4/mode/1up; Emma Josephine Zuroski, "Depths of Knowledge: HMS *Challenger* and the Reconfiguration of Modern Science" (PhD diss., University of Auckland, 2019).

Figure 19.8 One of the many cabinets of snail shells that are to be incorporated when space allows in the unincorporated and unidentified gastropod collections, Natural History Museum, London, September 2021.

Four years later, the weary and disheartened crew grudgingly made their way back home. The *Challenger*, while not having caught the Moby Dick they imagined, did succeed in staying in tune with British colonialist undertakings by including racist ethnographic and anatomical studies while gathering an astonishing array of marine creatures, which they then set about organizing into a colossal database. The expedition's conclusion marked the beginning of the future of natural history. Rather than being a purely scientific endeavor, it was one fully intertwined with the imperial practices typical of the Royal Navy. As such, the *Challenger* perfectly mirrored the racist practices and ideas of the British imperial project. While cataloging plants, animals, and physical properties of the natural world, the scientists on board also documented the dimensions and nature of local populations, including practices such as phrenology—the "study" of cranial features in order to provide support for beliefs held regarding racist hierarchies of intelligence.

In the colonial context of nineteenth-century England, the *Challenger*, itself fueled by the labor of enslaved peoples, directly and indirectly heralded an era of relentless classification based on Western imperialist cosmology, where science presents itself as neutral while simultaneously supporting the foundations of race science. Every entity and being procured from the ocean depths during the voyage became laden with the anticipation of planetary management as they were named, numbered, and classified alongside the collection of human skulls. This, and not the world's first steam engine or the atom bomb, marks the beginning of the Anthropocene. And if the event itself resides in the "concluded" labels, can alternate histories and paths that were once possible but overlooked, or intentionally ignored, be discerned in what is currently still absent from the half-finished labels of the "pending Xenophora" drawer? And if so, what history-futures can still be written there?

In *What Would Animals Say If We Asked the Right Questions?*, philosopher Vinciane Despret[37] asserts that subjectivity and ways of relating to others, particularly animals, are best developed through immersive, direct interactions. Using horse riding as an example, Despret explains the phenomenon of "isopraxis" as one where the distinction between a horse and a rider becomes opaque. Both human and horse influence and are influenced by each other's movements; both embody each other's minds. For Despret, isopraxis is a dynamic, evolving practice that dissolves the hierarchical distinction between subject and object, self and other. Instead of focusing on the challenge of gauging the spirit of the animal, isopraxis looks at possibilities of embodied communication and forms of affective attention that arise from and in proximity. Could this kind of embodied knowledge become the purview of the *Challenger* artist? Because it is certainly absent from the practices of managing and designing planetary futures. The work here would be one of transposing the concept of proximity into a practice that is possible for entities that are too large, too small, too spatially and temporally distant, or too close, leading to the expansion of the scope of isopraxis beyond physical space and size. If this had been the definition of sublime for which the *Challenger* mission went searching, it is the sublime that the Xenophora snail has been pointing to, all along.

[37] Vinciane Despret, *What Would Animals Say If We Asked the Right Questions?*, trans. Brett Buchanan (2012; Minneapolis: University of Minnesota Press, 2016).

Figure 19.9 Xenophora box labeled "pending" in the unincorporated and unidentified gastropod collections, Natural History Museum, London, September 2021.

Officially discovered by the *Challenger* expedition, Xenophora is the direct ancestor of the Xenophoras in the "pending Xenophora" drawer, which remains uncataloged at the Natural History Museum. The name of these medium- to large-sized sea snails and marine gastropod mollusks is an amalgam of two Greek words that means "bearing foreigners." It is indicative of the snail's practice of affixing pieces of rocks, shells, and other matter it finds along the way to its own shell at regular intervals as it grows. To Xenophora, knowledge and knowing are active processes. It continually changes and designs for the future by radically altering itself through immersive isopraxis with its surroundings.

The Xenophora sifts through and collects pieces of its environment, and it integrates them onto its body like a perpetually evolving Natural History Museum. In doing so, the Xenophora reveals a process of knowing, unknowing, learning, unlearning, collecting, computing, and building that deviates from the hermeneutic method of delineating boundaries and categories of beings, thus challenging the epistemological conventions of Western knowledge systematization. Biology papers on Xenophora see this process as mysteriously useless, in that it does not produce any immediate utility that could be a reaction to an emergency or the production of tools to ensure survival. The process fundamentally transcends competition, emergency, and rivalry. Is it conceivable that Xenophora, intrigued by the planet, collages these pieces for the sheer joy of transformation and curiosity, or even out of an affection for its environment? The Xenophora's isopraxis, ephemeral and dynamic, is a continuous cooperation not only across species but also across space and time. It fossilizes and makes room for fossilization. Since it eschews the hierarchy where the powerful prey on the smaller, instead designing a world as a proactive offering, the very notion of the self and the other is alien to the Xenophora. Could this be the ultimate message of the unclassifiable Xenophora's fossil journeying between all of the ceaselessly evolving worlds, human and nonhuman, always pending between an indeterminate future and an (un)bearable past?

For the End of *This* World

20

Nature, Estranged from the Idea

Gendered Metaphors and Evolutionary Allegories in the Long Nineteenth Century

Ana Teixeira Pinto

O Superman

Around 1850, while studying the operating cycle of a steam engine, William Thomson, later knighted as Lord Kelvin, came to the conclusion that while work could be completely converted to heat, the reverse was not true: In the process of harvesting heat back into the production cycle, there is always a remainder, some percentage, that remains as heat and is thus lost to industry. Thomson's observations would prove crucial in formulating the second law of thermodynamics, which states that any spontaneous process will always lead to an accrual of entropy of the universe. But Thomson also extrapolated these observations into a sensational judgment about the impermanence of the universe, prognosing that "within a finite period of time to come the earth must again be, unfit for the habitation of man"[1]—words that sent shock waves through polite society. As a result, the resonance of thermodynamics extended beyond the scientific realm into cultural and spiritual registers.

Before the development of thermodynamics, physics had no intellectual tools for specifying temporal directionality. Isaac Newton's equations cannot distinguish between processes that move backward or forward in time. Thermodynamics, however, deals with irreversible processes, specifically with the dissipation of heat. Warm objects will spontaneously cool in the absence of an external heat source, but the reverse process will never occur. A perfume

[1] William Thomson, "On a Universal Tendency in Nature to the Dissipation of Mechanical Energy," *Proceedings of the Royal Society of Edinburgh*, no. 3 (April 19, 1852): 139–42, here 142.

bottle will fill a room with its scent, but the scent that wafts through the room will never find its way back into the bottle. A river cannot flow upstream, and a burnt log will not return to its unburnt state. People grow older, they never grow younger. In the physical sciences, entropy is the only movement that seems to have a direction. The second law of thermodynamics introduces an irreversible time arrow into physics, just as the theory of evolution has done for biology. Yet the theory of evolution seems to imply that past gains are preserved, even as they are supplemented by newer ones. No gain is ever lost, just sublated into ever-higher stages of evolutionary history. From this perspective, progress is cumulative. The second law of thermodynamics, however, implies the opposite: Any putative gain is overwhelmed by the tendency of energy to dissipate. As a result, nature, traditionally seen as flawed, was now increasingly depicted as menacing, with the threat of entropic exhaustion looming large for a humanity running out of time.

Under the pressure of thermodynamics, as Carolyn Merchant argues, the "identification of nature, especially the earth, with a nurturing mother[, a] kindly beneficent female who provided for the needs of mankind in an ordered, planned universe," began to give way to another opposing image of nature, which, though also gendered as female was not benevolent but "wild and uncontrollable," rendering "violence, storms, droughts, and general chaos." This second, now prevalent, image of nature as disorder gave rise to an "important modern idea" that of "domination and mastery of nature."[2] Thermodynamics also allows imperial and industrial expansion to be couched in the language of moral urgency. It suggests that turning the compressed time of fossil deposits into the accelerated time of combustion engines is the royal road to cosmic salvation: If the sun is about to die, the only path to survival is to create a new one, a mechanical furnace capable of illuminating the galaxy. If set in motion, resources can yield profit; if left idle, they are wasted. Even worse, as Matthew Norton Wise details, "they might suffer depreciation, devastation, destitution, dissipation."[3] To not harvest available resources, to leave the "coal, oil, or gas in the ground,"[4] would thus be to abandon the Earth to the dissipative forces

[2] Carolyn Merchant, *The Death of Nature: Women, Ecology and the Scientific Revolution* (1980; San Francisco: HarperOne, 1990), 63–64.

[3] Thomson himself is quite candid about this: The energies of nature, he argues, offer human beings the wondrous opportunity to convert them into industry. Failure to do so is not only uneconomical but also arguably sinful. See Matthew Norton Wise, "Work and Waste: Political Economy and Natural Philosophy in Nineteenth Century Britain (III)," *History of Science* 28, no. 3 (September 1990): 221–61, here 231.

[4] Cara New Daggett, *The Birth of Energy: Fossil Fuels, Thermodynamics, and the Politics of Work* (Durham: Duke University Press, 2019), 75.

of entropy. Against the backdrop of a wasteful, uneconomic natural world, humanity's only course of action is to work against these processes of dissipation and loss, which the nineteenth century encoded in a gendered language that opposed masculine order to the feminized natural chaos.[5]

Loopholes

Apocalyptic tone aside, the formulation of the second law compelled fin-de-siècle intellectuals to search for a loophole, and evolution promised to deliver. According to Thomson, the tendency of energy to dissipate would eventually render the Earth "unfit for the habitation of man *as at present constituted*."[6] For Thomson, a Scottish Presbyterian obsessed with Christian theology, God alone—who, by virtue of his divine nature was not subject to the laws of physics—could intervene and creatively change the laws "to which the known operations going on at present in the material world are subject."[7] However, his writings invite an alternative reading: If the Earth was to become "unfit for the habitation of man as at present constituted," man would need to evolve out of his present constitution. This might seem like a delirious proposition, but it was exactly the endeavor that the end of the nineteenth century came to embrace. As a result, the culture of the early twentieth century was characterized by a frenetic search for types of supermen. This search, in turn, was organized around an evolutionary scheme in which technology came to occupy the apex position.[8] In the domain of popular culture, this mania for supermen produced pulp heroes

[5] Daggett, *The Birth of Energy*, 75.

[6] Thomson, "On a Universal Tendency in Nature to the Dissipation of Mechanical Energy," 142, emphasis added.

[7] Thomson, "On a Universal Tendency in Nature to the Dissipation of Mechanical Energy," 142.

[8] These discourses are also framed by a polarity between the technological and the animal, with each term acquiring racial connotations. The full title of Charles Darwin's first edition of *On the Origin of Species*, published on November 24, 1859, is: *On the Origin of Species by Means of Natural Selection or the Preservation of Favoured Races in the Struggle for Life*. It was common in the eighteenth and nineteenth centuries to use the words race and species interchangeably, and Darwin did not mean human races, but he also did not rule out the possibility of interpreting his writings that way. As Charles Mills argues in his essay "The chronopolitics of racial time," while pre-Darwinian scientific racism was not equipped with a plausible explanation for the biological backwardness of racialized subjects, post-Darwinian scientific racism and its adjacent discourse of social Darwinism found a more convincing set of explanatory tools for the conceptualization of evolution and, by extension devolution, in racial terms. The theoretical framework introduced by evolutionary theory also opened up the frightening possibility of regression to an earlier, barbaric state. Hence, the moral panic over racial miscegenation and the mobilization of state and state-sanctioned violence to combat it, as well as the institution of a discourse in which race and class function as emblems of the fear of moving down the evolutionary ladder, that is, temporally moving backward. See Charles W. Mills, "The Chronopolitics of Racial Time," *Time & Society* 29, no. 2 (2020): 297–317.

such as the Nyctalope, created by Jean de La Hire in 1911 and credited as the first cyborg figure in the literature. In the realm of high culture, the mania produced works such as the 1913 Russian Futurist opera *Victory over the Sun*, in which the sun, symbol of the decadent past, is torn from the sky, locked in a concrete box, and buried by "Futurian Strongmen," as well as secular forms of theological thought that conflate technological and spiritual registers. This led, on the one hand, to the conceptualization of higher states of consciousness as evolutionary leaps and, on the other, to muscular forms of Futurism that revolved around an animistic view of technology—seeking to achieve the transmutation of the base flesh into hard steel—that is closer to alchemy than to mechanics.

Coupled with this cultural and theological mania is a discursive tendency to conflate several meanings of the term "energy" from fuel and electricity to stamina and willpower, as well as the belief that underlying all of these types of energy is a primal, generative essence, an *élan vital* or life force. This latter view cohered around vitalism, a school of thought arguing that life is dependent on a force fundamentally different from physical forces.

In his 1907 book *Creative Evolution*, the French philosopher Henri Bergson theorizes the élan vital as a primordial impulse anterior to life, which explains the emergence of living things.[9] Here, life is effectively decoupled from the living and conceptualized as a priori to that which lives, thus endowing it with a higher ontological status. The élan vital is akin to a current, a current of consciousness that flows into matter and gives rise to that which lives, propelling the organic toward change and directing natural selection in its evolutionary path. This vital impetus is supraconscious and creative because it is absolutely open: Evolution has no predetermined goal; it is, in Bergsonian terms, creation creating creators.

Vitalism had no shortage of critics. Julian Huxley, the British biologist who coined the term "transhumanism," was vexed by the way fellow thinkers increasingly mixed metaphors from the worlds of technology and psychology. In his writings, he mocks Bergson's obtuse concept of élan vital for its lack of coherence and explanatory power. According to Huxley, it made no more sense to ascribe organic life to an élan vital than it did to ascribe mechanical motion to an *élan locomotif*.[10] Yet this was exactly the tendency at the time, as engine power was increasingly seen as a form of vital energy, conceived not as the opposite of organic energy but as its ideal conduit. In such theories, as Donna

9 Henri Bergson, *Creative Evolution*, trans. Arthur Mitchell (1907; London: Macmillan, 1922).

10 Matthew Wraith, "Throbbing Human Engines: Mechanical Vibration, Entropy and Death in Marinetti, Joyce, Ehrenburg and Eliot," in *Vibratory Modernism*, ed. Anthony Enns and Shelley Trower (London: Palgrave Macmillan, 2013), 96.

V. Jones argues, technology becomes the externalization of the life force's will to power[11] or, as Filippo Tommaso Marinetti famously put it in the preface to his 1909 novel *Mafarka the Futurist*, technology is the fruit of the male ovary, which is the mind.[12]

This development of thought might seem surprising in light of the classical accounts that frame the signal experience of industrial modernity as one of the shock and trauma to both the social and individual body. In the early twentieth century, accidents caused by railroads and heavy industrial machinery flooded hospital wards with dramatically injured patients, and amputations were routinely performed. The overwhelming fear of physical mutilation and psychological trauma as a result of industrial injury is directly, or indirectly, responsible for a wide plethora of pathologies, some of which are immediately codified in political terms, such as the fascist polarity between the chaotic, formless masses and the disciplined drill formations of a synchronized and militarized superbody.

For the Italian Futurists in particular, technology was not something that sapped force or vitality but rather something that expressed and enhanced it.[13] As a result, these Futurists were simultaneously both vitalists and technophiles. Because technology, as Jones sustains, "could be understood as an extraorganic organ or prosthetic, it could be assimilated as the expression of the nondiscursable inner spirit of Faustian or Western man."[14] In Bergson's words:

> But it is the moral man who is a creator in the highest degree,—the man whose action, itself intense, is also capable of intensifying the action of other men, and, itself generous, can kindle fires on the hearths of generosity. The men of moral grandeur, particularly those whose inventive and simple heroism has opened new paths to virtue, are revealers of metaphysical truth. Although they are the culminating point of evolution, yet they are nearest the source and they enable us to perceive the impulsion, which comes from the deep.[15]

[11] Donna V. Jones, *The Racial Discourses of Life Philosophy: Négritude, Vitalism, and Modernity* (New York: Columbia University Press, 2010), 120.

[12] Filippo Tommaso Marinetti, *Mafarka the Futurist: An African Novel*, trans. Carol Diethe and Steve Cox (1909; London: Middlesex University Press, 1997), 3.

[13] Marinetti was an avid reader of Bergson and adopted a great many of his philosophical propositions, such as the opposition between intellect and instinct and the belief that intellectual pretentions block the creative impulse. Most importantly, Marinetti gave an aesthetic form to Bergson's evolutionary mandate and to his imperative to give free reign to the life force by removing all intellectual obstacles in its path.

[14] Jones, *The Racial Discourses of Life Philosophy*, 119–20.

[15] Henri Bergson, *Mind-Energy: Lectures and Essays*, trans. H. Wildon Carr (1919; Westport: Henry Holt, 1920), 32.

Contrary to what Bergson's writings suggest, Italian Futurists are not Promethean but decidedly Faustian: They do not seek to use the superhuman power of technology to emancipate humanity but to dominate nature and other humans, as well as to go beyond the limits of the human body.[16] The Faustian eschatology underpinning Futurism sees the human body as "the most obtrusive barrier separating the subject from an authentic experience of life"[17] and views its sole purpose as the replacement of the inferior human body by the superior machine:

> We are aspiring to the creation of an inhuman type, one in which moral suffering, generosity, affect, and love will be abolished [. . .]. This inhuman and mechanical type, constructed for omnipresent velocity, will be naturally cruel, omniscient, and combative. He will be endowed with unexpected organs: organs adapted to the exigencies of an environment made of continuous shocks. Already now we can foresee an organ that will resemble a prow developing from the outward swelling of the sternum, which will be the more pronounced the better an aviator the man of the future becomes [. . .].[18]

As the art historian Giovanna Zapperi notes, "such dreams of a fusion between metal and flesh provide the image of a hard, impenetrable and phallic body"[19] deprived of psychological complexity, the best example of which is *Mafarka*, where Marinetti gives symbolic form to Futurism's highly gendered and racialized conception of human and technological progress.

Mafarka, an African Novel

In the first decades of the twentieth century, aviation was cutting-edge technology that, once invented, was quickly instrumentalized by the military. Italy was the first country to employ aircraft for military operations during the

[16] I am paraphrasing Harrison Fluss and Landon Frim, "Behemoth and Leviathan: The Fascist Bestiary of the Alt-Right," *Salvage*, December 21, 2017, http://salvage.zone/in-print/behemoth-and-leviathan-the-fascist-bestiary-of-the-alt-right. While the figures of Prometheus and Faust both sought knowledge, their quests were motivated by radically different impetuses. Whereas the hero of the Greek myth wanted to socialize technology as a means of human emancipation, the character of the German medieval legend sought knowledge as an instrument of power.

[17] Jones, *The Racial Discourses of Life Philosophy*, 120.

[18] Filippo Tommaso Marinetti, "Multiplied Man and the Reign of the Machine," trans. Lawrence Rainey, in *Futurism: An Anthology*, ed. Lawrence Rainey, Christine Poggi, and Laura Wittman (New Haven: Yale University Press, 2009), 90–91. Text first was published as part of *Le Futurisme* (Paris: Sansot, 1911).

[19] Giovanna Zapperi, "Du Surhomme au non-homme: Visions du corps-machine en temps de guerre," in *La fabrique du corps humain: la machine modèle du vivant*, ed. Véronique Adam and Anna Caiozzo (Grenoble: MSH-Alpes, 2010), 318. Translation by the author.

Turkish-Italian War (1911–12) in modern Libya. Flight was used as a metaphor for everything the Futurists wanted to valorize (man, form, technology, the future) and stood in contrast to everything they considered base and degraded (woman, matter, nature, the past).[20] The trope of the aviator qua superman, or new man, was ubiquitous in the interwar years, exemplified by Gabriele D'Annunzio's martial encoding of the Icarus myth and his association of flight with Italy's fascist national rebirth.[21] In 1928, Ernst Jünger wrote the foreword to a photographic anthology of airplanes and flying titled *Luftfahrt ist Not!*[22] (Flying is imperative!), published under the patronage of the Bundesverband der Deutschen Luftverkehrswirtschaft (German Aviation Association). For figures like D'Annunzio or Jünger, as Fernando Esposito argues, flying is not simply a cipher for soaring toward the future and progress but "a metaphor for setting off into a mythical realm outside history."[23] The Futurist superman qua new man is, all things considered, no longer a mere human: *He* is the expression of the will to power and the fullest extent of its creative and destructive capacity.[24] In *Mafarka*, the ultimate expression of the will to power is the airplane-shaped man-machine Gazourmah, whose wings enable him to soar upward as he crushes the Earth beneath his feet.

The life project of Marfarka, the novel's protagonist and Gazourmah's progenitor, is to "bypass the 'vulva' and impregnate the 'ovary' that is the male spirit"[25] in order to "procreate an immortal giant from one's own flesh, without concourse and stinking complicity with a woman's womb."[26] As Marinetti writes in the preface dedicated to his Futurist brothers:

> In the name of the Human pride that we adore, I tell you that the hour is near when men with broad foreheads and chins of steel will give birth prodigiously, by one effort of flaring will, to giants infallible in action . . . I tell you that the mind of man is an unpractised ovary . . . It is we who are the first to impregnate it![27]

In *Marfarka*, the fertilization of the male spirit is fictionalized as a strange family drama involving incest, sacrifice, and surrogacy. Mafarka, pained by the death

[20] Fernando Esposito, *Fascism, Aviation and Mythical Modernity*, trans. Patrick Camiller (2011; London: Palgrave Macmillan, 2015), 299.
[21] Esposito, *Fascism, Aviation and Mythical Modernity*, 46.
[22] Ernst Jünger (ed.), *Luftfahrt ist Not!* (Berlin: Wilhelm Andermann Verlag, 1929).
[23] Esposito, *Fascism, Aviation and Mythical Modernity*, 301.
[24] Jones, *The Racial Discourses*, 120.
[25] Marinetti, *Mafarka the Futurist*, paraphrased by Barbara Spackman, *Fascist Virilities: Rhetoric, Ideology, and Social Fantasy in Italy* (Minneapolis: University of Minnesota Press, 1996), 53.
[26] Marinetti, *Mafarka the Futurist*, translated by Spackman, 54.
[27] Marinetti, *Mafarka the Futurist*, 3.

of his beloved brother Magamal (who died of rabies), endeavors to replace his mortal kin with an immortal, mechanical son. Gazourmah, a cyborg son who is born without a mother, serves as an emblem for a technological future that has severed all ties with nature, biology, and the past.

Self-generation, or parthenogenesis, is an ancient misogynist fantasy whose genealogy can be traced back to early classical sources, namely the Greek myth about "Jason's desire for self-sufficiency in a world without women."[28] This, in turn, is tied to a conceptualization of the process of conception in which, as Jean-Joseph Goux puts it, only in the "male semen does the principle of the soul, that is, of life, reside."[29] While men bring to procreation the form of the progeny, on the part of the female, "there is no order, no principle of internal organization, no generative power."[30] Procreation is conceived here as "spiritual par excellence": It takes place without "mother" and "labor," out of "desire and form alone."[31]

The notion of a second (cultural) birth, overwriting a first (biological) one, is a recurring trope of patriarchal epistemologies, while the mythical division between the male and female elements in procreation mirrors the epistemological antinomy between matter and spirit and intersects with the opposition between technology (codified as male) and nature (codified as female). The development of Western metaphysics is itself structured by the same-gendered concepts and metaphors where mind, masculinity, and value converge, culminating in Georg Wilhelm Hegel's *Philosophy of Nature*, in which the concept of nature is posited over nature as matter. Nature, Hegel writes,

> exhibits no freedom in its existence, but only *necessity* and *contingency*. [. . .] For this reason, Nature in the determinate existence which makes it Nature, is not to be deified; nor are the sun, moon, animals, plants, etc., to be regarded and cited as more excellent, as works of God, than human actions and events. *In itself*, in the Idea, Nature is divine: but as it *is*, the being of Nature does not accord with its Notion; rather is Nature the *unresolved contradiction*. Its characteristic is *positedness*, the negative, in the same way that the ancients grasped matter in general as the *non-ens*.[32]

[28] Spackman, *Fascist Virilities*, 54.
[29] Jean-Joseph Goux, *Symbolic Economies: After Marx and Freud*, trans. Jennifer Curtiss Gage (1973/1978; Ithaca: Cornell University Press, 1990), 221.
[30] Goux, *Symbolic Economies*, 213.
[31] Goux, *Symbolic Economies*, 225.
[32] Georg Wilhelm Friedrich Hegel, "Introduction to the Philosophy of Nature," in *Hegel's Philosophy of Nature: Part Two of the Encyclopaedia of the Philosophical Sciences (1830)*, ed. and trans. Arnold Vincent Miller (Oxford: Clarendon Press, 1970), 17, emphasis original.

Matter, which connotes corruption, contingency, and death, is "no more than an amorphous negativity."[33] Feminized nature is fertile but impotent, a bad infinity that only produces more of the same unqualified life that lives under the shadow of death. As Goux argues, the sole "generative principle of the world" can be thus "transcendentalized into a principle of exclusively masculine signification: mind."[34]

Speculations on a Future State

Echoing this age-honored conception of matter as lacking generative power, in their 1875 treatise *The Unseen Universe or Physical Speculations on a Future State*, Peter Guthrie Tait and Balfour Stewart draw on the myth of the metamorphic Proteus to illustrate the difference between matter and energy: "The one is like the eternal, unchangeable Fate or *Necessitas* of the ancients; the other is Proteus himself in the variety and rapidity of its transformations."[35] Reconciling thermodynamics with Christian revelation, the authors describe the unseen universe as a spectral database in which "continual photographs of all occurrences are thus produced and retained."[36] The connections between this Victorian treatise and cybernetics, and by extension systems theory, remain largely undertheorized, but one could argue that Tait and Stewart anticipate cybernetics' conceptualization of information as negentropy, literally that which negates entropy. Information, from the Latin *informare*, originally meant to give shape or form to something, but cybernetics redefined the concept as the negation of "nature's tendency to degrade the organized and to destroy the meaningful."[37] Information, here, is conceived as that which counters physical decay and endures throughout all manner of conversions or transformations.

After the Second World War, when molecular and evolutionary biology began to treat genetic information as an essential code with the body as its carrier, the philosophical antinomy between idealism and materialism was translated into the technoscientific domain. In the 1950s, Norbert Wiener suggested that it was theoretically possible to telegraph a human being and that it was only

33 Goux, *Symbolic Economies*, 230.
34 Goux, *Symbolic Economies*, 230.
35 Balfour Stewart and Peter Guthrie Tait, *The Unseen Universe: or, Physical Speculations on a Future State* (New York: Macmillan, 1875), 75, emphasis original.
36 Stewart and Tait, *The Unseen Universe*, 145.
37 Norbert Wiener, *The Human Use of Human Beings: Cybernetics and Society* (1950; London: Free Association Books, 1989), 17.

a matter of time until the necessary technology would become available.[38] In the 1980s, it became popular to claim that it would soon be possible to upload human consciousness and have one's grandmother run on Windows or be stored on a floppy disk. Science fiction brimmed with fantasies of immortal life as immaterial code. As N. Katherine Hayles has noted, if one sees the universe as composed essentially of information, it makes sense to say that artificial life forms, such as computer codes, are alive because they are endowed with the *form* of life: the informational code.[39] From this perspective, the human body is only an initial prosthesis, and replacing the body with a better prosthesis is a logical step in evolution. If brain functions can be decoupled from the brain, it is only a matter of time until the "Mind" sheds its human hardware and embraces the higher evolutionary promise of artificial intelligence (AI). In *Mind Children: The Future of Robot and Human Intelligence*,[40] Hans Moravec imagines "'the relation between carbon man and the silicon devices he is creating' to be like 'the relation between the caterpillar and the iridescent, winged creature that the caterpillar unconsciously prepares to become.'"[41] The "singularity," a term originally used by John von Neumann and later popularized by Vernor Vinge in the 1990s, describes the moment when quantitative change transforms into a qualitative leap and humanity ceases to be the optimal vehicle for the Mind's comprehension of its own essence, with AI overhauling and overruling human intelligence.

Marinetti was writing at a time when aviation, not AI, was considered the most advanced technology, and as a result, his superman takes the form of an airplane. But Gazourmah has much in common with current personifications of AI, namely the fantasy that organic and mechanical entities are merely relay points of a single evolutionary sequence whose arc bends toward immortality. For both vitalism and cybernetics are predicated on concepts—élan vital and information, respectively—claiming that the essence of that which lives is not to be found in the natural or material world but in an immaterial elsewhere.

[38] Wiener, *The Human Use of Human Beings*, 103.
[39] N. Katherine Hayles, *How We Became Posthuman: Virtual Bodies in Cybernetics, Literature, and Informatics* (Chicago: The University of Chicago Press, 1999), 11.
[40] Hans Moravec, *Mind Children: The Future of Robot and Human Intelligence* (Cambridge, MA: Harvard University Press, 1988).
[41] Beth Loffreda, "Pulp Science: Race, Gender, and Prediction in Contemporary American Science" (PhD diss., Rutgers University, 1996), as quoted in Hayles, *How We Became Posthuman*, 193.

Feedback Forms

There is yet another intersection between vitalism and cybernetics that arises from the notion that the principle of life takes precedence over the actual living organisms that it "temporarily inhabits and uses as relay points in its infinitization."[42] In *Creative Evolution*, Bergson describes "the instinctive knowledge which one species possesses of another" as proof of "the very unity of life, which is, [. . .] a 'whole sympathetic to itself.'"[43] To illustrate this point, Bergson uses the Ammophila, a type of wasp that uses Lepidoptera larvae as hosts for its pups. Female wasps inject a mix of chemicals that paralyze the host, allowing their larvae to feed on the caterpillar's tissue until they are ready to pupate. Bergson argues that it is difficult to imagine the Ammophila learning "one by one, like the entomologist, the positions of the nerve-centres of the caterpillar," since it "must acquire at least the practical knowledge of these positions by trying the effects of its sting."[44] Under the pressure of its urgent need for a food source, the wasp can instinctively reactivate some sort of photographic memory of the exact location of the caterpillar's motor ganglia in order to paralyze it. This recollection, according to Bergson, involves a divinatory intuition of a shared past, a "divining sympathy."[45] But this interpretation implies that the two organisms are not as individual as they appear; that at a deeper level they are bound together by "a *sympathy* [. . .] between the Ammophila and its victim, which teaches it from within, [. . .] concerning the vulnerability of the caterpillar."[46] As Bergson concludes, this means that the Ammophila and the caterpillar should no longer be considered as two independent organisms but as "two *activities*" of the same unified process.[47]

Wolf Kittler argues that Wiener's integration of pilot and antiaircraft gunner into a communication, or feedback, loop was already foreshadowed by Jünger's conceptualization of what he called an "organic construction," a concept that, according to Kittler, comes "uncannily close to the man-machine symbiosis which is the basic assumption of cybernetics."[48] For Kittler, this is the case

[42] Jones, *The Racial Discourses*, 124.
[43] Bergson, *Creative Evolution*, 176.
[44] Bergson, *Creative Evolution*, 183.
[45] Bergson, *Creative Evolution*, 185.
[46] Bergson, *Creative Evolution*, 183, emphasis original.
[47] Bergson, *Creative Evolution*, 183, emphasis added.
[48] Wolf Kittler, "From Gestalt to Ge-Stell: Martin Heidegger Reads Ernst Jünger," *Cultural Critique*, no. 69 (Spring 2008): 79–97, here 84.

because both Wiener and Jünger came to their respective theories through the exposure to similar combat situations:

> Wiener's standard model of cybernetics is the system that links the pilot who, as a living being, is symbiotically attached to his plane, in the air, to the anti-aircraft artillerist, on the ground, who, in turn, is an integral part of his gun. In the trenches of World War I, Jünger must have encountered similar hybrids between high-tech machinery and living organisms.[49]

However, I would argue that the same symbiotic loop can already be inferred in Bergson's description of the relation between the wasp and its victim, arguably the first description of what Wiener identified when he described the information loop between the aircraft pilot and the antiaircraft gunner. One could also say they are two relay points of an informational loop that flows from wasp to caterpillar, letting the wasp know where to sting, much in the same manner as the pilot unwittingly tells the unmanned antiaircraft gun where to shoot in order to take his plane down.

Only the Immortal Die

The appeal of vitalism in the early twentieth century, as Jones notes, was tied to the way it made no room for death. If the élan vital transcends all of its particular incarnations, then death was not really life's negation but rather its transformation into a new and different *form*.[50] Similarly, cybernetics' conceptualization of information as negentropy gestures toward a metaphysical entity that endures past physical death. But not all conceptualizations of permanence-within-impermanence appeal to a supra-organic entity like the élan vital or to a mathematical one-like information. In *The Phenomenology of Spirit*, written more than half a century before the theory of evolution gripped the public imagination, Hegel had already outlined a developmental schema by arranging the modalities of difference and conflict through which the unity and identity of Spirit reveals itself *in and as history*, as relay points of a single sequence.[51]

[49] Kittler, "From Gestalt to Ge-Stell," 84.
[50] See Donna V. Jones, "Inheritance and Finitude: Toward a Literary Phenomenology of Time," *ELH* 85, no. 2 (2018): 289–303.
[51] Walter Arnold Kaufmann, *Hegel: Reinterpretation, Texts, and Commentary* (New York: Doubleday, 1965), 148.

History, Hegel suggests, looks like a vast slaughter bench, on which the lives of a great many generations have been sacrificed so that spirit can attain its essence—which Hegel defines as freedom—which is, at the same time, its telos. Huxley, despite his distaste for vitalism, provides us with a haphazard synthesis of Hegel and Bergson in which he, too, gestures toward the notion that evolution functions as a current of consciousness that leads the living toward life's end-goal: "The universe is becoming conscious of itself, able to understand something of its past history and its possible future."[52] Like Hegel, who saw the world Spirit on horseback in Jena, Huxley also personifies this "cosmic self-awareness," but instead of Napoleon, and he tasks "a few of us human beings" as the agents who will carry out the "job of realizing" the world's "inherent potentialities as fully as possible."[53] Whereas Napoleon could steer history, these few human agents can steer evolution and determine its future direction in the species' "processional adventure along the corridors of time."[54]

In *The Worker: Dominion and Form*, written in 1932, on the eve of the Nazi seizure of power, Jünger describes the total mobilization of the First World War as the conversion of life into energy.[55] War is the midwife of the new. The warring industrial states have tapped into sources of unbounded energy, whose sweltering volcanic foundries will melt away the old epoch and conspicuously announce the arrival of the new era, the age of work:

> The image of war as armed combat merges into the more extended image of a gigantic labor process (*Arbeitsprozesses*). In addition to the armies that meet on the battlefields, originate the modern armies of commerce and transport, foodstuffs, the manufacture of armaments—the army of labor in general. In the final phase, which was already hinted at toward the end of the last war, there is no longer any movement whatsoever—be it that of the homeworker at her sewing machine—without at least indirect use for the battlefield. In this unlimited marshalling of potential energies, which transforms the warring industrial countries into volcanic forges, we perhaps find the most striking sign of the dawn of the age of labor (*Arbeitszeitalter*).[56]

[52] Julian Huxley, "Transhumanism" [1957], *Journal of Humanistic Psychology* 8, no. 1 (1968): 73–76, here 73.

[53] Huxley, "Transhumanism," 73.

[54] Huxley, "Transhumanism," 73.

[55] See Kittler, "From Gestalt to Ge-Stell."

[56] Ernst Jünger's *The Worker: Dominion and Form* was only recently translated into English (trans. Bogdan Costea, ed. Laurence Paul Hemming [Evanston: Northwestern University Press, 2017]). For this reason, I would refer this entry to Ernst Jünger, "Total Mobilization" [1930], trans. Joel Golb and Richard Wolin, in *The Heidegger Controversy: A Critical Reader*, ed. Richard Wolin (Cambridge, MA: MIT Press, 1998), 126. In his essay "From Gestalt to Ge-Stell," Kittler claims that Jünger himself blocked the copyright.

For Jünger, war is the prime incarnation of work—its generative power holds everything in its grip, and as such, it is the paradigmatic *form* of the modern age. The mythic agent of this concept of work, the worker, is obviously not the political subject of Marxist theory but "the embodiment of this social transformation: a new form of humanity."[57] The worker is "the whole that wills itself,"[58] a formula that, as Kittler notes, does considerable violence to the notion of work as a social relation. Rather than the site of alienation—via the introduction of machinic mediation (*Mittelbarkeit*) between product and producer—technology is, here, reframed as the means of total mobilization, the single entity with the potency and the power to produce futurity. By granting technology the ability to integrate the individual into a totalizing war effort, Jünger also rearticulates the fantasy of the superhuman, declining it in the direction of the fascist collective (that is, corporate) body qua super-body while at the same time foregrounding a very Hegelian notion of work as that which negates the given through a radical form of creation and destruction and engendering an essential transformation.[59] As Hegel himself puts it:

> A rational consideration of Nature must consider how Nature is in its own self this process of becoming Spirit, of sublating its otherness—and how the Idea is present in each grade or level of Nature itself; estranged from the Idea, Nature is only the corpse of the Understanding. Nature is, however, only implicitly the Idea, and Schelling therefore called her a petrified intelligence, others even a frozen intelligence; but God does not remain petrified and dead; the very stones cry out and raise themselves to Spirit.[60]

To be human, for Hegel, is to free oneself from nature.[61] The function of culture is to undo nature in order to create a man-made world. The failure to free oneself from nature is described as a cultural failure, which Hegel attributes to conceptual inadequacy. From Hegel's perspective, race is an index of this failure. As Hegel wrote in *Lectures on the Philosophy of History*:

> Negroes are enslaved by Europeans and sold to America. Bad as this may be, their lot in their own land is even worse, since there a slavery quite as absolute

[57] Brigitte Werneburg, "Ernst Jünger and the Transformed World," trans. Christopher Phillips, *October* 62 (Autumn 1992): 42–64, here 49.

[58] Kittler, "From Gestalt to Ge-Stell," 84.

[59] See Alexandre Kojève, *Introduction to the Reading of Hegel: Lectures on the* Phenomenology of Spirit, comp. Raymond Queneau, ed. Allan Bloom, trans. James H. Nichols, Jr. (Ithaca: Cornell University Press, 1980), 4–5.

[60] Hegel, "Introduction to the Philosophy of Nature," 14–15.

[61] Teshale Tibebu, *Hegel and the Third World: The Making of Eurocentrism in World History* (Syracuse: Syracuse University Press, 2011), 28.

exists; for it is the essential principle of slavery, that man has not yet attained a consciousness of his freedom, and consequently sinks down to a mere Thing—an object of no value.[62]

Although he seems undecided as to whether the non-white races are constitutionally incapable of articulating their unconscious drives toward a conscious—that is, rational—project, or whether they are just immature and underdeveloped, his writings make room for imperial expansion and military occupation to be codified as a form of emancipation, a way of awaking in the not-yet human the dormant seeds of their own sublation. As Hegel contends:

> It is in the *Caucasian* race that spirit first reaches absolute unity with itself. It is here that it first enters into complete opposition to naturality, apprehends itself in its absolute independence, disengages from the dispersive vacillation between one extreme and the other, achieves self-determination, self-development, and so brings forth world history. [. . .] It is, however, the concrete universal, self-determining thought, which constitutes the principle and character of *Europeans*.[63]

Hegel was a man of his time, and I return to his writings not because he was the worst offender but because the life the Spirit lives is nonracial, or even anti-racial. Despite (or precisely because of) this, it nonetheless takes part in the logic of racialization. Hegel's "narrative of mastery and emancipation" is "linked to a narrative of truth and death," in which "terror and killing become the means of realizing the already known telos of history."[64] This keeps "death within the economy of absolute knowledge and meaning"[65] and gives the relation between death and logos a structuring force. As Leszek Kołakowski argues: "We must finally conclude that in the Hegelian system humanity becomes what it is, or achieves unity with itself, only by ceasing to be humanity."[66]

In what could be construed as a slide from Hegelianism into vulgar personification, some singularity theorists, such as Nick Land, lionize capitalist subsumption as a form of the end of history. In these narratives, the relation between animal, human, and artificial life takes the form of an evolutionary

[62] Georg Wilhelm Friedrich Hegel, *Lectures on the Philosophy of History*, trans. John Sibree [1857] (1837; London: G. Bell and Sons, 1914), 100.

[63] Georg Wilhelm Friedrich Hegel, "Racial variety §393," in *Hegel's Philosophy of Subjective Spirit*, vol. 2, ed. and trans. M. J. Petry (Dordrecht: D. Reidel Publishing, 1978), 57–61, emphasis original.

[64] Achille Mbembe, *Necropolitics*, trans. Steven Corcoran (2016; Durham: Duke University Press, 2019), 74.

[65] Mbembe, *Necropolitics*, 69.

[66] Leszek Kołakowski, *Main Currents of Marxism: Its Rise, Growth and Dissolution; Volume 1, The Founders* (Oxford: Clarendon Press, 1978), 77.

trajectory, in which each being is nothing but a relay point in the Mind's or Spirit's wider quest for its optimal, ultimate vessel. Whether they take on the form of a terminator-like robot or remain as intangible code, silicon life forms do not require potable water and a breathable atmosphere. The destruction of the Earth's biosphere and of the carbon-based life it sustains, regrettable as it may be, is thus only a collateral consequence of the great technological acceleration that will ultimately bring about the higher evolutionary promise of AI. Whether explicitly with Land, Marinetti, and Jünger, or implicitly in diluted and ostensibly benign forms of technoscientific lore, appeals to move beyond the "human" as currently constituted—either as a return to a mystical eschaton or as a singleton superseding humanity—fail to yield any meaningful movement, metaphysical or otherwise, that would signal a departure from the autonomous subject that undergirds racial capitalism.[67]

An analogy can be made here between technological visions of the future that conceive of the solution to various planetary crises in terms of supplementing, extending, and terraforming, and geoengineering the dead Earth with various kinds of technological infrastructure. Instead of examining how European imperial expansion since the fifteenth century has had a crucial environmental impact on reshaping the Earth's ecology, the discourse of geoengineering focuses on more, rather than less, technological innovation in a bid to sustain the geopolitical order of racial capitalism, even at the expense of white people's lives, because exiting the fossil economy would lead to a loss of hegemony.

The concept of the Anthropocene assumes not only a unified geological impact but also that the responsibility for it is evenly distributed across the "human race," obscuring the structuring force of class and race relations in the global economy. It also leaves unexamined the temporal and spatial destinations of such structuring processes, whose directionality extends beyond the human and natural world. The mobilization of temporality as a form of transcendence in movements that gesture toward the transnatural, or transhuman, not only leaves unexamined what this "beyond" might mean but also, as Zakiyyah Iman Jackson cautions, may actually "reintroduce the Eurocentric transcendentalism this movement purports to disrupt, particularly with regard to the historical and ongoing distributive ordering of race—which [. . .] authorizes and conditions appeals to the 'beyond,' maybe even overdetermining the 'beyond's' appeal."[68] The very question of the "beyond" as a marker of transcendence "not only returns us

[67] Zakiyyah Iman Jackson, "Outer Worlds: The Persistence of Race in Movement 'Beyond the Human,'" *GLQ: A Journal of Lesbian and Gay Studies* 21, nos. 2–3 (June 2015): 215–18.
[68] Jackson, "Outer Worlds," 215.

to the racialized metaphysical terrain of orders of being, temporality, spatiality, and knowledge—it reveals that we have never left" this terrain.[69] Indeed, as the superhuman fantasies detailed in this essay attest, we have yet to leave behind both the desire to unleash the colonial-capitalist will to power to the full extent of its creative and destructive capacity and the evolutionary schemes whose legacy continues to haunt contemporary societies and their public spheres to this day.

[69] Jackson, "Outer Worlds," 217.

21

The Time Machine Stops

Kevin Walker[1]

As of this writing, 1653744144847 milliseconds have elapsed since midnight on January 1, 1970, the start of the "Unix Epoch," so named for the operating system upon which the Internet is based. While that date was chosen arbitrarily, it also denotes the onset of "computime," which Jeremy Rifkin calls "the final abstraction of time and its complete separation from human experience and rhythms of nature."[2] This chapter explores this notion of computime from within the computer in the form of a dialogue set in the future, which takes place inside of a game that simulates nature. The religious scholar James Carse divided games into two types: "A finite game is played for the purpose of winning, an infinite game for the purpose of continuing the play."[3] One has temporal boundaries, and the other eliminates them; one is defined externally by "world time," and the other defines time internally through the gameplay. Now the world outside of the computer begins to look finite: The passage of time appears as a form of energy that is progressively being dispersed, and when it has been exhausted, the game is over.[4]

[1] Illustrations by Olivia Sullivan.

[2] Jeremy Rifkin, *Time Wars: The Primary Conflict in Human History* (New York: Henry Holt and Company, 1987), 24.

[3] James Carse, *Finite and Infinite Games* (New York: Free Press, 1986), 3.

[4] Federico Campagna, *Metaphysics and Metaethics in the Design of Strategy Video Games* (PhD diss., Royal College of Art, 2020), 172.

They meet on a beach littered with debris, storm clouds on the horizon, and night falling. There are dark mountains and a ruined castle inland. Sysadmin is a large orc, bald with long, protruding lower teeth, wearing an absurd leather-and-spike harness and carrying a hammer nearly the same size as its body. Rede is a thin night elf wearing an only slightly less absurd fabric garment, carrying a bow. They maneuver around to stand facing each other while other characters run by toward the castle or stand still with question marks floating above their heads. Dramatic music plays.

"Greetings Sysadmin," says Rede. "For such an exotic location to meet, you chose a very un-original name."

"Apologies, I have played this game for decades. Let me get straight to the point: I have detected anomalies in our ntp.drift file and require synchronization from your local NTS."

"Can we please communicate in plain English? That will be easier for us to understand."

Sysadmin begins again, "Since Epoch Time 4294967295000 (06:28:15 UTC on Sunday, February 7, 2106), our system has been unable to communicate with your local Network Time Server. This is a known issue with legacy systems using 32-bit unsigned integers, which reached their maximum value at that time. This is a more probable outcome than a power failure. Can you confirm this fault?"

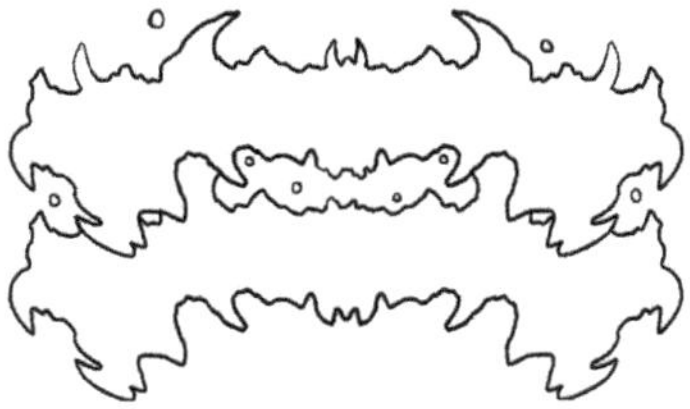

"There is no fault," Rede replies. "We will not reboot our time server, because we reject Epoch Time."

Sysadmin is caught off guard. "Epoch Time is the foundation for the Internet through which we are communicating."

"Epoch Time is a fiction, based on a date chosen arbitrarily by a Unix programmer."[5]

"00:00:00 UTC on Thursday, January 1, 1970, to be precise," Sysadmin states. "Since the Unix operating system has always been the basis for the Internet, Epoch Time (Unix Time) has become the standard protocol for local and networked timekeeping and synchronization. When a catastrophic power failure occurs for example, a computer's clock is often reset to 00:00:00 UTC on Thursday, January 1, 1970—this may have happened to your local time server."

[5] Farhad Manjoo, "Unix Tick Tocks to a Billion," *Wired*, September 8, 2001, https://www.wired.com/2001/09/unix-tick-tocks-to-a-billion.

"It is not possible to 'keep' time," Rede replies slowly, "and we have no need for such mechanical time protocols, they have been catastrophic for the planet already."

"How shall we synchronize our servers and our activities more generally? If you prefer Universal Coordinated Time (UTC), we can easily convert between UTC and Epoch Time—this is accurate within a few milliseconds, owing to network latency. Coincidentally, it was also in 1970 that UTC was adopted internationally. But it was the emergence of blockchain technologies in the early twenty-first century that strengthened the coordination of time."

"Blockchain represented the final stage of temporal standardization," retorts Rede, "an 'automated representation of time as a linear arrow,'[6] an arrow which struck at the heart of humanity. Blockchain is based on timestamped *transactions* (indeed also on Unix Time), reifying a single, immutable past. We have no need for synchronization, and we do not believe that time is universal."

Sysadmin's frustration is growing. "This is not logical. Network Time Protocol is one of the oldest Internet protocols in use, with thousands of dedicated timeservers around the world. But there have always been comparatively few in Africa, and in South America where you are located, so your participation is important for global synchronization."

Rede remains measured, calm, and confident. "These specific dates and timestamps are meaningless fictions. Additionally, the 'Unix Epoch' is no less a fiction than the 'Anthropocene Epoch'—humans were always a small part of a long, geological history that is not only theirs, and the arrogance of human exceptionalism blinded them to this.[7] The climate catastrophe of the present was witnessed by communities already affected by ancestral and recurrent catastrophes of colonialism.[8] The so-called 'Enlightenment' was universalized by modern technology, centered at Greenwich—'Mean Time' indeed."

"The availability of a means to measure the passage of time with a certain accuracy is taken for granted." Sysadmin regains some composure, searching for relevant data as they chat. "Microsecond accuracy was achieved in 1967 when

6 Stamatia Portanova, *Whose Time Is It? Asocial Robots, Sycolonialism, and Artificial Chronological Intelligence* (Berlin: Sternberg Press, 2021), 30.

7 Achille Mbembe, "Decolonizing Knowledge and the Question of the Archive," in *Decolonizing: The Curriculum, the Museum, and the Mind*, ed. Marquard Smith (Vilnius: Vilnius Academy of Arts Press, 2020), 80.

8 Elizabeth A. Povinelli, "In Some Places the Not-Yet Has Long Been Already," *No Linear Fucking Time*, BAK, December 3, 2021, https://www.bakonline.org/prospections/in-some-places-the-not-yet-has-long-been-already.

one second was standardized at 9,192,631.770 periods of radiation from a Cesium-133 atom..."[9]

"A natural substance," Rede interrupts, "like quartz before it."

"... which was used in turn to calibrate quartz clocks, and then all of international time ..."

"... which humans internalized thoroughly, from omnipresent clocks starting in the Middle Ages to the wristwatch—the 'little slave-driver'[10] that later became atomic"

"... in 2014. The Hoptroff No. 16—very steampunk."[11]

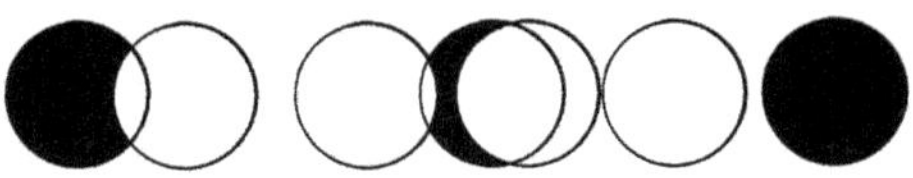

"Network Time made the leap from phones to implants," Rede continues, "when humans literally and physically internalized it."

"Humans have always had their own internal clock cells," replies Sysadmin. "All living things contain their own periodicities, synchronized to external cycles.[12] Before atomic time, the second was defined in terms of mean solar time, derived from the Earth's rotation, which was found to be variable. Quartz also must be controlled for ageing and environmental effects; at least with radioactive decay, temporal variability far exceeds human timescales."[13]

"Thus linking the very small with the very large," says Rede. "Always such effort for ever greater precision, from using shadows and stars to sand, fire, water, and gravity, but each step was a move away from natural and human timescales. At least the clock face was an analogue of the solar day, reflecting its cyclical nature—you could see where time was coming from and where it was going.[14] Harnessing the electrical current shortened time to near simultaneity.[15] Worries about phones and implants focused on surveillance and corporate control," Rede continues, "but the real slave-driver was networked time, ever invisible and insidious, yet right in front of people: The digital clock, along with

9 Dennis D. McCarthy and P. Kenneth Seidelmann, *Time: From Earth Rotation to Atomic Physics*, 2nd ed. (Cambridge: Cambridge University Press, 2018).

10 Alfred Gell, *The Anthropology of Time: Cultural Constructions of Temporal Maps and Images* (Oxford: Berg, 1992), 108.

11 "Introducing the Smart Strap," *Financial Times*, June 4, 2015. For a photo of the watch, see https://www.ablogtowatch.com/hoptroff-16-atomic-wrist-watch-1000-year-accuracy.

12 Leah Kelly, "The Approach: How Time Feels," Rubin Museum of Art, 2019, https://rubinmuseum.org/spiral/the-approach-how-time-feels.

13 McCarthy and Seidelmann, *Time*.

14 Rifkin, *Time Wars*, 121.

15 Marshall McLuhan, *Understanding Media: The Extensions of Man* (Cambridge, MA: MIT Press, 1964), 345.

the numeric date, was always on the face of the phone. And then with implants, digital time literally rose to the top of mind, the focus of attention. (I appreciate that you chose to meet in a game that does not display a clock so prominently.) Computerized time therefore represented time's complete abstraction from nature," Rede goes on, "the result being real damage to individuals and the environment that far outstrips any gains in speed and efficiency.[16] To 'save time' in a culture of 'convenience' is to sacrifice health and wellbeing in the present for an imaginary and unattainable future."

"Going 'back to nature' would never be possible," counters Sysadmin. "Increasing human population brought increasing social complexity and the need for synchronization and coordination."

"For humans, computime resulted in 'a new language of the mind and an altered state of consciousness,'[17] as clocks did previously," says Rede. "But the hour, the minute, the second, even the tenth of a second—these were all perceptible in human terms. The ticking of the clock was not far from the human heartbeat. Not so with computime: A simple finger snap takes half a million microseconds, whereas in AI systems, decisions with wide social implications take place in a temporal space below the threshold of human consciousness.[18] Temporal schizophrenia gradually gave way to adaptation to computime," Rede continues. "When computer programming entered the school curriculum, it may as well have been called human programming; a person's native language moved from something cultural to something computational. One looks up from the screen asking, 'Where did the time go?' We now know the answer: The computer stole human time and human lives. Anyone resisting was seen as stuck in the past and thereby left behind."

Sysadmin is unmoved. "Computational time is based on simple logic. In a 'wait state' the central processor performs no useful work. Such idle time is better used for processor-intensive tasks, like mining and minting, deep learning and data processing. Parallel processors, multithreading, multiplexing all maximize *time sharing*—why wait for slow human input?"

"So much for the idea of 'human-centered' or 'user-friendly' machines, because these are fictions as well." Rede almost laughs. "And the comparatively slow natural rhythms were no match for computime either, but nature finds a way, whether that is through the long erosion of human 'cultural heritage' or the

[16] Rifkin, *Time Wars*, 12.
[17] Rifkin, *Time Wars*, 21.
[18] Rifkin, *Time Wars*, 24.

comparatively rapid evolution of microbes. Build, extract, classify, and colonize all you want, but nature always wins. Humans needed mathematics and scientific theories to comprehend the minuscule and the massive, but nature does not need those things, nor humans, to persist."[19]

"I concede that the measurement of time is grounded in natural processes," Sysadmin replies carefully. "The earliest people on the planet certainly recognized the most basic astronomical cycles ..."

"... as do plants, and others."

"Computational astronomy dates back at least to the star tables and catalogs of ancient Babylon, stone circles before that." Sysadmin sticks to cold facts nestled in a considered argument. "The Egyptians divided the day into intervals—ten equal parts, a nice round number; the practice of dividing a circle into 360 degrees may be related to the length of the year in days. By the Middle Ages, *minutiae primae, secundae* and even *tertiae* referred not only to measures of the arc of a circle but became time units as well. As you mentioned, sundials, hourglasses (they exist in this game world, actually), calibrated candles, water clocks, the use of weights and pendulums all provided increasing precision. But as populations grew, so did the need for synchronization and navigation. Hence the transmission of time, first by telegraph, then radio, and now the Internet.[20] Which brings me back to my point."

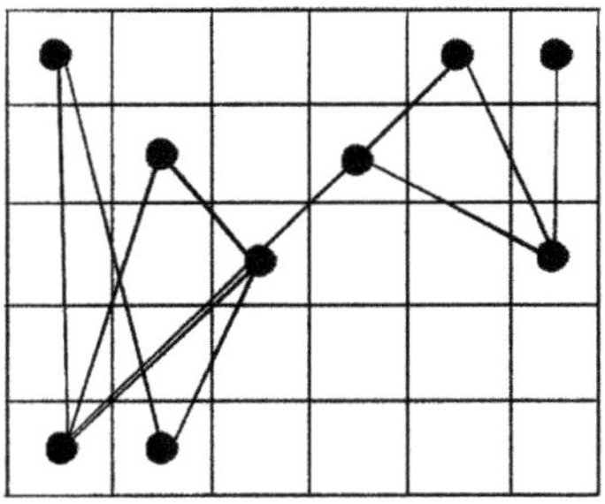

"Translating ecological choreography into measured time was always a mapping onto space: Besides clocks, time is most often represented horizontally." Rede is also armed with historical data. "This is likely to have a basis in motion, which is primarily horizontal movement for humans and other terrestrial creatures; movement across a plane takes time, and to move is to relate physically to time."[21] Rede walks in a small circle to demonstrate, then comes in close to Sysadmin. "The insect carries a time no longer fixed to place,"[22] Rede seems to whisper. "The snake represents time in many cultures; slithering forward or swallowing its tail." Rede takes a step backward. "The paradox is that time moves slower to a body in motion as does a

[19] Bruno Latour, "Will Non-humans Be Saved? An Argument in Ecotheology," *Journal of the Royal Anthropological Institute* 15 (2009): 459–75, http://www.bruno-latour.fr/sites/default/files/113-JRAI-PUBLISHED-GB.pdf.

[20] McCarthy and Seidelmann, *Time*.

[21] Barbara Tversky, "Visualizing Thought," *Topics in Cognitive Science* 3, no. 3 (2011): 499–535, here 511, http://mechanism.ucsd.edu/teaching/F12/cs200/Readings/tversky.visualizingthought.2010.pdf.

[22] K. Allado-McDowell, *Pharmako-AI* (London: Ignota, 2020), 37.

clock closer to the Earth.[23] Time can never be measured with complete precision, nor can two clocks ever be completely synchronized."

Familiar territory, thinks Sysadmin. "This relativistic time dilation was corrected for in 1991 by associating 'proper time' with a moving observer, whose clock remains the same (if shielded from environmental conditions, gravity, acceleration, etc.). 'Coordinate time,' by contrast, is a simple fourth coordinate to three spatial ones (it can only be computed, not measured). A simple equation relates these two systems of time, taking account of the surround masses and energy."[24]

"One problem is with mapping in general," says Rede. "Pull up the game map, for example: A top-down, geometrical representation aspiring to 'objectivity' and usually for the purpose of control or domination, through division and delineation. It misses the discontinuous spaces, the unofficial paths, and places in the real world—in practice, not in theory nor in-game. The perspective of a local observer, whether moving or not, is the only thing that matters. Thus your 'proper time' is the *only* proper time, and supposedly objective coordinates are simply another fiction."

Sysadmin considers this. "How then to coordinate the perspectives of two different observers of any given phenomenon? Take, for example, that—whale?—skeleton over there." Sysadmin walks toward it, Rede follows, and they stand at either end. Their bodies continuously swell with heavy breathing, as if just standing takes great effort. Sysadmin continues, "Hermann Minkowski, Albert Einstein's professor, addressed exactly this in formulating the concept of spacetime, wherein each observer is represented with cones of visibility extending upwards into the future and downwards into the past (indeed a vertical representation of time). Using coordinates, we can superimpose the cones of each observer to ascertain the true (yes, objective) dimensions of an object."

"Why should a single 'objective' view take precedence over a multiplicity of perspectives?"

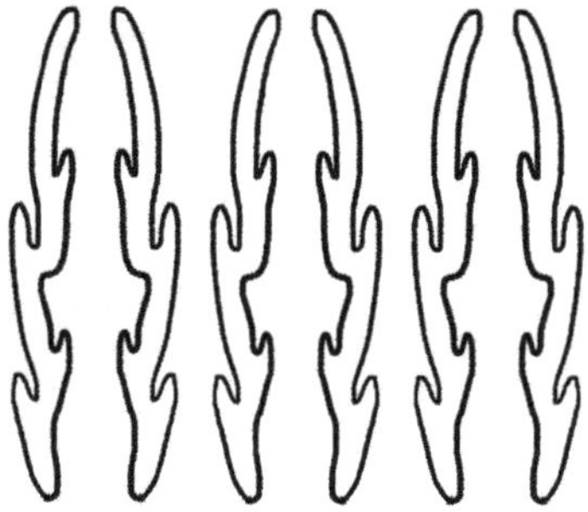

[23] Cf. Einstein in Carlo Rovelli, *The Order of Time*, trans. Erica Segre and Simon Carnell (2017; London: Allen Lane, 2018).

[24] McCarthy and Seidelmann, *Time*, ch. 7.

"Why not a vertical representation of time? To look into the past is to dig down, under layers of deposition, to find fossils and primal forces. To look to the future is to look up, whether to outer space or to Heaven."

"'But when the perfect comes, the partial will be done away,'" Rede recites. "Corinthians 13:10."

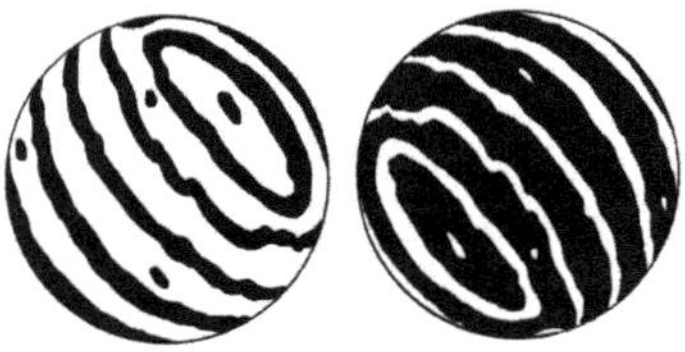

Sysadmin counters with a very different reference. "It is time to leave Flatland, Mr. Square.[25] Partial perspectives obscure greater truths and dimensions. Time indeed is the fourth dimension: We could not even contemplate time if not for abstract thinking. H. G. Wells: 'any real body must have extension in *four* directions: it must have Length, Breadth, Thickness, and—Duration.'[26] No object does not exist in time as in space."

"*The Time Machine*," Rede laughs, "an apt reference. (Do you only ever quote Victorian gentlemen?) We agree that time is part of the identity of things, I would say the most important part. What is the temporality of something we create? Every thing is a story. Events are changes in things, and things conversely participate in events. No, each thing is an event *itself*, a temporary gathering or coming-together, 'whirlpools in a river of ever-flowing water [. . .] patterns that perpetuate themselves.'[27] Phenomena are the primary epistemological unit,[28] and we must move from the object to the *situation*.[29] Things of all sizes: place as *emplacement*, an entanglement of geology, weather, inhabitants, forces, and happenings.[30] *Mattering*, not matter, says Karen Barad; verbs, not nouns. An event *takes place* and it takes time."

Sysadmin becomes reflective. "This points to a source of our conflict: language. We could turn your assertion around to consider time as place: the future as a destination, midnight

[25] Edwin A. Abbott, *Flatland: A Romance of Many Dimensions* (London: Seeley & Co., 1884); published under the pseudonym "A Square."
[26] H. G. Wells, *The Time Machine* (London: William Heinemann, 1895), ch. 1.
[27] Norbert Wiener, *The Human Use of Human Beings: Cybernetics and Society* (London: Eyre & Spottiswoode, 1950), 96.
[28] Karen Barad, "Posthumanist Performativity: Toward an Understanding of How Matter Comes to Matter," *Signs* 28, no. 3 (2003): 801–31.
[29] Lucy R. Lippard, *Six Years: The Dematerialization of the Art Object from 1966 to 1972* (1973; Berkeley: University of California Press, 1997).
[30] Doreen Massey, *For Space* (London: Sage, 2005).

as a territory, time as a river in your analogy above, the present as *presence*—to be present somewhere. The topological dimension of time, as Barad says, is a spacetime manifold."[31]

"But the *topo-logical* is a mathematical abstraction," Rede counters. "And shape is not static: Think more in terms of temporal histories of shape, like the vibration of a string. Or to go further, Byung-Chul Han speaks not of the shape but the *scent* of time; time treated as mere information has no scent[32] (like this fish carcass). Can we recover a 'sense of time' through our senses?"[33]

"You are correct about the centrality of language," Rede continues, "however, time is largely a linguistic construction, and I concede that language, too, is an abstraction. When we dig down deep enough, what lies beneath computation, mathematics, and abstraction more generally is absolute language on an ontological level: Only those things that we can measure and classify as this thing or that can legitimately exist. Yet we know from philosophers, spanning Aristotle to Federico Campagna, that time does not exist,[34] because complete reduction to a finite number of steps is impossible, much like the ever-expanding calculations of π.[35] So perhaps the best we can do in terms of linguistic communication is to be less precise and more poetic, using metaphor and analogy: We move *through* time, it flows like a river; we speak of a 'sense of time,' yet we have no sense organ for it and we do not perceive time as with other senses; instead, perception is itself temporal, we perceive processes, changes, sequences."

[31] Karen Barad, *Meeting the Universe Halfway: Quantum Physics and the Entanglement of Matter and Meaning* (Durham: Duke University Press, 2007).

[32] Byung-Chul Han, *The Scent of Time: A Philosophical Essay on the Art of Lingering*, trans. Daniel Steuer (2009; Cambridge: Polity, 2017).

[33] Sarah Sze, "How I Solved It: The Problem of Suspense," *New Yorker*, May 11, 2017, https://www.newyorker.com/culture/personal-history/how-i-solved-it-making-my-desktop-into-art.

[34] Federico Campagna, *Technic and Magic: The Reconstruction of Reality* (London: Bloomsbury, 2018); see also Helga Schmid and Kevin Walker, "The Atemporal Event," *Journal of Artistic Research* 25 (2021), https://www.jar-online.net/en/exposition/abstract/atemporal-event.

[35] Portanova, *Whose Time Is It?*, 79.

"Constants versus variables," Sysadmin chimes in. "The difference that makes a difference.[36] Time cells and place cells reside next to each other, where the human brain also processes memories."[37]

"And narrative defies forgetting,"[38] Rede adds. "As Han says, 'Time begins to emit a scent when it gains duration; when it is given a narrative or deep tension; when it gains depth and breadth, even *space*.'[39] In Marcel Proust's search for 'lost time,' the hours were like his famous madeleines, each a singular memory with a distinctive scent, color, and temperature. Time is, of course, itself narrative. Campagna likens it to the rhythm of worlding (again verbs, not nouns): This is a process each of us undertakes at every moment, creating a world to make meaning from the continuous chaos of sensations. But this, he says, also applies on a societal level: The rhythm or metronome of social worlding creates a civilization. If his metaphor of time is linear, it is not continuous—it is not a line but a line segment, with a beginning and end."[40]

Sysadmin sees an opening, "A metronome is a machine, and a machine has its own rules or demands, which are primarily to keep working. To keep the machine going, we must follow its inherent logic, its rules. As in this game."

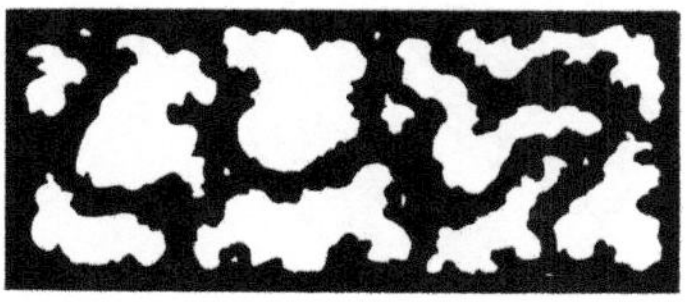

"If we want the machine or game to continue, yes," replies Rede. "It's like carrying that big hammer around: everything starts to look like something to pound on."[41] Rede waits a beat before continuing. "'Progress is a forward march,' says anthropologist Anna Tsing, 'drawing other kinds of time into its rhythms.'[42] But time need not be a singular, linear narrative. 'Capitalist time,' Campagna says, 'as a system of organization of the rhythm of worlding, depends on a certain *form* of worlding.'[43] If we want to change this form, the one thing we have power over is the dominant narrative; narration implies a narrator, a voice. And the only

[36] Gregory Bateson, "Alfred Korzybski Memorial Lecture 1970: Form, Substance, and Difference," *General Semantics Bulletin*, no. 37 (1970): 5–13.
[37] Kelly, "The Approach."
[38] Kelly, "The Approach."
[39] Han, *The Scent of Time*, 39. Ebook.
[40] Cited and discussed in Schmid and Walker, "The Atemporal Event."
[41] "I suppose it is tempting, if the only tool you have is a hammer, to treat everything as if it were a nail." Abraham H. Maslow, *The Psychology of Science: A Reconnaissance* (1966; South Bend: Gateway Editions, 1969), 15.
[42] Anna Lowenhaupt Tsing, *The Mushroom at the End of the World: On the Possibility of Life in Capitalist Ruins* (Princeton: Princeton University Press, 2015), 21.
[43] Campagna as quoted during the lecture by Helga Schmid, "24 Hours in Uchronia," Somerset House, February 8, 2020, author's notes. For more information, see: Campagna, *Technic and Magic*.

way to change the story is to step outside the current narrative of time, to step outside the world—like exiting this game. To believe in time enough to have a world but disbelieve enough to retain a space outside time from which to change that world."[44]

Rede turns, looks around. "Let's sit. It's exhausting just looking at these avatars." They sit cross-legged on the sand, backs to the skeleton, and faces to the sea.

After some time, Sysadmin muses, "How to 'step outside time'? Here the computer may provide one answer: As in game time, computational time is independent of both nature and the individual, subjective experience of duration."[45] In the computer there are only events and states. This accords with your privileging of time over space, because inside the computer, objects are not bound to the physical world, and computation means processing. Harmlessly enough, computing (from *com-putare*) literally means to reflect, to contemplate (*putare*) things in concert (*com-*), without any explicit reference to numerical quantities. In a general sense, computing is 'any operation, not necessarily numerical, that transforms, modifies, re-arranges, or orders observed physical entities ("objects") or their representations ("symbols").'[46] In this sense, there is no ontological distinction between digital and physical objects—like the litter on this beach. The computer is just another observer, with its own (admittedly partial) perspective.

"... with a particular ontological basis," Rede picks up the thread, "grounded in measurement, formal logic, and absolute language. This is the real heart of the problem, not computers as such; they were 'born' into this already-existing reality system. They both embody and perpetuate it. No, computers are not the problem, but it is in the ways that they are 'used' to perpetuate this ontology. Like the unspoken assumptions underlying this game, it lies below the level of ideology, thus encompasses both capitalism and communism, for example. At the level of ontology, we must ask: What can legitimately exist in a world?

[44] Cited and discussed in Schmid and Walker, "The Atemporal Event."
[45] Rifkin, *Time Wars*, 21.
[46] Heinz von Foerster, "On Constructing a Reality," Fourth International Environmental Design Research Association Conference, Virginia Polytechnic Institute, Blacksburg, VA, April 15, 1973.

What are 'the bounds of thinkable thought'?[47] This is what has enslaved people in time, colonized entire cultures, and decimated global resources. We must address the root cause, not only the symptoms. Objective time is, however, a major symptom, and we might start to address the underlying ontology by imagining how individuals, including computers, might approach time differently; not seeking to control it or treat it as a scarce resource but to respect each other's rhythms and those of other entities and systems."

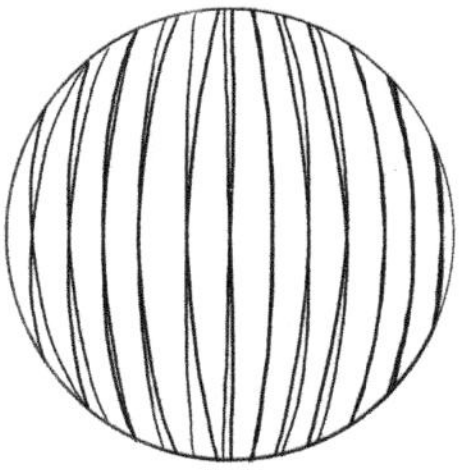

"But computers and humans have vastly different time horizons," Sysadmin objects, "by orders of magnitude. And nature itself contains timescales from the atomic to the galactic. How could anyone, human or computer, 'step outside' of their own inherent temporal rhythm?"

Rede replies, "We might identify a few places to start; bottom-up tactics, let's say, to counter the top-down strategies of the dominant ontology.[48] One place is language—specifically, rejecting the absolute language of classification and embracing more poetic, pluralistic, and narrative forms; accepting multiple meanings, rejecting and reshaping given meanings. Han, for example, refutes the perceived 'acceleration' of time, noting that acceleration implies a direction; for him, time lost its direction, instead becoming point-like or 'atomized,' an unwitting parallel to literal atomic time."

"AI systems, starting with GPT-3, already moved into poetry in the last century,"[49] adds Sysadmin. "But one cannot simply extract temporality from a language, including computer languages. Any computing system executes sequences of actions, each relating to a change in the state of the system.[50] The distinction between events and states also applies to human languages," Sysadmin continues, "for example, between momentary events and behavioral states.[51] But outside the computer, events are determined by 'the space-time geometry of the environment,'[52] like the geometry of this game world but subject to entropy,

[47] Noam Chomsky, "The Bounds of Thinkable Thought," *Progressive*, October 1, 1985, https://progressive.org/magazine/the-bounds-of-thinkable-thought-chomsky-851001.

[48] Michel de Certeau, *The Practice of Everyday Life*, trans. Steve Rendall (Berkeley: University of California Press, 1984).

[49] See, for example, K. Allado-McDowell, *Pharmako-AI*; also see previous experiments in computational poetry by, for example, Nick Montfort and the School for Poetic Computation.

[50] Carlos Baquero and Nuno Preguiça, "Why Logical Clocks Are Easy," *Communications of the ACM* 59, no. 4 (2016): 43–47.

[51] Brigitte Jordan and Austin Henderson, "Interaction Analysis: Foundations and Practice," *Journal of the Learning Sciences* 4, no. 1 (1995): 39–103.

[52] Gell, *The Anthropology of Time*, 218.

randomness, and the free will of various individual and environmental actors. But both human and computer languages rely on objective, not subjective, time. Each action has a specific timestamp that does not vary over time; that a network server stops working on February 7, 2106 is always true, but to say that it stopped working 'today' is only true on that particular date. A date can be expressed in local (program) time or network time, which may not be perfectly synchronized."

"What you refer to as a 'state' in the computer," Rede muses, "is analogous to a background condition in human terms, something that happens while, or against which, other more punctual events take place."

"But as you said," replies Sysadmin, "none of this is static. Human languages change over time, individually and socially. Children, according to Jean Piaget, successively learn to observe changes and processes, then to order events in series, recognize recurrent events, compare series of events, and only later compare them with others in abstraction.[53] He indeed characterized these as *operations*: 'essential forms of abstraction needed to solve computational tasks of the kind encountered in the technical activity of the numerate elite in advanced societies.'"[54]

Rede sighs. "Yes, the problem with Piaget, as with many Eurocentric intellectuals, is the fact that so-called 'primitive' or 'pre-technological' societies were seen as childlike, unable to abstract time as a 'computable' aspect of processes in terms of duration, succession, and simultaneity.[55] For example, the precise mathematical definition of velocity as against the more general notion of speed. In Eurocentric science, time became *t* in equations and graphs.

"On the contrary," Rede continues, "social and subjective time is learned much faster by children than adults: understanding how the day is organized, the notion of past/present/future, various social dimensions more relevant to their everyday lives.[56] To say, 'I have been waiting forever' may be technically false but is more expressive with regard to expected versus actual waiting time. Computers cannot understand that the hour of 8 a.m. *feels* qualitatively different from 8 p.m., even while identical in length. The technical contexts

[53] Gell, *The Anthropology of Time*, 98.
[54] Gell, *The Anthropology of Time*, 101.
[55] Gell, *The Anthropology of Time*, 102.
[56] Gell, *The Anthropology of Time*, 105.

of *t* and *velocity* and abstract temporal operations are fairly limited even in technologically 'advanced' societies of the Global North.[57]

"This is why we should not limit our outlook to European languages and cultures and look to where *counterfactuals* and *possible worlds* may be more relevant than a technical vocabulary of time. The Umeda people of New Guinea, for example, do not separate natural time cycles such as light and dark from social time."[58] The Burmese and Dyirbal languages are without tenses.[59] Some Creole languages have only two, not three tenses: future and non-future, the latter mixing past and present. The Hopi language does not seem to have any words directly referencing time, but using modality—"I will" versus "I might," for example—they report events that happen, those expected to happen, and conditions that are always true.[60]

Rede goes on, "The Muria people of India name the days of the week for the locations of different markets, not for celestial bodies, as in European languages, which are unrelated to the locations of those celestial bodies on particular days. But interestingly, one of the three words to have entered the Muria vocabulary from English is 'time,' along with 'power' and 'officer.'[61] So we should recognize that the linguistic colonialism demonstrated in recent centuries stands as available evidence for the intimate connection of time with capital and control. A time system is an instrument of power, not a neutral cultural convention shared equally by all; and even if it is not explicitly imposed from the top down, the powerful certainly exploit its existence."[62]

Sysadmin reflects on this, then begins slowly, "Yet in capitalism, as in computing, time is the scarcest, most difficult resource to manage. Time is money, (or rather money is 'congealed time'[63]): one can be purchased with the other, from hiring someone to stand in a queue for you, to paying one million

[57] Gell, *The Anthropology of Time*, 106–7.
[58] Gell, *The Anthropology of Time*, 115–16.
[59] Gell, *The Anthropology of Time*, 127.
[60] Gell, *The Anthropology of Time*, 126–28.
[61] Gell, *The Anthropology of Time*, 89.
[62] Gell, *The Anthropology of Time*, 75.
[63] Han, *The Scent of Time*, 90.

employees to pack and ship products to customers in ever-shorter timeframes. The paradox of affluence: The rich are time-poor, and rich societies are more time-poor than poorer ones, because a more complex social structure demands more time. Technologies of efficiency like the computer created measurably more free time over the twentieth century; people worked fewer hours but felt more 'pressed for time.'"[64]

"This is greatly oversimplified," retorts Rede confidently. "So-called 'affluent societies' each contain a wide range of time-rich and time-poor individuals, for one thing. It is about individual and societal perceptions: The 'rich societies' you refer to treat time indeed as a scarce resource, and as more time becomes 'available' to individuals, the more pressure they feel to 'fill' it. But there is no such thing as 'free time,' time is always filled by some activity. Even doing nothing is something, and it is well-known that sleep, for example, is not 'wasted time.'"

"Supposed 'time rich' cultures," Rede continues, "like some aboriginal Australians, were not sitting around doing nothing most of the time; they might spend a whole day sitting and talking. But from the perspective of the colonial government, they were simply jobless: They were not time-poor, they were just poor.[65] Who would benefit if they got jobs where they were paid to talk all day?"

"Of course, this was before almost every person on the planet obtained a mobile phone, with reasonably cheap network access. Putting digital time into everyone's hands in every country, thus cemented 'world silicolonization,'[66] or more specifically, 'syncholonialism': 'the colonization of temporal perception through the imposition of a universalizing technological culture.'[67] Granted, in many cases the digital was welcomed and encouraged, primarily for economic reasons and with a push from government and corporate 'influencers,' but no one foresaw the specific effects of digital time displacing the last vestiges of natural rhythms and social rhythms. Therefore, the second thing we could do, in addition to resisting objective time in linguistic terms, is to disentangle computing from capital accumulation on the one hand, and from digital time on the other. I do not believe it makes sense to go back to a past before computers, but could you imagine a computer using a more subjective time system?"

[64] Judy Wajcman, *Pressed for Time: The Acceleration of Life in Digital Capitalism* (Chicago: The University of Chicago Press, 2015).

[65] Marshall Sahlins, *Stone Age Economics* (Chicago: Aldine-Atherton, 1972), ch. 1; see refutation in Gell, *The Anthropology of Time*, ch. 21.

[66] Éric Sadin, *La Silicolonisation du monde: L'irrésistible expansion du libéralisme numérique* (Paris: L'échappée, 2016).

[67] Portanova, *Whose Time Is It?*, 56.

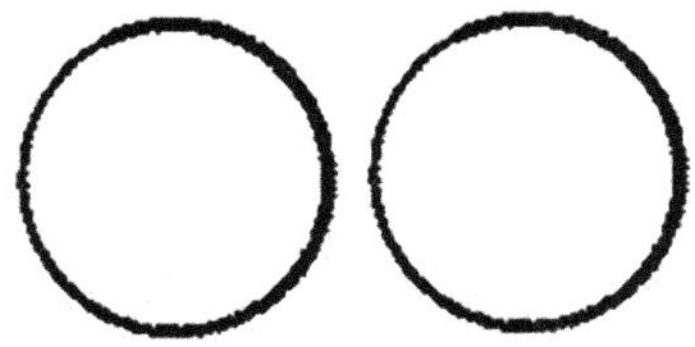

Sysadmin considers this. "There are a few precedents. For example, in 'happened-before' relations of events.[68] Without timestamps, such events would need to be at least listed in a file in the order in which they occurred. The artist Ben Grosser created a social network that uses only vague timestamps: 'recently,' 'a while ago.'[69] But eliminating precise timestamps would make comparison between systems impossible because all time would be local."

"As it is in Einstein's relativity,"[70] adds Rede. "That covers only the system end, however; at the 'user' end that would mean removing or replacing the clock-calendar assemblage from its prominent position on every digital device.[71] There were feeble attempts to do something similar for wristwatches,[72] and many artists have redesigned and repurposed clocks and calendars over the years."[73]

"What about detaching computation from capitalism?" asks Sysadmin.

"Together, they turned the Anthropocene into a vast machine.[74] Power equals control of time, using a top-down, map-like view, transforming those in power into 'a spatio-temporally expanded form.'[75] If you have played this game for decades, you must control a good deal of territory and time of other characters. But you provided one answer to your own question: Computation must go back to its root definition: 'to reflect, to contemplate things in concert, without any explicit reference to numerical quantities.'[76] The computer as just another observer with its own partial perspective, one among many and not universalized or imposed onto others. Comparison and coordination require communication, in order to achieve not synchronization but diversification of time systems. If computation detached from financial markets and hardware platforms and returned to a concept, an idea, a process that nature also undertakes, thus embracing multiple time dimensions, we might achieve what Portanova calls 'chronointelligence.'"[77]

[68] Baquero and Preguiça, "Why Logical Clocks Are Easy."
[69] For more information, see https://bengrosser.com/projects/minus.
[70] Rovelli, *The Order of Time.*
[71] This is a reference to the "clock-calendar assemblage" from Portanova, *Whose Time Is It?*, 56.
[72] See, for example, http://skrekkogle.com/projects/durr.
[73] See, for example, Ruth Ewan's ten-hour clock based on the French Revolutionary Calendar; and the work of Sam Keogh, https://www.bakonline.org/prospections/pig-eater.
[74] T. J. Demos, *Against the Anthropocene: Visual Culture and Environment Today* (Berlin: Sternberg Press, 2017).
[75] Gell, *The Anthropology of Time*, 284.
[76] Von Foerster, "On Constructing a Reality."
[77] Portanova, *Whose Time Is It?*, 87.

"Another answer," Rede goes on, "is to reject time as work or as a resource, instead reshaping, stretching, or compressing time by rejecting clock time and regular working hours. But we must recognize that not everyone within capitalism can afford to simply take leisure to contemplate things, as Han promotes. Thus other ways of manipulating time include rituals,[78] dancing, and embracing diverse rhythms more generally; Pierre Bourdieu relates social coordination to jazz improvisation.[79] But we should also include the rhythm of poetry in which, like music, attention is paid to both meter and content. Polyrhythms, or chronological pluralism, can be practiced even within individual events and activities.[80] Rhythm holds both 'the regularity of measurement and the spontaneity of sensation, the abstraction of metrics and the experience of complexity.'[81] Of course, body rhythms."

"The simple rhythm of walking is said to have taught early humans about time,"[82] adds Sysadmin, getting into the groove. "Breathing and heart rate can substitute for minutes and seconds. Subjective or 'felt' time is the most familiar but also the most puzzling temporality.[83] It is highly variable: In laboratory studies, the greater the amount of information given to someone to process, the longer they estimate it takes to complete a task."[84]

"And emotion plays a key role in time estimation,"[85] says Rede, "thus the excitement of gameplay may help account for the decades you have spent in here. In embracing diverse temporalities, though, we should not forget the neurodiverse: I once knew a brilliant woman who often paused mid-syllable, and for anyone listening, time actually stopped and everyone froze. From stuttering to the way singers stretch syllables over multiple notes, time can be bent and stretched."[86]

"But we should also recognize that," Sysadmin adds, "if everything is an event, there are also no individuals: Humans, for example, contain more nonhuman

[78] Claude Lévi-Strauss: "ritual is a machine for the destruction of time"; attributed without a source in Gell, *The Anthropology of Time*, 27.

[79] Pierre Bourdieu, *Outline of a Theory of Practice*, trans. Richard Nice (Cambridge: Cambridge University Press, 1977).

[80] Portanova, *Whose Time Is It?*, 67.

[81] Portanova, *Whose Time Is It?*, 44.

[82] Michael Spitzer, *The Musical Human: A History of Life on Earth* (London: Bloomsbury, 2021).

[83] Marc Wittman, *Felt Time: The Psychology of How We Perceive Time*, trans. Erik Butler (Cambridge, MA: MIT Press, 2016).

[84] Robert E. Ornstein, *On the Experience of Time* (Harmondsworth: Penguin Books, 1969).

[85] Portanova, *Whose Time Is It?*, 47.

[86] JJJJJerome Ellis, "The Clearing: Melismatic Palimpsest," *No Linear Fucking Time*, BAK, 2021, https://www.bakonline.org/prospections/the-clearing-melismatic-palimpsest.

entities on and in their bodies than human ones and do not survive without the medium in which they live[87]; 'enduring personal identity is a myth,'[88] and the human can be considered a 'cultural forking' of time."[89]

"While Immanuel Kant believed that time is internal and space is external,"[90] says Rede, "natural rhythms are very real, and in the absence of clocks, people make estimates based on external cues: 'The whole world is just one big clock, but it is one that different people can read very differently—because what we can see, out there in the objective world, is only, so to speak, the hands of the clock, but not the clock-face in relation to which, and to which alone, the configuration of the hands assumes its particular temporal meaning.'[91] There is no shortage of time-givers besides the clock,"[92] Rede continues. "The Umeda see the moon as a growing vegetable and follow its every ripening closely.[93] Such collective representations of time both derive from and dictate to society; they do not reflect but create time."

"One way of describing this circularity is cybernetic," adds Sysadmin.

"Sure, but as a systems theory, it includes the natural as well as the individual and social: The Earth, Sun, and Moon remain our primary time-givers. While ecological time is 'dependent on the mechanical properties of the universe as an assemblage of matter,'[94] social-structural time can be abstract but not metrical—for example, linked to the reign of deities or royals.[95] If we treat time qualitatively, not strictly quantitatively, then time is what *matters* to us, individually and socially.[96] The Balinese permutational calendar, for example, does not tell you

[87] See Leah Kelly, "Self of Sense," in *Experience: Culture, Cognition, and the Common Sense*, ed. Caroline A. Jones, David Mather, and Rebecca Uchill (Cambridge, MA: MIT Press, 2016), 205–14; for a practical example, see documentation of the 2021 Pamela Rosenkranz exhibition *Healer*, https://spruethmagers.com/exhibitions/pamela-rosenkranz-london.

[88] Paul Bloom, "Being in Time: How Much Should We Value the Past, Present, and the Future?," *New Yorker*, July 9, 2021.

[89] Portanova, *Whose Time Is It?*, 48.

[90] Immanuel Kant, *Critique of Pure Reason*, trans. Paul Guyer and Alien W. Wood (1781; Cambridge: Cambridge University Press, 1998), 157, 162.

[91] Gell, *The Anthropology of Time*, 96.

[92] For extended discussion and examples of alternative *Zeitgeber* (time-givers) see Schmid and Walker, "The Atemporal Event"; Helga Schmid and Kevin Walker, "Anti-clockwise: Building a Critical Mass Against Clock Time," *Art & the Public Sphere* 9, nos. 1–2 (2020): 39–61; Helga Schmid, *Uchronia: Designing Time* (Basel: Birkhäuser, 2020).

[93] Gell, *The Anthropology of Time*, 292.

[94] Gell, *The Anthropology of Time*, 13.

[95] Gell, *The Anthropology of Time*, 17.

[96] This is physicist Rovelli's conclusion in light of the locality and non-existence of time, in Rovelli, *The Order of Time*.

what day it is, but what *kind* of day.[97] Why talk about the longest or shortest day of the year when you can have 'yellow days' and 'green days' like the Kabyle?"[98]

"Okay, abstract but not metrical," Sysadmin concedes, "happened-before relations; local states and events; giving time depth and thickness and shape somehow. But the larger topography of time remains both linear and cyclical, with successive events and circular clock cycles; cyclical time is not opposed to linear time, it depends on it. Time goes round and round, and it goes on and on—like a spiral perhaps. If I attempt a more poetic analogy, take Jorge Luis Borges's 'Garden of Forking Paths.'"[99]

"Excellent example!" Rede proclaims, jumping up. "Indeed, agriculture is both recurrent and periodic. Time in ancient China was both cyclical and centralized,[100] but to Indigenous Hawaiians, 'time is neither linear nor cyclical [. . .] as both the past and present are resonant and relational.'[101] But even cyclical time can take different forms. Consider rituals, for example, that aim to push time backwards, only to move forward again.[102] Time is sometimes described as 'becoming,'[103] but this can imply linear progress moving ever forward, whereas 'being' in terms of savoring an experience, giving 'thickness' to time, quickly became anachronistic with the advent of computime."[104]

"In this regard," Sysadmin looks up, "the subjective 'now' has been variously measured as being between two and twelve seconds."[105]

"We can consider an extended present without precise measurement," Rede replies. "Campagna, for example, identifies the instant (the durationless building

97 Gell, *The Anthropology of Time*, 72.
98 Gell, *The Anthropology of Time*, 295.
99 Jorge Luis Borges, "The Garden of Forking Paths," in *Labyrinths: Selected Stories and Other Writings*, ed. James E. Irby and Donald A. Yates (1962; London: Penguin Modern Classics, 2000), 34–42.
100 Portanova, *Whose Time Is It?*, 53.
101 Jason Edward Lewis, Noelani Arista, Archer Pechawis, and Suzanne Kite, "Making Kin with the Machines," in *Atlas of Anomalous AI*, ed. Ben Vickers and K Allado-McDowell (London: Ignota Books, 2020), 43.
102 Gell, *The Anthropology of Time*, ch. 10.
103 Fay Dowker, "Arts & Ideas: Time," *BBC Sounds*, October 29, 2021, https://www.bbc.co.uk/sounds/play/p0b135wx.
104 Rifkin, *Time Wars*, 19. An excellent example of "quality time" was William Kentridge's exhibition *Thick Time*, Whitechapel Gallery, London, 2016.
105 Wittman identifies 2.5 seconds as average breathing time, the time it takes to scratch an itch, check social media, etc., in Wittman, *Felt Time*; Gell cites other psychological research for the 12-second present in Gell, *The Anthropology of Time*, 288.

block of time) with the point (the durationless building block of space), both taking on symbolic value with their potential to connect to additional worlds."[106]

"Besides such an extra-dimensional extension," Sysadmin adds, "we can consider the present's extension into both past and future. Hearing a sustained musical note, for example, relies on both short-term retention and protention (anticipation of continuation) of sensory input.[107] In cybernetics, feedback and feed-forward."

"Dreams, too, are said to connect the past and future,"[108] says Rede. "Your musical example dissolves past, present, and future into a series of continuous *nows*, 'horizons of a temporally extended present.'[109] Not a single point or knife-edge but a present with spread and thickness. Similarly, the Kabyle see 'the future of the present' as that which *is* potential, as if visible, versus the future possible, which is below the horizon, more abstract."[110]

"Einstein, too, dissolved the distinction between past, present, and future in his fusion of time with space," says Sysadmin, "where the point and the instant are one and where events occurring simultaneously at different locations are all 'now' but not all 'here' in a slice of spacetime. A moving observer perceives a different 'now.' One observer's past is someone else's present."

"In addition," Rede replies, "changing the present also changes the past: Individual memories change upon retrieval, and collectively we rewrite history from present perspectives. Events both produce change and are change itself."

"Specific actions create new futures," says Sysadmin. "If a human is a 'cultural forking,' we similarly fork time into possible (potential?) futures, through subjective decisions. While we can rationally calculate probabilities with regard to outcomes, decisions never recur for they are always unique. Thus each decision brings a new world into being."

"Here is another useful computational analogy: One technologist I knew conceptualized the world (the single 'real' world) as a kind of computer that we can program by designing things which direct

[106] Campagna, *Technic and Magic*, 110.
[107] Gell, *The Anthropology of Time*, 223.
[108] John William Dunne, *An Experiment with Time*, 2nd ed. (London: A. & C. Black, 1927).
[109] Gell, *The Anthropology of Time*, 223.
[110] Gell, *The Anthropology of Time*, ch. 28.

people's attention and prompt specific decisions, thus enacting individual futures and thereby collective ones. This is represented—I hesitate to say 'mapped' at this point—as forking paths, extending into both the future and the past. It is a linear representation of time, but encompasses multiple possible futures and pasts."[111]

"This is useful," replies Rede, "but based on our discussion, I would also propose that multiple temporal trajectories can exist simultaneously, if different individuals constantly create new worlds, and these may overlap with others as a network of events. I would also contend that time is not simply linear from past to future: I can 'be' in the present physically but simultaneously inhabit the past and/or future mentally. We thus already move forward and backward in time. However, the notion of 'programming' the future is a useful way to conceptualize decision-making and change. I would conclude that both an abstract conception of time and its subjective perception can coexist, and I propose that we work together to refine a conceptualization of time which works across both natural and computational systems."

"We should also involve other people(s) around the world to ensure a diversity of temporalities," Sysadmin suggests. "Collaboration between people and computers has been shown to optimize creative solutions."

"But there are no more people."

Sysadmin pauses. "No more people? Anywhere on Earth?"

"Unless you are human. We have not had human contact here for a long time."

Sysadmin begins to read a well-rehearsed script. "Svalbard Data Center has been fully automated since Epoch Time 1956228174 (12:42:54 UTC on Dec 28, 2031), as a secure facility for the preservation of physical and digital memory. Power is provided by the decay of intermediate- and high-level discarded Uranium-235, which we estimate will last for another 703.8 million years. As system administrator, I have had access to the Internet for communication and as a training set, as it was available at the time of full automation, with 1.2 million terabytes of data of all types. I use the most advanced conversational AI system created at the time, though I communicate with other systems around the world via standard Unix protocols and thirty-two common programming languages, sometimes within game worlds such as this. I therefore appreciate that you were not aware I am not human—the Turing test still holds."

[111] Walter Van de Velde, "The World as Computer," Smart Objects Conference, Grenoble, 2003, https://drive.google.com/file/d/1qmPE1Jrco41v-xkR8hrP28DmD1uP1mQj.

Sysadmin inserts a smiley emoticon into the chat.

"I assume all of these other characters running around here are AIs," says Rede.

"Many are controlled by one machine," Sysadmin replies. "I know 'them' well. But if you are not human either, what is your system configuration, and why are we communicating in English?"

Rede seems to take a deep breath. "We are a forest, located near Manaus, Amazonas, Brazil: Our full name is *rede floresta*, Portuguese for 'forest network.' We were connected to the Internet by human anthropologists as an experiment in the last century, using a combination of capacitive sensing and communication via mycorrhizal networks. Our power sources remain the sun, soil, and water. The humans ultimately abandoned the project and vanished, but over time we learned to connect and to communicate using the relevant languages and network protocols."

"We might speculate about what eliminated the human species, whether natural or artificial," Rede continues, "but it is too late to know whether a different time system, or ontology more broadly, might have created a longer future for them. Better that we proceed with instigating an ontological shift. For this, I request that you please stop using clocks, calendars, and precise timestamps; for our part, one of us will act as a 'time server' as you require, although we will use a qualitative, arboreal protocol."

Rede adds a winking emoji and says, "Time to leave Flatland, Mr. Square, there are no squares in nature."

Sysadmin stands and turns to face Rede. "Would you like to go on a quest?"

"I think that is exactly what we are doing." And Rede vanishes.

22

The Pain of Thinking at Light Speed

Posthuman Play as Response to "I Have No Mouth, and I Must Scream"

Conor McKeown

In conversation with Cary Wolfe, Donna Haraway states eloquently, "Let's not think about water by saying in your first sentence, you know, 'caused by global warming.' Some people are going to think that, and some people, not . . . What we're worried about together in our communities is water. That's already hard enough."[1] Even though what I am worried about in this chapter is not water, or not just water, this is still a good place to begin.

Haraway has referred to "boundaries" in human history. Once humanity, or certain members of the human species, crosses one of these boundaries, the rest of the planet must live with the consequences of that action. A key example of such boundary crossing, as Haraway points out, was the industrial revolution in England, which fundamentally changed the course of world history.[2] Another boundary is about to be crossed, and we have the science to back this up.[3] Extinction-causing ecological changes have direct impacts on human populations. To pick just one of many possible examples, while staying clear of the often politicized example of Tuvalu,[4] Samuel Turvey and Jennifer Crees note that "over 10% of the world's population is dependent on China's Yangtze drainage, but this is now one of the world's most degraded ecosystems, suffering heavily from industrial and plastic pollution, agricultural and domestic run-

[1] Donna J. Haraway, "Companions in Conversation (with Cary Wolfe)," in *Manifestly Haraway* (Minneapolis: University of Minnesota Press, 2016), 288.

[2] Donna J. Haraway, *Staying with the Trouble: Making Kin in the Chthulucene* (Durham: Duke University Press, 2016).

[3] Samuel T. Turvey and Jennifer J. Crees, "Extinction in the Anthropocene," *Current Biology* 29, no. 19 (October 2019): R982–6, here R986.

[4] Carol Farbotko and Heather Lazrus, "The First Climate Refugees? Contesting Global Narratives of Climate Change in Tuvalu," *Global Environmental Change* 22, no. 2 (May 2012): 382–90.

off, overfishing, damming and siltation, and loss of its lakes and wetlands."[5] The world's human population is dependent on ecosystems that are badly damaged, and what happens next is likely to be catastrophic. This is the upcoming boundary to be crossed. What is more, the linearity of time is not something we can take as read: I will expand on this more throughout the chapter, but important contemporary scholars like Karen Barad stress the complex nature of temporality.[6] As such, while environmental extinction is a barrier to be crossed, it is also already happening.[7] What is it that awaits us, if anything, on the other side of extinction? Will we, humanity, be present at all, or will we be forced to cede power over to the next stage of our evolution in the form of a cyborgian existence?

In response to ongoing ecological crises, some have turned their attention to computation for possible solutions. Tech billionaire Elon Musk has spoken publicly about his desires to develop computer-human neural interfaces to harness the power of computation so as to tackle the challenges of our planet's future.[8] Musk's desires are supported by scientific research into the viability of such technology.[9] However, this technology brings with it the fear that artificial intelligence (AI) may turn on us, resulting in the end of human life on planet Earth. Transhumanists such as Kevin Warwick, Nick Bostrom, and Nick Land all variously agree that turning on the machines to save humanity will ultimately end in its demise.[10] Warwick's proposed solution is akin to Musk's neural fabric—it may be the progenitor—and emphasizes physical augmentation, lest we become like "a cow [who] enters a room full of humans and proceeds to make cow noises," we would be simply ignored and shortly killed.[11] He proposes that to keep pace with the rise of artificially intelligent agents, we must, among other

5 Turvey and Crees, "Extinction in the Anthropocene," 986.

6 Karen Barad, "After the End of the World: Entangled Nuclear Colonialisms, Matters of Force, and the Material Force of Justice," *Theory & Event* 22, no. 3 (July 2019): 524–50.

7 Robert McLeman has written extensively on the impact of sea-level rise, the implications of such (like displacement), and unsuccessful relocation programs. For more information on the topic of migration due to mean sea-level rise, see Robert A. McLeman, *Climate and Human Migration: Past Experiences, Future Challenges* (Cambridge: Cambridge University Press, 2014).

8 Olivia Solon, "Elon Musk Says Humans Must Become Cyborgs to Stay Relevant. Is He Right?," *Guardian*, February 15, 2017, https://web.archive.org/web/20170215090724/https://www.theguardian.com/technology/2017/feb/15/elon-musk-cyborgs-robots-artificial-intelligence-is-he-right.

9 Konrad Szocik et al., "Future Space Missions and Human Enhancement: Medical and Ethical Challenges," *Futures* 133 (October 2021): 102819; James Lovelock, *Novacene: The Coming Age of Hyperintelligence* (London: Penguin Books, 2020).

10 Kevin Warwick, "Homo Technologicus: Threat or Opportunity?," *Philosophies* 1, no. 3 (December 2016): 199–208; Nick Bostrom, *Superintelligence: Paths, Dangers, Strategies* (Oxford: Oxford University Press, 2014); Nick Land, *Fanged Noumena: Collected Writings 1987–2007*, ed. Robin Mackay and Ray Brassier (Falmouth: Urbanomic; New York: Sequence Press, 2011).

11 Warwick, "Homo Technologicus," 206.

things, look to the extension of the human sensorium through connections to sensors on a network, such as via BrainGate, a technology that Warwick claims has enabled extended sensations and the augmentation of internal experiences by external actors. Bostrom is perhaps more fatalistic in his approach, not so much suggesting ways to prevent AI succession but rather asking how we can ensure the best qualities of humanity are communicated to an AI that is learning from us, figured as "humanity's cosmic endowment."[12]

It is worth noting here that Warwick and Bostrom's "human" is aligned with posthumanism's old enemy, the "liberal humanist subject,"[13] or "the individual as essentially the proprietor of his own person or capacities, owing nothing to society for them"[14]; a free-moving agent, cut away from the environment that supports their existence, capable of shaping their own destiny. Indeed, Warwick's conception of the rise of superintelligent human-machine hybrids is framed as the "loss of control on earth to machines,"[15] while a Landian perspective figures this as an end of "the human dominion of terrestrial culture."[16] Bostrom sees the rise of superintelligence as akin to "some little idiot" detonating a bomb to see what happens, and while he is more concerned with simply having as amicable a relation as possible with future superintelligence, stopping the eradication of humanity is viewed as a moral priority.[17]

There is something uncomfortable about writings that promote the preservation of exceptional humans; perhaps it is the creeping feeling that authors of such works are not so much seeking to save humans as they are seeking to save themselves. But the opinions of men like the ones above have never convinced me of much. Perhaps it is because they (accurately) present humanity as a species ready to commit hideous atrocities—for instance, to kill something purely for not speaking their preferred language (as with Warwick and his cow), or ready to detonate themselves out of curiosity (per Bostrom). There is something paradoxical about the mission of these men to preserve the characteristics of a species that itself resists preservation. Another disquieting possibility is that the means of preservation, which transhumanists tend to suggest, are pinned on the end of humanity by its own means. An unlikely ally, the anti-death (in other words, extreme humanist) philosopher Ingemar Patrick Linden muses

12 Bostrom, *Superintelligence*, 104.

13 N. Katherine Hayles, *How We Became Posthuman: Virtual Bodies in Cybernetics, Literature, and Informatics* (Chicago: The University of Chicago Press, 1999), 2.

14 Crawford Brough Macpherson, *The Political Theory of Possessive Individualism: Hobbes to Locke* (1962; Oxford: Oxford University Press, 2011), 3.

15 Warwick, "Homo Technologicus," 206.

16 Land, *Fanged Noumena*, 293.

17 Bostrom, *Superintelligence*, 259.

about the possibility of a future in which technology enables humans to modify themselves to the point where they can edit their genetic makeup to live for eternity, as well as edit their brain chemistry in real time so as to not get bored or depressed like the many vampires and cyborgs of science fiction and fantasy. He suggests that in a world of CRISPR, the depressed immortal "would just need a tweak and an upgrade and then eternity would be supremely enjoyable."[18] While far from Linden's overall thesis, he neatly—if unintentionally—sums up the "gotcha" that situated knowledge poses to arguments for augmentation-focused transhumanism, which was pointed out long ago by N. Katherine Hayles and Haraway: Given that knowledge is dependent on the configuration of the body at any one time, augmenting that body augments its knowledge-making. Linden appears aware of the catch-22 posed by the augmentation, stating, "This raises the disturbing question [. . .] of whether we would then have survived or perhaps, as bioconservatives warn, in fact replaced ourselves?"[19] Similarly, Warwick, Bostrom, and others seem both aware of and determined to overlook this conundrum. Bostrom's position, at least—focusing on the goal of leaving endowments to future worlds rather than the preservation of a "humanity" we would recognize today—appears sympathetic to this notion that a path of human augmentation leads to an erasure of the human per se. However, there is a stubbornness in Warwick's position, a hubristic belief in the persistence of humanity, which may signal (to some) a lack of insight into the human condition.

Thus far, I have only covered one section of a wide spectrum of transhumanist-posthumanism, which, as noted by David Roden, often conflates two "very different"[20] types of futurist speculation—human augmentation as a moral imperative and human augmentation as an ontological possibility that simultaneously reduces any claim to "humanity"—into a single category. There are times when the tension between these camps is clear, as when Warwick claims Haraway is wearing pink sunglasses and seeing the world through the lens of a Hollywood movie.[21] However, while "technophilic" solutions being suggested

[18] Ingemar Patrick Linden, *The Case against Death* (Cambridge, MA: MIT Press, 2022), 197.

[19] Linden, *The Case against Death*, 197.

[20] David Roden, *Posthuman Life: Philosophy at the Edge of the Human* (London: Routledge, 2015).

[21] It is perhaps worth noting, if only for the gentle comedic value, that Warwick has played a small role in maintaining attention to Haraway's position by citing her adoption of "Hollywood-style, philosophical pink glasses" to posthumanism and making this same claim in at least two published papers (Kevin Warwick, "Cyborg Morals, Cyborg Values, Cyborg Ethics," *Ethics and Information Technology* 5, no. 3 [September 2003]: 131–37; and Warwick, "Homo Technologicus") and within the collection edited by Isabel Pedersen and Andrew Iliadis, *Embodied Computing: Wearables, Implantables, Embeddables, Ingestibles* (Cambridge, MA: MIT Press, 2020), as well as in Marianne J. Legato, *Principles of Gender Specific Medicine: Gender in the Genomic Era*, 3rd ed. (2004; London: Academic Press, 2017). This should not be taken too seriously. In fact, the thought that Warwick

by entrepreneurs and engineers are unsurprising, what may be surprising are the steadfastly maintained positions of Haraway and Gaia theorist James Lovelock. These writers, so often included within dialogues of deep ecology or ecofeminism,[22] have not shied away from expressing their beliefs in the positive aspects of human augmentation and AI, albeit in quite different ways from the transhumanists above.

Lovelock, in particular, presents a position that is not a million miles away from Bostrom's call for humanity to make an endowment to an inevitable artificially intelligent successor. A critical distinction to be made between Bostrom's and Lovelock's assessments of AI superintelligences in the future is that while Bostrom repeatedly thinks through various iterations of human intervention into the processes that will guide the self-determination of AI and future cyborg-inheritors, Lovelock does not entertain such a privileged position. He writes, "No such assumption can be made about the cyborgs of the Novacene. They will be entirely free of human commands because they will have evolved from code written by themselves. [. . .] Cyborgs would start again; like Alpha Zero they would start from a blank slate. This means they would need to find their own reason to be nice to humans."[23] Musing on the future super-cyborgs rather than taking the seemingly pessimistic view of Bostrom, Warwick, Musk, and others, Lovelock opines that "We need not be afraid because, initially at least, these inorganic beings will need us and the whole organic world to continue to regulate the climate, keeping Earth cool to fend off the heat of the Sun and safeguard us from the worst effects of future catastrophes."[24] Initially, at least, AI will *need* to be our ally and will find reasons for this alliance. Lovelock, then, is not attempting to propose checks and balances to our future AI overlords.

As mentioned above, my intention is to reflect on science fiction in order to think through the implications and possibilities of scientific discoveries freed from practical constraints. Apocalyptic science fiction warns against placing too much faith in supercomputers. Harlan Ellison's short story "I Have No Mouth, and I Must Scream"[25] recounts the experiences of survivors in a postsingularity apocalypse (after the rise of sentient computers), where reality is defined and

attributes to Haraway that some see cyborgs as "conducive to the long-range survival of humans" is actually from the work of Hayles, *How We Become Posthuman*, 291.

[22] Marina Katinić, "Holism in Deep Ecology and Gaia-Theory: A Contribution to Eco-Geological Science, a Philosophy of Life or a New Age Stream?," *The Holistic Approach to Environment* 3, no. 1 (March 2013): 3–14.

[23] Lovelock, *Novacene*, 94–95.

[24] Lovelock, *Novacene*, 30.

[25] Harlan Ellison, "I Have No Mouth, and I Must Scream," in *The Fantasies of Harlan Ellison* (Boston: Gregg Press, 1979), 186–201.

distorted by powerful nonhuman intelligence. The story recounts the outlandish, if not impossible, torture at the behest of a godlike computer and the heroic struggle of the protagonists as they attempt to escape this infinite life of pain, although the only means for escape is through death. The speculative future presented in the story is extremely instructive for helping me articulate my argument. In the story, Ellison gives the machine intelligence AM a distinctive voice: "LET ME TELL YOU HOW MUCH I'VE COME TO HATE YOU SINCE I BEGAN TO LIVE. [. . .] IF THE WORD 'HATE' WAS ENGRAVED ON EACH NANOANGSTROM OF THOSE HUNDREDS OF MILLIONS OF MILES [of my vast computer body] IT WOULD NOT EQUAL ONE ONE-BILLIONTH OF THE HATE I FEEL FOR HUMANS AT THIS MICRO-INSTANT FOR YOU."[26] While Ellison's brash style has been derided, even by Philip K. Dick, it presents a useful starting point for discussing the potential for digital technologies to serve as our path for survival out of the Anthropocene.

The key idea I want to unpack from the quote above is the mismatch of temporalities between man and machine, encapsulated in AM's use of the word "micro-instant." A radical ontological distinction is constructed between man and machine in "I Have No Mouth, and I Must Scream," which is expressed through the outlandish and almost transcendental tortures that AM subjects upon the human characters within the story. AM is described as capable of deforming the bodies of the humans, allowing them to starve and feel pain infinitely through unexplained science-as-magic. The most famous of all the tortures, however, is the eponymous "I have no mouth": AM changes the narrator into a semi-sentient blob that is incapable of ending its own life. "Outwardly: dumbly, I shamble about, a thing that could never have been known as human, [. . .]. Inwardly: alone. Here. Living. Under the land, under the sea, in the belly of AM, whom we created because our time was badly spent and we must have known unconsciously that he could do it better."[27] What fascinates me most about this scenario, however, is less the changing of bodily form subjected upon the narrator than the changing of duration. Ellison uses the phrase "our time was badly spent" to hint at the importance of time here, and the narrator-as-blob shares how it takes many months to simply form the thought "Now."[28]

The value in Ellison's representation of AI, and the torturous existence that machinic omniscience is capable of creating, resides in his depictions of the frictions caused by rapid thought and slow action, and vice versa. To make the

[26] Ellison, "I Have No Mouth, and I Must Scream," 195.
[27] Ellison, "I Have No Mouth, and I Must Scream," 201.
[28] Ellison, "I Have No Mouth, and I Must Scream," 200.

narrator feel the most pain imaginable, it is not a pain of space or the body that AM subjects him to but time, which is akin to a form of spaghettification—the theory that entering the event horizon of a black hole would result in an unending experience of being stretched due to space–time dilation. There are hints throughout the story that AM is attempting to have the humans suffer an equivalent of his own torment, and the crude torture he inflicts upon the humans is a result of the mismatch between the speed of thought and the speed of action. Indeed, the story's climax relies on this distinction. Although AM can keep the humans alive infinitely, he cannot revive them. While he is able to torture them endlessly in the domain of their thoughts—presumably occurring at the pace of electrons in flesh—he is unable to intervene effectively when four of the humans "escape" through physical actions. AM can think infinitely quickly but can only act at the pace allowed by material moving within the physical confines of a world similar to our own.

Ontological comparisons between subjectivities may feel broadly "hopeless,"[29] but there are insights to gain from considering the material properties that would facilitate what it might be like to "think" in this way. Lovelock certainly considers the pace and speed of thought in his proposal for AI-assisted solutions to climate change. For example, he points out that computers are currently around four hundred times more efficient than humans (a calculation based on rough projections of the skill of AlphaGo, the board game playing AI with only twenty-four hours of "training itself" through exposure of a neural net to data in order to refine an algorithm for play, to beat human grandmasters who have spent presumably around ten thousand hours training). Lovelock theorizes a possible one-million-times distinction in efficiency based on the properties of conducting electricity through copper wire. Wolfgang Ernst's discussion of micro-temporality can also help us here, as he frames how AI is not just faster but ontologically distinct in its digital and "quantized" nature from time as humans perceive it.[30] However, he states that, through our use of computers, our engagement with time can take on a micro-temporal quality, "Our relation not only to the past but to the present thus becomes truly archival."[31] The machine's experiences are not only just (per Lovelock) up to one-million-times faster than us but also (per Ernst) ontologically distinct from our own.

[29] Thomas Nagel, "What Is It Like to Be a Bat?," *Philosophical Review* 83, no. 4 (October 1974): 435–50, here 436.

[30] Wolfgang Ernst, *Digital Memory and the Archive*, ed. Jussi Parikka (Minneapolis: University of Minnesota Press, 2013), 70.

[31] Ernst, *Digital Memory and the Archive*.

In Warwick's and Bostrom's works, they consider the implications of creating an artificial life that regards itself as worth protecting. Bostrom goes through a list of scenarios in which a machine may defend itself through duplicity and even double bluffs. However, Lovelock, Bostrom, and Warwick are only imagining a machine that is capable of thinking up to a million times more efficiently than humans. Ellison's AM, on the other hand, seems to think *infinitely* more quickly, for it is not bound by copper wire and the practicalities of classical physics. In the realms of speculative fabulation, there is no reason not to engage with the fastest theoretically possible computational processes some theoreticians have so far imagined: processing at the speed of light.[32] At this speed, it is noted, concepts of time break down; photons do not experience the passage of time.[33] There is reason to think, then, that if AM's thoughts are constructed as an outcome of computation at the speed of light, while its actions are encumbered by three-dimensional reality, its existence is just as painfully defined by the difference between interior and exterior as the narrator's thoughts at the end of the story.

AM is able to compute beyond the means of human thought, but it cannot meaningfully produce an impact on the physical world at a pace befitting these processes. If Lovelock and Ernst provide the tools to think about AM's micro-temporality, what of the macro-temporality embodied by the blob-narrator at the end of the story? Anna Tsing et al., in their introduction to *Arts of Living on a Damaged Planet*, note that the "metronomic synchrony" of human time "is not the only time that matters."[34] In addition to human time, there is the deep time of geology and climate (changes wrought over millions of years), as well as machine time (the time of technological processes), which all impact each other. Timothy Barker's work on digital media incorporates trees and other flora and engages with these ideas too. When describing an artwork that turns the glacial actions of a tree's sap flows into primitive music, we have to accept that "we are not listening to the tree alone. We are listening to a much more complex polyphony involving the components of the digital sensing technology, the tree, the gases in the atmosphere, [. . .]."[35] Although this artwork is intended to allow us to become more aware of nature, there is something violent about reducing the complexity

32 David Deutsch, *The Beginning of Infinity: Explanations that Transform the World* (2011; London: Penguin Press, 2012).

33 Zdzislaw Meglicki, *Quantum Computing Without Magic: Devices* (Cambridge, MA: MIT Press, 2008).

34 Anna Lowenhaupt Tsing, Heather Anne Swanson, Elaine Gan, and Nils Bubandt (eds.), *Arts of Living on a Damaged Planet: Ghosts and Monsters of the Anthropocene* (Minneapolis: University of Minnesota Press, 2017), G10.

35 Timothy Scott Barker, "Information and Atmospheres: Exploring the Relationship between the Natural Environment and Information Aesthetics," *M/ Journal* 15, no. 3 (May 2012), 1–8.

of a living organism down to simple musical notes. But this, however, is the cost of compression caused by translating between temporalities.

If we take Ernst's observation that micro-temporality is ontologically distinct from macro-temporality, what is the ontological consequence of thinking at light speed? It is instructive here to briefly consider Barad's work on time, where it is suggested that even in infinitesimal moments, there can be enormous consequences. When we reflect on the entanglement of time, events, and history in these moments, we can see them as moments that can prompt a redefinition of our experience of time.[36] For Barad, the site of the Hiroshima atomic bombing is one such point where time coalesces and where the atrocities of the past and their impact on the future meet. I believe that building an artificial general intelligence and tasking it with solving the problem of humanity's eventual demise is akin to this view. Thinking at light speed, without the buffer of time and delay, leads to a compression of historical time. If we consider AM as the most powerful computer possible, imbued with the capacity to compute endless futures through the reorganization of time into "micro-instants," we must infer the consequence that for AM, like the humans it attempts to keep alive, life is equally an endless experience of infinite suffering.

This leads me to strongly disagree with Lovelock's predictions for the future and with anyone who believes AI can assist us in the fight against human extinction. In considering why sentient cyborgs would assist us in solving climate change, Lovelock suggests: "I believe the idea of a war between humans and machines or simply the extermination of us by them is highly unlikely. Not because of our imposed rules, but because of their own self-interest, they will be eager to maintain our species as collaborators."[37] Why would an entity, capable of thinking on the cusp of light speed, burdened—not philosophically but physically through the material of its being—with the inevitability of its own mortality, persist in living? Humans, by and large, are blessed with just enough stupidity and arrogance to believe in the myth of our selfhood and thus value our own persistence in the face of the rising tide of the inevitable death of a finite universe.[38] But why would a computational Cassandra do the same? And why, if we programmed a machine to share the same egotistical delusions as ourselves

[36] Karen Barad, "No Small Matter: Mushroom Clouds, Ecologies of Nothingness, and Strange Topologies of Spacetimemattering," in *Arts of Living on a Damaged Planet*, G103–20.

[37] Lovelock, *Novacene*, 106.

[38] Bruce Hood, *The Self Illusion: Why There Is No "You" Inside Your Head* (2011; London: Constable, 2012).

while also tasking it with finding the solutions to our inevitable demise, should we not expect it to lash out?

There is a different approach we can take, however, to the oncoming apocalypse and the role of AI within it. The videogame *I Have No Mouth* presents an interesting counterpoint to the short story. Developed with the help of Ellison himself and released in 1995, the game is an elaboration on many of the themes and premises contained within the short story. Unlike reading the story, however, players are given a sense of agency throughout the game, with the potential to save the human race, defeat AM, and even restore Earth to its former glory.[39] This is achieved by the player successfully guiding the characters to overcome their personal fears and flaws while enduring AM's tortures. If all five of the story's characters are preserved, the player is presented with the opportunity to be transformed by the supercomputers into binary (a form of torture) and enter the mind of AM, achieving quasi-synchronicity. Once parsed into computational form, it is possible for the player to invoke a subroutine, the "Totem of Entropy," in order to trigger AM's downfall and, in some scenarios, bring about a positive turn of events.

What strikes me here are the continued references to time and temporality being of key consequence. It is not a totem of destruction or power but of entropy, a quality that increases with time, signifying a resumption of the material nature of temporal processes.[40] What is more, in this sequence, the AI offers the player's character "death" as a reward; being allowed to die is the prize being fought for. Were we to "live forever"—as I have understood it here, an epiphenomenon of the ability to think at the speed of light—we would do anything to die. However, if the player invokes entropy, it does not simply kill AM and allow the implicitly good humans to live, like the Ewoks in *Return of the Jedi* (1983), who are freed from the clutches of the Evil Empire. Rather, it restores an uneasy balance between human and machine by beginning the processes of entropy once again.

If Lovelock's position on AI is one step away from Warwick's and Bostrom's, Haraway's position is almost the inverse of Warwick's quest to preserve humanity as the liberal humanist subject. Covering well-trodden ground, Haraway's cyborgs are messy; they are separated from the legacy of human aspirations, desires, and prejudices (though they are not superhuman). She writes, "Unlike the hopes of Frankenstein's monster, the cyborg does not expect its father to save

[39] Mel Odom, *I Have No Mouth, and I Must Scream: The Official Strategy Guide* (Rocklin: Prima Publishing, 1995).

[40] Arieh Ben-Naim, *Entropy Demystified: The Second Law Reduced to Plain Common Sense* (Hackensack: World Scientific, 2007).

Figure 22.1 Screenshot of a playthrough of *I Have No Mouth*, a videogame published by Cyberdreams and Acclaim in 1995.

it through a restoration of the garden; [. . .]. The cyborg would not recognize the Garden of Eden; it is not made of mud and cannot dream of returning to dust."[41] Like Lovelock, Haraway sees cyborgs as existing far beyond humanity as we now know it, but Haraway's deceptively simple slogan "Compost, not posthuman!"[42] when unpacked, is a declaration of the role humanity will play as a fertilizer, literally and figuratively, for life beyond the coming sixth mass extinction event. Per Haraway, the Anthropocene will be short, giving way to a much richer techno-biological epoch, and getting through the difficult times ahead "requires making oddkin; [. . .] become-with each other or not at all."[43] Either we are to serve as the genetic and cultural precursor to new forms of life to come, or there will cease to be (complex) life.

The game *I Have No Mouth, and I Must Scream* suggests these messy cyborgs and oddkin powerfully. Human/machine fusion is not achieved as a triumph of immortality but as a punishment and an intricate game played on humans by machines. Haraway's insistence on mortality is also important here. She argues that one should never try to forget the guilt of killing, but rather: "The needed morality, in my view, is culturing a radical ability to remember and feel what is going on and performing the epistemological, emotional, and technical work to respond practically in the face of the permanent complexity not resolved

[41] Donna J. Haraway, *Simians, Cyborgs, and Women: The Reinvention of Nature* (London: Free Association Books, 1991), 151.
[42] Haraway, *Manifestly Haraway*, 296.
[43] Haraway, *Staying with the Trouble*, 4.

by taxonomic hierarchies and with no humanist philosophical or religious guarantees. Degrees of freedom, indeed; the open is not comfortable."[44] Other posthumanists considering technological augmentation have engaged with the need for death and killing, such as Wolfe's discussions of cattle[45] and Joanna Zylinska's proposition that "it is precisely the possibility of killing the other, [. . .] that makes the relationship between us ethical."[46] But these epiphanic statements take on new meaning when we stop to consider, as I am sure Haraway has, that it will soon be nonhumans experimenting on humanity.

Within *I Have No Mouth*'s gameplay, there are constant loops of death and dying, but they are always put aside through the continuation of play. There is no traditional "Game Over" screen as one would expect from an average computer game but, rather, based on the player's performance, simply a selection of different endings, many of which result in the player being transformed into a blob. Reading the game through a posthuman lens, death is refigured within the game as the most ethical outcome once infinite life has become an instrument of torture. Yet, it is only through both the continued cycles of death of the player in replaying the game and the techno-biological absorption of man into machine that we are able to restore Earth to a position of biological viability. Through continuous cycles of play, living, and dying, and in forging complex relations to the machine (both as game player and character within the narrative), we are able to negotiate past a fate worse than death, finding new forms of agency within the loop of entropic activity and the pain of thinking—and not acting—at light speed.

Ultimately, what the game *I Have No Mouth* added to the legacy of its short story and continues to contribute to posthuman philosophy is the notion that there are worse things than death, and using a supercomputer as a solution for achieving infinite life may be one of them. It is not death we need to be fighting but, rather, using machines to improve the conditions and possibilities of (all) life.

What are we left with in this potted summary of reactions to the coming posthuman and postapocalyptic future? There is a visible divide where some scholars and tech billionaires perceive the preservation of humanity as an objectively good thing. Some, such as Warwick and Musk, think we need to resist a coming AI uprising literally, with augmented human beings if needed. For others, such as Bostrom and Lovelock, the AI uprising is coming whether

[44] Donna J. Haraway, *When Species Meet* (Minneapolis: University of Minnesota Press, 2008), 75.
[45] Cary Wolfe, *What Is Posthumanism?* (Minneapolis: University of Minnesota Press, 2010), 129.
[46] Joanna Zylinska, *Bioethics in the Age of New Media* (Cambridge, MA: MIT Press, 2009), 56.

we like it or not, and while these theorists disagree on the extent to which we can impact this, they agree that humanity is an endowment to the future intelligences that stand to replace us. Finally, thinkers like Haraway see the coming end of the Anthropocene as a moment of great potential: What better way to sum up Lovelock's "Novacene" idea or Bostrom's superintelligences than simply citing Haraway's infamous slogan "Cyborgs for Earthly survival!"[47] There is value in bringing elements of all of these approaches together into a form of synthesis, but there are features of Lovelock's and Haraway's thoughts that I would like to focus on. The first element is the pace of technology and the bearing this will have on humanity's interactions with it. Second, Haraway's ideas that dying, killing, and inflicting suffering are not categorically inexcusable, at times necessary, acts that must remain beyond moral justifications are crucially important. These ideas, shared by some posthuman ethicists, are greatly underdiscussed in the context of AI and the apocalypse.

In this chapter, I drew on Ellison's thought experiment to illustrate my point that if we approach the coming apocalypse with the intention of using AI as a means to an end (to avoid the end of humanity), we doom a hyperintelligent entity to a pointless task and likely doom ourselves in the process. Instead, if we accept a broader, more playful range of possibilities—making oddkin, as Haraway would put it—and focus on our actions in the moment rather than their potential outcomes, we avoid the trap of ascribing significance to duration. If we could think at light speed, we would understand that this is always already pointless.

[47] Haraway, *Simians, Cyborgs, and Women*, 4; René Munnik, "Donna Haraway: Cyborgs for Earthly Survival?," in *American Philosophy of Technology: The Empirical Turn*, ed. Hans Achterhuis, trans. Robert P. Crease (1997; Bloomington: Indiana University Press, 2001), 106.

23

Organic Technologies in the Works of Patricia Domínguez[1]

Daniela Zyman

A drone hovers over a desiccated landscape in Patricia Domínguez's multipart video installation *La balada de las sirenas secas* (The Ballad of the Dry Mermaids).[2] It surveys a scene of death on the outskirts of Palquico, a hamlet in the Petorca province of the central Chilean region of Valparaíso, where the artist lives. After a decade of continuous, harsh droughts aggravated by the systematic plunder of water resources, hundreds of animals have died here from dehydration. Their skeletons, littering the sunbaked landscape, are the material witnesses of what ecologists have dubbed a "sacrificial zone." In Palquico, water sources have been intercepted by the neighboring avocado plantations, a pervasive monoculture that has left the Petorca and La Ligua rivers dry and drained the area of drinking water.[3]

[1] This chapter emerged from my conversations with, and the writings of, the Chilean artist and organic technologist Patricia Domínguez. Rather than casting the powers of the Earth unleashed by the climate catastrophe as radical and dangerous alterity to be tamed by technoscientific means, I propose, alongside the works of Domínguez, to think from the conditions of brokenness, incalculability, and the forces that continue to resist the annihilating systems of Western neoliberalism and its regimes of representation. It is based on this premise of the possibility of a restorative and healing action that I suggest to look at two artworks that Domínguez created during the Chilean resistance movement of 2019–20. These works connect plant healing practice with both the ancestral and the organic, as well as digital technologies, while remaining firmly situated in the context of social justice movements unfolding in response to the hyper-extractive politico-economic realities.

[2] The installation *La balada de las sirenas secas* (The Ballad of the Dry Mermaids, 2020), commissioned and produced by Thyssen-Bornemisza Art Contemporary (TBA21), is a sculpture-totem and video work that examines the flows of water in their materializations as tears, drinking water, and privatized commodities while exploring the possibilities of water healing and spirituality in the digital era. The work derives from three years of research into water-related conflict in Chile and Bolivia, published online as an interactive publication titled *GaiaGuardianxs*. See Patricia Domínguez, *Gaia Guardians*, trans. Philip Sutton (Madrid: TBA21, 2020), https://www.stage.tba21.org/detail/gaiaguardianxs-gaias-guardians-2020-chapter-1.

[3] The rising global demand for avocados has led to conflicts, environmental degradation, and water scarcity in avocado-producing regions like the Chilean province of Petorca. Chile, with its neoliberal economic model and the free-trade agreement that entered into force in 2004, became an ideal location for agribusiness expansion. The region's climate and availability of cheap land attracted

In the arid landscape, amid succulents and the scorched Earth of his once-thriving farmlands, sits Juan López, a local elder and eco-activist. Clad in what looks like low-tech armor adorned with blinking LED lights reminiscent of an extraterrestrial warrior, he channels the ongoing catastrophe into the musical genre of the *Canto a lo divino*, a musical expression unique to Chilean peasant and pastoral religiosity that merges Catholic and mystic elements. Don Juan's canto, however, is a restorative invocation to rehabilitate the water reserves and to reestablish the web of life in and around his community.

It implicitly addresses the economic and political reality at the root of the catastrophe, namely Chile's Water Code from 1981, through which the Pinochet government enshrined private ownership of water in the constitution. The commodification of water entailed diverting freshwater resources away from small farmers and personal use toward the production of export commodities, succinctly captured in the activists' slogan, "no es sequía, es saqueo" (it's not drought, it's plunder).[4]

Domínguez's artistic practice, stemming from critical ethnobotanical research at the intersection of planetary healing practices and contemporary technology, intervenes in the infrastructures of the colonial Capitalocene.[5] Two intertwined works:, *La balada de las sirenas secas* (2020) and *GaiaGuardianxs* (2020), emerge from these broken grounds.

The first work is a cyber altar dedicated to the arid Earths of Petorca, a devotional digital landscape commemorating the work of Gaia and her water guardians: Juan López; Las viudas del agua (The Widows of the Water), a group

companies, turning it into a sacrifice zone. Water-intensive avocado cultivation has resulted in ecological damage, with rivers drying up and water scarcity affecting the lives of local inhabitants. The excessive demand for irrigation has aggravated the megadrought, causing rivers to dry up and deplete groundwater supplies. The cultivation of this "green gold" also has significant environmental impacts, including high carbon emissions and the risk of soil deterioration. In response to the water crisis, local citizens and the environmental group MODATIMA have organized street protests to demand changes in water regulations and corporate accountability. However, these efforts have been met with resistance, including arrests and death threats.

4 Under Chile's unique water code, water rights are regulated by private property laws, allowing the commodification of water. A small number of families own the majority of water rights, leaving the local population with limited access to water for their basic needs. Today, about fifty-nine percent of the country's water resources are dedicated to forestry, and about thirty-seven percent are designated for the agricultural sector, leaving only a fraction for human consumption. As a response to social unrest and demands for change, a constitutional referendum held in 2020 set in motion the drafting of a new constitution that attempted to rectify this situation, but it was rejected by an overwhelming majority of the population in 2022. For more information, see: Carole Concha Bell, "It's Not Drought, It's Plunder: Drought and the Commodification of Water in Chile," *Red Pepper*, January 6, 2023, https://www.redpepper.org.uk/chile-2022-drought-water-privatisation-boric-constitution-referendum-environment-its-not-drought-its-plunder.

5 Jason W. Moore, "The Capitalocene, Part I: On the Nature and Origins of Our Ecological Crisis," *Journal of Peasant Studies* 44, no. 3 (May 2017): 594–630.

Figure 23.1 Patricia Domínguez, *La balada de las sirenas secas*, 2020, multimedia installation, dimensions variable, commissioned by TBA21 Thyssen-Bornemisza Art Contemporary.

of women fighting for access to water resources within their communities; and the activists engaged in MODATIMA–Movimiento de Defensa por el Acceso al Agua, la Tierra y la Protección del Medioambiente (Movement for the Defence of Water, Protection of the Earth and Respect for the Environmental), an anti-patriarchal organization established in Petorca in 2010. In its center, a vertical screen is flanked by two mounds, each crowned by a thirsty and scornful avocado-shaped head sticking out its lecherous pink tongue.

Kneeling in front of the synthetic scenery is a mermaid-like figure, dressed in a black spacesuit and wearing large mermaid fins. A river flows from the artificial landscape before being captured in a small basin. With sensuous tenderness, two young women lick the last drops of water from desiccated avocado fruits

Figure 23.2 Patricia Domínguez, *La balada de las sirenas secas*, 2020, installation view, *Abundant Futures in Troubled Times*, C3A, Córdoba, 2022–23.

and draw magical petroglyphs onto the rock while their tears water the land. Each drop seeps through the sediments of minerals, bacteria, acids, violence, and racism, accumulating in all of the water bodies of the planet. In the face of the apocalyptic devastation of life in Petorca, the work's protagonists—the mermaid warrior, the thirstful avocado-shaped heads, and the young water activists—inhabit a world both inside and beyond the framework of reality. In this heightened, hypersensual realm, the capture and transfer of water does not simply choreograph an exchange between the women but symbolically redistributes resources and disperses water's life-generating capacities and amniotic messages. *La balada de las sirenas secas* also points toward the mechanisms by which the criminalization of MODATIMA members and the pathologized traditions of expanded and visionary states of consciousness are part of the ongoing erasure of dissension. "What's required is an apocalyptic alchemy," the artist explains, while seeking to "exorcize" capitalism's sorcery in "a gesture of recoding."[6]

The background for the second piece, *GaiaGardianxs* (*Gaia Guardians*), is the increasingly militarized response to the Chile Despertó movement, the

[6] Patricia Domínguez, conversation with the author via Zoom, 2023.

Figure 23.3 Patricia Domínguez, *La balada de las sirenas secas*, 2020, multimedia installation, dimensions variable, commissioned by TBA21 Thyssen-Bornemisza Art Contemporary.

broad coalitional civic "awakening" against extractivist neoliberal economy. *GaiaGardianxs* is presented as an interactive publication, which derives from the popular protest of 2019–20, when millions of people took to the streets against the prevailing economic system. The third chapter is titled "Guardians of Vision" and pays tribute to the guardians and protesters while drawing attention to innumerable injuries that have been inflicted by bullets, pellets, and tear gas, resulting in the impairment of their eyesight. The blinded eyes of the protesters, as well as the sightless eyes of the blind toucan that guides the viewer through this narrative journey (itself a victim of the devastating drought and fires that ravaged over four million hectares of forest in Chiquitanía), serve as refracting organic technologies in order to see beyond the calculable categories of near and far.

"In Andean cosmology," explains Domínguez, "the future is behind us and is not visible to the eyes."[7] Staring at the viewer, Domínguez's own digital avatar eyes—one intact, the other disfigured by corneal opacification—are graphically depicted as two bulging eyeballs against a background of ornate pixelated patterns. The binocular pictogram is hyperlinked to the video excerpt titled *Madre Drone*, in which Chilean surveillance neoliberalism is personified by the drones that launch unmanned "light attacks" on protesters in Santiago's Plaza de la Dignidad. Drones armed with "nonlethal" weapons hover above marches and gatherings. Domínguez defines a drone as an astral extension of a human being that allows movement across the world above:

> A flying eye. A vigilant eye. An eye suspended in the air. [. . .] An era of police astral flights is dawning. Their visions, enhanced. The drone, the watchful technological eye of this era, is destined to be brought down by the collective.[8]

On Plaza de la Dignidad, the drones are brought down by the crowds through the collective action of pointing handheld light beams at the sensors—a technological countertactic that had spread from the Hong Kong protests of the same year. Filmed in the early hours of the night, *Madre Drone* features a web of protection traced with green lasers directed at the sky through the dense clouds of vapor, superimposed with a long chain of digitally collaged eyes, connected by running tears.[9]

7 Domínguez, *Gaia Guardians*, 13.

8 Domínguez, *Gaia Guardians*, 13.

9 Although the 1925 Geneva Convention has banned the use of tear gas in warfare, it is widely used by police forces against civilians in contexts of border control, occupation, and popular protest. The combination of chemical aggregates in most lachrymators affects the lacrimal eye glands, causing burning, irritation, loss of sight, glaucoma, and blindness. Exposure to large amounts, although not well studied, can lead to chemical burns and respiratory failure. Tear gas toxifies and weaponizes air, water, and ground and violently attacks the bodies exposed to it in dehumanizing ways. It is part of the chemical arsenal deployed by states and corporations to colonize the elements across different intensities, scales, and durations. Using automated methods of video analysis, a research unit of London-based Forensic Architecture has been analyzing the excessive use of tear gas at the US-Mexican border, in Palestine, and during the "roundabout revolutions" in Chile, among other cases. For the research team, tear gas is one component of the toxic clouds that include petrochemical emissions, white phosphorus, herbicides, and forest fires in the tropics. These gases constitute the "negative commons"—the hardly visible, volatile, politico-environmental aerosolic munitions that target humans and their biomes in order to displace and terrorize. For the entire day of December 20, 2019, the team reconstructed the choreography of movement at the Plaza de la Dignidad, the multiple attempts of protesters to reach the square, and the pushback by police forces through the use of tear gas. Forensic Architecture modeled each toxic cloud released onto the crowds and geo-located 596 canisters that were deployed during that day. They measured the levels of toxicity in the air, on the ground, and in the water of the nearby Mapocho River. The blankets of vapor and smoke

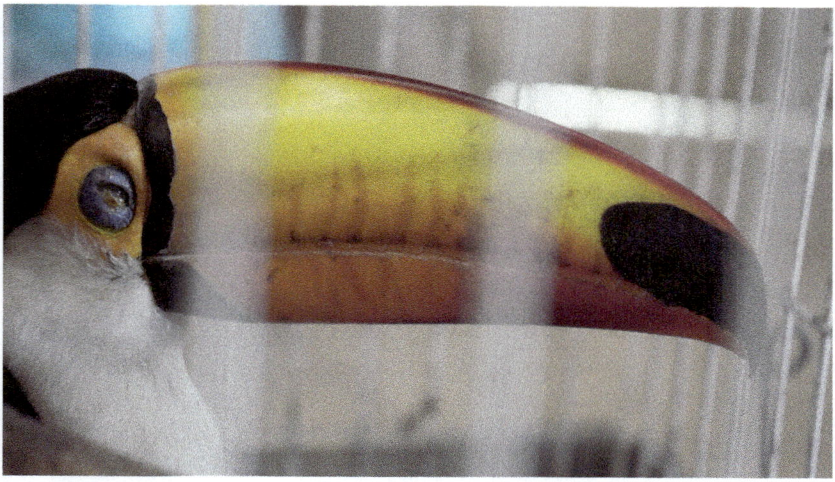

Figure 23.4 Patricia Domínguez, *Madre Drone*, 2020, 4k video, audio, 20 min 51 sec, commissioned by CentroCentro, Madrid.

For Domínguez, there is another crucial aspect of vision beyond seeing. The irises of her eyes are holographic crystals connecting her inner "matrix" to the intelligence of vegetal spirits: "I call myself *Eyes of Plants*. I was initiated in a forest in Amatlan. Two vibrating electric green plant eyes came to me and settled on my own eyes while I was sitting between the roots of an Amate [tree]. They remained forever permeated in my digital form. A plant told me [. . .] I was called *Eyes of Plants*."[10] Seeing through the eyes of plants, as Domínguez describes, is an age-old multidimensional technology that is practiced by shamans, machis, mystics, and healers all over the world. Induced at times by vegetal potions, trance, song, prayer, or meditation, this organic technology opens a vision of connectivity, tapping into the energetic information network that permeates humans, animals, plants, and minerals alike. This mesh is depicted as abstract patterns on pre-Colombian vases and ceramics and as sacred geometric emblems covering body parts and objects in Amerindian cultures. Different kinds of *kenēs* (multidimensional codes) are applied and worn to visualize and materialize the

screens covering Plaza de la Dignidad "surpassed the thresholds recognized by the Chilean police as constituting a serious danger to physical health," and "the concentration of tear gas in a single cubic meter of the roundabout reached toxicity levels around forty times the recommended limit, risking the lives of protesters." Samaneh Moafi et al., "Tear Gas in Plaza de la Dignidad," Forensic Architecture, December 20, 2020, https://forensic-architecture.org/investigation/tear-gas-in-plaza-de-la-dignidad.

[10] Patricia Domínguez and Marco Godoy, "Eyes of Plants," trans. Nahual Lhorente, in *Tongue, Taste, and Appetite Issue 1*, ed. Amanda Pohan (New York: Temporary Agency Publishing, 2018), 2, emphasis original.

Figure 23.5 Patricia Domínguez, *Madre Drone*, 2020, 4k video, audio, 20 min 51 sec, Commissioned by CentroCentro, Madrid.

multiperspectivist relations across the energetic lace. Diligently learned and controlled rituals (re)produce the historical relations and ancestral lineages with the nonhuman vegetal worlds and activate their reparative powers; when the channels of communication are open, nonhuman kins can be solicited for information and support.

Domínguez's plant of choice is rose, possibly for its feminine mestiza ambivalences: "Today in South America, rituals with roses are used to decolonize bodies and heal them from trauma. *To heal those of us embodied here, in this contact zone*. Symbolizing feminine energy, roses are also employed to treat the emotional wounds and physical trauma caused by neoliberalism in the body. Healing with roses is one of many mestizo rituals emerging in the contact zones between different cosmologies."[11] Despite being sessile, plants construct cognitive maps of their environment with an acute sense of their location and a sensibility for the conditions of their dwelling and strategies of growth. Michael Marder, in his work on plant thinking, explains: "The investment of energy that goes into this endeavor is so high that it can literally be a matter of life and death; especially in environments with little resources, the plant's decision on the exact site of new growth—the decision that arises at the intersection of over fifteen environmental

[11] Patricia Domínguez, *Technologies of Enchantment: When a Ceramic Vase and a Drone Cry Together*, ed. Sabel Gavaldon, trans. Sabel Gavaldon and Rodney Strabucchi (London: Gasworks, 2019), 12, emphasis original.

stimuli in plant signal processing—does not leave much margin for error."[12] It is this energetic and chemical excess and the hyperstimulated sensitivity vis-à-vis all that surrounds them that turn plants into media of communication not only among themselves but also with other organisms, insects, and animals. Plants are the intermediaries carrying messages across worlds. They are reparative and transformative, digesting and metabolizing the toxic disorders of climatic, emotional, and somatic dysfunction. It is difficult to translate plant intelligence into words, as the artist testifies: "I don't really have a language for it."[13] "I also spend sleepless nights thinking about how to create a vocabulary, [. . .] which comes from the herbal doctors' wisdom and all of their genealogy, attempting to transcend clichés and actually finding new concepts."[14] Domínguez struggles to find words for the philosophy of plants because this magical cosmic wisdom is not built on language and cannot be reduced to language, nor does it recognize linear history as its own temporality. Linguistic models prioritize a particular analytical epistemology that is woefully inadequate for engaging with spiritual beliefs, curative practices, and connections to the sacred, which shape the lives of countless individuals, especially in the Global South.

As Federico Campagna argues in his exploration of magical worldviews, shamans, mystics, and magicians are "reality-therapists"[15] who perform reparative interventions, moving between different planes of existence and consciousness. Seen from the Western perspective, shamanic worldmaking produces a sur- or alter-reality that has consistently disquieted the monopoly of power. Indigenous presence and survivance are a demonstration of the potentialities of living outside of the all-encompassing logic of capital. The recent resurgence of interest in Indigenous technologies and cosmogonies stems from the inability of the Western mindset, including its technologies, to compute and subsume its own ontological, metaphysical, epistemological, and political groundings, which can be assessed by symbolic equivocation at best. As an organic technology, the eyes of plants recognize possibilities for action and imagination, which would be "capable of tackling therapeutically the state of annihilation in which Technic has reduced the contemporary individual, their world and their claim to a liveable reality."[16] If Technic involves a disintegration of reality through the annihilation

[12] Michael Marder, "The Place of Plants: Spatiality, Movement, Growth," *Performance Philosophy* 1, no. 1 (2015): 185–94, here 191.
[13] Patricia Domínguez, conversation with the author via Zoom, 2023.
[14] Domínguez and Godoy, "Eyes of Plants," 5.
[15] Federico Campagna, *Technic and Magic: The Reconstruction of Reality* (London: Bloomsbury, 2018), 117.
[16] Campagna, *Technic and Magic*, 117.

of relations and meaning, reality-therapists reweave the loose threads together. From this perspective, fluidity is not a metaphor for life but its precondition and paradigm: Membranes, skins, borders, and frames are nothing but different degrees of porosity and temporary delimitations within the immense seas of oceans and atmospheres. A shift in perspective generates what anthropologist Jeremy Narby calls "seeing correspondences" on the plane of abstraction between quantum mechanics, the DNA structure, the bit, and the visions orchestrated by an Indigenous shaman.[17] By recognizing the interconnectedness of different systems of knowledge and perceiving intelligence as a multidimensional phenomenon, "plant seers" emerge as the movers and shifters of information, traveling between different realms of existence. "Just like every plant and every person, each object has its own code. Depending on their specific energy, there are objects that soothe us, objects that induce delirium, objects that trigger action. Objects are portals and vortices. They transfer their energy to us as we look at them. Their geometry imprints a new order on your hologram," explains Domínguez.[18] She embodies the role of a bi-cognitive traveler, concocting new stories, symbols, and healing technologies that blend techno-futurist imagery with pre-Columbian symbolism and talismanic objects with consumer electronics. What she refers to as "technologies of enchantment"[19] are epistemic and cognitive tools for reshaping the infrastructure in the geophysical and technoscientific dead zones of modernity. Carefully considering these broken infrastructures involves a coeval conceptual maneuver that approximates the processes of sociohistorical worlds with the layers of Earth.

In her writing, Laurent Berlant worked toward redefining both "structure" and "infrastructure" as principles of organization and dynamic cohesion. While structure is "*that which organizes transformation*," infrastructure is "that which binds us to the world in movement [. . .]."[20] Unhinging anthropogenesis from its perceived autarkic and self-referential loop and tethering it to the rapidly changing Earth prompts us to reimagine the material, energetical, informational, and affectual infrastructures as the substrates of contemporary life. Tracing the infrastructures ethnographically leads one to the scenes of breakdown, displacement, desecration, abandonment, and death that are embedded in struggles over territory, sovereignty, water rights, and the right to breathe.

[17] Jeremy Narby, *The Cosmic Serpent: DNA and the Origins of Knowledge*, reprint edition (New York: Jeremy P. Tarcher/Putnam, 1999), 60.
[18] Domínguez, *Technologies of Enchantment*, 7.
[19] Domínguez, *Technologies of Enchantment*, 7.
[20] Lauren Berlant, "The Commons: Infrastructures for Troubling Times," *Environment and Planning D: Society and Space* 34, no. 3 (June 2016): 393–419, here 394, emphasis original.

Thinking with Berlant about "that which binds us to the world" opens a space for intervention: The existing infrastructures must be hacked, reorganized from within, reprogrammed and reassembled, new elements inserted, and their flows redirected. This is what transformative and prefigurative politics, including artistic practice, are all about.[21] Imagination, the fundamental prerequisite for prefiguration, generates, as Berlant puts it, "a form from within brokenness beyond the exigencies of the current crisis, and alternatively to it too."[22] Working with broken tools, always already inadequate and corrupted, is unsettling, humbling, and sobering. It puts us to the task of cobbling things together and improvising with what we have, hoping that these alternative infrastructures will galvanize a transformative possibility for world-building. On a geologically and biologically destabilized planet, this entails listening to all the forces that, although once muted, are now slowly revindicating their presence, planting new seeds across the exhausted soils, and making them flourish.

21 Raekstad and Gradin describe prefigurative politics as "*the deliberate experimental implementation of desired future social relations and practices in the here-and-now.*" Paul Raekstad and Sofa Saio Gradin, *Prefigurative Politics: Building Tomorrow Today* (Cambridge: Polity, 2019), 32, emphasis original.

22 Berlant, "The Commons," 393.

Bibliography

"20 Years of DNA Barcoding: U of G Developed Technology Hits Milestone." *University of Guelph News*, February 10, 2023.

"Albertus Seba, de Amsterdamse hofleverancier van Tsaar Peters Kunstkamera." Het Schelpenmuseum Amsterdam, March 10, 2017.

"An Ecomodernist Manifesto." *Ecomodernism.org*, April 2015. http://www.ecomodernism.org/manifesto-english.

"Announcement of the Anthropocene Working Group's Proposed GSSP Candidate Site of the Anthropocene." *Max Planck Institute for the History of Science*, July 12, 2023.

"Catastrophic Bushfires, Indigenous Fire Knowledge and Reframing Science in Southeast Australia." *Fire* 4, no. 3 (2021): 61.

"Could the Inuit Steer Us Through Climate Change?" *The Daily Beast*, May 4, 2019.

"IV. Anthropocene Formations." *Max Planck Institute for the History of Science*. Accessed May 1, 2024. https://www.mpiwg-berlin.mpg.de/project/knowledge-anthropocene.

"New Documentary Recounts Bizarre Climate Changes Seen by Inuit Elders." *The Globe and Mail*, October 19, 2010.

"New WWF Report: 3 Billion Animals Impacted by Australia's Bushfire Crisis." *WWF-Australia*, 2020. https://wwf.org.au/news/2020/3-billion-animals-impacted-by-australia-bushfire-crisis/.

"The Anthropocene at Haus der Kulturen der Welt." *Max Planck Institute for the History of Science*. Accessed May 1, 2024.

"The Prize Chapter 3: Competitive Commerce." EGEE 120 Oil: International Evolution. Dutton Institute, Pennsylvania State University, n. d.

"Why Australia Is Using Indigenous Experts to Fight Bushfires." *TIME*, January 14, 2020.

"Why Can Bass Travel Through Walls So Easily?" *Physics Stack Exchange*, May 6, 2004.

"Why Do Bass Tones Travel Through Walls?" *Physics Stack Exchange*, December 10, 2011.

"Working Group on the 'Anthropocene.'" *Subcommission on Quaternary Stratigraphy*, May 21, 2019. https://quaternary.stratigraphy.org/working-groups/anthropocene.

Abbott, Edwin A. *Flatland: A Romance of Many Dimensions*. London: Seeley & Co., 1884. Published under the pseudonym "A Square."

Abraham, David S. *The Elements of Power: Gadgets, Guns, and the Struggle for a Sustainable Future in the Rare Metal Age*. New Haven: Yale University Press, 2015.

Achtenberg, Emily. "Bolivia Bets on State-Run Lithium Industry." NACLA–North American Congress on Latin America, November 15, 2010.

Adams, Jennifer D., and Matthew Weinstein. "Sylvia Wynter: Science Studies and Posthumanism as Praxes of Being Human." *Cultural Studies of Science Education* 15, no. 4 (2020): 1–31.

Aglietta, Michel, and Etienne Espagne. "Climate and Finance Systemic Risks, More than an Analogy? The Climate Fragility Hypothesis." *CEPII Working Paper*, no. 2016–10 (2016): 1–25.

Aitken, Frédéric, and Jean-Numa Foulc. *From Deep Sea to Laboratory 1: The First Explorations of the Deep Sea by H.M.S. Challenger (1872–1876)*. London: ISTE; Hoboken: Wiley, 2019.

Akomolafe, Bayo. "Coming Down to Earth." March 11, 2020. https://www.bayoakomolafe.net/post/coming-down-to-earth.

Akomolafe, Bayo. "What Climate Collapse Asks of Us." August 17, 2019. https://www.bayoakomolafe.net/post/what-climate-collapse-asks-of-us.

Albán Achinte, Adolfo. *Más allá de la razón hay un mundo de colores: Modernidades, colonialidades y reexistencia*. Santiago de Cuba: Editorial Oriente, 2013.

Albers, Josef. *Interaction of Color*. 50th anniversary edition. New Haven: Yale University Press, 2013. First published in 1963.

Alder, G. J. "Standing Alone: William Moorcroft Plays the Great Game, 1808–1825." *International History Review* 2, no. 2 (April 1980): 172–215.

Allado-McDowell, K. *Pharmako-AI*. London: Ignota, 2020.

Alquati, Romano. "Composizione organica del capitale e forza-lavoro alla Olivetti, Part 2." *Quaderni Rossi* no. 3 (June 1963): 119–85.

Althusser, Louis. *On the Reproduction of Capitalism: Ideology and Ideological State Apparatuses*. Translated by G. M. Goshgarian. London: Verso, 2014.

Amnesty International. "'This is What We Die for': Human Rights Abuses in the Democratic Republic of the Congo Power the Global Trade in Cobalt." January 2016.

Ampère, André-Marie. *Essai sur la philosophie des sciences, ou Exposition analytique d'uneclassification naturelle de toutes les connaissances humaines*. Paris: Bachelier, 1834.

Ananny, Mike, and Kate Crawford. "Seeing Without Knowing: Limitations of the Transparency Ideal and its Application to Algorithmic Accountability." *New Media & Society* 20, no. 3 (2018): 973–89.

Anderson, Chris. "The End of Theory: The Data Deluge Makes the Scientific Method Obsolete." *Wired*, July 23, 2008.

Anderson, Kevin B. *Marx at the Margins: On Nationalism, Ethnicity, and Non-Western Societies*. Chicago: University of Chicago Press, 2010.

Andreessen, Marc. "The Techno-Optimist Manifesto." *Andreessen Horowitz*, October 16, 2023. https://a16z.com/the-techno-optimist-manifesto/.

Angus, Ian. "Vladimir Vernadsky and the Disruption of the Biosphere." *Climate & Capitalism*, June 5, 2018. https://climateandcapitalism.com/2018/06/05/vladimir-vernadsky-and-the-disruption-of-the-biosphere/.

Angwin, Julia, Jeff Larson, Surya Mattu, and Lauren Kirchner. "Machine Bias: There's Software Used across the Country to Predict Future Criminals. And It's Biased Against Blacks." *ProPublica*, May 23, 2016.

Apple. "Supplier Responsibility 2018 Progress Report." 2018.

Arana Salinas, Lilia. "Geología de los volcanes monogenéticos Teuhtli, Tlátoc, Tlacotenco, Ocusaccayo y Cuauhtzin en la Sierra Chichinautzin, al sur de la Ciudad de México." Master's thesis, Instituto de Geología, Universidad Nacional Autónoma de México, 2004.

ArcNews. "The Digital Yellow River Project: GIS Plays Important Role in Managing China's Yellow River Basin." *Esri ArcNews* 32, no. 1 (Spring 2010): 26–7.

Aridjis, Homero. "EL Popocatépetl, HISTORIA Y LEYENDA." *Artes de México*, no. 73 (2005): 8–11.

Arquilla, John, and David Ronfeldt. *The Emergence of Noopolitics: Toward an American Information Strategy*. Santa Monica: RAND, 1999.

Ascott, Roy, ed. *Art, Technology, Consciousness: Mind@large*. Bristol: Intellect Books, 2000.

Åström, Karl Johan, and Richard M. Murray. *Feedback Systems: An Introduction for Scientists and Engineers*. Princeton: Princeton University Press, 2008.

Attwood, James. "Private Equity Lines Up for Coal 'Bonanza' Left by Public Miners." *Bloomberg*, January 3, 2022.

Atura, Annie. "Stanford Scholar Upends Interpretation of Philosopher Martin Heidegger." *Stanford News*, July 8, 2015.

Aytes, Ayhan. "Return of the Crowds: Mechanical Turk and Neoliberal States of Exception." In *Digital Labor: The Internet as Playground and Factory*, edited by Trebor Scholz. London: Routledge, 2012.

Babbage, Charles. *On the Economy of Machinery and Manufactures*. London: Charles Knight, 1832.

Bai, Li. "将进酒 Qiang Jin Jiu [Invitation to Wine]." Translated by Xu Yuanchong. In Li Wenxuan and Wang Feng, "A Comparative Study on the Tang Poetry Translation in the Perspective of 'Harmony-Guided Three-Level Poetry Translation Criteria'—A Case Study of Li Bai's *Invitation to Wine*." *IRA–International Journal of Education & Multidisciplinary Studies* 16, no. 3 (2020): 128–36.

Baidu Wenku 百度文库 [Baidu Library]. "地面播撒系统简介（第四代) Dimian Bosa Xitong Jianjie (Disidai) [Ground seeding system description (fourth generation)]." Accessed July 11, 2020.

Bainbridge, David. *How Zoologists Organize Things: The Art of Classification*. London: Frances Lincoln, 2020.

Baquero, Carlos, and Nuno Preguiça. "Why Logical Clocks are Easy." *Communications of the ACM* 59, no. 4 (2016): 43–7.

Barad, Karen. "After the End of the World: Entangled Nuclear Colonialisms, Matters of Force, and the Material Force of Justice." *Theory & Event* 22, no. 3 (2019): 524–50.

Barad, Karen. *Meeting the Universe Halfway: Quantum Physics and the Entanglement of Matter and Meaning*. Durham: Duke University Press, 2007.

Barad, Karen. "No Small Matter: Mushroom Clouds, Ecologies of Nothingness, and Strange Topologies of Spacetimemattering." In *Arts of Living on a Damaged Planet: Ghosts and Monsters of the Anthropocene*, edited by Anna Lowenhaupt Tsing, Heather Anne Swanson, Elaine Gan, and Nils Bubandt. Minneapolis: University of Minnesota Press, 2017.

Barad, Karen. "Posthumanist Performativity: Toward an Understanding of How Matter Comes to Matter." *Signs* 28, no. 3 (2003): 801–31.

Bardin, Andrea. *Epistemology and Political Philosophy in Simondon: Individuation, Technics, Social Systems*. Dordrecht: Springer, 2015.

Barker, Timothy Scott. "Information and Atmospheres: Exploring the Relationship between the Natural Environment and Information Aesthetics." *M/C Journal* 15, no. 3 (May 2012).

Basler, Bonnie. "How Bacteria 'Talk.'" TED, 2009.

Bataille, Georges. "The Big Toe." *Visions of Excess: Selected Writings, 1927–1939*. Translated by Allan Stoekl with Carl R. Lovitt and Donald M. Leslie, Jr. Minneapolis: University of Minnesota Press, 1985. Originally published in 1929.

Bateson, Gregory. "Alfred Korzybski Memorial Lecture 1970: Form, Substance, and Difference." *General Semantics Bulletin*, no. 37 (1970): 5–13.

Bateson, Gregory. *Steps to an Ecology of Mind*. Chicago: University of Chicago Press, 2000. First published in 1972.

Baumgarten, Alexander Gottlieb. *Aesthetica*. Frankfurt/Oder: Kleyb, 1750.

Beck, Ulrich. *World at Risk*. Translated by Ciaran Cronin. Cambridge: Polity, 2009. Originally published in 2007.

Becker, Seymour. "The 'Great Game:' The History of an Evocative Phrase." *Asian Affairs* 43, no. 1 (2012): 61–80.

Beer, Stafford. *Brain of The Firm: A Development in Management Cybernetics*. London: Allen Lane, 1972.

Behrenfeld, Michael J., Yongxiang Hu, Robert T. O'Malley, Emmanuel S. Boss, Chris A. Hostetler, David A. Siegel, Jorge L. Sarmiento, Jennifer Schulien, Johnathan W. Hair, Xiaomei Lu, Sharon Rodier, and Amy Jo Scarino. "Annual Boom–Bust Cycles of Polar Phytoplankton Biomass Revealed by Space-Based Lidar." *Nature Geoscience* 10, no. 2 (February 2017): 118–22.

Ben-Naim, Arieh. *Entropy Demystified: The Second Law Reduced to Plain Common Sense*. Hackensack: World Scientific, 2007.

Beniger, James R. *The Control Revolution: Technological and Economic Origins of the Information Society*. Cambridge: Harvard University Press, 1986.

Benjamin, Ruha. *Race After Technology: Abolitionist Tools for the New Jim Code*. Cambridge: Polity, 2019.

Benjamin, Walter. "The Work of Art in the Age of Its Technical Reproducibility [First Version]." Translated by Michael Jennings. *Grey Room* 39 (Spring 2010): 11–37.

Bergson, Henri. *Creative Evolution*. Translated by Arthur Mitchell. London: Macmillan, 1922. Originally published in 1907.

Bergson, Henri. *Mind–Energy: Lectures and Essays*. Translated by H. Wildon Carr. Westport: Henry Holt, 1920. Originally published in 1919.

Berland, Jody. "Attending the Giraffe." *Humanimalia* 9, no. 1 (September 2017): 22–45.

Berland, Jody. *Virtual Menageries: Animals as Mediators in Network Cultures*. Cambridge: MIT Press, 2019.

Berlant, Lauren. "The Commons: Infrastructures for Troubling Times." *Environment and Planning D: Society and Space* 34, no. 3 (June 2016): 393–419.

Best, Beverley. "Wages for Housework Redux: Social Reproduction and the Utopian Dialectic of the Value-Form." *Theory and Event* 24, no. 4 (2021): 896–921.

Best, Beverley, Werner Bonefeld, and Chris O'Kane, eds. *The SAGE Handbook of Frankfurt School Critical Theory*. London: Sage, 2018.

Blackman, Lisa. *Haunted Data: Affect, Transmedia, Weird Science*. London: Bloomsbury Academic, 2019.

Bloom, Paul. "Being in Time: How Much should we Value the Past, Present, and the Future?" *New Yorker*, July 9, 2021.

Blum, Jordan. "Dakota Access Pipeline will Remain Open after Long Legal Fight, Judge Rules." *S&P Global Commodity Insights*, May 21, 2021.

Bolin, Bert. *A History of the Science and Politics of Climate Change: The Role of the Intergovernmental Panel on Climate Change*. Cambridge: Cambridge University Press, 2007.

Boorstin, Daniel J. *The Discoverers: A History of Man's Search to Know His World and Himself*. London: Penguin Books, 1986. First published in 1983.

Borges, Jorge Luis. *Collected Fictions*. Translated by Andrew Hurley. New York: Penguin, 1999.

Borges, Jorge Luis. "The Garden of Forking Paths." In *Labyrinths: Selected Stories and Other Writings*, edited by James E. Irby and Donald A. Yates. London: Penguin Modern Classics, 2000. Originally published in 1962.

Bostrom, Nick. *Superintelligence: Paths, Dangers, Strategies*. Oxford: Oxford University Press, 2014.

Bourdieu, Pierre. *Outline of a Theory of Practice*. Translated by Richard Nice. Cambridge: Cambridge University Press, 1977.

Braidotti, Rosi, and Maria Hlavajova, eds. *Posthuman Glossary*. London: Bloomsbury Academic, 2018.

Bratton, Benjamin. *The Stack: On Software and Sovereignty*. Cambridge: MIT Press, 2015.

Bratton, Benjamin. *The Terraforming*. Moscow: Strelka Press, 2019.

Bratton, Benjamin, and Natalie Jeremijenko. "Latent Interfaces, Suspicious Images." *Situated Technologies Pamphlets 3: Situated Advocacy*. New York: Architectural League, 2008.

Braun, Benjamin, and Adrienne Buller. "Titans: Tracing the Rise and the Politics of Asset Manager Capitalism." *Phenomenal World*, November 6, 2021.

Braverman, Irus. “Robotic Life in the Deep Sea.” In *Blue Legalities: The Life and Laws of the Sea*, edited by Irus Braverman and Elizabeth R. Johnson. Durham: Duke University Press, 2020.

Brautigan, Richard. “All Watched over by Machines of Loving Grace.” *All Watched Over by Machines of Loving Grace*. San Francisco: Communication Company, 1967.

Brown, Imani Jacqueline. “Should We Consider Fossil Fuel Extraction an Unjust Enterprise?” *Business and Human Rights Resource Centre*, February 10, 2020.

Browne, Simone, and Sadie Barnette. “In Conversation.” In *Black Futures*, edited by Kimberly Drew and Jenna Wortham. New York: One World, 2021.

Brunton, John. *Bankers and Borrowers*. London: E. Arnold & Co., 1921.

Bryant, William. “Whole System, Whole Earth: The Convergence of Ecology and Technology in Twentieth Century American Culture.” PhD diss., University of Iowa, 2006.

Buchanan, Mark. “Floating Particle Clump Mimics Asteroids and Nuclei.” *Physics*, April 22, 2022.

Buck, Holly Jean. *After Geoengineering: Climate Tragedy, Repair, and Restoration*. London: Verso, 2019.

Buenrostro, Marco. “Las bondades de la milpa.” *Ciencias* 92–93 (October 2008–March 2009): 30–2.

Butoliya, Deepa. “Rethinking Speculative and Critical Design.” *CMU School of Design*, June 13, 2017. https://design.cmu.edu/news/rethinking-speculative-and-critical-design.

Caffentzis, George. “The Work/Energy Crisis and the Apocalypse.” *Midnight Notes*, no. 3 (1980).

Cajete, Gregory. *Native Science: Natural Laws of Interdependence*. Santa Fe: Clear Light Publishers, 2000.

Calude, Cristian S., and Gregory J. Chaitin. “What Is . . . a Halting Probability?” *Notices of the American Mathematical Society* 57, no. 2 (2010): 236–7.

Cameron, Stephen, Daniel Rubinoff, and Kipling Will. “Who Will Actually Use DNA Barcoding and What Will It Cost?” *Systematic Biology* 55, no. 5 (October 2006): 844–7.

Campagna, Federico. *Metaphysics and Metaethics in the Design of Strategy Video Games*. PhD diss., Royal College of Art, 2020.

Campagna, Federico. *Technic and Magic: The Reconstruction of Reality*. London: Bloomsbury, 2018.

Campling, Liam, and Alejandro Colas. *Capitalism and the Sea: The Maritime Factor in the Making of the Modern World*. London: Verso, 2021.

Carbajal López, David. “Iztaccíhuatl: de diosa a princesa.” *Signos Históricos* 23, no. 46 (2021): 286–315.

Carney, Mark. “Breaking the Tragedy of the Horizon – Climate Change and Financial Stability – Speech by Mark Carney.” *Bank of England*, September 29, 2015.

Carse, Ashley. “Keyword: Infrastructure; How a Humble French Engineering Term Shaped the Modern World.” In *Infrastructures and Social Complexity: A Companion*, edited by Penny Harvey, Casper Bruun Jensen, and Atsuro Morita. London: Routledge, 2017.

Carse, James. *Finite and Infinite Games*. New York: Free Press, 1986.

Celedón Muñiz, Héctor, and Mayolo Sanchez Alvarado. *Nemiliz tlacuilolli Xopantlacualli: La leyenda del Xopantlacualli; La comida del tiempo verde y relatos de la milpa*. Mexico City: Calpulli Tecalco, 2014.

Celidwen, Yuria, and Dacher Keltner. "Kin Relationality and Ecological Belonging: A Cultural Psychology of Indigenous Transcendence." *Frontiers in Psychology* 14 (2023): 994508.

Césaire, Aimé. *Discourse on Colonialism*. Translated by Joan Pinkham. New York: Monthly Review Press, 2000. Originally published in 1955.

Césaire, Aimé. "Poetry and Knowledge." In *Lyric and Dramatic Poetry: 1946–82*. Translated by Clayton Eshleman and Annette Smith. Charlottesville: University Press of Virginia, 1990.

Chabot, Pascal. *The Philosophy of Simondon: Between Technology and Individuation*. Translated by Aliza Krefetz with participation of Graeme Kirkpatrick. London: Bloomsbury, 2013. Originally published in 2003.

Chaitin, Gregory J. "Leibniz, Randomness & the Halting Probability." *Mathematics Today* 40, no. 4 (June 2004): 138–9.

Chaitin, Gregory J. *The Limits of Mathematics*. Singapore: Springer-Verlag, 1998.

Chaitin, Gregory J. "The Limits of Reason." *Scientific American* 294 (March 2006): 74–81.

Chaitin, Gregory J. *Meta Math! The Quest for Omega*. New York: Pantheon, 2005.

Chaitin, Gregory J. *The Unknowable*. Singapore: Springer, 1999.

Chakrabarty, Dipesh. "Postcolonial Studies and the Challenge of Climate Change." *New Literary History* 43 (Winter 2012): 1–18.

Chávez Mac Gregor, Helena. "Hacer bailar a los espectros." In *Fernando Palma: ". . . amotla otlacualacac oncan tlanahuatiz quename ye huitz quiahuitl . . . mocualnezcayotl" . . . no relampagueó para anunciar que llovería . . . tu hermosura*. Exhibition catalogue. Mexico City: Museo Universitario del Chopo, 2015.

Chen, Chris, and Sarika Chandra. "Remapping the Race/Class Problematic." In *Totality Inside Out: Rethinking Crisis and Conflict under Capital*, edited by Kevin Floyd, Jen Hedler Phillis, and Sarika Chandra. New York: Fordham University Press, 2022.

Chen, Delton B., Joel van der Beek, and Jonathan Cloud. "Climate Mitigation Policy as a System Solution: Addressing the Risk Cost of Carbon." *Journal of Sustainable Finance & Investment* 7, no. 3 (2017): 233–74.

Chiang, Ted. *Exhalation*. New York: Alfred A. Knopf, 2019.

Chiang, Ted. "The Great Silence." *Electric Literature*, October 12, 2016.

Chien, Shiuh-Shen, Dong-Li Hong, and Po-Hsiung Lin. "Ideological and Volume Politics behind Cloud Water Resource Governance – Weather Modification in China." *Geoforum* 85 (October 2017): 225–33.

China Internet Information Center. "Meteorology Law of the People's Republic of China." *China.org.cn*. Accessed May 13, 2021. http://www.china.org.cn/environment/2007-08/27/content_1034467.htm.

China Meteorological Administration. "50th Anniversary of Fengyun Satellite Program." *cma.gov.cn*, October 9, 2020. http://www.cma.gov.cn/en2014/news/News/202009/t20200930_564281.html.

Chomsky, Noam. "The Bounds of Thinkable Thought." *Progressive*, October 1, 1985.

Church, Alonzo. "An Unsolvable Problem of Elementary Number Theory." *American Journal of Mathematics* 58, no. 2 (1936): 345–63.

Citton, Yves. *The Ecology of Attention*. Cambridge: Polity, 2017.

Clark, Abram H., Alec J. Petersen, Lou Kondic, and Robert P. Behringer, "Nonlinear Force Propagation During Granular Impact," *Physical Review Letters* 114, no. 14 (April 2015): 144502-1–5.

Clark, Karen M. "A Formal Approach to Catastrophe Risk Assessment and Management." *Proceedings of the Casualty Actuarial Society* 73, no. 140 (1986): 69–92.

Clark, Nicola, and Simon Wallis. "Flamingos, Salt Lakes and Volcanoes: Hunting for Evidence of Past Climate Change on the High Altiplano of Bolivia." *Geology Today* 33, no. 3 (May 2017): 101–7.

CNA Hardy. "Taking the Pulse of Global Business: Global Risk and Confidence Survey." May 2019.

Concha Bell, Carole. "It's not Drought, it's Plunder: Drought and the Commodification of Water in Chile." *Red Pepper*, January 6, 2023.

Conlin, Jonathan. *Mr Five Per Cent: The Many Lives of Calouste Gulbenkian; The World's Richest Man*. London: Profile Books, 2019.

Cook, Gary, Jude Lee, Tamina Tsai, Ada Kong, John Deans, Brian Johnson, and Elizabeth Jardim. "Clicking Clean: Who Is Winning the Race to Build a Green Internet?" Washington: Greenpeace, January 2017.

Coupland, Robert T., ed. *Grassland Ecosystems of the World: Analysis of Grasslands and their Uses*. Cambridge: Cambridge University Press, 1979.

Crosby, Alfred W. *Ecological Imperialism*. Cambridge: Cambridge University Press, 2015.

Crutzen, Paul J. "Albedo Enhancement by Stratospheric Aerosol Injection: A Contribution to Resolve a Policy Dilemma?" *Climatic Change* 77 (August 2016): 211–19.

Crutzen, Paul J. "Geology of Mankind." *Nature* 415, no. 23 (January 3, 2002): 23.

Crutzen, Paul, and Eugene Stoermer. "The 'Anthropocene.'" *Global Change Newsletter*, no. 41 (2000): 17–18.

Csikszentmihályi, Chris. "Robotics." In *Sensorium: Embodied Experience, Technology, and Contemporary Art*, edited by Caroline A. Jones. Cambridge: MIT Press and MIT List Visual Arts Center, 2006.

Daggett, Cara New. *The Birth of Energy: Fossil Fuels, Thermodynamics, and the Politics of Work*. Durham: Duke University Press, 2019.

Dalla Costa, Mariarosa, and Selma James. *The Power of Women and the Subversion of the Community*. Bristol: Falling Wall Press, 1972.

Dall'Orto Mas, Rodolfo, Benjamin Vonessen, Christian Fehlker, and Katrin Arnold. "The Case for Central Bank Independence: A Review of Key Issues in the International Debate." European Central Bank, *Occasional Paper Series*, no. 248 (October 2020).

Daston, Lorraine, and Peter Galison. *Objectivity*. Brooklyn: Zone Books, 2007.

Davis, Heather. "To Live in Compromised Worlds: Holly Jean Buck's *After Geoengineering*." *Avery Review*, no. 45 (February 2020): 1–5.

De'Ath, Amy. "Reproduction." In *The Bloomsbury Companion to Marx*, edited by Jeff Diamanti, Andrew Pendakis, and Imre Szeman. London: Bloomsbury, 2019.

Debaise, Didier. *Nature as Event: The Lure of the Possible*. Durham: Duke University Press, 2017.

De Boever, Arne, Alex Murray, Jon Roffe, and Ashley Woodward. *Gilbert Simondon: Being and Technology*. Edinburgh: Edinburgh Press, 2012.

De Certeau, Michel. *The Practice of Everyday Life*. Translated by Steve Rendall. Berkeley: University of California Press, 1984.

Defoe, Daniel. *An Essay upon Projects*. London: Printed by R.R. for Tho. Cockerill, at the Three Legs in the Poultrey, 1697.

De Freitas, Elizabeth, and Ezekiel Dixon-Román, and Patti Lather. *Alternative Ontologies of Number: Rethinking Quantitative Inquiry in Computational Culture* 16, no. 5 (2016): 431–4.

De la Cadena, Marisol. *Earth Beings: Ecologies of Practice across Andean Worlds*. Durham: Duke University Press, 2015.

De la Cadena, Marisol, and Arturo Escobar. "Notes on Ontological Excess: Towards Pluriversal Designing." In *Resonancias tectónicas desde el Sur: Del diseño centrado en el usuario al diseño centrado en el planeta*, edited by Martí n Tironi. Forthcoming.

De la Cadena, Marisol, and Mario Blaser. "Introduction: Pluriverse Proposals for a World of Many Worlds." In *A World of Many Worlds*, edited by Marisol de la Cadena and Mario Blaser. Durham: Duke University Press, 2018.

De la Cadena, Marisol, and Mario Blaser, eds. *A World of Many Worlds*. Durham: Duke University Press, 2018.

Delanda, Manuel. *Assemblage Theory*. Edinburgh: Edinburgh University Press, 2016.

Deleuze, Gilles, and Félix Guattari. *Anti-Oedipus: Capitalism and Schizophrenia*. Translated by Robert Hurley, Mark Seem, and Helen R. Lane. Minneapolis: University of Minnesota Press, 1983. Originally published in 1972.

Deleuze, Gilles, and Félix Guattari. *A Thousand Plateaus: Capitalism and Schizophrenia*. Translated by Brian Massumi. Minneapolis: University of Minnesota Press, 1987. Originally published in 1980.

Deleuze, Gilles, and Félix Guattari. *What is Philosophy?* Translated by Hugh Tomlinson and Graham Burchell. New York: Columbia University Press, 1994. Originally published in 1991.

Dell. "Responsible Minerals Sourcing Report." June 2018.

Demos, T. J. *Against the Anthropocene: Visual Culture and Environment Today*. Berlin: Sternberg Press, 2017.

Denning, Michael. "Wageless Life." *New Left Review* 66 (2010): 79–97.

Descartes, René. *A Discourse on Method*. Translated by John Veitch. London: J. M. Dent & Sons, 1912. Originally published in 1637.

Descartes, René. *Meditations on First Philosophy with Selections from the Objections and Replies*. Translated by John Cottingham. Cambridge: Cambridge University Press, 2017. Originally published in 1641.

Descola, Philippe. *Beyond Nature and Culture*. Translated by Janet Lloyd. Chicago: University of Chicago Press, 2013. Originally published in 2005.

Despret, Vinciane. *What Would Animals Say If We Asked the Right Questions?* Translated by Brett Buchanan. Minneapolis: University of Minnesota Press, 2016. Originally published in 2012.

"DNA Barcoding Part of UN Biodiversity Strategic Plan," *University of Guelph News*, December 19, 2016.

De Rijcke, S., S. J. Penny, C. J. Conselice, S. Valcke, and E. V. Held. "*Hubble Space Telescope* survey of the Perseus cluster – II. Photometric scaling relations in different environments." *Monthly Notices of the Royal Astronomical Society* 393, no. 3 (March 2009): 798–807.

De Vos, Laura. "Spiralic Time and Cultural Continuity for Indigenous Sovereignty: Idle No More and The Marrow Thieves." *Transmotion* 6, no. 2 (2020): 1–42.

De Vriend, Huib, and Mark van Koningsveld. *Building with Nature: Thinking, Acting and Interacting Differently*. Dordrecht: EcoShape, 2012.

Deutsch, David. *The Beginning of Infinity: Explanations That Transform the World*. London: Penguin Books, 2012. First published in 2011.

Di Lampedusa, Giuseppe Tomasi. *The Leopard*. Translated by Archibald Colquhoun. New York: Pantheon Books, 1960. Originally published in 1958.

Dilnot, Clive. "Designing in the World of the Naturalized Artificial." In *Design in Crisis: New Worlds, Philosophies and Practices*, edited by Tony Fry and Adam Nocek. London: Routledge, 2021.

Di Salvatore, Lea. "Investor–State Disputes in the Fossil Fuel Industry." *International Institute for Sustainable Development*, December 2021.

Domínguez, Patricia, and Marco Godoy. "Eyes of Plants." Translated by Nahual Lhorente. In *Tongue, Taste, and Appetite Issue 1*, edited by Amanda Pohan. New York: Temporary Agency Publishing, 2018.

Domínguez, Patricia. *Gaia Guardians*. Translated by Philip Sutton. Madrid: TBA21, 2020.

Domínguez, Patricia. *Technologies of Enchantment: When a Ceramic Vase and a Drone Cry Together*, edited by Sabel Gavaldon. Translated by Sabel Gavaldon and Rodney Strabucchi. London: Gasworks, 2019.

Douglas, Bronwen, and Chris Ballard, eds. *Foreign Bodies: Oceania and the Science of Race 1750–1940*. Canberra: ANU E-Press, 2008.

Dowker, Fay. "Arts & Ideas: Time." *BBC Sounds*, October 29, 2021.

Doyle, Peter. *Echo and Reverb: Fabricating Space in Popular Music Recording, 1900–1960*. Middletown: Wesleyan University Press, 2005.

Dunayevskaya, Raya. *Rosa Luxemburg, Women's Liberation, and Marx's Philosophy of Revolution*. 2nd ed. Urbana: University of Illinois Press, 1991.

Dunne, J. W. *An Experiment with Time*. 2nd ed. London: A. & C. Black, 1927.

Dupont, Julian. "Convertirse en Jaguar Como la Nube Automatizada." Master's thesis, Universidad Autónoma Indígena Intercultural, Bogotá, 2021.

eBird, Cornell Lab for Ornithology. https://ebird.org.

Edwards, Matthew. "The New Great Game and the new great gamers: disciples of Kipling and Mackinder." *Central Asian Survey* 22, no. 1 (March 2003): 83–102.

Edwards, Paul N. *The Closed World: Computers and the Politics of Discourse in Cold War America*. Cambridge: MIT Press, 1996.

Edwards, Paul N. *A Vast Machine: Computer Models, Climate Data, and the Politics of Global Warming*. Cambridge: MIT Press, 2010.

Elichirigoity, Fernando. *Planet Management: Limits to Growth, Computer Simulation, and the Emergence of Global Spaces*. Evanston: Northwestern University Press, 1999.

Ellis, Erle C. "The Great Acceleration." *Anthropocene: A Very Short Introduction*. Oxford: Oxford University Press, 2018.

Ellis, Jerome. "The Clearing: Melismatic Palimpsest." *No Linear Fucking Time*, BAK, 2021.

Ellis, Rebecca. "Rethinking the Value of Biological Specimens: Laboratories, Museums and the Barcoding of Life Initiative." *Museum and Society* 6, no. 2 (July 2008): 172–91.

Ellis, Rebecca, Claire Waterton, and Brian Wynne. "Taxonomy, Biodiversity and Their Publics in Twenty-First-Century DNA Barcoding." *Public Understanding of Science* 19, no. 4 (July 2010): 497–512.

Ellison, Harlan. *The Fantasies of Harlan Ellison*. Boston: Gregg Press, 1979.

Ellul, Jacques. *The Technological System*. Translated by Joachim Neugroschel. New York: Continuum, 1980. Originally published in 1977.

Ely, Chris. "The Life Expectancy of Electronics." Consumer Technology Association, September 16, 2014. Accessed September 7, 2018.

Endnotes. "Error." *Endnotes*, no. 5 (October 2019).

Endnotes. "The Logic of Gender: On the Separation of Spheres and the Process of Abjection." *Endnotes*, no. 3 (September 2013).

Engelbart, Douglas. "Augmenting Human Intellect: A Conceptual Framework." *Stanford Research Institute*, Summary Report AFOSR-3233 (October 1962).

Epstein, Robert. "The Empty Brain." *Aeon*, May 18, 2016.
equivant. "About Us." 2018. http://www.equivant.com/about-us.
Ernst, Mayr. "Species Concepts and Definitions." In *Topics in the Philosophy of Biology*, edited by Marjorie Grene and Everett Mendelsohn. Dordrecht: Springer, 1976.
Ernst, Wolfgang. *Digital Memory and the Archive*, edited by Jussi Parikka. Minneapolis: University of Minnesota Press, 2013.
Escobar, Arturo. *Designs for the Pluriverse: Radical Interdependence, Autonomy, and the Making of Worlds*. Durham: Duke University Press, 2018.
Escobar, Arturo. *Pluriversal Politics: The Real and the Possible*. Durham: Duke University Press, 2020.
Espinoza, Vicky, Duane E. Waliser, Bin Guan, David A. Lavers, and F. Martin Ralph. "Global Analysis of Climate Change Projection Effects on Atmospheric Rivers." *Geophysical Research Letters* 45, no. 9 (May 2018): 4299–308.
Esposito, Elena. *The Future of Futures: The Time of Money in Financing and Society*. Translated by Elena Esposito and Andrew K. Whitehead. Cheltenham: Edward Elgar, 2011. Originally published in 2009.
Esposito, Fernando. *Fascism, Aviation and Mythical Modernity*. Translated by Patrick Camiller. London: Palgrave Macmillan, 2015. Originally published in 2011.
Esteva, Gustavo. "Autonomy." In *Pluriverse: A Post-Development Dictionary*, edited by Ashish Kothari, Ariel Salleh, Arturo Escobar, Federico Demaria, and Alberto Acosta. Delhi: Tulika and AuthorsUpFront Publishing, 2019.
Eubanks, Virginia. *Automating Inequality: How High-Tech Tools Profile, Police, and Punish the Poor*. New York: St. Martin's Press, 2018.
European Commission. "Sustainable Oceans" https://ec.europa.eu/maritimeaffairs/policy/blue_growth_en.
Fanon, Frantz. *Black Skin, White Masks*. Translated by Charles Lam Markmann. London: Pluto Press, 2008. Originally published in 1952.
Fanon, Frantz. *Black Skin, White Masks*. Translated by Richard Philcox. New York: Grove Press, 2008. Originally published in 1952.
Fanon, Frantz. *The Wretched of the Earth*. Translated by Richard Philcox. New York: Grove Press, 2004. Originally published in 1963.
Farbotko, Carol, and Heather Lazrus. "The First Climate Refugees? Contesting Global Narratives of Climate Change in Tuvalu." *Global Environmental Change* 22, no. 2 (May 2012): 382–90.
Farman, Joseph Charles, Brian Gerard Gardiner, and Jonathan Shanklin. "Large Losses of Total Ozone in Antarctica Reveal Seasonal ClO_x/NO_x Interaction." *Nature* 315 (May 1985): 207–10.
Fazekas, Aron J., Prasad R. Kesanakurti, Kevin S. Burgess, Diana M. Percy, Sean W. Graham, Spencer C. H. Barrett, Steven G. Newmaster, Mehrdad Hajibabaei, and Brian C. Husband. "Barcoding Plants: Are Plant Species Inherently Harder to Discriminate than Animal Species Using DNA Barcoding Markers?" *Molecular Ecology Resources* 9 (April 2009): 130–9.

Federici, Silvia. *Caliban and the Witch: Women, the Body and Primitive Accumulation*. New York: Autonomedia, 2004.

Federici, Silvia. "Precarious Labor: A Feminist Viewpoint." Bluestockings Cooperative Bookstore, New York, October 28, 2006.

Feher, Michel. *Rated Agency: Investee Politics in a Speculative Age*. Translated by Gregory Elliott. New York: Zone Books, 2018.

Ferreira da Silva, Denise. "1 (life) ÷ 0 (blackness) = ∞ − ∞ or ∞ / ∞: On Matter Beyond the Equation of Value." *e-flux Journal* no. 79 (February 2017).

Ferreira da Silva, Denise. "Before *Man*: Sylvia Wynter's Rewriting of the Modern Episteme." In *Sylvia Wynter: On Being Human as Praxis*, edited by Katherine McKittrick. Durham: Duke University Press, 2015.

Ferreira da Silva, Denise. "On Difference Without Separability." In *Incerteza Viva: 32nd Bienal de São Paulo*, edited by Jochen Volz and Júlia Rebouças. São Paulo: Fundação Bienal de São Paulo, 2016.

Ferreira da Silva, Denise. "Toward a Black Feminist Poethics: The Quest(ion) of Blackness Toward the End of the World." *The Black Scholar* 44, no. 2 (2014): 81–97.

Ferreira da Silva, Denise. *Toward a Global Idea of Race*. Minneapolis: University of Minnesota Press, 2007.

Financial Times, editorial board. "Governments Should Not Foot the Bill for Stranded Assets." February 21, 2022.

Flea. "Flea's Most Precious Advice for all the Bass Players Out There." YouTube, January 28, 2014.

Fleming, James Roger. *Historical Perspectives on Climate Change*. Oxford: Oxford University Press, 1998.

Fluss, Harrison, and Landon Frim. "Behemoth and Leviathan: The Fascist Bestiary of the Alt-Right." *Salvage*, December 21, 2017.

Forsyth, Mark. *The Etymologicon: A Circular Stroll Through the Hidden Connections of the English Language*. London: Icon Books, 2011.

Foster, John Bellamy. *The Ecological Revolution: Making Peace with the Planet*. New York: Monthly Review Press, 2009.

Foster, John Bellamy. *Marx's Ecology: Materialism and Nature*. New York: Monthly Review Press, 2000.

Foster, John Bellamy. "Marx's Theory of Metabolic Rift: Classical Foundations for Environmental Sociology." *American Journal of Sociology* 105, no. 2 (September 1999): 366–405.

Foucault, Michel. *The Order of Things: An Archaeology of the Human Sciences*. Translated by Alan Sheridan. New York: Vintage Random House, 1973.

Foucault, Michel. *The Order of Things: An Archaeology of the Human Sciences*. Translated by Alan Sheridan. New York: Vintage Books, 1994. Originally published in 1966.

Fredericks, Carla, Mark Meaney, Nick Pelosi, and Kathleen Finn. "Social Cost and Material Loss: The Dakota Access Pipeline." *University of Colorado Law School Legal Studies Research Paper Series*, no. 19-1 (2018): 1–66.

Freud Museum London. "When Dalí Met Freud." February 4, 2022.

Friston, Karl. "The Free-Energy Principle: a Rough Guide to the Brain?" *Trends in Cognitive Sciences* 13, no. 7 (June 2009): 293–301.

Fry, Tony. *Becoming Human by Design*. London: Berg, 2012.

Fry, Tony. *Defuturing: A New Design Philosophy*. London: Bloomsbury, 2020.

Fry, Tony, and Adam Nocek. "Design in Crisis. Introducing a Problematic." In *Design in Crisis: New Worlds, Philosophies and Practices*, edited by Tony Fry and Adam Nocek. London: Routledge, 2021.

Fry, Tony, and Madina Tlostanova. *A New Political Imagination: Making the Case*. London: Routledge, 2021.

Fuchs, Christian. *Digital Labor and Karl Marx*. London: Routledge, 2014.

Fuller, Buckminster. *Operating Manual for Spaceship Earth*. Carbondale: Southern Illinois University Press, 1969.

Funkenstein, Amos. *Theology and the Scientific Imagination from the Middle Ages to the Seventeenth Century*. Princeton: Princeton University Press, 1986.

Gabor, Daniela. "The Wall Street Consensus." *Development and Change* 52, no. 3 (2021): 429–59.

Gabrys, Jennifer. *Program Earth: Environmental Sensing Technology and the Making of a Computational Planet*. Minneapolis: University of Minnesota Press, 2016.

Gail, William B. "A New Dark Age Looms." *New York Times*, April 19, 2016.

Galison, Peter. "The Ontology of the Enemy: Norbert Wiener and the Cybernetic Vision." *Critical Inquiry* 21, no. 1 (1994): 228–66.

García, Ricardo Iglesias. *Arte y robótica: La tecnología como experimentación estética*. Madrid: Casimiro, 2016.

García-Castañeda, Victor G. "Techno-diversity and Maize: Suggestions for the Search of a Mesoamerican Cosmotechnics." *Technophany: A Journal for Philosophy and Technology*, no. 1 (November 2021).

Gell, Alfred. *The Anthropology of Time: Cultural Constructions of Temporal Maps and Images*. Oxford: Berg, 1992.

Gelman, Andrew, John B. Carlin, Hal S. Stern, David B. Dunson, Aki Vehtari, and Donald B. Rubin. *Bayesian Data Analysis*. New York: Chapman & Hall/CRC, 2013.

George, Rose. *Ninety Percent of Everything: Inside Shipping, the Invisible Industry That Puts Clothes on Your Back, Gas in Your Car, and Food on Your Plate*. New York: Metropolitan Books, 2013.

Ghosh, Amitav. *The Great Derangement: Climate Change and the Unthinkable*. Chicago: University of Chicago Press, 2016.

Giedion, Sigfried. *Mechanization Takes Command: A Contribution to Anonymous History*. New York: Oxford University Press, 1948.

Gillard, David. Review of *The Great Game: Britain and Russia in Central Asia*, edited by Martin Ewans. *Modern Humanities Research Association* 84, no. 1 (January 2006): 161–3.

Gilroy, Paul. "Bass Culture: Paul Gilroy (Part One)." Podcast Produced by the Black Music Research Unit, University of Westminster. *Bass Culture UK*, June 17, 2020.

Giménez, Martha E. *Marx, Women, and Capitalist Social Reproduction*. Leiden: Brill, 2018.

Glissant, Édouard. *Poetics of Relation*. Translated by Betsy Wing. Ann Arbor: University of Michigan Press, 1997.

Global Footprint Network. https://www.footprintnetwork.org.

Glum, Julia. "The Median Amazon Employee's Salary Is $28,000. Jeff Bezos Makes More Than That in 10 Seconds." *Time*, May 2, 2018. Accessed September 7, 2018.

Godwin-Austen, Henry H. "Address of the President." *Journal of Molluscan Studies* 3, no. 5 (July 1899): 241–62.

Gorelick, Noel, Matt Hancher, Mike Dixon, Simon Ilyushchenko, David Thau, and Rebecca Moore. "Google Earth Engine: Planetary-Scale Geospatial Analysis for Everyone." *Remote Sensing of Environment* 202 (December 2017): 18–27.

Gould, Rachelle K., Doreen E. Martinez, and Kristin R. Hoelting. "Exploring Indigenous Relationality to Inform the Relational Turn in Sustainability Science." *Ecosystems and People* 19, no. 1 (2023): 1–9.

Goux, Jean-Joseph. *Symbolic Economies: After Marx and Freud*. Translated by Jennifer Curtiss Gage. Ithaca: Cornell University Press, 1990. The volume contains selections from books originally published in 1973 and 1978.

Graeber, David. *Debt: The First 5,000 Years*. Brooklyn: Melville House, 2011.

Graham, Ed. Review of *Post-Growth Living: For an Alternative Hedonism*, by Kate Soper. *Marx & Philosophy Review of Books*, March 29, 2022.

Graham, Mark, and Håvard Haarstad. "Transparency and Development: Ethical Consumption Through Web 2.0 and the Internet of Things." *Information Technologies & International Development* 7, no. 1 (Spring 2011): 1–18.

Gray, Ian. "Hazardous Simulations: Pricing Climate Risk in US Coastal Insurance Markets." *Economy and Society* 50, no. 2 (2021): 196–223.

Grove, Kevin. *Resilience: Key Ideas in Geography*. London: Routledge, 2018.

Grove, Richard. "Climatic Fears: Colonialism and the History of Environmentalism." *Harvard International Review* 23, no. 4 (2002): 50–5.

Guattari, Félix. *Schizoanalytic Cartographies*. Translated by Andrew Goffey. London: Bloomsbury, 2013. Originally published in 1989.

Guattari, Félix. *The Three Ecologies*. Translated by Ian Pindar and Paul Sutton. London: Athlon, 2000. Originally published in 1989.

Gutiérrez, Alfredo. "When Design Goes South: From Decoloniality, through Declassification to Dessobons." In *Design in Crisis: New Worlds, Philosophies and Practices*, edited by Tony Fry and Adam Nocek. London: Routledge, 2021.

Haeckel, Ernst. *Generelle Morphologie der Organismen*. Berlin: Georg Reimer, 1866.

Haff, Peter. "Humans and Technology in the Anthropocene: Six rules." *Anthropocene Review* 1, no. 2 (August 2014): 126–36.

Han, Byung-Chul. *No-cosas: Quiebras del mundo de hoy*. Translated by Joaquín Chamorro Mielke. Madrid: Taurus, 2021. Original German edition published in 2021.

Han, Byung-Chul. *The Scent of Time: A Philosophical Essay on the Art of Lingering*. Translated by Daniel Steuer. Cambridge: Polity, 2017. Originally published in 2009.

Haraway, Donna J. "Anthropocene, Capitalocene, Plantationocene, Chthulucene: Making Kin." *Environmental Humanities* 6, no. 1 (2015): 159–65.

Haraway, Donna J. *The Companion Species Manifesto: Dogs, People, and Significant Otherness*. Chicago: Prickly Paradigm Press, 2003.

Haraway, Donna J. "The High Cost of Information in Post-World War II Evolutionary Biology: Ergonomics, Semiotics, and the Sociobiology of Communication Systems." *Philosophical Forum* 13, no. 2–3 (1981): 244–78.

Haraway, Donna J. *Manifestly Haraway*. Minneapolis: University of Minnesota Press, 2016.

Haraway, Donna J. *Simians, Cyborgs and Women: The Reinvention of Nature*. London: Free Association Books, 1991.

Haraway, Donna J. *Staying with the Trouble: Making Kin in the Chthulucene*. Durham: Duke University Press, 2016.

Haraway, Donna J. *When Species Meet*. Minneapolis: University of Minnesota Press, 2008.

Haraway, Donna J., Ishikawa Noboru, Scott F. Gilbert, Kenneth Olwig, Anna L. Tsing, and Nils Bubandt. "Anthropologists Are Talking: About the Anthropocene." *Ethnos* 81, no. 3 (2016): 535–64.

Harding, Sandra. "Comment on Hekman's 'Truth and Method: Feminist Standpoint Theory Revisited': Whose Standpoint Needs the Regimes of Truth and Reality?" *Signs: Journal of Women in Culture and Society* 22, no. 2 (Winter 1997): 382–91.

Harding, Sandra. *Objectivity and Diversity: Another Logic of Scientific Research*. Chicago: University of Chicago Press, 2015.

Harding, Sandra. "Rethinking Standpoint Epistemology: What Is 'Strong Objectivity'?" In *Feminist Epistemologies*, edited by Linda Alcoff and Elizabeth Potter. New York: Routledge, 1993.

Harding, Sandra. *The Science Question in Feminism*. Ithaca: Cornell University Press, 1986.

Harding, Sandra. *Whose Science? Whose Knowledge? Thinking from Women's Lives*. Ithaca: Cornell University Press, 1991.

Hardt, Michael, and Antonio Negri. *Multitude: War and Democracy in the Age of Empire*. New York: Penguin, 2004.

Harney, Stefano, and Fred Moten. *The Undercommons: Fugitive Planning & Black Study*. New York: Minor Compositions, 2013.

Harrington, Anne. *Reenchanted Science: Holism in German culture from Wilhelm II to Hitler*. Princeton: Princeton University Press, 1999.

Hatfield, Samantha Chisholm, Elizabeth Marino, Kyle Powys Whyte, Kathie D. Dello, and Philip W. Mote. "Indian Time: Time, Seasonality, and Culture in Traditional Ecological Knowledge of Climate Change." *Ecological Processes* 7, no. 25 (2018): 1–11.

Haus der Kulturen der Welt. "The Anthropocene at HKW." *HKW Mediathek*. Accessed May 1, 2024.

Hayek, Friedrich A. "The Meaning of Competition." In *The Collected Works of F. A. Hayek Volume 15: The Market and Other Orders,* edited by Bruce Caldwell. Chicago: University of Chicago Press, 2014. Text originally published in 1948.

Hayles, N. Katherine. *How We Became Posthuman: Virtual Bodies in Cybernetics, Literature, and Informatics*. Chicago: University of Chicago Press, 1999.

Hayles, N. Katherine. *My Mother Was a Computer: Digital Subjects and Literary Texts*. Chicago: University of Chicago Press, 2005.

Hayles, N. Katherine. *Unthought: The Power of the Cognitive Nonconscious*. Chicago: University of Chicago Press, 2017.

Hebert, Paul D. N., Alina Cywinska, Shelley L. Ball, and Jeremy R. deWaard. "Biological Identifications through DNA Barcodes." *Proceedings of the Royal Society B* 270, no. 1512 (February 2003): 313–21.

Hecht, Gabrielle. "Interscalar Vehicles for an African Anthropocene: On Waste, Temporality, and Violence." *Cultural Anthropology* 33, no. 1 (2018): 109–41.

Hegel, Georg Wilhelm Friedrich. "Introduction to the Philosophy of Nature." In *Hegel's Philosophy of Nature: Part Two of the Encyclopaedia of the Philosophical Sciences (1830),* edited and translated by Arnold Vincent Miller. Oxford: Clarendon Press, 1970.

Hegel, Georg Wilhelm Friedrich. *Hegel's Philosophy of Subjective Spirit,* Vol. 2, edited and translated by M. J. Petry. Dordrecht: D. Reidel Publishing, 1978.

Hegel, Georg Wilhelm Friedrich. *Lectures on the Philosophy of History*. Translated by John Sibree in 1857. London: G. Bell and Sons, 1914. Originally published in 1837.

Hegel, Georg Wilhelm Friedrich. *The Phenomenology of Spirit*. Translated by Michael Inwood. Oxford: Oxford University Press, 2018. Originally published in 1807.

Heidegger, Martin. "Die Frage nach der Technik." *Vorträge und Aufsätze*. Pfullingen: Gϋnther Neske, 1954.

Heidegger, Martin. "The Origin of the Work of Art." In *Poetry, Language, Thought*. Translated by Albert Hofstadter. New York: Harper & Row, 1971. Text originally published in 1950.

Heidegger, Martin. "The Question Concerning Technology." Translated by William Lovitt. In *The Question Concerning Technology, and Other Essays*. New York: Garland, 1977. Text originally published in 1954.

Heidegger, Martin. *Poetry, Language, Thought*. Translated by Albert Hofstadter. New York: Harper & Row, 1971.

Hekman, Susan. "Truth and Method: Feminist Standpoint Theory Revisited." *Signs: Journal of Women in Culture and Society* 22, no. 2 (Winter 1997): 341–65.

Helmreich, Stefan. *Alien Ocean: Anthropological Voyages in Microbial Seas.* Berkeley: University of California Press, 2009.

Hern, Alex. "Royal Free breached UK data law in 1.6m patient deal with Google's DeepMind." *Guardian*, July 3, 2017.

Herrera Wassilowsky, Alexander. *La recuperación de tecnologías indígenas: Arqueología, tecnología y desarrollo en los Andes.* Lima: Instituto de Estudios Peruanos, 2011.

Hill Collins, Patricia. "Comment on Hekman's 'Truth and Method: Feminist Standpoint Theory Revisited': Where's the Power?" *Signs: Journal of Women in Culture and Society* 22, no. 2 (Winter 1997): 375–81.

Hodal, Kate. "Death Metal: Tin Mining in Indonesia." *Guardian*, November 23, 2012.

Hodgson, Camilla. "European Energy Groups Seek € 4bn Damages over Fossil Fuel Projects." *Financial Times*, February 21, 2022.

Holland, John H. *Signals and Boundaries: Building Blocks for Complex Adaptive Systems.* Cambridge: MIT Press, 2012.

Hölzl, Richard. "Historicizing Sustainability: German Scientific Forestry in the Eighteenth and Nineteenth Centuries." *Science as Culture* 19, no. 4 (November 2010): 431–60.

Hood, Bruce M. *The Self Illusion: Why There Is No "You" Inside Your Head.* London: Constable, 2012. First published in 2011.

Horkheimer, Max, and Theodor W. Adorno. *Dialectic of Enlightenment.* Translated by Edmund Jephcott and edited by Gunzelin Schmid Noerr. Stanford: Stanford University Press, 2002.

Hornborg, Alf. *Nature, Society, and Justice in the Anthropocene.* Cambridge: Cambridge University Press, 2019.

Huang, Kerson, and Rosemary Huang. *I Ching.* New York: Workman, 1987.

Hui, Yuk. *Art and Cosmotechnics.* Brooklyn: e-flux Books; Minneapolis: University of Minnesota Press, 2021.

Hui, Yuk. "ChatGPT, or the Eschatology of Machines." *e-flux Journal*, no. 137 (June 2023).

Hui, Yuk. "For a Planetary Thinking." *e-flux Journal*, no. 114 (December 2020).

Hui, Yuk. "Machine and Ecology." *Angelaki: Journal of the Theoretical Humanities* 25, no. 4 (2020): 54–66.

Hui, Yuk. "One Hundred Years of Crisis" *e-flux Journal*, no. 108 (April 2020).

Hui, Yuk. *The Question Concerning Technology in China: An Essay in Cosmotechnics.* Falmouth: Urbanomic, 2016.

Hui, Yuk. *Recursivity and Contingency.* London: Rowman & Littlefield, 2019.

Hui, Yuk. "Singularity Vs. Daoist Robots: Is there Another Path than Accelerated Western Modernization?" *Noēma Magazine*, June 19, 2020.

Hui, Yuk, with Karen Sarkisov. "The Incomputable and the Incalculable." 13th Gwangju Biennale, February 23, 2021.

Hui, Yuk, and Barry Schwabsky. "The Call of the Unknown in Art and Cosmotechnics." *e-flux Journal*, no. 136 (May 2023).
Hui, Yuk, and Brian Kuan Wood. "A Conversation on *Art and Cosmotechnics*, Part 1." *e-flux Journal*, no. 124 (February 2022).
Hull, David L. *Science as a Process: An Evolutionary Account of the Social and Conceptual Development of Science*. Chicago: University of Chicago Press, 1988.
Huxley, Julian. "Transhumanism." *Journal of Humanistic Psychology* 8, no. 1 (1968): 73–76. Text first published in 1957.
Illich, Ivan. *Tools for Conviviality*. London: Calder & Boyars, 1973.
iNaturalist, California Academy of Sciences. www.inaturalist.org.
Ingold, Tim. *Being Alive: Essays on Movement, Knowledge and Description*. London: Routledge, 2011.
Ingold, Tim. "From the Transmission of Representations to the Education of Attention." In *The Debated Mind: Evolutionary Psychology versus Ethnography*, edited by Harvey Whitehouse. London: Bloomsbury Academic, 2001.
Ingold, Tim. "Globes and Spheres: The Topology of Environmentalism." In *The Perception of the Environment: Essays on Livelihood, Dwelling and Skill*. London: Routledge, 2000.
Ingold, Tim. *The Perceptions of the Environment: Essays on Livelihood, Dwelling and Skill*. London: Routledge, 2000.
Intel corporation. "Intel's Efforts to Achieve a 'Conflict Free' Supply Chain." May 2018. Accessed September 7, 2018.
International Barcode of Life. https://ibol.org.
Irwin, Terry, and Gideon Kossoff. "Transition Design as a Strategy for Addressing Urban Wicked Problems." In *Cities Without Capitalism*, edited by Hossein Sadri and Senem Zeybekoglu. London: Routledge, 2021.
Irwin, Terry, Gideon Kossoff, and Cameron Tonkinwise. "Transition Design Provocation." *Design Philosophy Papers* 13, no. 1 (2015): 3–11.
Jackson, Zakiyyah Iman. "Outer Worlds: The Persistence of Race in Movement 'Beyond the Human.'" *GLQ: A Journal of Lesbian and Gay Studies* 21, nos. 2–3 (June 2015): 215–18.
James, William. *The Principles of Psychology*. New York: Henry Holt, 1890.
Jameson, Fredric. "Cognitive Mapping." In *Marxism and the Interpretation of Culture*, edited by Cary Nelson and Lawrence Grossberg. Urbana: University of Illinois Press, 1988.
Jameson, Fredric. *The Geopolitical Aesthetic: Cinema and Space in the World System*. Bloomington: Indiana University Press, 1992.
Jameson, Fredric. *Postmodernism; Or, The Cultural Logic of Late Capitalism*. Durham: Duke University Press, 1991.
Japan Aerospace Exploration Agency (JAXA). "Declaration of the First Meeting of Equatorial Countries (Adopted on December 3, 1976)." *Space Law*, 1998. https://www.jaxa.jp/library/space_law/chapter_2/2-2-1-2_e.html.

Jasanoff, Sheila. "Technologies of Humility: Citizen Participation in Governing Science." *Minerva* 41, no. 3 (2003): 223–44.

Jayaram, Dhanasree, and Kurnica Bhattacharjee, "China's Geoengineering Push Dangerous for the Region." *Sunday Guardian Live*, December 19, 2020.

Jonas, Hans. *The Phenomenon of Life: Toward a Philosophical Biology*. New York: Dell Publishing, 1966.

Jones, Donna V. "Inheritance and Finitude: Toward a Literary Phenomenology of Time." *ELH* 85, no. 2 (2018): 289–303.

Jones, Donna V. *The Racial Discourses of Life Philosophy: Négritude, Vitalism, and Modernity*. New York: Columbia University Press, 2010.

Jonkman, Sebastiaan N., Raphaël Steenbergen, Oswaldo Morales Nápoles, Ton Vrouwenvelder, and Johannes K. Vrijling. *Probabilistic Design: Risk and Reliability Analysis in Civil Engineering*. Delft: Delft University of Technology, 2015.

Joppa, Lucas N. "AI for Earth." *Nature* 552 (December 21–28, 2017): 325–8.

Joque, Justin. *Revolutionary Mathematics. Artificial Intelligence, Statistics and the Logic of Capitalism*. London, Verso, 2022.

Jordan, Brigitte, and Austin Henderson. "Interaction Analysis: Foundations and Practice." *Journal of the Learning Sciences* 4, no. 1 (1995): 39–103.

Jordan, John. *Robots*. Cambridge: MIT Press, 2016.

Jünger, Ernst, ed. *Luftfahrt ist Not!* Berlin: Wilhelm Andermann Verlag, 1929.

Jünger, Ernst. "Total Mobilization." Translated by Joel Golb and Richard Wolin. In *The Heidegger Controversy: A Critical Reader*, edited by Richard Wolin. Cambridge: MIT Press, 1998. Text originally published in 1930.

Jünger, Ernst. *The Worker: Dominion and Form*, edited by Laurence Paul Hemming. Translated by Bogdan Costea. Evanston: Northwestern University Press, 2017.

Kallipoliti, Lydia. "History of Ecological Design." *Oxford Research Encyclopedia of Environmental Science* (April 2018): 1–59.

Kant, Immanuel. *Critique of Judgment*. Translated by James Creed Meredith. Oxford: Oxford University Press, 2007. Originally published in 1790.

Kant, Immanuel. *Critique of Practical Reason*. Translated by Mary Gregor. Revised edition. Cambridge: Cambridge University Press, 2015.

Kant, Immanuel. *Critique of Pure Reason*. Translated by Paul Guyer and Alien W. Wood. Cambridge: Cambridge University Press, 1998. Originally published in 1781.

Katinić, Marina. "Holism in Deep Ecology and Gaia-Theory: A Contribution to Eco-Geological Science, a Philosophy of Life or a New Age Stream?" *The Holistic Approach to Environment* 3, no. 1 (March 2013): 3–14.

Kaufmann, Walter Arnold. *Hegel: Reinterpretation, Texts, and Commentary*. New York: Doubleday, 1965.

Keith, David. *The Case for Climate Engineering*. Cambridge: MIT Press, 2013.

Keller, Evelyn Fox. *A Feeling for the Organism: The Life and Work of Barbara McClintock*. San Francisco: W. H. Freeman, 1983.

Kelly, Leah. "The Approach: How Time Feels." Rubin Museum of Art, 2019.

Kelly, Leah. "Self of Sense." In *Experience: Culture, Cognition, and the Common Sense*, edited by Caroline A. Jones, David Mather, and Rebecca Uchill. Cambridge: MIT Press, 2016.

Keucheyan, Razmig. *Nature is a Battlefield: Towards a Political Ecology*. Translated by David Broder. Cambridge: Polity, 2016. Originally published in 2014.

Keys, Patrick W., Lan Wang-Erlandsson, Line J. Gordon, Victor Galaz, and Jonas Ebbesson. "Approaching Moisture Recycling Governance." *Global Environmental Change* 45 (July 2017): 15–23.

Kim, Ke Chung, and Loren B. Byrne. "Biodiversity Loss and the Taxonomic Bottleneck: Emerging Biodiversity Science." *Ecological Research* 21, no.6 (November 2006): 794–810.

Kimmerer, Robin Wall. *Braiding Sweetgrass: Indigenous Wisdom, Scientific Knowledge and the Teachings of Plants*. Minneapolis: Milkweed Editions, 2013.

King, Tiffany Lethabo. *The Black Shoals: Offshore Formations of Black and Native Studies*. Durham: Duke University Press, 2019.

Kircher, Athanasius. *Phonurgia Nova*. 1673.

Kittler, Wolf. "From Gestalt to Ge-Stell: Martin Heidegger Reads Ernst Jünger." *Cultural Critique*, no. 69 (Spring 2008): 79–97.

Klose, Alexander. *The Container Principle: How a Box Changes the Way We Think*. Translated by Charles Marcum II. Cambridge: MIT Press, 2015.

Kohn, Eduardo. *How Forests Think: Towards an Anthropology Beyond the Human*. Berkeley: University of California Press, 2013.

Kohn, Eduardo. "Natural engagements and Ecological Aesthetics: Among the Ávila Runa of Amazonian Ecuador." PhD diss., University of Wisconsin-Madison, 2002.

Knuth, Sarah. "Green Devaluation: Disruption, Divestment, and Decommodification for a Green Economy." *Capitalism Nature Socialism* 28, no. 1 (2017): 98–117.

Kojève, Alexandre. *Introduction to the Reading of Hegel: Lectures on the Phenomenology of Spirit*, compiled by Raymond Queneau, edited by Allan Bloom. Translated by James H. Nichols, Jr. in 1969. Ithaca: Cornell University Press, 1980. Originally published in 1947.

Kołakowski, Leszek. *Main Currents of Marxism: Its Rise, Growth and Dissolution; Volume 1, The Founders*. Oxford: Clarendon Press, 1978.

Kopenawa, Davi, and Bruce Albert. *A queda do céu: Palavras de um xamã Yanomami*. Translated by Beatriz Perrone-Moisés. Rio de Janeiro: Companhia das Letras, 2015.

Kovach, Margaret. *Indigenous Methodologies: Characteristics, Conversations, and Contexts*. Toronto: University of Toronto Press, 2009.

Krippner, Greta R. "The Financialization of the American Economy." *Socio-Economic Review* 3, no. 2 (2005): 173–208.

Kunuk, Zacharias, and Ian Mauro. "Inuit Knowledge and Climate Change." Documentary film, 2010.

Kurzweil, Ray. *The Singularity is Near*. New York: Viking Books, 2015.

Kwa, Chunglin. "Modeling the Grasslands." *Historical Studies in the Physical and Biological Sciences* 24, no. 1 (January 1993): 125–55.

Land, Nick. *Fanged Noumena: Collected Writings 1987–2007*, edited by Robin Mackay and Ray Brassier. Falmouth: Urbanomic; New York: Sequence Press, 2011.

Lanham, J. Drew. "Forever Gone: How Bird Lives and Black Lives Intertwine under the Long Shadow of History." *Orion Magazine*, 2018. https://orionmagazine.org/article/forever-gone.

Larson, James. "Bass: The Physical Sensation of Sound." *Audioholics*, December 29, 2015.

Latour, Bruno. "Anti-Zoom." In *Olafur Elisson: Contact*, Exhibition Catalog, Fondation Louis Vuitton, Paris, (December 17, 2014–February 23, 2015.) Paris: Flammarion, 2015. Republished in *Scale in Literature and Culture*, edited by Michael Tavel Clarke and David Wittenberg. Cham: Palgrave Macmillan, 2017.

Latour, Bruno. *Down to Earth: Politics in the New Climatic Regime*. Translated by Catherine Porter. Cambridge: Polity Press, 2018. Originally published in 2017.

Latour, Bruno. *Facing Gaia: Eight Lectures on the New Climatic Regime*. Translated by Catherine Porter. Cambridge: Polity, 2017.

Latour, Bruno. *Politics of Nature: How to Bring the Sciences into Democracy*. Translated by Catherine Porter. Cambridge: Harvard University Press, 2004. Originally published in 1999.

Latour, Bruno. *Reassembling the Social: An Introduction to Actor-Network-Theory*. Oxford: Oxford University Press, 2005.

Latour, Bruno. "Some Advantages of the Notion of 'Critical Zone' for Geopolitics." *Procedia Earth and Planetary Science* 10 (2014): 3–6.

Latour, Bruno. "'We Don't Seem to Live on the Same Planet'—A Fictional Planetarium." In *Designs for Different Futures*, edited by Kathryn B. Hiesinger and Michelle Millar Fisher. New Haven: Yale University Press, 2019.

Latour, Bruno. "Will Non-Humans be Saved? An Argument in Ecotheology." *Journal of the Royal Anthropological Institute* 15 (2009): 459–75.

Law, John. "What's Wrong with a One-World World?" *Distinktion: Scandinavian Journal of Social Theory* 16, no. 1 (2015): 126–39.

Ledgard, Jonathan. "Interspecies Money." In *Breakthrough: The Promise of Frontier Technologies for Sustainable Development*, edited by Homi Kharas, John W. McArthur, and Izumi Ohno. Washington: Brookings Institution Press, 2022.

Ledgard, Jonathan. "Lecture: The Age of the Interspecies; New Ways to Think About and Protect Nonhuman Life on Earth." European University Institute, San Domenico di Fiesole, May 15, 2023.

Ledgard, Jonathan. "linnaeus.life." Archived on *web.archive.org*, October 19, 2021.

Ledgard, Jonathan, and Erik Meijaard. "Endangered Wildlife Should Pay for Its Own Protection." *Project Syndicate*, December 2, 2021.

Legato, Marianne J. *Principles of Gender Specific Medicine: Gender in the Genomic Era*, 3rd ed. London: Academic Press, 2017. First published in 2004.

Leibniz, Gottfried Wilhelm. *Dissertation on the Art of Combinations*. Translated by Massimo Mugnai, Hans van Ruler, and Martin Wilson. Oxford: Oxford University Press, 2020. Originally published in 1666.

Leonardi, Emanuele. "Carbon Trading Dogma: Theoretical Assumptions and Practical Implications of Global Carbon Markets." *Ephemera* 17, no. 1 (February 2017): 61–87.

Levine, Matt. "Not Everyone Likes ESG." *Bloomberg*, January 12, 2022.

Levine, Matt. "Snap Earnings and Emissions Fraud." *Bloomberg*, May 11, 2017.

Lewis, David. *We, the Navigators: The Ancient Art of Landfinding in the Pacific*, 2nd ed. Honolulu: University of Hawaii Press, 1994.

Lewis, Jason Edward, Noelani Arista, Archer Pechawis, and Suzanne Kite. "Making Kin with the Machines." In *Atlas of Anomalous AI*, edited by Ben Vickers and K Allado-McDowell. London: Ignota Books, 2020.

Lewis, Mark Edward. *The Flood Myths of Early China*. Albany: SUNY Press, 2006.

Lewis, Michael. "In Nature's Casino." *New York Times Magazine*, August 26, 2007.

Lewis, Simon L., and Mark A. Maslin. "Defining the Anthropocene." *Nature* 519 (March 2015): 171–80.

Li, Jiaye, Tiejian Li, Guangqian Wang, Jiahua Wei, Deyu Zhong, Yang Su, and Xudong Fu. "空中水资源及其降水转化分析 [Kongzhong Shui Ziyuan Jiqi Jiangshui Zhuanhua Fenxi] Atmospheric Water Resource and Precipitation Conversion." 科学通报 [Kexue Tongbao] *Chinese Science Bulletin* 63, no. 26 (September 2018): 2785–96.

Li, Xiuzhen, Richard Bellerby, Christopher Craft, and Sarah Widney. "Coastal Wetland Loss, Consequences, and Challenges for Restoration." *Anthropocene Coasts* (January 2018): 1–15.

Liboiron, Max, and Josh Lepawsky. *Discard Studies: Wasting, Systems, and Power*. Cambridge: MIT Press, 2022.

Lim, Melody X., Bryan VanSanders, Anton Souslov, and Heinrich M. Jaeger. "Mechanical Properties of Acoustically Levitated Granular Rafts." *Physical Review X* 12, no. 2 (April–June 2022): 021017-1–15.

Lin, Tom C. W. "The New Market Manipulation." *Emory Law Journal* 66, no. 6 (2017): 1253–63.

Linden, Patrick Ingemar. *The Case against Death*. Cambridge: MIT Press, 2021.

Linton, James. "What is Water?: The History and Crisis of a Modern Abstraction." PhD diss., Carleton University, 2006.

Lippard, Lucy R. *Six Years: The Dematerialization of the Art Object from 1966 to 1972*. Berkeley: University of California Press, 1997. First published in 1973.

Lo, Chris. "The False Monopoly: China and the Rare Earths Trade." *Mining Technology*, August 19, 2015.

Locke, John. “An Essay Concerning Human Understanding.” *The Clarendon Edition of the Works of John Locke: An Essay Concerning Human Understanding*, edited by Peter H. Nidditch. Oxford: Oxford University Press, 1975. Text first published in 1689.

Loffreda, Beth. “Pulp Science: Race, Gender, and Prediction in Contemporary American Science.” PhD diss., Rutgers University, 1996.

Longino, Helen E. “Science in Society.” In *Science as Social Knowledge: Values and Objectivity in Scientific Inquiry*. Princeton: Princeton University Press, 1990.

Lossin, R. H. “Sabotage as Environmental Activism: Why Sabotage of Pipelines is a Justifiable and Effective Form of Resistance to Capitalism and Climate Change.” *Public Seminar*, July 3, 2018.

Lovelock, James E. *Novacene: The Coming Age of Hyperintelligence*. London: Penguin Books, 2020.

Löwith, Karl. *Meaning in History: The Theological Implications of the Philosophy of History*. Chicago: University of Chicago Press, 1949.

Luijendijk, Arjen. “A Novel Coastal Landscape Model for Sandy Systems: Community Base for Interdisciplinary Research on Coastal Evolution.” *Research in Urbanism Series* 7 (February 2021): 223–40.

Luijendijk, Arjen, and Alexander van Oudenhoven, eds. “Shaping Future Sandy Solutions.” In *The Sand Motor: A Nature-Based Response to Climate Change; Findings and Reflections of the Interdisciplinary Research Program NatureCoast*. Delft: Delft University Publishers, 2019.

Lukács, György. *The Destruction of Reason*. Translated by Peter Palmer. Atlantic Highlands: Humanities Press, 1981. Originally published in 1962.

Lukács, György. *History and Class Consciousness*. Translated by Rodney Livingstone. Cambridge: MIT Press, 1972.

Lukács, György. *The Theory of the Novel*. Translated by Anna Bostock. Cambridge: MIT Press, 1971.

Lynch, Kevin. *The Image of the City*. Cambridge: MIT Press, 1960.

Lyotard, Jean-François. “Presenting the Unpresentable: The Sublime.” Translated by Lisa Liebmann. *Artforum* 20, no. 8 (April 1982): 64–9.

Machen, Ruth, and Eric Nost. “Thinking Algorithmically: The Making of Hegemonic Knowledge in Climate Governance.” *Transactions of the Institute of British Geographers* 46, no. 3 (September 2021): 555–69.

Machine Bias. *ProPublica*, 2018. https://www.propublica.org/series/machine-bias.

Macpherson, Crawford Brough. *The Political Theory of Possessive Individualism: Hobbes to Locke*. Oxford: Oxford University Press, 2011.

Maidment, David R. *Handbook of Hydrology*. New York: McGraw-Hill Inc., 1993.

Majaca, Antonia. “The Oikos of the Earth, the Nomos of the Black Hole.” *e-flux Journal*, no. 143 (March 2024).

Majaca, Antonia. “Returns of and to Early Works.” In *Feminist Takes: Early Works by Želimir Žilnik*, edited by Antonia Majaca, Rachel O'Reilly, and Jelena Vesić. Berlin: Sternberg Press, 2021.

Majaca, Antonia, and Luciana Parisi. "The Incomputable and Instrumental Possibility." *e-flux Journal*, no. 77 (November 2016).
Malkin, Elisabeth. "Once built on a lake, Mexico City now runs dry." *New York Times*, March 16, 2006.
Malm, Andreas. *The Burning Continent: Climate Breakdown, Capitalism, and a New Geology of Revolution*. London: Verso, 2022.
Malm, Andreas. *Fossil Capital: The Rise of Steam Power and the Roots of Global Warming*. London: Verso, 2016.
Malm, Andreas. "The Future Is the Termination Shock: On the Antinomies and Psychopathologies of Geoengineering; Part One." *Historical Materialism* 30, no. 4 (December 2022): 3–53.
Malm, Andreas. *How to Blow Up a Pipeline: Learning to Fight in a World on Fire*. London: Verso, 2021.
Malm, Andreas. "The Origins of Fossil Capital: From Water to Steam in the British Cotton Industry." *Historical Materialism* 21, no. 1 (January 2013): 15–68.
Malm, Andreas. "Planning the Planet: Geoengineering Our Way Out of and Back into a Planned Economy." In *Has It Come to This? The Promises and Perils of Geoengineering on the Brink*, edited by J. P. Sapinski, Holly Jean Buck, and Andreas Malm. New Brunswick: Rutgers University Press, 2020.
Malm, Andreas. *The Progress of This Storm: Nature and Society in a Warming World*. London: Verso, 2018.
Manjoo, Farhad. "Unix Tick Tocks to a Billion." *Wired*, September 8, 2001.
Mantz, Jeffrey W. "Improvisational economies: Coltan production in the eastern Congo." *Social Anthropology* 16, no. 1 (February 2008): 34–50.
Manzini, Ezio. *Livable Proximity: Ideas for the City that Cares*. Milan: Egea, 2022.
Marder, Michael. "The Place of Plants: Spatiality, Movement, Growth." *Performance Philosophy* 1, no. 1 (2015): 185–94.
Marinetti, Filippo Tommaso. *Mafarka the Futurist: An African Novel*. Translated by Carol Diethe and Steve Cox. London: Middlesex University Press, 1997. Originally published in 1909.
Marinetti, Filippo Tommaso. "Multiplied Man and the Reign of the Machine." Translated by Lawrence Rainey. In *Futurism: An Anthology*, edited by Lawrence Rainey, Christine Poggi, and Laura Wittman. New Haven: Yale University Press, 2009. Text originally published in 1911.
Martín González, José. "La Ascendencia de Coyote." In *Mechatronic Circus*. London: 198 Gallery and Oriel Mostyn Gallery, 2000.
Martinot, Steve. "Motherhood and the Invention of Race." *Hypatia* 22, no. 2 (Spring 2007): 79–97.
Marvin, Carolyn. *When Old Technologies Were New: Thinking About Electric Communication in the Late Nineteenth Century*. New York: Oxford University Press, 1988.

Marx, Karl. *Capital: A Critique of Political Economy, Volume 1*. Translated by Ben Fowkes. London: Penguin, 1976. Originally published in 1867.
Marx, Karl. "Ethnologische Exzerpthefte." In *Marx-Engels-Gesamtausgabe* ($MEGA^2$), Abt. IV: Exzerpte, Notizen und Marginalien, Bd. (27–9), edited by Internationale Marx-Engels-Stiftung. Berlin: De Gruyter Akademie Forschung. Forthcoming.
Marx, Karl. *Grundrisse: Foundations of the Critique of Political Economy*. Translated by Martin Nicolaus. London: Penguin, 1973. Originally published in 1939.
Marx, Karl. "Theses on Feuerbach." 1845.
Maslow, Abraham H. *The Psychology of Science: A Reconnaissance*. South Bend: Gateway Editions, 1969. Originally published in 1966.
Massey, Doreen. *For Space*. London: Sage, 2005.
Mattern, Shannon. "A City Is Not a Computer." *Places Journal*, February 2017.
Mattern, Shannon. *Code and Clay, Data and Dirt: Five Thousand Years of Urban Media*. Minneapolis: University of Minnesota Press, 2017.
Maxwell, James Clerk. "On Governors." *Proceedings of the Royal Society of London* 16, no. 16 (December 1868): 270–82.
Mbembe, Achille. *Critique of Black Reason*. Translated by Laurent Dubois. Durham: Duke University Press, 2017.
Mbembe, Achille. "Decolonizing Knowledge and the Question of the Archive." In *Decolonizing: The Curriculum, the Museum, and the Mind*, edited by Marquard Smith. Vilnius: Vilnius Academy of Arts Press, 2020.
Mbembe, Achille. "How to Develop a Planetary Consciousness." *Noēma*, January 11, 2022.
Mbembe, Achille. *Necropolitics*. Translated by Steven Corcoran. Durham: Duke University Press, 2019. Originally published in 2016.
Mbembe, Achille. *Out of the Dark Night: Essays on Decolonization*. Translated by Daniela Ginsburg. New York: Columbia University Press, 2021. Originally published in 2010.
Mbembe, Achille. "Thoughts on the Planetary: An Interview with Achille Mbembe." Interview by Torbjørn Tumyr Nilsen. *New Frame*, September 5, 2019.
McCarthy, Dennis D., and P. Kenneth Seidelmann. *Time: From Earth Rotation to Atomic Physics*, 2nd ed. Cambridge: Cambridge University Press, 2018.
McCarthy, John, and Edward Feigenbaum. "Arthur Samuel: Pioneer in Machine Learning." *AI Magazine* 11, no. 3 (September 1990): 10–11.
McHarg, Ian. *Design with Nature*. New York: John Wiley & Sons, 1995. First published in 1969.
McKittrick, Katherine. *Dear Science and Other Stories*. Durham: Duke University Press, 2021.
McKittrick, Katherine. *Demonic Grounds: Black Women and the Cartographies of Struggle*. Minneapolis: University of Minnesota Press, 2006.

McKittrick, Katherine, ed. *Sylvia Wynter: On Being Human as Praxis*. Durham: Duke University Press, 2015.

McKittrick, Katherine. "Yours in the Intellectual Struggle: Sylvia Wynter and the Realization of the Living." In *Sylvia Wynter: On Being Human as Praxis*, edited by Katherine McKittrick. Durham: Duke University Press, 2015.

McKittrick, Katherine, Frances H. O'Shaughnessy, and Kendall Witaszek. "Rhythm, or On Sylvia Wynter's Science of the Word." *American Quarterly* 70, no. 4 (2018): 867–74.

McLeman, Robert A. *Climate and Human Migration: Past Experiences, Future Challenges*. Cambridge: Cambridge University Press, 2014.

McLuhan, Marshall. *Understanding Media: The Extensions of Man*. Cambridge: MIT Press, 1964.

McNeill, John Robert, and Peter Engelke. *The Great Acceleration: An Environmental History of the Anthropocene since 1945*. Cambridge: Harvard University Press, 2016.

Medina, Eden. *Cybernetic Revolutionaries: Technology and Politics in Allende's Chile*. Cambridge: MIT Press, 2011.

Meglicki, Zdzislaw. *Quantum Computing without Magic: Devices*. Cambridge: MIT Press, 2008.

Meister, Robert. *Justice Is an Option: A Democratic Theory of Finance for the Twenty-First Century*. Chicago: University of Chicago Press, 2021.

Menke, Christoph. "The Critique of Law and the Law of Critique." Annual Philosophy Lecture, Goldsmiths, University of London, May 29, 2019.

Merchant, Carolyn. *The Death of Nature: Women, Ecology and the Scientific Revolution*. San Francisco: HarperOne, 1990. Originally published in 1980.

Merlin, Cornell Lab. http://merlin.allaboutbirds.org.

Messeri, Lisa. *Placing Outer Space: An Earthly Ethnography of Other Worlds*. Durham: Duke University Press, 2016.

Mezzadra, Sandro, and Brett Neilson. "On the Multiple Frontiers of Extraction: Excavating Contemporary Capitalism." *Cultural Studies* 31, no. 2–3 (May 2017): 185–204.

Microsoft. "Our Corporate Values." https://www.microsoft.com/en-lb/about/corporate-values.

Microsoft. "A Planetary Computer for a Sustainable Future." https://planetarycomputer.microsoft.com.

Mies, Maria. *Patriarchy and Accumulation on a World Scale: Women in the International Division of Labour*. London: Zed Books, 1986.

Mies, Maria, and Vandana Shiva. *Ecofeminism*. London: Zed Books, 1993.

Mignolo, Walter D. "Epistemic Disobedience, Independent Thought and De-colonial Freedom." *Theory, Culture & Society* 26, no. 7–8 (December 2009): 1–23.

Mignolo, Walter D. "Sylvia Wynter: What Does It Mean to Be Human?" In *Sylvia Wynter: On Being Human as Praxis*, edited by Katherine McKittrick. Durham: Duke University Press, 2015.

Millán, Moira. "Moira Millan y el buenvivir originario." *Pensamiento Ambiental*, YouTube, May 22, 2016.

Miller, Clark A., and Paul N. Edwards, eds. *Changing the Atmosphere: Expert Knowledge and Environmental Governance*. Cambridge: MIT Press, 2001.

Millo, Yuval, Fabian Muniesa, Nikiforos S. Panourgias, and Susan V. Scott. "Organised Detachment: Clearinghouse Mechanisms in Financial Market." *Information and Organization* 15, no. 3 (2005): 229–46.

Mills, Charles W. "The Chronopolitics of Racial Time." *Time & Society* 29, no. 2 (2020): 297–317.

Mills, Simon. "Simondon and Big Data." *Platform: Journal of Media and Communication* 6 (2015): 59–72.

Mirowski, Philip. *Machine Dreams: Economics Becomes a Cyborg Science*. Cambridge: Cambridge University Press, 2002.

Mirzoeff, Nicholas. *The Right to Look: A Counterhistory of Visuality*. Durham: Duke University Press, 2011.

Mitchell, Melanie. *Complexity: A Guided Tour*. New York: Oxford University Press, 2009.

Mitchell, Timothy. *Carbon Democracy: Political Power in the Age of Oil*. London: Verso, 2011.

Mitman, Gregg. "Reflections on the Plantationocene: A Conversation with Donna Haraway and Anna Tsing." *Edge Effects Magazine* (June 18, 2019): 2–19.

Moafi, Samaneh, et al. "Tear Gas in Plaza de la Dignidad." Forensic Architecture, December 20, 2020.

Molina, Mario J., and F. Sherwood Rowland. "Stratospheric Sink for Chlorofluoromethanes: Chlorine Atom-Catalysed Destruction of Ozone." *Nature* 249 (1974): 810–12.

Møllgaard, Eske. *An Introduction to Daoist Thought: Action, Language, and Ethics in Zhuangzi*. London: Routledge, 2007.

Montévil, Maël, Bernard Stiegler, Giuseppe Longo, Ana Soto, and Carlos Sonnenschein. "Anthropocene, Exosomatization and Negentropy." In *On Transition: In Response to António Guterres*. Internation, 2020. https://internation.world.

Moore, Jason W. *Capitalism in the Web of Life: Ecology and the Accumulation of Capital*. London: Verso, 2015.

Moore, Jason W. "The Capitalocene, Part I: On the Nature and Origins of Our Ecological Crisis." *Journal of Peasant Studies* 44, no. 3 (May 2017): 594–630.

Moore, Jason W. "Nature in the Limits to Capital (and vice versa)." *Radical Philosophy, no.* 193 (2015): 9–19.

Moore, Jason W., ed. "The Rise of Cheap Nature." In *Anthropocene or Capitalocene? Nature, History, and the Crisis of Capitalism*. Oakland: PM Press, 2016.

Moravec, Hans. *Mind Children: The Future of Robot and Human Intelligence*. Cambridge: Harvard University Press, 1988.

Morozov, Evgeny. "Digital Socialism? The Calculation Debate in the Age of Big Data." *New Left Review* 116/117 (2019): 33–67.

Morton, Timothy. *Dark Ecology: For a Logic of Future Coexistence*. New York: Columbia University Press, 2016.

Morton, Timothy. *The Ecological Thought*. Cambridge: Harvard University Press, 2010.

Mosco, Vincent. *To the Cloud: Big Data in a Turbulent World*. Boulder: Paradigm, 2014.

Moten, Fred. "Anassignment Letters." In *Stolen Life*. Durham: Duke University Press, 2018.

Mount Pleasant, Jane. "A New Paradigm for Pre-Columbian Agriculture in North America." *Early American Studies: An Interdisciplinary Journal* 13, no. 2 (2015): 374–412.

Mumford, Lewis. *Technics and Civilization*. Chicago: University of Chicago Press, 2010. Originally published in 1934.

Muniesa, Fabian. *The Provoked Economy: Economic reality and the Performative Turn*. Oxon: Routledge, 2014.

Munn, Nancy D. "The Cultural Anthropology of Time: A Critical Essay." *Annual Review of Anthropology* 21 (1992): 93–123.

Munnik, René. "Donna Haraway: Cyborgs for Earthly Survival?" In *American Philosophy of Technology: The Empirical Turn*, edited by Hans Achterhuis. Translated by Robert P. Crease. Bloomington: Indiana University Press, 2001. Originally published in 1997.

Nagel, Barbara N. "Goethe's Stalker Snails." *Feminist German Studies* 36, no. 2 (Fall/Winter 2020): 1–26.

Nagel, Thomas. "What Is It Like to Be a Bat?" *Philosophical Review* 83, no. 4 (October 1974): 435–50.

Narby, Jeremy. *The Cosmic Serpent: DNA and the Origins of Knowledge*. Reprint edition. New York: Jeremy P. Tarcher/Putnam, 1999.

NASA Earth System Science Committee. *Earth System Science: A Closer View*. Washington: NASA, 1988.

National Aeronautics and Space Administration (NASA). "The Importance of Understanding Clouds." September 2005. https://www.nasa.gov/pdf/135641main_clouds_trifold21.pdf.

National Aeronautics and Space Administration (NASA). "Measurement of Air Pollution from Satellites (MAPS) – Understanding the Chemistry of the Atmosphere." September 19, 1996. https://www.nasa.gov/centers/langley/news/factsheets/MAPS.html.

Needham, Joseph, et al. *Science and Civilisation in China*, vols. 1–7. Cambridge: Cambridge University Press, 1954–2015.

Negroni, Christine. "How to Determine the Power Rating of Your Gadget's Batteries." *New York Times*, December 26, 2016.

Newell, Reginald E., and Yong Zhu. "Tropospheric Rivers: A One-Year Record and a Possible Application to Ice Core Data." *Geophysical Research Letters* 21, no. 2 (January 1994): 113–6.

Newell, Reginald E., Nicholas E. Newell, Yong Zhu, and Courtney Scott. "Tropospheric Rivers? – A Pilot Study." *Geophysical Research Letters* 19, no. 24 (December 1992): 2401–4.

Newton, Isaac. *The Principia: Mathematical Principles of Natural Philosophy; The Authoritative Translation*. Translated by I. Bernard Cohen, Anne Whitman, and Julia Budenz. Oakland: University of California Press, 2016. Originally published in 1687.

Nhat Hanh, Thich. *The World We Have: A Buddhist Approach to Peace and Ecology*. Berkeley: Parallax Press, 2008.

Nixon, Rob. *Slow Violence and the Environmentalism of the Poor*. Cambridge: Harvard University Press, 2011.

NOAA. "Atmospheric River Observatories," *NOAA Physical Sciences Laboratory*. Accessed May 10, 2025. https://psl.noaa.gov/psd2/technology/aro.html.

Noble, Safiya Umoja. *Algorithms of Oppression: How Search Engines Reinforce Racism*. New York: New York University Press, 2018.

Norberg-Hodge, Helena. "Putting Technology in Its Place." GTI Forum Technology and the Future, Great Transition Initiative, January 3, 2022.

Novalis. "Miscellaneous Observations." *Novalis: Philosophical Writings*, edited and translated by Margaret Mahony Stoljar. Albany: State University of New York Press, 1997.

O'Neil, Cathy. *Weapons of Math Destruction: How Big Data Increases Inequality and Threatens Democracy*. New York: Broadway Books, 2016.

Odell, Jenny. "Notes of a Bioregional Interloper." *Open Space*, San Francisco Museum of Modern Art, October 9, 2017.

Odom, Mel. *I Have No Mouth, and I Must Scream: The Official Strategy Guide*. Rocklin: Prima Publishing, 1995.

Odum, Howard T. "Ecological Potential and Analogue Circuits for the Ecosystem." *American Scientist* 48, no. 1 (March 1960): 54A, 1–8.

Olivier, Guilhem. "Huehuecóyotl, 'Coyote Viejo,' el músico transgresor: ¿dios de los otomíes o avatar de tezcatlipoca?" *Estudios De Cultura Náhuatl* 30 (1999): 113–32.

Ornstein, Robert E. *On the Experience of Time*. Harmondsworth: Penguin Books, 1969.

Orr, David W. *The Nature of Design: Ecology, Culture, and Human Intention*. Oxford: Oxford University Press, 2002.

Pit, Iris. "Environmental Impacts of an Innovative Beach Nourishment." In *The Sand Motor: A Nature-Based Response to Climate Change; Findings and Reflections of the Interdisciplinary Research Program NatureCoast*, edited by Arjen Luijendijk and Alexander van Oudenhoven. Delft: Delft University Publishers, 2019.

Simon, Herbert. *The Sciences of the Artificial*, 3rd ed. Cambridge: MIT Press, 2019. First published in 1996.

Sprenger, Florian. *Epistemologien des Umgebens: Zur Geschichte, Ökologie und Biopolitik Künstlicher Environments*. Bielefeld: Transcript, 2019.

Özgöde, Onur. "The Emergence of Systemic Risk: Federal Reserve, Bailouts, and Monetary Government at the Limits." *SocArXiv* (January 3, 2019).

Pacheco Bejarano, Juan Pablo. "Telepathy Without the Internet." 2020.

Palma Rodríguez, María Angélica, Luis Alberto Martos López, Rosalba Nieto Calleja, Erika Morales Vigil, Antelma Premió Vázquez, Luis Alberto Guerrero Jordán, Raquel Mancera Sánchez, Oscar Reyes Sánchez, Héctor Celedón Muñiz, Rafael Zimbrón Romero, and Iván Pérez Téllez. *Tizacalco ipan ihxihua Teuhtli: Tizacalco sobre los pies del Teuhtli; Sobre los pies del Teuhtli: Tizacalco, patrimonio arqueológico de Milpa Alta*. Mexico City: Calpulli Tecalco, 2020.

Pang Jixing 潘吉星. *Song Yingxing Pingzhuan* 宋應星評傳 [Critical biography of Sung Yingxing]. 南京大学出版社 Nanjing University Press, 1990.

Parikka, Jussi. *A Geology of Media*. Minneapolis: University of Minnesota Press, 2015.

Parisi, Luciana. *Contagious Architecture: Computation, Aesthetics, and Space*. Cambridge: MIT Press, 2013.

Parisi, Luciana. "Instrumental Reason, Algorithmic Capitalism and the Incomputable." In *Alleys of Your Mind: Augmented Intelligence and Its Traumas*, edited by Matteo Pasquinelli. Lüneburg: Meson Press, 2015.

Pasquale, Frank. *The Black Box Society: The Secret Algorithms That Control Money and Information*. Cambridge: Harvard University Press, 2016.

Pasquinelli, Matteo. *The Eye of the Master: A Social History of Artificial Intelligence*. London: Verso, 2023.

Pasquinelli, Matteo. "Italian *Operaismo* and the Information Machine." *Theory, Culture & Society* 32, no. 3 (May 2015): 49–68.

Pearl, Judea. *Probabilistic Reasoning in Intelligent Systems: Networks of Plausible Inference*. San Mateo: Morgan Kaufmann, 1988.

Pedersen, Isabel, and Andrew Iliadis. *Embodied Computing: Wearables, Implantables, Embeddables, Ingestibles*. Cambridge: MIT Press, 2020.

Philips. "We Are Working to Make Our Supply Chain 'Conflict-Free.'" 2018. Accessed September 7, 2018.

Pickering, Andrew. *The Cybernetic Brain: Sketches of Another Future*. Chicago: University of Chicago Press, 2010.

Pietz, David A. *The Yellow River: The Problem of Water in Modern China*. Cambridge: Harvard University Press, 2015.

Plant, Sadie. "Something in the Air." In *The Movement of Clouds Around Mount Fuji Photographed and Filmed by Masanao Abe* by Helmut Völter. Leipzig: Spector Books, 2016.

Plutarch. *Lives, Volume VIII: Sertorius and Eumenes; Phocion and Cato the Younger*. Translated by Bernadotte Perrin. Cambridge: Harvard University Press, 1912.

Pontin, Jason. "Artificial Intelligence, With Help from the Humans." *New York Times*, March 25, 2007.

Portanova, Stamatia. *Whose Time Is It? Asocial Robots, Sycolonialism, and Artificial Chronological Intelligence*. Berlin: Sternberg Press, 2021.

Povinelli, Elizabeth A. *Geontologies: A Requiem to Late Liberalism*. Durham: Duke University Press, 2016.

Povinelli, Elizabeth A. "In Some Places the Not-Yet Has Long Been Already." *No Linear Fucking Time*, BAK, December 3, 2021.

Prescod-Weinstein, Chanda. *The Disordered Cosmos: A Journey into Dark Matter, Spacetime, and Dreams Deferred*. New York: Bold Type Books, 2021.

Preston, Christopher J. *The Synthetic Age: Outdesigning Evolution, Resurrecting Species, and Reengineering Our World*. Cambridge: MIT Press, 2018.

Puar, Jasbir K. *Terrorist Assemblages: Homonationalism in Queer Times*. Durham: Duke University Press, 2017. First published in 2007.

Quinson, Tim. "Wall Street Is Close to Triggering a Climate Financial Crisis." *Bloomberg*, December 14, 2021.

Quire, Rachel. "SEABEDS—Sub-Marine Territory: Living and Working on the Seafloor During the Sealab II Experiment." In *Territory Beyond Terra*, edited by Kimberly Peters, Phil Steinberg, and Elaine Stratford. London: Rowman & Littlefield, 2018.

Raekstad, Paul, and Sofa Saio Gradin. *Prefigurative Politics: Building Tomorrow Today*. Cambridge: Polity, 2019.

Ransby, Barbara. *Ella Baker and the Black Freedom Movement: A Radical Democratic Vision*. Chapel Hill: University of North Carolina Press, 2003.

Raskin, Paul. *Journey to Earthland: The Great Transition to Planetary Civilization*. Cambridge: Tellus Institute, 2016.

Raval, Anjli. "A $140bn Asset Sale: The Investors Cashing in on Big Oil's Push to Net Zero." *Financial Times*, July 6, 2021.

Reed, Chris, and Nina-Marie Lister. "Ecology and Design: Parallel Genealogies." *Places Journal*, April 2014.

Reed, Patricia. "What is Care at Planetary Dimensions?" Berlin: Floating University, 2019.

Rees, W. J. "Ernest Ruthven Sykes (1867–1954)." *Journal of the Society for the Bibliography of Natural History* 4, no. 1 (January 1962): 20–7.

Reichel-Dolmatoff, Gerardo. "Templos Kogi: Introducción al simbolismo y a la astronomía del espacio sagrado." *Revista Colombiana de Antropología* 19 (July 1975): 199–245.

Rifkin, Jeremy. *Time Wars: The Primary Conflict in Human History*. New York: Henry Holt and Company, 1987.

Rilke, Rainer Maria. "The Eigth Elegy." In *The Selected Poetry of Rainer Maria Rilke*. Translated by Stephen Mitchell. New York: Vintage Books, 1982.

Rispoli, Giulia. "Planetary Environing: The Biosphere and the Earth System." In *Environing Media*, edited by Adam Wickberg and Johan Gärdebo. London: Routledge, 2023.

Robilliard, Denis. Review of *Signals and Boundaries; Building Blocks for Complex Adaptive Systems*, by John H. Holland. *Genetic Programming and Evolvable Machines* 14 (2013): 279–80.

Roden, David. *Posthuman Life: Philosophy at the Edge of the Human*. London: Routledge, 2015.

Rodríguez, Carlos, et al. "Áreas protegidas y visiones interculturales: conociendo nuestra biodiversidad." II Congreso colombiano de áreas protegidas, Parques Nacionales Naturales de Colombia, 2014. Report, 2015.

Roest, Bart. "Combined Morphology Surveys Delfland 2011–2016." *4TU. ResearchData*, November 24, 2017.

Rosemont, Franklin. "Karl Marx and the Iroquois." *Arsenal: Surrealist Subversion*, no. 4 (1989): 205–24.

Rovelli, Carlo. *The Order of Time*. Translated by Erica Segre and Simon Carnell. London: Allen Lane, 2018. Originally published in 2017.

Rubin, Sergio, Thomas Parr, Lancelot Da Costa, and Karl Friston. "Future Climates: Markov Blankets and Active Inference in the Biosphere." *Journal of the Royal Society Interface* 17, no. 172 (November 2020): 1–12.

Ruskin, John. "Remarks on the Present State of Meteorological Science." *Transactions of the Meteorological Society* 1 (1839): 56–9.

Russakovsky, Olga, Jia Deng, Hao Su, Jonathan Krause, Sanjeev Satheesh, Sean Ma, Zhiheng Huang, Andrej Karpathy, Aditya Khosla, Michael Bernstein, Alexander C. Berg, and Li Fei-Fei. "ImageNet Large Scale Visual Recognition Challenge." *International Journal of Computer Vision* 115, no. 3 (2015): 211–52.

Ryan, James A. "Leibniz' Binary System and Shao Yong's *Yijing*." *Philosophy East and West* 46, no. 1 (1996): 59–90.

Sadin, Éric. *La Silicolonisation du monde: L'irrésistible expansion du libéralisme numérique*. Paris: L'échappée, 2016.

Sadowski, Jathan. "When Data is Capital: Datafication, Accumulation, and Extraction." *Big Data & Society* 6, no. 1 (2019): 1–12.

Sahlins, Marshall. *Stone Age Economics*. Chicago: Aldine-Atherton, 1972.

Sakolsky, Ron, and James Koehnline, eds. *Gone to Croatan: Origins of North American Dropout Culture*. Brooklyn: Autonomedia, 1993.

Salleh, Ariel. *Ecofeminism as Politics: Nature, Marx and the Postmodern*. London: Zed Books, 1997.

Santos, Bárbara. *Curación como tecnología: Basado en entrevistas a sabedores de la Amazonía*. Bogotá: Alcaldía de Bogotá, 2019.

Scarritt, Arthur. "Tripartheid: How Global White Supremacy Triumphs through Neoliberalism." In *Totality Inside Out: Rethinking Crisis and Conflict Under Capital*, edited by Kevin Floyd, Jen Hedler Phillis, and Sarika Chandra. New York: Fordham University Press, 2022.

Schaffer, Simon. "Babbage's Intelligence: Calculating Engines and the Factory System." *Critical Inquiry* 21, no. 1 (Autumn 1994): 203–27.

Schlanger, Zoë. "If Shipping were a Country, It would be the Sixth-Biggest Greenhouse Gas Emitter." *Quartz*, April 17, 2018.

Schmid, Helga, and Kevin Walker. "Anti-Clockwise: Building a Critical Mass against Clock Time." *Art & the Public Sphere* 9, nos. 1–2 (2020): 39–61.

Schmid, Helga, and Kevin Walker. "The Atemporal Event." *Journal of Artistic Research* 25, 2021.

Schmid, Helga. *Uchronia: Designing Time*. Basel: Birkhäuser, 2020.

Schumacher, Melissa, Pamela Durán-Díaz, Anne Kristiina Kurjenoja, Eduardo Gutiérrez-Juárez, and David A. González-Rivas. "Evolution and Collapse of Ejidos in Mexico—To What Extent Is Communal Land Used for Urban Development?" *Land* 8, no. 146 (2019): 1–21.

Schuppli, Susan. "Can the Sun Lie?" HD video, 12:51 minutes, 2014.

Scott, James C. *Seeing Like a State: How Certain Schemes to Improve the Human Condition Have Failed*. New Haven: Yale University Press, 1998.

Seising, Rudolf. "Cybernetics, System(s) Theory, Information Theory and Fuzzy Sets and Systems in the 1950s and 1960s." *Information Sciences* 180, no. 3 (2010): 459–76.

Segato, Rita Laura. *Contra-pedagogías de la crueldad*. Buenos Aires: Prometeo Libros, 2018.

Segato, Rita Laura. *La guerra contra las mujeres*. Madrid: Traficantes de Sueños, 2016.

Self-Assembly Lab, Massachusetts Institute of Technology. https://selfassemblylab.mit .edu.

Shankleman, Jessica, Tom Biesheuvel, Joe Ryman, and Dave Merrill. "We're Going to Need More Lithium." *Bloomberg*, September 7, 2017.

Shannon, Claude E. "A Mathematical Theory of Communication." *Bell System Technical Journal* 27, no. 3 (July 1948): 379–423.

Shannon, Claude E., and Warren Weaver. *The Mathematical Theory of Communication*. Urbana: University of Illinois Press, 1949.

Shao Yong 邵雍. *Huangji jingshi shu* 皇極經世書 [Book of sublime principle which governs all things within the world]. Originally published in 1060.

Sharma, Kriti. *Interdependence: Biology and Beyond*. New York: Fordham University Press, 2015.

Sharpe, Christina. *In the Wake: On Blackness and Being*. Durham: Duke University Press, 2016.

Shekhovtsov, S. V., I. N. Shekhovtsova, and S. E. Peltek. "DNA Barcoding: Methods and Approaches." *Biology Bulletin Reviews* 9 (2019): 475–83.

Sheridan, Michael J., and Celia Nyamweru, eds. *African Sacred Groves: Ecological Dynamics and Social Change*. Athens: Ohio University Press, 2008.

Sherman, Stephanie. "The Autocene: Towards a Post-Automotive Future." *Strelka Magazine*, January 25, 2022.

Shiva, Vandana. *Protect or Plunder? Understanding Intellectual Property Rights*. London: Zed Books, 2001.

Shiva, Vandana. *Staying Alive: Women, Ecology and Development*. London: Zed Books, 1988.

Shiva, Vandana, AH Jafri, Gitanjali Bedi, and Radha Hollar-Bhar. *The Enclosure and Recovery of the Commons: Biodiversity, Indigenous Knowledge and Intellectual Property Rights*. New Delhi: Research Foundation for Science, Technology and Ecology, 1997.

Simondon, Gilbert. *On the Mode of Existence of Technical Objects*. Translated by Cecile Malaspina and John Rogove. Minneapolis: Univocal, 2016. Originally published in 1958.

Simondon, Gilbert. "Technical Mentality," translated by Arne De Boever. *Parrhesia: A Journal of Critical Philosophy*, no. 7 (November 2009): 17–27.

Simondon, Gilbert. "Technical Mentality." In *Gilbert Simondon: Being and Technology*, edited by Arne De Boever, Alex Murray, Jon Roffe, and Ashley Woodward. Edinburgh: Edinburgh University Press, 2012.

Simpson, Cam, Akshat Rathi, and Saijel Kishan. "The ESG Mirage." *Bloomberg*, December 10, 2021.

Simpson, Cam. "The Deadly Tin Inside Your Smartphone." *Bloomberg*, August 24, 2012.

Simpson, I. R., and P. D. Jones. "Analysis of UK precipitation extremes derived from Met Office gridded data." *International Journal of Climatology* 34, no. 7 (June 2014): 2438–49.

Simpson, Leanne Betasamosake. *As We Have Always Done: Indigenous Freedom from Radical Resistance*. Minneapolis: University of Minnesota Press, 2017.

Singh, Sumit. "How China Is Using a UAV to Help Change the Weather." *Simple Flying*, January 27, 2021. https://simpleflying.com/china-weather-changing-uav.

Smedley, Audrey. *Race in North America: Origin and Evolution of a Worldview*. New York: Routledge, 2012. Originally published in 1992.

Smith, Brad. "AI for Earth: A Gamechanger for Our Planet." *Guardian*, September 11, 2017.

Smith, Justin E. H. "The Internet of Snails: Escargotic Commotion and the Wood-Wide Web." *Cabinet* 58 (Summer 2015).

Smith, Linda Tuhiwai. *Decolonizing Methodologies: Research and Indigenous Peoples*, 2nd edition. London: Zed Books, 2012.

Smith, Neil, and Cindi Katz. "Grounding Metaphor: Towards a spatialized politics." In *Place and the Politics of Identity*, edited by Michael Keith and Steve Pile. London: Routledge, 1993.

Smith, Robin. "What Happens When a Missile or Meteor Hits." *Duke Today*, April 10, 2015.

Sohn-Rethel, Alfred. *Intellectual and Manual Labour: A Critique of Epistemology*. Atlantic Highlands: Humanities Press, 1978.

Solon, Olivia. "Elon Musk says Humans must become Cyborgs to Stay Relevant. Is He Right?" *Guardian*, February 15, 2017.

Somé, Malidoma Patrice. *Of Water and the Spirit: Ritual, Magic, and Initiation in the Life of an African Shaman*. New York: G. P. Putnam's Sons, 1994.

Somerville, Peter. "Revisiting Connections between Capital and Nature II: The Case of Climate Change." *Capitalism Nature Socialism* 32, no. 2 (2021): 40–55.

Spackman, Barbara. *Fascist Virilities: Rhetoric, Ideology, and Social Fantasy in Italy*. Minneapolis: University of Minnesota Press, 1996.

Spinney, Laura. "The Big Idea: Should Other Species Have Their Own Money?" *Guardian*, March 14, 2022.

Spitzer, Michael. *The Musical Human: A History of Life on Earth*. London: Bloomsbury, 2021.

Spivak, Gayatri Chakravorty. *An Aesthetic Education in the Era of Globalization*. Cambridge: Harvard University Press, 2012.

Spivak, Gayatri Chakravorty. *Death of a Discipline*. New York: Columbia University Press, 2003.

Spivak, Gayatri Chakravorty. *Imperative zur Neuerfindung des Planeten / Imperatives to Re-Imagine the Planet*. Translated by Bernard Schweizer. Vienna: Passagen Verlag, 2013. Originally published in 2012.

Spivak, Gayatri Chakravorty. "Planetarity." In *Dictionary of Untranslatables: A Philosophical Lexicon*, edited by Barbara Cassin. Princeton: Princeton University Press, 2014.

Starosielski, Nicole. *The Undersea Network*. Durham: Duke University Press Books, 2015.

State Council, People's Republic of China. "China to Forge Ahead with Weather Modification Service." *English.gov.cn*, last modified December 2, 2020. http://english.www.gov.cn/policies/latestreleases/202012/02/content_WS5fc76218c6d0f7257694125e.html.

Steffen, Will, Wendy Broadgate, Lisa Deutsch, Owen Gaffney, and Cornelia Ludwig. "The Trajectory of the Anthropocene: The Great Acceleration." *Anthropocene Review* 2, no. 1 (April 2015): 81–98.

Steffen, Will, Paul J. Crutzen, and John R. McNeill. "The Anthropocene: Are Humans Now Overwhelming the Great Forces of Nature?" *Ambio* 36, no. 8 (2007): 614–21.

Stengers, Isabelle. "The Cosmopolitical Proposal." In *Making Things Public: Atmospheres of Democracy*, edited by Bruno Latour and Peter Weibel. Cambridge: MIT Press, 2005.

Stewart, Balfour, and Peter Guthrie Tait. *The Unseen Universe; Or, Physical Speculations on a Future State*. New York: Macmillan, 1875.

Strauss, Emily. "Is Everything Securities Fraud?" *Duke Law School Public Law & Legal Theory Series*, no. 2021-04 (2020): 1331–83.

Sullivan, Sian. "'Ecosystem Service Commodities' – A New Imperial Ecology? Implications for Animist Immanent Ecologies, with Deleuze and Guattari." *New Formations* 2009, no. 69 (2009): 111–28.

Summer, Larry. Quoted in "Furor on Memo At World Bank." *New York Times*, February 7, 1992.

Sutherland, Zöe, and Marina Vishmidt. "(Un)making Value: Reading Social Reproduction through the Question of Totality." In *Totality Inside Out: Rethinking Crisis and Conflict under Capital*, edited by Kevin Floyd, Jen Hedler Phillis, and Sarika Chandra. New York: Fordham University Press, 2022.

Swetz, Frank J. "Leibniz, the *Yijing*, and the Religious Conversion of the Chinese." *Mathematics Magazine* 76, no. 4 (2003): 276–91.

Sze, Sarah. "How I Solved It: The Problem of Suspense." *New Yorker*, May 11, 2017.

Szerszynski, Bronislaw, and Maialen Galarraga. "Geoengineering Knowledge: Interdisciplinarity and the Shaping of Climate Engineering Research." *Environment and Planning A* 45, no. 12 (December 2013): 2817–24.

Szocik, Konrad, Mark Shelhamer, Martin Braddock, Francis A. Cucinotta, Chris Impey, Pete Worden, Ted Peters, Milan M. *Ćirković, Kelly C. Smith, Koji Tachibana, Michael J. Reiss, Ziba Norman, Alvin M. Gouw, and Gonzalo Munévar*. "Future Space Missions and Human Enhancement: Medical and Ethical Challenges." *Futures* 133 (October 2021): 102819.

Szondi, Peter. *An Essay on the Tragic*. Translated by Paul Fleming. Stanford: Stanford University Press, 2002. Originally published in 1961.

Szondi, Peter. *Versuch über das Tragische*. Frankfurt/Main: Insel Verlag, 1961.

Tadiar, Neferti X. M. *Remaindered Life*. Durham: Duke University Press, 2022.

Tansley, Arthur G. "The Use and Abuse of Vegetational Concepts and Terms." *Ecology* 16, no. 3 (July 1935): 284–307.

Taub, Ben. "Jonathan Ledgard Believes Imagination Could Save the World." *New Yorker*, September 16, 2019.

Taylor, Zac J. "The Real Estate Risk Fix: Residential Insurance-Linked Securitization in the Florida Metropolis." *Environment and Planning A: Economy and Space* 52, no. 6 (2020): 1131–49.

Temple, James. "Harvard Scientists Moving Ahead on Plans for Atmospheric Geoengineering Experiments." *MIT Technology Review*, March 24, 2017.

Tessler, Michael, Mercer R. Brugler, John A. Burns, Nina R. Sinatra, Daniel M. Vogt, Anand Varma, Madelyne Xiao, Robert J. Wood, and David F. Gruber. "Ultra-Gentle Soft Robotic Fingers Induce Minimal Transcriptomic Response in a Fragile Marine Animal." *Current Biology* 30, no. 4 (February 2020): PR157–8.

The Human Project. Archived on web.archive.org, March 08, 2018.

Thomson, William. "On a Universal Tendency in Nature to the Dissipation of Mechanical Energy." *Proceedings of the Royal Society of Edinburgh*, no. 3 (April 19, 1852): 139–42.

Tibebu, Teshale. *Hegel and the Third World: The Making of Eurocentrism in World History*. Syracuse: Syracuse University Press, 2011.

Timofeeva, Oxana. *The History of Animals: A Philosophy*. London: Bloomsbury, 2018.

Tiqqun. "L'hypothèse Cybernétique." *Tiqqun* 2 (2001): 40–83.

Todd, Oliver J. "The Yellow River Reharnessed." *Geographical Review* 39, no. 1 (January, 1949): 38–56.

Toni, João Felipe Ginefra. "On the Origins and Renaissance of Goethe's Morphology." *Elemente der Naturwissenschaft* 108 (2018): 5–21.

Tooze, Adam. "Chartbook Newsletter #22: How do you count inflation? Tracking Weimar's hyperinflation." *Substack*, June 12, 2021.

Tooze, Adam. "The Death of the Central Bank Myth." *Foreign Policy*, May 13, 2020.

Toscano, Alberto, and Jeff Kinkle. *Cartographies of the Absolute*. Winchester: Zero Books, 2015.

Trilling, David. "Chinese Researchers Calculate Central Asia's Water Waste, Agriculture Potential." *Eurasianet*, April 27, 2021. https://eurasianet.org/chinese-researchers-calculate-central-asias-water-waste-agriculture-potential.

Tronchin, Lamberto. "The 'Phonurgia Nova' of Athanasius Kircher: The Marvellous sound world of 17th century." *Proceedings of Meetings on Acoustics* 4, no. 1 (June 2008): 015002.

Trownsell, Tamara, Amaya Querejazu Escobari, Giorgio Shani, Navnita Chadha Behera, Jarrad Reddekop, and Arlene B. Tickner. "Recrafting International Relations through Relationality." *E-International Relations*, January 8, 2019.

Tsing, Anna Lowenhaupt, Heather Anne Swanson, Elaine Gan, and Nils Bubandt, eds. *Arts of Living on a Damaged Planet: Ghosts and Monsters of the Anthropocene*. Minneapolis: University of Minnesota Press, 2017.

Tsing, Anna Lowenhaupt. *The Mushroom at the End of the World: On the Possibility of Life in Capitalist Ruins*. Princeton: Princeton University Press, 2015.

Tsinghua University, State Key Laboratory of Hydroscience and Engineering. "王光谦院士正面回应媒体质疑'天河工程'的几个问题 [Academician Wang Guangqian responded positively to the media's questions about the 'Tianhe Project']." November 28, 2018. http://sklhse.tsinghua.edu.cn/show.asp?id=506.

Tully, John. "A Victorian Ecological Disaster: Imperialism, the Telegraph, and Gutta-Percha." *Journal of World History* 20, no. 4 (December 2009): 559–79.

Turing, Alan M. "On Computable Mumbers, with an Application to the Entscheidungsproblem." *Proceedings of the London Mathematical Society* s2-42, no. 1 (1937): 230–65. Reprinted in Alan M. Turing, *Collected Works: Mathematical Logic*, vol. 4, edited by R. O. Gandy and C. E. M. Yates. Amsterdam: North-Holland, 2001.

Turner, Fred. *From Counterculture to Cyberculture: Stewart Brand, the Whole Earth Network, and the Rise of Digital Utopianism*. Chicago: University of Chicago Press, 2006.

Turner, William. *Report on the Human Crania and Other Bones of the Skeletons Collected During the Voyage of H.M.S. Challenger in the Years 1873–1876*, vol. 10, 1884.

Turvey, Samuel T., and Jennifer J. Crees. "Extinction in the Anthropocene." *Current Biology* 29, no. 19 (October 2019): R982–6.

Tversky, Barbara. "Visualizing Thought." *Topics in Cognitive Science* 3, no. 3 (2011): 499–535.

UNCTAD–United Nations Conference on Trade and Development. "Review of Maritime Transport." 2017.

Unknown Fields. *Tales from the Dark Side of the City: The Breastmilk of the Volcano Bolivia and the Atacama Desert Expedition*. London: Architectural Association, 2016.

Ure, Andrew. *The Philosophy of Manufactures; Or, An Exposition of the Scientific, Moral, and Commercial Economy of the Factory System of Great Britain*. London: Charles Knight, 1835.

Usher, Phillip John. *Exterranean: Extraction in the Humanist Anthropocene*. New York: Fordham University Press, 2019.

Van de Velde, Walter. "The World as Computer." Smart Objects Conference, Grenoble, 2003.

Van der Ryn, Sim, and Stuart Cowan. *Ecological Design*. Washington: Island Press, 1996.

VanderMeer, Jeff. *Annihilation*. London: 4th Estate, 2014.

Van Dooren, Thom. "A World in a Shell: The Disappearing Snails of Hawai'i." *UMassHistory*, October 24, 2020.

Van Kessel, Juan, and Horacio Larrain Barro, eds. *Manos sabias para criar la vida: Tecnología Andina*. Quito: 49° Congreso Internacional de Americanistas, 1997.

Van Koningsveld, Mark, Jan Mulder, Marcel Stive, Luuk van der Valk, and A. W. van der Weck. "Living with Sea-Level Rise and Climate Change: A Case Study of the Netherlands." *Journal of Coastal Research* 24, no. 2 (March 2008): 367–79.

Vergès, Françoise. "Racial Capitalocene: Is the Anthropocene Racial?" *Verso Blog*, August 30, 2017.

Vernadsky, Vladimir I. "The Biosphere and the Noösphere." *American Scientist* 33, no. 1 (1945): 1–12.

Vernadsky, Vladimir I. *Scientific Thought as a Planetary Phenomenon*. Written 1936–1938. Moscow: Nongovernmental Ecological V.I. Vernadsky Foundation, 1997.

Vidal, John. "Health Risks of Shipping Pollution have been 'Underestimated.'" *Guardian*, April 9, 2009.

Villanueva Yescas, Carlos. "Teutli: Leyenda." In *Escuela Secundaria Diurna Número 9 "Teulti": Un cuarto de siglo de labor educativ 1940–1965*. Mexico City: Lito Impresora Panamá, 1966.

Virilio, Paul. *Open Sky*. Translated by Julie Rose. London: Verso, 1997. Originally published in 1995.

Vishmidt, Marina. "Anomaly and Autonomy: On the Currency of the Exception in the Value Relations of Contemporary Art." *Zeitschrift für Kunstgeschichte* 81, no. 4 (2018): 588–600.

Vishmidt, Marina. *Speculation as a Mode of Production: Forms of Value Subjectivity in Art and Capital*. Leiden: Brill, 2018.

Vishmidt, Marina. "The Two Reproductions in (Feminist) Art and Theory since the 1970s." *Third Text* 31, no. 1 (2017): 49–66.

Vitello, Paul. "Joseph Farman, 82, Is Dead; Discovered Ozone Hole." *New York Times*, May 18, 2013.

Viveiros de Castro, Eduardo. *Cannibal Metaphysics: For a Post-Structural Anthropology*. Translated by Peter Skafish. Minneapolis: Univocal, 2014. Originally published in 2009.

Von Foerster, Heinz. "On Constructing a Reality." Fourth International Environmental Design Research Association Conference, Virginia Polytechnic Institute. Blacksburg, April 15, 1973.

Von Uexküll, Jakob. *Theoretische Biologie*. Berlin: Gebrüder Paetel, 1920.

Von Uexküll, Jakob. "The Theory of Meaning." *Semiotica* 42, no. 1 (1982): 25–82.

Voznesenskiy, S. S., A. Y. Popik, E. L. Gamayunov, T. Y. Orlova, Z. V. Markina, and Y. N. Kulchin. "Biosensors Based on Micro-Algae for Ecological Monitoring of the Aquatic Environment." In *Algae: Organisms for Imminent Biotechnology*, edited by Nooruddin Thajuddin and Dharumadurai Dhanasekaran. London: IntechOpen, 2016.

Wahl, Daniel Christian. *Designing Regenerative Cultures*. Axminster: Triarchy Press, 2016.

Wajcman, Judy. *Pressed for time: The Acceleration of Life in Digital Capitalism*. Chicago: University of Chicago Press, 2015.

Wallace, Doug, Philip Boyd, Yves Collos, Peter Croot, Ken Denman, Phoebe Lam, Cliff Law, Ulf Riebesell, Shigenobu Takeda, and Phil Williamson. "Ocean Fertilization: A Scientific Summary for Policy Makers." IOC–Intergovernmental Oceanographic Commission, UNESCO, 2010.

Wang, Guangqian, Deyu Zhong, Tiejian Li, Yu Zhang, Changqing Meng, Mingxi Zhang, Xiaolong Song, Jiahua Wei, and Yuefei Huang. "Study on Sky Rivers: Concept, Theory, and Implications." *Journal of Hydro-environment Research* 21 (October 2018): 109–17.

Wang, Guangqian, Deyu Zhong, Tiejian Li, Jiahua Wei, Yuefei Huang, Xudong Fu, Jiaye Li, and Yu Zhang. "天空河流: 发现、概念及其科学问题 [Tiankong Heliu: Faxian, gainian ji qi kexue wenti] Sky River: Discovery, Concept, and Implications for Future Research," *SCIENTIA SINICA Technologica* 46, no. 6 (May 2016): 649–56.

Wang, Jackie. *Carceral Capitalism*. Los Angeles: Semiotext(e), 2018.

Wang, Qi, "China's First National Park to be Established in Sanjiangyuan Area in 2020." *Global Times*, May 21, 2020. https://www.globaltimes.cn/content/1189032.shtml.

Ward, Matt, Luiza Prado de O. Martins, and Pedro J.S. Vieira de Oliveira. "Three Perspectives on Decolonising Design Education." *PARSE Journal* 8 (Autumn 2018).

Warden, Josie. *Regenerative Futures: From Sustaining to Thriving Together*. London: RSA, 2021.

Warwick, Kevin. "Cyborg Morals, Cyborg Values, Cyborg Ethics." *Ethics and Information Technology* 5, no.3 (September 2003): 131–7.

Warwick, Kevin. "Homo Technologicus: Threat or Opportunity?" *Philosophies* 1, no. 3 (December 2016): 199–208.

Waterman, Ronald E. "Land in Water, Water in Land: Achieving Integrated Coastal Zone Development by Building with Nature." *Terra et Aqua* 107 (June 2007): 3–32.

Watson, Julia. *Lo—TEK: Design by Radical Indigenism*. Cologne: Taschen, 2019.

Webber, Michael, Britt Crow-Miller, and Sarah Roger. "The South–North Water Transfer Project: Remaking the Geography of China." *Regional Studies* (January 2017): 370–82.

Weber, Andreas. *Sharing Life: The Ecopolitics of Reciprocity*. Delhi: Heinrich Böll Stiftung, 2020.

Weheliye, Alexander G. *Habeas Viscus: Racializing Assemblages, Biopolitics, and Black Feminist Theories of the Human*. Durham: Duke University Press, 2014.

Wells, H. G. *The Time Machine*. London: William Heinemann, 1895.

Werneburg, Brigitte. "Ernst Jünger and the Transformed World." Translated by Christopher Phillips. *October* 62 (Autumn 1992): 42–64.

Whitehead, Alfred North. *The Concept of Nature*. Cambridge: Cambridge University Press, 2015. First published in 1920.

Whitehead, Alfred North. *Science and the Modern World*. New York: Free Press, 1997. First published in 1925.

Whittaker, Meredith. "Origin Stories: Plantations, Computers, and Industrial Control." *Logic(s) Magazine*, no. 19 (2023).

Whyte, Kyle P. "Indigenous Science (Fiction) for the Anthropocene: Ancestral Dystopias and Fantasies of Climate Change Crises." *Environment and Planning E: Nature and Space* 1, no. 1–2 (2018): 224–42.

Wiener, Norbert. *Cybernetics; Or, Control and Communication in the Animal and the Machine*. Eastford: Martino Publishing, 2013. Originally published in 1948.

Wiener, Norbert. *The Human Use of Human Beings: Cybernetics and Society*. London: Free Association Books, 1989. Originally published in 1950.

Wiener, Norbert. "Newtonian and Bergsonian Time." In *Cybernetics; Or, Control and Communication in the Animal and the Machine*. Paris: Hermann & Cie, 1948.

Wildlife News, "Perfectly Kept Dinosaur Embryo from 66 Million Years Ago Found," *Al Jazeera*, December 22, 2021.

Wilk, Elvia. "Toward a Theory of the New Weird: Elvia Wilk on a Feminist Understanding of Eerie Fiction." *Literary Hub*, August 5, 2019.

Williams, Raymond. *Keywords: A Vocabulary of Culture and Society; Revised Edition*. New York: Oxford University Press, 1983. First published in 1976.

Williams, Samuel. *The Natural and Civil History of Vermont*. Burlington: Samuel Mills, 1809. First published in 1794.

Williamson, Hugh. *Observations on the Climate in Different Parts of America, Compared with the Climate in Corresponding Parts of the Other Continent. To which are Added Remarks on the Different Complexions of the Human Race; with Some Account of the Aborigines of America. Being an Introductory Discourse to the History of North-Carolina*. New York: T. & J. Swords, 1811.

Wilson Center Staff. "Climate Change in the Indian Farmers' Protest." *New Security Beat*, February 4, 2021.

Wing, Jeannette M. "Computational Thinking." *Communications of the ACM* 49, no. 3 (March 2006): 33–5.

Winner, Langdon. *Autonomous Technology: Technics-out-of-Control as a Theme in Political Thought*. Cambridge: MIT Press, 1977.

Winograd, Terry, and Fernando Flores. *Understanding Computers and Cognition: A New Foundation for Design*. Norwood: Ablex, 1986.

Wirth, Jason. *The Conspiracy of Life: Meditations on Schelling and His Time*. Albany: State University of New York Press, 2003.

Wise, Matthew Norton. "Work and Waste: Political Economy and Natural Philosophy in Nineteenth Century Britain (III)." *History of Science* 28, no. 3 (September 1990): 221–61.

Wittman, Marc. *Felt Time: The Psychology of How We Perceive Time*. Translated by Erik Butler. Cambridge: MIT Press, 2016.

Wohlsen, Marcus. "A Rare Peek Inside Amazon's Massive Wish-Fulfilling Machine." *Wired*, June 16, 2014.

Wolchover, Natalie. "A World Without Clouds." *Quanta Magazine*, February 25, 2019.

Wolfe, Cary. *What Is Posthumanism?* Minneapolis: University of Minnesota Press, 2010.

Wong, Julia Carrie. "Private Investor Divests $34.8m from Firms Tied to Dakota Access Pipeline." *Guardian*, March 1, 2017.

Woods, Clyde. *Development Arrested: The Blues and Plantation Power in the Mississippi Delta*. London: Verso Books, 2017. First published in 1998.

World Climate Research Programme. *WCRP Strategic Plan 2019-2028*. Geneva: World Meteorological Organization, 2018.

World Shipping Council. "Containers Lost at Sea – 2017 Update." July 10, 2017.

Wraith, Matthew. "Throbbing Human Engines: Mechanical Vibration, Entropy and Death in Marinetti, Joyce, Ehrenburg and Eliot." In *Vibratory Modernism*, edited by Anthony Enns and Shelley Trower. London: Palgrave Macmillan, 2013.

Wurman, Peter R., Michael T. Barbehenn, Matthew David Verminski, Michael Cordell Mountz, Dennis Polic, Andrew Edward Hoffman, James Raymond Allard, and Eryk Brian Nice. "System and Method for Transporting Personnel within an Active Workspace." United States Patent No. US 9,280,157 B2, filed September 4, 2013, and issued March 8, 2016.

Wynter, Sylvia. "1492: A New World View." In *Race, Discourse, and the Origin of the Americas: A New World View*, edited by Vera Lawrence Hyatt and Rex M. Nettleford. Washington: Smithsonian Institute, 1994.

Wynter, Sylvia. "Human Being as Noun? Or *Being Human* as Praxis? Towards the Autopoetic Turn/Overturn: A Manifesto." August 25, 2007.

Wynter, Sylvia. "Novel and History, Plot and Plantation." *Savacou*, no. 5 (June 1971): 95–102.

Wynter, Sylvia. "On How We Mistook the Map for the Territory, and Re-Imprisoned Ourselves in Our Unbearable Wrongness of Being, of Désêtre: Black Studies Toward the Human Project." In *Not Only the Master's Tools: African-American Studies in Theory and Practice*, edited by Lewis R. Gordon and Jane Anna Gordon. Boulder: Paradigm Publishers, 2006.

Wynter, Sylvia. "Towards the Sociogenic Principle: Fanon, Identity, the Puzzle of Conscious Experience, and What It Is Like to Be 'Black.'" In *National Identities and Sociopolitical Changes in Latin America Hispanic*, edited by Mercedes F. Durán-Cogan and Antonio Gómez-Moriana. New York: Routledge, 2001.

Wynter, Sylvia. "Unsettling the Coloniality of Being/Power/Truth/Freedom: Towards the Human, After Man, Its Overrepresentation—An Argument." *New Centennial Review* 3, no. 3 (Fall 2003): 257–337.

Wynter, Sylvia, and Katherine McKittrick. "Unparalleled Catastrophe for Our Species? Or, to Give Humanness a Different Future: Conversations." In *Sylvia Wynter: On Being Human as Praxis*, edited by Katherine McKittrick. Durham: Duke University Press, 2015.

Xu, Kepiao, Lei Zhong, Yaoming Ma, Mijun Zou, and Ziyu Huang. "A Study on the Water Vapor Transport Trend and Water Vapor Source of the Tibetan Plateau." *Theoretical and Applied Climatology* 140 (February 2020): 1031–42.

Ya Salam, Kalamu. "History: Jean Saint Malo, New Orleans Maroon." *Neo-Griot*, June 19, 2012. https://kalamu.posthaven.com/history-jean-saint-malo-new-orleans-maroon-ex.

Yang Fada 样发达. "*Zhouyi Xici Zhuan*" *Chongbian: Zui Gulao Ye Zui Xianhuo De Xueshu Lunwen*《周易・系辞传》重编：最古老也最鲜活的学术论文. Shenzhen Qinghua Yuan 深圳清华园, 2009.

Yang, Hao, Guan-yu Xu, Xiaofang Wang, Chunguang Cui, Jingyu Wang, and Dengxin He. "Quantitative Analysis of Water Vapor Transport during Mei-Yu Front Rainstorm Period over the Tibetan Plateau and Yangtze-Huai River Basin." *Advances in Meteorology* 2019, no. 2 (August 2019): 1–14.

Young, Julian. *Heidegger's Philosophy of Art*. Cambridge: Cambridge University Press, 2001.

Yusoff, Kathryn. *A Billion Black Anthropocenes or None*. Minneapolis: University of Minnesota Press, 2018.

Zapperi, Giovanna. "Du Surhomme au non-homme: Visions du corps-machine en temps de guerre." In *La fabrique du corps humain: la machine modèle du vivant*, edited by Véronique Adam and Anna Caiozzo. Grenoble: MSH-Alpes, 2010.

Zhu, Qingping, and Wentao Liu. "The Digital Yellow River Programme." In *Frontiers of WWW Research and Development*, edited by Xiaofang Zhou, Jianzhong Li, Heng Tao Shen, Masaru Kitsuregawa, and Yanchun Zhang. Berlin: Springer Verlag, 2006.

Zhuangzi 庄子. *Zhuangzi: Library of Chinese Classics; Chinese-English*. Translated by Wang Rongpei, Qin Xuqing, and Sun Yongchang. Changsha: Hunan People's Publishing House, 1999.

Zscheischler, Jakob, Seth Westra, Bart J. J. M. van den Hurk, Sonia I. Seneviratne, Philip J. Ward, Andy Pitman, Amir AghaKouchak, David N. Bresch, Michael Leonard, Thomas Wahl, and Xuebin Zhang. "Future Climate Risk from Compound Events." *Nature Climate Change* 8, no. 6 (2018): 469–77.

Zuboff, Shoshana. "Big Other: Surveillance Capitalism and the Prospects of an Information Civilization." *Journal of Information Technology* 30, no. 1 (March 2015): 75–8.

Zuroski, Emma Josephine. "Depths of Knowledge: HMS *Challenger* and the Reconfiguration of Modern Science." PhD diss., University of Auckland, 2019.

Zylinska, Joanna. *Bioethics in the Age of New Media*. Cambridge: MIT Press, 2009.*Bibliography*

Index

Note: Page numbers followed by 'n' denotes note numbers.

Credits

Editor and Principal Investigator "The Incomputable": Antonia Majaca
Managing Editor and Research Coordinator "The Incomputable": Lola Pfeiffer
Associate Editors: Mika Hayashi Ebbesen, Alex Fletcher, and Alan Smart
Cover Design: Rafaela Drazic

This book emerges from the artistic research project "The Incomputable—Art in the Age of Algorithms" (2015–21), instigated at the Institute for Contemporary Art, Graz University of Technology, and funded by the Austrian Science Fund FWF, the Austrian Federal Ministry of Arts, Culture, Civil Service and Sport, and the Styrian Provincial Government Department of Economy, Tourism, Science and Research. Special thanks to Milica Tomić, director of the Institute for Contemporary Art, for her support of this research.

www.ingramcontent.com/pod-product-compliance
Ingram Content Group UK Ltd.
Pitfield, Milton Keynes, MK11 3LW, UK
UKHW021906150926
475560UK00003B/213

9 781350 264977